THE HISTORY OF THE COMMERCIAL CRISIS 1857-1858

AND THE STOCK EXCHANGE PANIC OF 1859

All Published in

Reprints of Economic Classics

by D. Morier Evans

The Commercial Crisis 1847-1848 [1849]

Facts, Failures and Frauds [1859]

Speculative Notes and Notes on Speculation [1864]

THE HISTORY OF THE COMMERCIAL CRISIS 1857 - 1858 AND THE STOCK EXCHANGE PANIC OF 1859

BY

D. MORIER EVANS

[1859]

REPRINTS OF ECONOMIC CLASSICS

AUGUSTUS M. KELLEY · PUBLISHERS
NEW YORK 1969

First Edition 1859

(London: Groombridge & Sons, *5 Paternoster Row*, 1859)

Reprinted 1969 by

AUGUSTUS M. KELLEY · PUBLISHERS

New York New York 10010

SBN 678 00517 6

Library of Congress Catalogue Card Number

74-84577

PRINTED IN THE UNITED STATES OF AMERICA

by SENTRY PRESS, NEW YORK, N. Y. 10019

THE HISTORY

OF THE

COMMERCIAL CRISIS, 1857-58.

AND THE

STOCK EXCHANGE PANIC OF 1859.

THE HISTORY

OF THE

COMMERCIAL CRISIS,

1857-58,

AND THE

STOCK EXCHANGE PANIC

OF

1859.

BY

D. MORIER EVANS,

AUTHOR OF "THE COMMERCIAL CRISIS, 1847-48," "FACTS, FAILURES, AND FRAUDS,"
ETC. ETC.

LONDON:

GROOMBRIDGE AND SONS, 5, PATERNOSTER ROW.

MDCCCLIX.

TO

ARTHUR ANDERSON, ESQ.,

AS ONE OF THE PRINCIPAL REPRESENTATIVES

OF THE MERCANTILE AND MARITIME INTERESTS

OF THE UNITED KINGDOM,

THE HISTORY OF THE COMMERCIAL CRISIS, 1857-58,

IS

Respectfully Inscribed

BY

THE AUTHOR.

TO THE READER.

THE "History of the Commercial Crisis of 1857-58" is now presented after some short delay, the labour entailed in the arrangement of the work, and the care requisite in seeing it through the press, having been considerable. While the volume may be regarded as exhibiting, in particular, the special career of the great revulsion experienced at the close of 1857, it also contains a retrospective glance at the several similar "dread visitations" which have occurred since the remarkable epoch of 1825. It may, therefore, be said to trace the financial history of the country for more than the last quarter of a century. The connecting link between the Crisis of 1847-48 and that of 1857-58 has been distinctly preserved, and the materials collected in the Supplementary Chapter and the General Appendix will, it is thought, be found useful to all who may be engaged in banking and mercantile pursuits.

The causes and effects of the Crisis of 1857-58, which is admitted to have been the most severe that England, or any other nation, has ever encountered, are set forth in a plain and intelligible manner, the Report of the Parliamentary Committee clearly demonstrating the rise and progress of the system of business which first caused the inflation, and which, when it could be no longer supported, terminated in the ordinary, and on this occasion fearful, collapse. Whatever defects may exist in the banking system of the United Kingdom—and its most ardent supporters will allow that modifications may be essential—the operation of the Charter of 1844 cannot be held responsible for the great mischief which followed the over-trading of 1853 and the four succeeding years. But although the Crisis produced such injurious results, and the consequences were very alarming, the material wealth of the country was not endangered—a circumstance since established by the recovery and steady revival in business. The basis, also, on which trade was carried out appears to have been more substantial, the final liquidation of many of the estates being less unsatisfactory than could have been anticipated.

During the progress of the book, and when the writer imagined his task was nearly at an end, the commercial world was surprised by the Panic at the Stock Exchange in April and May, 1859. In accordance with the suggestion received from more than one friend, it was considered advisable to introduce a short chapter on that topic, accompanied by the extraordinary fluctuation in prices which then took place, in order to preserve, in a succinct form, a record of those startling events. This has been done in a manner which exhibits the frightful transition then experienced, the facts having been obtained from the best sources.

In the presence of such a mass of statistics, it may be imagined that some few typographical errors will be discovered. It is not, however, believed that any which shall become apparent will iuterfere with the integrity of the narrative, or detract from the general usefulness of the work.

The writer has only to add that he believes this volume, together with his previous works, the "Commercial Crisis, 1847-48," and "Facts, Failures, and Frauds," afford more combined information in relation to financial and commercial progress than has hitherto been published.

Birchin Lane, Lombard Street,
November 9, 1859.

CONTENTS.

SECTION THE SIXTH.

APPENDIX.

SECTION THE FIRST.

The Origin and Antecedents of Panics—The Course of the Several Great Events in Connection therewith—The Leading Features of the Particular Convulsions of 1825-26, and those occurring at the Subsequent Dates of 1836-37, 1847-48, and 1857-58—Description of Various Panics.

WITHIN the last sixty years, at comparatively short intervals, the commercial world has been disturbed by a succession of those terrible convulsions that are now but too familiar to every ear by the expressive name, "panic." Each separate panic has had its own distinctive features, but all have resembled each other in occurring immediately after a period of apparent prosperity, the hollowness of which it has exposed. So uniform is this sequence, that whenever we find ourselves under circumstances that enable the acquisition of rapid fortunes, otherwise than by the road of plodding industry, we may almost be justified in auguring that the time for panic is at hand.

First in the melancholy list stands the panic of 1825-26. This was preceded by a speculative mania of such a fantastic kind, that the very names of the "Bubble Companies," as they were called, if now quoted, would look like a sarcasm upon speculation in general. Associations had been formed without the slightest regard to the distinction that ought to be preserved between those enterprises that may be left to individual competition, and those that require the aid of subscribed capital, and not only failure, but ridicule, was the natural result. The history of subsequent panics, with all their attendant evils, will reveal nothing so ludicrously lamentable as the various companies for the pretended sale of milk, bread, fish, etc., that started into existence before the panic of 1825-26, by which they were utterly annihilated. Large sums also had been invested in foreign loans, and trade with Spanish America and other distant countries had been followed out to an unjustifiable extent. A sudden increase in imports was the result of the commercial operations, mostly carried on by means of capital obtained on credit; and so far did these exceed the usual amount, that the rates of exchange turned against this country, and the Bank of England, to check the efflux of bullion, raised her rate of discount and diminished her issues. The glut of money, and the consequently low rate of interest that had prevailed during the greater part of the year 1824 and at the beginning of 1825, had induced private bankers to advance money on securities not readily to be realized; and hence, when the merchants applied to them for assistance, they were unable to afford it, and several failures occurred in the commercial world, as the commencement of a state of distress which soon reached the bankers themselves. Alarm ensued in the most complete sense of the word, and might be plainly read in the countenances of the multitudes who thronged Lombard Street, to draw out their balances.* The apprehensions of the general public had been caused, in the first instance, by the failure of several great banking establishments in the country, and the consequent doubts respecting the stability of the London bankers were confirmed by two or three failures in the metropolis. By the middle of 1825, the panic reached its height, and subsequently expended its force, upwards of seventy banking-houses having failed or suspended payment. The year had commenced with an aspect of the most complete prosperity, and money had been readily advanced, in the previous twelve months, to carry out every project, however wild and abortive. But the sums thus invested had yielded no return, and, by a natural sequence of cause and effect, the year closed amidst general anxiety and distress.†

* A curious anecdote is related of the cause of "a run" upon one of the Lombard Street establishments. A poor woman, having met with a slight accident, seated herself, to recover strength, at the door of one of the banks; a crowd immediately collected, which terminated in a report that the house was unsafe, and hence the "run" upon it.

† The number of banks that stopped payment in 1825 was seventy-nine, with fifty-eight branches; the estimated liabilities reached £14,100,000, and the assets were taken at £10,800,000, showing a deficiency of £3,300,000. In 1826, the number was twenty-five, with two branches; liabilities taken at £4,650,000, and the assets £3,300,000; deficiency, £1,350,000.

The disasters of this year (1825), commencing in September, 1824, and reaching

The spirit of speculation that preceded the panic of 1836-37 was totally different in its character to that which had convulsed

their height by the close of March, are generally attributable to the speculations of the preceding twelve months, aided by excessive paper and credit, and a more than usual facility on the part of bankers in making advances to individuals; but specifically they may be considered to have had their origin—*1st.* In desperate attempts to outstrip the natural course of commerce. All the great articles of importation became objects of speculation, carried so much beyond reasonable bounds, that the excess in the imports of indigo above the averages of 1822-23-24, amounted to 38 per cent., cotton 48, wine 56, wool 90, silk 98 per cent., and other chief articles of commerce in similar proportions. The total comparative increase amounted in value to £115,000,000. *2nd.* In increased issues by the country banks and the Bank of England. A redundancy of capital prevailed, and the credit of the country assumed an unwonted aspect of advancement. In the spring of the year money was worth only 2½ per cent., and the Three per Cent. Consols, which, in April, 1823, stood at 73, in April, 1824, had advanced to 94. This state of dubious prosperity, while it gave an artificial stimulus to adventure, encouraged the Ministers, in its earliest stages, to effect a reduction of interest upon Government Securities, in which also they were further assisted by inordinate issues on the part of the Bank of England. This reduction of interest on stock was followed by a proportionate fall of interest on private securities. Under such circumstances, men, accustomed to draw large incomes from their capital, were forced to content themselves with diminished incomes from the same source, or to try new and more adventurous modes of obtaining additional profits. In this way a powerful impulse was given to the spirit of daring speculation, exhibited in loans, in joint-stock bubbles, and in vain attempts to raise prices, and to elevate the supplies beyond the demands of commerce. Hence, also, the spread of a like spirit throughout the provinces, and the wild activity of the banks, not merely in speculating on their own accounts, but in accommodating, by thoughtless issues of their notes, the gambling dispositions of all around them.

The increase on the country and the Bank of England issues is shown in the following returns:—

The country bank circulation amounted—

In 1823	To	£4,479,448
„ 1824	„	6,724,069
„ 1825	„	8,755,307

The Bank of England circulation amounted—

In 1823, April 5	To	£16,845,840
„ 1824, „ 3	„	19,313,989
„ 1825, „ 2	„	20,328,979

The advance, in less than three years, was, therefore, nearly £8,000,000, without a corresponding increase in the trade or industry of the country, sufficient to have justified the issue of a single additional pound. *3rd.* In an enormous increase of private paper and of transactions on credit. *4th.* In a further increase of foreign loans to the amount of £8,700,000, and in a drain of money for payments on foreign mines, which this year amounted to £3,997,000.

On these accounts alone, the nation bound itself to pay, within the year, many millions beyond its ordinary annual disbursements; but as paper was plentiful, and credit obtainable, not on experience, but "introduction," neither squalls nor breakers were contemplated. All was thought to be fair, and the risks of a great the commercial world eleven years before. Joint-stock banks, which had been made legal, as far back as 1826, by the repeal of an old law, forbidding the formation, in England and Wales, of any banking establishment having more than six members, suddenly became a favourite object of enterprise; and whereas, during the same years that immediately followed the alteration of the law, the number of joint-stock banks established in England and Wales had not exceeded thirty-four, a number which had increased to sixty by the end of December, 1835, no less than forty-two additional establishments were organized in the course of 1836, making a total of 102, of which more than three-fourths were banks of issue. This number, which may be easily raised to nearly 200, if the various branches be taken into consideration, refers to England and Wales only, and is therefore exclusive of thirteen joint-stock banks established in Ireland.

The quantities of paper-money which flooded the country, in consequence of this mania for joint-stock banking enterprise, became the natural cause of new speculations, commercial

---

commerce were regarded only as the excitements of a paradisaic state. But when the productiveness of mines came to be doubted, when the premiums of £30 per share on "Anglo-Mexicans," and £700 per share on "Real del Montes," came to be viewed as ruinous decoys—when foreign loans were looked upon with suspicion—when "calls" ceased to be met, and bills were more frequently dishonoured than paid—then paper shrivelled and credit collapsed, and then were depressions and embarrassments communicated to all the branches of trade and adventure, which an ephemeral currency had stimulated into unnatural activity. Then, also, the Bank of England contracted its issues. Up to the hour of commercial dissolution it had continued to augment them, but no sooner was help needed, than it contracted them to the extent of £3,500,000. The country banks followed the example; and, along with a rapidly diminishing circulation, the refusal of discount became general. Under the joint influence of the causes enumerated, capital was absorbed and dissipated, and the redundancy of money, which oppressed the market in 1824, converted into absolute scarcity. The funds fell, shares sank in value, the rate of interest advanced, goods became unsaleable, pawnbrokers and money-lenders were resorted to till their capitals were exhausted, book-debts ceased to be collectable, bills were rendered nugatory by the stoppage of drawer, acceptor, and endorser, and confidence vanished as gold disappeared. Then came the consummation of rebuke. Seventy-nine banks, with a nearly equal number of branches, broke. They had identified themselves with popular speculation, and, as far as customers' money would float them, they went; but, when that failed, they sank in the common flood. When the deluge was at its height, the Bank of England once more enlarged its issues; but it came like an act of repentance, when regrets are unavailing, and did but little in repairing the breaches which its floodings and its droughts had done so much to widen.—*British Losses by Bank Failures*, 1820 to 1857.

adventures of the most specious and dangerous character became once more the order of the day, and, in 1836, a heavy drain of bullion continued from April to the end of August, when it was arrested by the restrictive measures adopted by the Bank of England. These measures, which commenced in June, were further extended in August, and had the effect of checking the expansion of the mania in London; but the provincial joint-stock establishments, instead of reducing their issues, went on steadily extending them. However, the reaction that had occurred in the metropolis created a diminution of confidence throughout the kingdom, and a general run took place on the provincial joint-stock banks, who endeavoured to raise money by a forced sale of securities in London, and thus mainly contributed to the drain of bullion from the Bank of England vaults to the provinces and Ireland, which was one of the most striking phenomena of the year 1836. The alarm commenced in Ireland, when the Agricultural (joint-stock) Bank, an enormous establishment, with no less than forty-five branches, stopped payment,* and soon extended to the northern part of England, where, however, great disaster was prevented by timely assistance.†

Another important feature in the financial history of 1836-37 was the crisis that took place in American commerce, and revealed a system of over-trading between England and the United States, that had been carried on for something like two or three years. The great importing houses in the United States had adopted the plan of establishing agents in the foreign countries with which they were connected, and these were allowed, by some of the principal firms concerned in the American trade, to draw upon them for a certain stipulated amount, by bills of four months. Paper of this description was regarded with much favour in the English market, because it was calculated that the goods purchased by them could easily be sent to America, and sold within the space of three months, and that, therefore, there was sufficient time to remit funds to England before the acceptances became due. Indeed, so great was the confidence

* The liabilities of this establishment were estimated at £1,350,000, and the loss to the shareholders at about £400,000.

† At Carlisle there were some heavy failures. In Manchester, also, distress was occasioned by the suspension of of the Commercial Bank of England and the Northern and Central Bank of England, the latter with thirty eight branches.

in American bills, that the houses which granted the credits soon dropped the precaution of requiring security; and thus many of the New York and other houses were carrying on a large trade upon the mere basis of English acceptances. It should be observed, that the business of making these advances, as far as England was concerned, was almost confined to six houses in Liverpool, and two or three in London, whose acceptances at times amounted to no less than £15,000,000 or £16,000,000, while they had not capital equal to the sixth part of that sum to meet their obligations.

The measures that the Bank of England had adopted in order to employ her general resources, compelled her to lend a part of her funds to those dealers in money whose business it is to discount paper of unexceptionable character, and, as the paper purchased at this period consisted in a great measure of American bills drawn in the manner already described, the system that had been followed became known to the Bank directors, who, it is affirmed, ordered their agent at Liverpool to reject the paper of certain houses.* This injunction, it is supposed, was intended to be secret, but, as might have been expected, it soon oozed out and became generally known, and the firms specified were brought into discredit, though they still carried on their operations till March, 1837, when things came to a stand, and three of the leading establishments chiefly concerned in transactions with American acceptances were forced to suspend their payments. The bills of these houses alone—the houses of Wildes, Wiggins, and Wilson, called, in the phraseology of the day, the "three W.'s," amounted to £5,500,000,† while the aggregate liabilities of the American houses, generally, represented at least £12,000,000, the quantity of paper having been swelled to a most unjustifiable extent by every description of abuse. The assistance of the Bank of England enabled the three firms to continue their operations, but still the whole system of commercial transactions with America had received an irreparable shock, while the

* It is now very common, whenever there is a speculation in cotton or corn, and any pressure follows, to hear a rumour to the effect that the Bank has rejected certain paper, and, if it can be traced to an authentic source, it always exercises an unfavourable influence.

† These houses, when they at length wound up, paid good dividends, some even to the extent of 20s. in the pound. They were firms of great importance, and the suspensions at the period created much anxiety.

stoppage of the banks of the United States by President Jackson was an additional cause of embarrassment on the other side of the Atlantic.*

The crisis of 1847-48 was partly caused by circumstances similar to those that had operated so unfortunately in 1825-26. As on the former occasion, a rage for the formation of joint-stock companies, in every conceivable branch of trade, had interested the whole community, so that in 1845 the speculation mania was directed to the construction of railways in every possible locality; and again "bubbles were bursting" in all directions. The Irish famine, owing to the potato disease in the autumn of 1846, and the political disturbances that shook Europe in 1848, were, however, special causes peculiar to this period of commercial depression, and from which, for a lengthened period, there was no recovery.

By the gold discoveries in California and Australia, to which

---

* In 1837, all the banks of the United States suspended specie payments, and from that time to 1839, exceedingly injurious and, for the most part, fruitless efforts were made to effect their re-establishment in public confidence; for it was found that the theory and practice of the American banks were essentially false. Yet, notwithstanding the adoption of every description of expedient, while all the banks stopped, nine-tenths of them became in a short time insolvent. It could not be otherwise—for in most of them the needy bank directors had absorbed more than half the discounts, and a few favoured but defunct schemes the remainder. It could not be otherwise—for the yearly balances of trade had for some time been cleared off only by postponements of payment or renewal of bills. It could not be otherwise—for while debts increased, imports were, for the sake of keeping up appearances, allowed each year to become more and more extensive. It could not be otherwise—for although the great fire at New York, in 1834, had inflicted on the States a loss of 18,000,000 dollars, the people, led on by the banks, plunged, in the course of two succeeding years, into debts of 47,267,618 dollars', amount, for gambling land purchases, and the further sum of 111,693,519 on account of speculative foreign loans, forming together a mass of loss and liability to the amount of 176,961,137 dollars. It could not be otherwise—for, in less than five years, the banks, by fraud and failure, occasioned a loss to the Government of £3,813,000, and to the public at large of £91,362,625. Finally, it could not be otherwise—for the banks, instead of controlling and giving right direction to adventurous enterprise, identified themselves with speculation, descended from their high station as conservators of capital, and, while they enriched a few corrupt associations, ruined the community, and entailed permanent dishonour on the nation. To give the history of the institution entitled, in the accompanying list, "The United States Bank,"† would be to furnish an account of the conduct of nearly every one of the delinquent banks of America. They were mostly of the same genus—mania begat, and panic extinguished them. A brief epitome of its follies and its frauds may not,

† This bank suspended in 1839.

attention will hereafter be directed, an unhealthy stimulus was given to business, and in 1851 and 1852 signs of a reaction against the over-trading that had been carried on in both those countries began to appear. Nevertheless, the large profits that had been made during the years that immediately succeeded the auriferous discoveries caused losses to be felt less heavily than they would have been under other circumstances; and it was not till a more advanced date that the worst consequences of excessive enterprise were felt. The political troubles that had operated so mischievously during the years 1848 and 1849 were, in a great measure, terminated by the elevation of Louis Napoleon to the throne of France; but a new evil arose through the impetus which was given to French speculation by the so-called "Credit Mobilier," "Credit Foncier," and other establishments, and the Bourse of Paris presented a scene of frenzy that recalled the old days of the Rue Quincampoix.* On the other side of

---

however, be out of place, especially as its fall has been attributed in some quarters to the unreasonableness of the English creditors. Under its original charter, the first step of the United States Bank was to become the parent of paper issues to an unparalleled extent. This led, of course, to embarrassment. It had apparently made much money; but "the paper" had been parted with too freely. To retrieve this error, post notes were issued to an enormous amount, payable twelve months after date. But the remedy came too late. A suspension of specie payment followed. To this succeeded speculations in cotton, under the pretence of assisting trade. Then came false proclamations of prosperity, supported by deceitful reports, and tempting programmes of future usefulness and gain. Encouraged by the success which attended these manœuvres, and by the gullibility of the public on this side the Atlantic, the game of borrowing and of raising the wind in every form and shape was urged on with greater spirit than ever. Post notes and bonds were scattered all over England and the Continent, until the money-market of every capital in Europe was full to saturation. But during all this time no detailed accounts of the affairs of the bank were ever published. The world refused, however, to walk in darkness. Hope paid no dividends. Accounts were demanded. The pressure of the times, and not the honesty of the conductors, produced them, when it was found that the mighty establishment was a contemptible sham, a decoy, which had robbed its managers of their honour, and its supporters of their money. Nevertheless, a convulsive effort was made in England to keep the concern alive, and, for a while, the ricketty abortion went on, until at length it sank, more deeply indebted than before to the misplaced confidence of the British people. Great Britain is reckoned to have lost by this establishment capital to the extent of £10,000,000.—*British Losses by Bank Failures.*

* Enormous premiums in joint-stock companies were at this date realized, all of French origin. Combinations of capitalists were organized, who undertook all descriptions of works, not only in France, but also in Germany, Spain, and other countries, and no sooner was a project announced than the shares brought high prices.

the Atlantic, an extravagant style of living, that had become in vogue among the higher classes, led to the most desperate competition in trade, and the enormous expansion of American commerce in 1856-57 may be regarded as one of the most remarkable forerunners of the panic of the later date.

The crisis of 1847 was followed by a period of uneasy suspense in the commercial world, chiefly caused by the political convulsions that, in 1848 and 1849, shook nearly all the States on the Continent, and were felt in this country like the shock of a distant earthquake. When the abortive attempts made from time to time, in the interval, by a few noisy demagogues,* to gain so much as a passing notice, are remembered, and the ultimate position of the same persons is taken into account, it is felt that the famous 10th of April brought the question between order and anarchy in England to a settlement that is not likely to be speedily again disturbed. But while the struggle was yet pending, there was quite sufficient to render people uncomfortable about the stability of the Government, and, in a mercantile sense, this alarm was upheld by the very slow recovery of business that succeeded the great panic of the previous year. The general despondency would no doubt have produced serious results, had not the discovery of gold, first in California and afterwards in Australia, encouraged the hope that, whatever might happen, there would still be abundance of precious metal to meet the exigencies of commerce.† Sometimes this expectation, carried to an extreme, became a fear that gold would lose its value altogether, and that the fund-holder might be brought to ruin by the unexpected payment of his debt in comparatively worthless coin. Many political economists even talked of a future time, when the employment of bullion as a material of manufacture would be infinitely increased, and European tenements would be stocked with golden furniture, after the fashion of the Incas, and ordinary ornaments and utensils would be of the most massive auriferous manufacture.

The elevation of Louis Napoleon to the Presidency, and the en-

---

* The fate of the principal delinquents was settled by the Chartist trials in England and the proceedings adopted by the Government in Ireland.

† It was a trite remark, made by an influential joint-stock bank manager at the height of the excitement occasioned by the gold discoveries, that there would never be another *gold* panic; but he did live to see and experience the effects of one, viz., 1857-58.

suing conversion of the French republic into an empire, dissipated the clouds that overspread the political horizon; and, whatever may be the opinion repecting the *coup d'état* of December, 1851, there is no doubt that, for the time at least, heavy and desolating as was the sacrifice entailed, it assisted to arrest revolution. With partially restored confidence, the power of speculation began again to revive, and the gold discoveries, which had exercised a wholesome effect, while they mitigated despondency, now grew into a source of evil.* The awakened spirit took the ordinary form of a rage for joint-stock enterprise, and numerous companies were formed, in connection with the newly-unclosed treasure, which promised fairly at first, but which in the end were disastrous failures. At the same time, the lands of gold, California and Australia, seemed to offer a fresh and highly remunerative field for mercantile enterprise, and the trade of England and America, in the new direction, soon became so excessive, that, without the gift of prophecy, a mischievous result could easily be predicted. However, the storm, though seen, was distant; for, although several firms failed in 1851 and 1852,† they were so far from having any connection with the modern movements of the commercial world, that they chiefly consisted of old worn-out houses, who had long maintained a spurious existence on credit, and naturally crumbled away when a pressure arose. Still, even at this period, heavy losses were incurred in both the new branches of trade, though the large profits made in the first instance by the principal firms engaged in it enabled them to surmount immediate difficulties. In a word, at this period there was a pressure, not a panic, though some of the elements threatening a future extensive disruption of credit were already visible.

A short time before the final revulsion, the low state to which commercial morality had sunk was revealed by isolated cases,

---

* The Australian merchants, when the discoveries were first made, were quite alarmed for the result; wool growing and sheep farming, it was said, would be abandoned, and deputations waited on the Government regarding the encouragement of emigration.

† *Vide* list of suspensions, 1851-52. The reports of the meetings of creditors exhibited the results of speculation. Old-established firms entering into colliery and other adventures; houses trading on insufficient capital; neglected book-keeping, etc.; some utterly insolvent for years, and yet carrying on active business with high and respected names.

that by turns became leading topics of the day. Such were the transactions of Cole, Gordon, and Davidson, the failure of the Tipperary Bank, the suicide of Mr. John Sadleir, the disclosures of the Westminster Improvement Company, and the explosion of the Royal British Bank. But it was not till the end of 1857 that the utter rottenness of a commercial system, as carried on in Scotland, Liverpool, and London, was revealed to its full extent, proving that the enormous figures in the Board of Trade returns, supposed to indicate high prosperity, were based on an overwhelming quantity of accommodation bills, negotiated through country banks, and kept in circulation by every sort of pretext and expedient.* As exposure followed exposure, it seemed as though England and New York were vying with each other in barefaced fraud. Infamously bad management on the part of joint-stock bank directors was shown in the cases of the establishments in Scotland and the North of England that failed at this juncture, causing utter ruin to the shareholders. Never did wickedness and commerce seem so intimately allied as during the great panic of 1857; for what was shown to have occurred in the States—for instance, the disgraceful exposure connected with the Ohio Trust and Assurance Company—was found to have been repeated in England. Nor was the evil confined to the inhabitants of one country, or the members of one family of the human race; it spread like a pestilence over the northern part of Europe, and the anxieties of London were but weak, in comparison with the terror that absolutely paralyzed the Hanseatic Towns, and Denmark, Sweden, and Norway.

However, the evil of a panic is not altogether unmitigated. Not only is the commercial atmosphere cleared by the explosion of a number of establishments that have existed on a false basis—not only are wrong principles exposed, and therefore checked in their power of producing mischief, but the necessity of a reform in a state of laws under which such vast evils have been allowed to flourish is deeply imprinted on the public mind. The

* The cases of the Macdonalds, Monteiths, etc., will not be forgotten for a long period; the heavy liabilities, with the small assets, occasioned through the explosion of the accommodation bill system, recklessly supported even by the payment of a commission to men of straw for their signatures, showing the doubtful basis of business. The bankruptcy reports of the day illustrated the truth of this allegation in the most vivid colours.

moralist may draw ample materials for reflection from the cupidity and dishonesty which are brought to light under the influence of an extended system of credit; and in France, novelists and others have turned the ruling vice to account, by stigmatizing it in satires and dramas, that even philosophical writers have recognized as signs of the times. But in an age, like the present, of great luxury and ostentation, mere moral warnings will prove weak in their effect, when the desires that necessitate wealth are many, and riches appear accessible without toil. A revision of the law that will cause the miseries of panic to fall most heavily on the parties who have most deserved to suffer them, can alone prevent a recurrence of those terrible events that constitute, in the record of the last sixty years, such a chequered and unclean page in the history of the progress of modern commerce.

It is at all times, if possible, desirable to furnish corroborative views of facts or opinions expressed on such important topics as those referred to in the present pages, and, therefore, the writer has not failed to provide evidence to bear his statements out, selected, as they have been, from the best sources. In the first place, therefore, he presents, from an interesting pamphlet, entitled "Observations on the Currency, etc.," by Mr. S. Sandars, the annexed correct views on the general description of monetary panics, which even refer to a date antecedent to those the outlines of which have been described:—

ON MONETARY PANICS.

There are three kinds of panics—those arising from an abstraction of gold from the Bank of England—those from a contraction of credit capital—and those arising from the combined results of the contraction of credit capital and the scarcity of bullion in the Bank vaults.

1. The panic of 1793 and 1797 arose from the abstraction of gold from this country, as it was sent abroad for subsidies and loans advanced to our allies. These were gold panics, and were partially relieved by the advance of exchequer bills on the deposit of merchandise, but ultimately by the Restriction Act of 1797 making the Bank of England notes a legal conventional tender. This Act continued in force until Peel's bill of 1819, enacting a return to cash payments.

2. The panic of 1810, when failures to the extent of 10 millions occurred. Such panics arose from over-trading speculations in produce and excess of exports, and the Bank raised their discounts from 13 millions, in 1808, to 20 millions in 1810, or an increase of 45 per cent. This was a credit panic, as the Bank had the power of issuing an unlimited amount of inconvertible notes.

3. The panic of 1815, when 80 banks suspended payment, arose from the inflation of mercantile credit and vast speculations in cotton and other products. This was a credit panic, as the Bank still continued to issue inconvertible notes.

4. The panic of 1822 was caused by the Bank reducing its discounts from 15 millions to 3 millions, in order to secure the convertibility of the Bank note, and by the contraction of Bank of England and local notes from 48 millions, in 1818, to 26 millions in 1822. This was a gold panic, lowering the price of our funded and other properties, the stocks of goods held by the mercantile and trading interests, and agricultural products, to the extent of some 200 millions, and allowing that only one-fifth of such investments were forced on the markets during the panic, we may assume the total loss, including the failure of banks and mercantile houses, the partial closing of mills, and non-employment of workmen, caused an absolute loss of some £30,000,000.

5. The panic of 1826 arose from excessive speculations in foreign mines, in cotton and other products, and over-trading, and from the increased issue of notes, our circulating medium, which gradually rose from 33½ millions, in 1823, to 41 millions in 1826. This panic lowered the prices of funded and railway property, the stock of goods held by the mercantile interest and others, to the extent of 300 millions, and the assumed loss by the different parties may be estimated at some 45 millions, and, during such panic, the gold in the Bank was reduced below one million, and the credit of the Bank was saved only by the issue of some two millions of one pound bank notes accidentally found in the Bank.

6. The panic of 1832 lowered the price of public securities, etc., to the extent of 100 millions, causing a loss to the trading and other interests of some 15 millions.

7. The panics of 1836-37, called the American panics, as they originated and were chiefly confined to houses connected with the United States. The loss was estimated by the Manchester Chamber of Commerce, for that city alone, at 40 millions. The amount of gold in the Bank was decreased to so great an extent that its credit was maintained only by a loan of 2 millions from the Bank of France.

8. The panic of 1847 was caused by a bad harvest and great speculations in corn and railways. This panic was brought to a crisis from the surplus purchases of corn by the French Government being sent to this country in the months of June and July, to the extent of some 70,000 quarters, which were weekly forced upon the market at prices much below the current rates. For several successive weeks there was a fall of 8*s.* or 10*s.* a quarter, finally reducing prices from 96*s.* to 56*s.* Such a fall in the value of wheat brought to a state of bankruptcy a large number of houses connected with the trade in corn, and was followed by many other houses engaged in general trade. This brought on a money panic. The amount of gold in the Bank was reduced to 8 millions, and the Bank reserve of notes to 1½ millions, and its credit was only saved by the Government authorizing an addition to its issue of 2 millions. We thus see the anomalous effect of the Act of 1844, that brought the Bank to a state of insolvency, with 8 millions of sovereigns in its vaults. This panic lowered the price of funded and other properties 400 millions, and the loss to the funded and other interests was 60 millions.

9. The panic of 1857 arose from a money panic in America, that brought down a great number of our large merchants connected with America, followed by the failure of several large joint-stock banks in the North of England, arising from improvident advances by the managers of such banks. The gold in the Bank vaults was reduced to 6½ millions, and the notes in reserve to £957,000, and the solvency of the Bank was only sustained by another authorized issue of 2 millions of notes. This panic lowered the prices of funded and other properties 400 millions, and the

loss to the fund-holders and other interests, and the failures of banks and other houses, was some 80 millions. Thus the combined loss to the community generally, caused by the panics and the effects of the Acts 1819-44-45, may fairly be assumed to be 250 to 300 millions.

Of the progress of the particular epochs of 1825-26 and 1836-37, the annexed passages are extracted from Wade's "British History," published many years since by Mr. Effingham Wilson, a work of extensive research, which should be in the library of every professional and trading establishment. It is only to be regretted that it has not been brought down to a much later date.

FOREIGN LOANS AND JOINT-STOCK COMPANIES, A.D. 1825.

The general prosperity was such, that money or credit could be obtained for every enterprise, and the natural consequence of universal confidence was a general tendency to over-trading and speculation. Besides an infinite number of domestic undertakings in which the country had embarked, and which enhanced the prices of labour and commodities, a vast field of adventure had opened in South America. This partly arose from the determination of the English ministry to recognize the independence of the Transatlantic States, and to appoint *chargés des affaires* to Columbia, Mexico, and Buenos Ayres. Peru, Chili, and Guatemala, it was anticipated, would also be soon similarly treated, and raised to the rank of independent nations. The conduct of Government inspired confidence in individuals, who ceased to have any hesitation in advancing loans to the new governments, or in embarking in mining and other ventures. The money sent out of the country, a large portion of which was lost, was immense. The instalments paid on foreign loans, mining shares, and other speculations, in 1825, were estimated to amount to £17,582,773. Those on foreign loans only amounted to £11,304,623, and were as follow :—

£1,000,000	Brazilian loan of 1824	£350,000
2,000,000	Ditto, 1825	1,500,000
3,500,000	Danish ditto	2,625,000
2,000,000	Greek ditto	1,130,000
1,428,571	Guatemala ditto	357,143
400,000	Guadalajara	240,000
3,200,000	Mexican	2,872,000
2,500,000	Neapolitan	1,750,000
616,000	Peruvian	480,000

The mania for joint-stock companies was incredible. In 1824 and the beginning of the present year, 276 companies had been projected, of which the aggregate capital, on paper, was £174,114,050. Of these companies 33 were for canals and docks; 48 railroads; 42 gas; 6 milk; 8 supply of water; 4 coal mines; 34 metal mines; 20 insurances; 23 banking; 12 navigation and packets; 3 fisheries; 2 newspapers; 2 tunnels under the Thames; 3 for the embellishment of London; 2 sea-water baths; the rest miscellaneous. However absurd many of these projects, the shares of several rose to enormous premiums, especially the mining adventures in South America. The madness prevailing, and the sanguine anticipations of inordi nate gain, will be manifest from the following statement of the market-prices of the shares in five of the principal mining companies, at two periods, December 10, 1824, and January 11, 1825 :—

	December 10.	*January* 11.
Anglo-Mexican	£33 pr.	£158
Brazilian	10*s*. dis.	66
Colombian	£19 pr.	82
Real del Monte	550	1,350
United Mexican	35	1,550

On all the shares only £10 had been paid, except the Real del Monte, on which £70 had been paid. The adventurers obviously anticipated as rich a harvest as Pizarro and Cortes, and that without fighting, merely by the power of British capital, skill, and machinery.

THE COMMERCIAL CRISIS, A.D. 1826.

The mercantile reaction, which appeared at the close of the past year, continued with unabated force during the early part of the present. In November the number of bankrupts gazetted was 188; in December, 220; in January, 321; in February, 380; in March, 315. The number of bankrupts on March 4th was 93, which was the greatest number that had appeared; and from that time the plague may be said to have abated. As this was the most overwhelming revulsion in commerce that had ever happened, the causes in which it had originated were narrowly scrutinized; and the general inference seemed to be, that a wild spirit of speculation, springing, in the first instance, from the temptation of low prices, and fostered by the multiplication of paper-money and transactions on credit, was the primary source of the disorder. That over-trading was the origin, and the means indiscreetly afforded for over-trading accessories to the mischief, were facts clearly established from the returns obtained of the vast increase of imports, of the issues of the banks, and the number of bills of exchange in circulation. An excess of mercantile confidence, which opened the door to thoughtless enterprises, with fictitious capital, originated nine-tenths of the evil. The general prosperity encouraged the country banks and the Bank of England freely to make advances for almost every undertaking : they rapidly increased their issues of notes; but these, though powerful auxiliaries, were insufficient to account for the enormous redundancy of capital that marked the year 1824 and the summer of 1825. This could only be supplied by the vast extension of private credit by bills, promissory-notes, and open account. Such was the general confidence, that real money was hardly needed; credit was the universal currency; and hence was generated that redundancy of means which depressed the rate of interest, and induced individuals to seek profitable employment for their resources in foreign loans, foreign mines, and every imaginable domestic expedient. Alarmed at the speculative spirit abroad, the Bank of England were the first to adopt precautions, by contracting their circulation; and the example was followed by the country banks. This pulling up was soon felt by a pressure in the money-market. Some of the banks—that of Elford's, in the West of England, and of Wentworth's in the North—that had been extremely incautious in their advances to individuals, were unable to meet their engagements, and the fall of these houses involved the London firms with which they corresponded. Commercial confidence was destroyed, and a panic ensued. The Bank of England made strenuous efforts to mitigate pecuniary distress; and Government pursued a steady and judicious course. With the consequences of the folly and cupidity of individuals it could not properly intferere, but it sought to remedy some of the evils of the banking system. As a too great facility in the power of creating fictitious money had been a main ingredient in producing the mischief, Ministers sought to abridge the power of banks in issuing paper-money. For this purpose the circulation of one-pound notes was prohibited, and corporate bodies, or partnerships of more than six persons, were allowed to carry on the business of banking. Both these measures were improvements on the existing system, but not preventives of mercantile reaction; they afforded no effective guarantee against future panics, nor the over-issue and insolvency of bankers, nor against over-trading on baseless credit. The last are desiderata that can only emanate from individual prudence, and more general knowledge of the principles that govern the periodical vicissitudes of the commercial cycle.

REVIVAL OF INDUSTRY, A.D. 1827.

The present year was pregnant with events, and opened with a more cheering aspect than the preceding. Employment was generally to be had by the working classes, and though wages were still low, they enabled them to gain a livelihood. The different moneyed and manufacturing interests were recovering from the confusion of the last eighteen months by a progress which, though slow, was sure, and which by its slowness, perhaps, justified the belief that it did not proceed from factitious scheming, but was the natural return of mercantile health. The atmosphere had been cleared by the monetary crisis of 1825-26, and an entire decomposition of commercial elements effected. Masses of fictitious property were dispersed, and much of the real capital of the country distributed into new channels. Had this been the only result, the useful lesson over-speculation had taught might have been more instructive than injurious. Unfortunately an immense loss was sustained from the destruction of property, occasioned by the fluctuation in prices, and the sudden change in the employment of capital and industry. A check was thus given to internal improvement; and in consequence of the blight on mercantile confidence the legitimate movements of commerce long continued to be impeded. It is in the nature of great changes to involve the innocent with the guilty; and this was the worst result of the late revulsion; it not only swept away the delusive projects of the adventurers, but paralyzed for a season the operations of real business and commendable enterprise. In domestic politics the prominent topic of interest, in the history of the year, was the termination of the Liverpool Ministry, and the efforts to supply the vacancy occasioned by the sudden illness of that nobleman. Abroad the political horizon was auspicious. The arms of Russia and Persia were encountering each other on the banks of the Araxes, but the sound was too distant to disturb the repose of Europe. Our armament in the Tagus had guaranteed the Portuguese constitution against the machinations of the absolutists of Paris and Madrid. Even the interference of the great powers in behalf of the Greeks, which led to the battle of Navarino, failed to disturb the tranquillity of Europe.

PROSPEROUS STATE OF THE KINGDOM, A.D. 1836.

At the close of the past and commencement of the present year, the United Kingdom exhibited unusual signs of internal contentment and general prosperity. With the exception of partial depression in agriculture, all the great branches of national industry were unusually prosperous. In the great clothing districts of

Yorkshire and Lancashire the times were never known to be more favourable. In spite of the great development of the cotton-trade, it still continued to expand, and its utmost bounds seemed illimitable. It was the same with the woollen manufacture of Leeds and Huddersfield, the stuff manufacture of Bradford and Halifax, the linen manufacture of Barnsley and Knaresborough, the blanket and flannel manufactures of Dewsbury and Rochdale, they were all thriving. Even in the silk trade of Macclesfield, Coventry, and Spitalfields there were no complaints; no more than in the hosiery and lace trades of Nottingham, Derby, and Leicester. The potteries of Staffordshire continued prosperous, and the iron trade, in all its branches, was unusually flourishing. While manufacturing industry was in a state of energetic activity in the interior of the kingdom, it is almost superfluous to remark that the shipowners in the outports of London, Liverpool, Bristol, Glasgow, and Hull were not quiescent. One fact testifies to the prosperousness of commerce and existence of mercantile confidence, namely, the low rate of interest. Although there had been during the last twelvemonth several demands on the resources of moneyed men, the funds maintained a steady buoyancy; and the numerous projects on foot for improving the great lines of travelling and conveyance, at once attested abundance of pecuniary means and a lively spirit of improvement. That the general prosperity rested on stable grounds, there were solid reasons for concluding. A spirit of enterprise was abroad, but not of wild speculation. Except the mania for railways, which raged in England in common with other nations, there was no other abroad; and the avidity with which shares were bought up in these undertakings was justified by the actual success which had attended those of Liverpool and Manchester, Stockton and Darlington, Leeds and Selby. In 1824 the case was different; it was then pure castle-building; credit afforded unlimited means, and no project was too extravagant for support. At present, there was no want of commercial confidence, but it was a confidence indulged under a salutary reminiscence of former disasters. If anything could tend to its undue development, it was the state of the monetary system, which continued the most defective branch of industrial polity, and required unceasing watchfulness. It is as much a function of State to provide a safe and uniform currency, as a uniform standard of weights and measures, or a uniform and impartial course of judicial administration. None of the numerous provincial joint-stock banks of issue that had been established under an Act of the last reign (7 Geo. 4, c. 46), appear to have a subscribed capital exceeding two millions, with a paid-up capital of half a million. For one bank with so large a capital, there were many which did not possess a capital of a quarter of that amount, and as they frequently extended into branches in various parts of the country, the liabilities and consequent dangers of the parent bank were increased. One bank, with a capital of £600,000, had nearly forty branches in Bristol, Birmingham, Liverpool, Leeds, Manchester, Nottingham, and other places. In some banks, neither the amount of subscribed nor paid-up capital was known, which carried on, nevertheless, extensive business and had numerous branches. The subject, in the ensuing session, drew the attention of Parliament, and, at the instance of Mr. Clay, a committee of inquiry was appointed. Its investigations were not completed during the session, but enough was discovered to show the great irregularities and inconsistencies in the management of joint-stock banks; that they were not conducted on uniform and systematic principles; that the functions of the managing directors were not sufficiently defined and often irresponsibly exercised; and that, partly from this cause, and partly from the vague provisions of the partnership deeds, neither the interests of the shareholders nor of the public were adequately protected. In this state of things there was obviously cause for circumspection, though none, perhaps, for general alarm. One ground of confidence—at least in the old banking firms of the kingdom—is the better knowledge which the disastrous experience of former years had afforded, of the principles which ought to regulate banking associations in their advances to individuals, and in their issues of paper-money. The withdrawal of the small notes, too, is a guarantee against a popular, if not a commercial panic; and, as the obligation imposed on private bankers compels them to make periodical returns of their average circulation, timely notice is thereby afforded of the approach of the plague of over-issue, which the Bank of England would be culpably remiss in not checking on the first symptom of a redundant currency. Upon the whole, much of the machinery, as well as the material of commercial and manufacturing prosperity seemed safe and sound, and the natural result of lengthened peace at home and abroad, conjoined with a succession of the most favourable seasons. On the other hand there were, as before observed, complaints of agricultural distress. Farmers and landlords looked at their diminished incomes, not outgoings; they thought of the great sums they received during the war, not of the great sums they paid.

SPECULATION AND OVER-TRADING, A.D. 1837.

The state of general prosperity which was noticed at the beginning of the past year was followed in the summer by those symptoms of mercantile reaction that have been usually observed periodically to succeed periods of great industrial activity and commercial adventure. The present revulsion had a similar origin, and was marked by similar characteristics as those previously assigned to 1825-26; but its effects were more limited in this country, and less enduring than the ruinous pecuniary desolation which signalized the former period. The chief distinction between the elemental causes of the two was the more limited agency of private credit and the greater share banking and over-trading had in producing the crisis of 1836-37. In the existing difficulties, banking and speculation, especially of the American houses and of the Americans themselves, seem to have been the chief, if not the exclusive, sources of embarrassment. The recent partiality for joint-stock banks in England and Ireland grew out of the disasters which befel the banking firms in 1825; their destruction in that and former periods of commercial difficulty was considered to have arisen from the narrow basis on which they had been established, and that, by increasing their capital and the number of persons interested in their stability, their strength would be augmented. It was with the intention of carrying out these views that the Government prevailed upon the Bank of England to surrender some of their immunities. Joint-stock banks were no longer restricted to six, but were allowed, like the banks of Scotland, to have any number of partners with direct agencies in London. Either from the absence of enterprise, however, or the torpor that naturally followed the convulsion of 1825, these encouragements produced little immediate effect; and from that year to 1833, only thirty joint-stock banks had been established. But in 1833, the charter of the Bank being renewed, divested of most of its exclusive privileges, either from this cause or more probably the revival of commercial enterprise, joint-stock banks rapidly multiplied. In 1833 there was an addition of ten; in 1834 of eleven; in 1835 of nine; and in the first ten months of 1836 there was the enormous increase of forty-five joint-stock banks. In Ireland, from 1834 to the end of 1836, ten joint-stock banks had been established, making an aggregate of eighty-two, exclusive of their branches, which are equivalent to so many banks, in all the chief towns of the two kingdoms. The connection between these banking associations and the commercial difficulties of the present year formed a subject of controversy between Mr. Horsley Palmer, the leading advocate of the Bank of England, and Messrs. Loyd, Salomons, and Ricardo, who leaned to the side of the country banks. In considering their respective statements, there seems to have been little more than the old degree of *particeps criminis* that distinguished former periods of pecuniary pressure. The crisis of the present year was the counterpart, as before remarked, in its leading features, of that which ten years had preceded it: in both the coming storm was preluded by a wild spirit of mercantile venture; but the embarrassments created were neither so generally diffused in, nor exclusively limited to, England—they extended to Ireland and the United States, where a scene of monetary disorder presented itself wholly unexampled; bankers, importers, merchants, traders, and the Government being commingled in one mass of temporary insolvency. On both sides of the Atlantic difficulties, however, had a common origin—an inordinate thirst of gain; in America sought to be realized by land-jobs and over-trading in British produce; in England from excess of exports, railway projects, joint-stock companies for insurance, distilleries, cemeteries, newspapers, sperm-oil, cotton-twist, and zoological gardens. The mania for these share-undertakings was not limited to London, but was equally rife in Liverpool, Manchester, and Leeds. The banks fed the flame, though they did not kindle it. The first light, as Mr. Tooke has shown, always comes from the temptation of low prices and a tendency to higher, which, generating increasing consumption and demand, rouse into action the mercantile classes. No sooner are these symptoms abroad than the banks let go their paper, and instantly the commercial world is in a blaze. With the example of greater confidence in the banks, of readiness to afford advances to individuals, the sphere of private credit by bills and open account is instantly distended to an enormous size. The scene is changed as if by magic. Mistrust, stagnation, and inertness are converted into boundless confidence, mercantile activity, and speculative enterprise. Money, or what passes for money, is everywhere abundant; a community of sellers becomes one of buyers, and the wits of speculators and adventurers of every denomination set to work to absorb the seeming capital that overflows in every channel. That this was the cycle of the last as of former mercantile revolutions, is established by the state of prices and the issues of the banks. From 1834 to the summer of 1836 prices were on the advance, and speculation active. During the same period the provincial banks, both of England and Ireland, augmented their issues; and, though the Bank of England did not contemporaneously increase its circulation of notes, it was enabled to aid individual enterprise by the vast amount of private deposits at its disposal, and of which of late years it has become the great reservoir. It is private balances, not an increase of its issues, that has, since 1826, constituted the active trading resources of the Bank. In the use of their circulating capital, the directors have been accused of either undue eagerness to profit by its employment, or indiscreet precipitancy in their banking operations; of having afforded too much accommodation to individuals from August, 1835, to April, 1836, in order to facilitate the working of the West India loan of 1835; and then, when their own turn in that speculation had been served, of suddenly narrowing their discounts either to stop the efflux of gold to Ireland and the United States, or to enable them better to support the northern and central banks, to which they were committed, and the American houses. In the interval mentioned, advances could be readily obtained on stock and other approved securities; but as the summer advanced, discounts were abruptly refused to the largest and hitherto most respectable houses of Liverpool and London; trade in consequence became paralyzed; prices suddenly dropped from 30 to 40 per cent., and the various share bubbles floated on the tide of the previous pecuniary redundancy rapidly collapsed from want of dupes or instalments. This is one view that has been taken of the monetary pressure. But it is just to observe that there is always a period in the movements of commerce when it is incumbent on the banks to interfere for their own safety and that of the community; and that this point is, when commerce has obviously degenerated into unprincipled adventures, not founded on the regular demands of trade, either present or future, but solely on the command of unstinted resources. It is often only by withholding the means that the speculative furor can be arrested; that traffic can be prevented degenerating into mere gambling and monopoly, by which all pecuniary bargains and contracts are deranged, and prices forced up to an extravagant height, destructive of internal consumption and foreign commerce. At the same time the exercise of this wholesome check is sure to be inconvenient to some, and will assuredly incur the censure of those parties whose miscalculations or mercantile avidity have tempted them into undertakings beyond their available resources. In the production of present difficulties private credit participated, conjointly with the issues and advances of the banks, both in town and the country. These quicken into life, but, after that is done, private credit, by the multiplication of bills of exchange and the extension of current accounts, forms the great machinery of commercial operations. Of the expansive power of these agents and the mighty fulcrum they afford for speculation, the disclosures made by the great American houses of London—the "three W's," as they were termed—are a demonstration. The following account of these firms, published in June, 1837, presents features in the history of commerce deserving to be recorded. They are the amount of bills payable from June to December:—

Wilson and Co.	£936,300
Wiggin and Co.	674,700
Wildes and Co.	505,000
Total of acceptances	£2,116,000

An aggregate of acceptances to the amount of £2,116,000 is upwards of one-sixth part of the aggregate circulation of the private and joint-stock banks of England and Wales, and about one-eighth part of the average circulation of the Bank of England. Bills of exchange are not cash, but, when accepted by houses of undoubted credit, possess almost equal active force in the commercial world. The following are the amount of their shipments to America, which attest a not less speculative avidity in the United States than had prevailed in England:—

Wiggin and Co.	£1,118,900
Wildes and Co.	623,000
Wilson and Co. (dry goods account)	364,900

If this is not over-trading, it is certainly audacious enterprise. It shows that there is in British and American merchants, as well as in the seamen of the two countries, a spirit of hardy adventure that can be matched in no other nations. The above instances have been chronicled as examples of the commercial spirit of the age, and of the magnitude of individual transactions, aided by the resources of private credit. In conclusion, it is satisfactory to mention that, within two years after, almost the entire of the pecuniary difficulties of 1836-7 had passed away, commerce had resumed its wonted channels and activity, and that the great houses mentioned above were in a condition to meet all demands against them, chiefly in consequence of the banks of the United States, the whole of which had stopped payment, having again resumed payments in specie and the regular transaction of banking business with their customers.

SECTION THE SECOND.

The Gold and Silver Productions of South America—The Mines of Peru and Mexico, their Development and Resources—Other Producing Countries—Brazil, Chile, New Granada, the United States, and Russia—The Discovery of Gold in California, and its Effects—The Australian Production and Progress.

BEFORE the discoveries of gold in California and Australia, which have constituted such a distinctive feature of the last ten years, Spanish and Portuguese America were regarded as the chief source of the precious metals. From the days of Pizarro and Cortez, the names of Peru and Mexico have been associated, in many minds, with the notion of inexhaustible wealth; and, though the productiveness of the mines in those countries is no more to be compared with that which excited the cupidity of the old "Conquistadores," still they will always be famed in the history of gold and silver, those powerful agents in the moral and commercial fortunes of mankind.

The highest celebrity long appertained to Peru, where Potosi, worked as far back as 1545, maintained a sort of classical renown, like some of the fabulous mountains in the antique world. It is from Potosi that more silver has been brought than from any other spot in the world, and from this point, therefore, the price of commodities in Europe has often been influenced in a serious manner. Towards the end of the 16th century, the yield of Potosi might be estimated at £2,000,000 sterling, and at that period the precious metals were so much more rare than at the present day, that the effect produced by the influx of so much silver upon Spain must have been fully equal to that which would now be produced by six times the amount. The fact is curious, that all the treasure drawn from Potosi has been yielded by a single mountain, called the Ilatum Potocchi, or, for brevity, the Potosi, which stands in the midst of a vast uncultivated country, like an enormous sugar-loaf. The silver ore, to which it owes its reputation, was first discovered by an Indian in 1545, and the discovery brought together such a concourse of people, that the town of Potosi was founded, and so rapidly filled, that, towards the end of the 16th century, it is said to have contained no less than 160,000 inhabitants. After the first quarter of the 17th century, the production of the mines of Potosi began to decline, and amounted, at the end of that century, to about £700,000 sterling, while the contents of the ore were likewise diminishing. During the first half of the 18th century, a further decline took place; but a reaction afterwards occurred, and, in 1789, the yield amounted to £800,000 sterling. The insurrection of the colonies in Spanish America again caused a decrease, and at one time, during the struggle for independence, the production of Potosi was almost reduced to zero. In the year preceding the recent discoveries of gold, the average of the yield was under £200,000.

Notwithstanding the high celebrity of Peru, its reputation as the chief source of silver was ultimately eclipsed by the superior productiveness of Mexico. The mines of Zacatecas and Guanaxuato were worked before the end of the 16th century, and from that period the history of Mexico, as a mining country, is a record of increasing wealth. At the commencement of the 18th century, the gold and silver amounted to about £1,800,000, but in fifteen years this sum was increased to £2,600,000. The discovery of the mine of Valencia, shortly afterwards, brought with it an important augmentation of precious metal. In 1775, the produce of Mexico was £3,400,000; in 1788, it was £4,280,000; and in 1795 it was £6,400,000; at which point it remained nearly stationary until 1810, when the war of independence broke out.

The yield of the mines in Spanish America is generally reckoned in piastres, and, as nearly all the silver used in the civilized world came from that country, the Spanish piastre soon became the universal coin of commerce. The primitive "title" (or standard proportion) of the coin was of "11-12ths fine," or 917 in 1000; and there was a gold piece of the same weight, and containing the same proportion of precious metal, which was called a quadruple. To this high standard the Spanish Government rigidly adhered till the year 1772, when the Cabinet of Madrid violated its faith towards the whole world, and forfeited a long-established character by the reduction of the proportion of precious metal from 917 to 903 parts in 1000. All sorts of precautions were taken to prevent the discovery of the fraud; false weights were given to the assayers, and the agents employed by Government were bound by oath not to reveal a state secret, which had been discovered by every merchant and money-changer in the world. In 1786, the gold coinage became the subject of tampering, and the proportion of precious metal in the quadruple was lowered to 875 parts in 1000. No further depreciation, it should be observed, has taken place under the republics of Spanish America, which, to their honour be it recorded, have retained the standard which they fixed. Even to the present day the piastre retains its value. It settles accounts in India and China, it is used in Algiers, and is preferred by the Arab to all other moneys. The Turkish sultans have adopted it, but have reduced it to about a twentieth part of its original value by an admixture of base metal. As for the dollar of the United States, it is only the same coin under another name. In English money the silver Spanish piastre represents about 4*s*. 4*d*.

During the time that elapsed between the conquest and the year 1810, when the war of independence broke out, the quantity of precious metal extracted from the Mexican mines considerably augmented; but, as large quantities were smuggled out of the country to avoid payment of duty, the entire aggregate of production could not be clearly ascertained. Originally the Crown of Spain claimed a pound of silver, or gold, for every five pounds produced. This tax, which was afterwards reduced from one-fifth to one-tenth, was called "quint," and was levied in addition to other imposts, which fell with especial weight upon auriferous silver. Thus, there was a strong inducement to smuggling, which, in the case of gold, where a large value is contained in a small bulk, could be carried on with comparative facility. The productiveness of these two celebrated countries, about the time of the recent discoveries, may, however, be shown by the returns for the years 1846 and 1850. In the former year, Mexico yielded the value of £249,753 in gold, and £3,457,029 in silver; and Peru £96,241 in gold, and £1,000,583 in silver. In the latter year, Mexico yielded £382,901 in gold, and £5,883,333 in silver, and Peru similar amounts as in 1846.

At the commencement of the present century, two-thirds of the precious metals brought from the other side of the Atlantic came from Mexico and Peru. Other productive countries were Brazil, New Granada, Chile, and the United States. The discovery of gold in the Russian mountains obtained for the empire of the Czar a pre-eminence even above the old sources of metallic wealth, and shortly before California earned her sudden celebrity, the Ural chain was regarded with increasing interest by the monetary world of Europe.* However, it was not till within the last ten years that the notion of a discovery of gold seriously influencing the currency entered the public mind. This notion, at first viewed by many as a mere crotchet, has since been developed into a serious cause of alarm by the French political economist, M. Chevalier. With the discovery of the valuable deposits in California in the year 1848, a new epoch begins, and the excitement produced in the 16th century by the conquests of the Spaniards in America, and the consequent influx of precious metal into Europe, seems to be revived anew. Indeed, considering the enormous increase suddenly occasioned in the amount of gold and silver available for commercial purposes, any other result could not have been expected.

A comparison between the annual produce of fine gold and silver in the years 1846 and 1850, that is to say, at periods taken two years before and two years after the discovery, will at once make the force of these remarks intelligible. In 1846, the total produce of North and South America was to the value of £6,563,179, of which £1,301,560 represented gold, and £5,261,619 silver. In 1850, this amount was more than trebled, being raised to £20,601,813, that is to say, £13,341,989 of gold, and £7,259,824 of silver. The quantity of precious metal produced in Europe, Asia, and Africa, at these two periods, amounted respectively to £5,799,498 and £6,840,975, which, added to the American produce, give the very different totals of £12,800,000 and £27,500,000. The grand totals, in Troy weight, for the produce of the whole known world were 114,674 lbs. gold, and 1,979,084 lbs. silver, for 1846; and 365,930 lbs.

* A curious statement was made in several of the Continental journals some time ago, viz., that the late Emperor Nicholas regarded with alarm the discovery, on American soil, of such an enormous extent of riches, and that his apprehension was seriously increased when the colonial dependencies of England furnished similar wealthy resources.

gold, and 2,663,386 lbs. silver for 1850. It may be observed, that from all returns of the kind China and Japan have been excluded, because, although both these countries are highly productive, the amount of their produce is not to be ascertained. During the four years comprised in this comparative statement, there was a large increase in the silver production of Mexico, the yield for 1846 having been £3,457,029, while that for 1850 was £5,383,333.

The world had scarcely recovered from the surprise occasioned by the gold of California, when it was again astounded by the revelation that gold-fields of still greater promise had been found in Australia. Towards the end of May, 1851, it was discovered that New South Wales yielded gold superior in quality to that of California, and, in October, this district was in its turn eclipsed by the diggings in Port Philip. The conversion of a whole industrious population into a mob of miners was one of the earliest consequences of the revelation, and even those who remained at home talked hungrily of the "prize nuggets" that were to be secured in the new El Dorado. The same scene of excitement which had been witnessed in the United States was repeated in England. At first, attempts were made to cast doubt upon the authenticity of the statements. It was asserted that the Sydney accounts were much exaggerated, and that the produce of gold, if any, would not be large. Merchants and others deplored the embarrassments that would be occasioned in business through the rush to the diggings, and prophesied ruin and devastation to the colonies. Still, the movement went forward. Gold was produced, and gold was shipped to the mother country; and hardly had the earlier remittances been despatched from Port Jackson, before Victoria claimed to be ranked with New South Wales, as possessing auriferous deposits, which had only to be partially developed to prove that they were almost exhaustless.

Then came the organization of Californian and Australian gold mining companies, and the attendant symptoms of a mania on a small scale, which terminated, as usual, in the loss of the few millions of subscribed capital, and the exposure of the swindlers who concocted these schemes, and the dupes who suffered by them. Shares and scrip, which had been, through illegitimate dealing, run to a high premium, soon returned to par, then descended to a discount, as the advices from the residents abroad were more doubtful, and finally became utterly worthless. France deserves the credit of having inaugurated these enterprises at the outset of the discoveries in California; but even considered in the light of "a speculation," they were altogether unsuccessful, and it was not until British talent was directed to the introduction of these projects, that they in the slightest degree flourished.*

But, although English capital and English skill did not succeed in obtaining profits from either Californian or Australian mining adventure, the "alluvial deposits," as the surface workings were termed, constituted sufficient for large exports, and these, when different plans of operating were introduced, were supported by the yield secured from the lower depths by "hole sinking," and the other rude methods which the digger and untutored miner afterwards adopted. Every week brought intelligence from California of the large total forwarded to San Francisco from Sutter's Fort, Sierra Nevada, and the several fields which enjoyed an early reputation; but it was through the labour of the private adventurer, and not the associated company. Every month news was received of the increasing aggregate of the quantity sent by escort from Ballarat and Bendigo to Port Philip, and from Mount Alexander to Sydney; but here, again, it was found that the rough pan and the cradle evoked greater results than the recognized survey and the employment of regular mining implements. Steadily the produce and the shipments augmented, and it was, before long, acknowledged that the "clan" and "apportionment" system of the diggers defeated, in the most astounding manner, the combined efforts of machinery and directed resources.

With England and America placed in this position, the old country vying with its kindred relation in the production of wealth, the influence to be exercised by the change could not be disregarded. France, in an isolated position in this respect, endeavoured to make discoveries which would give her character and importance as a gold-producing nation, but without any success; notwithstanding that it is well understood

---

* Eccentric, indeed, were the names under which several of these companies were ushered into notice. One of Parisian extraction was entitled "L'Aigle d'Or" mining company, which promised great profits, but eventuated in complete failure. When the concern dissolved, and the directors were *non est*, they were facetiously termed in the market the "legal doers."

colonial as well as local explorations were effected for that purpose.* During the period the vast mutations made by these discoveries have been in progress, it was not unnatural that English as well as French political economists should eliminate their theories, and follow out such deductions as they believed would ensue from the course of events; but, admitting much that they have written to be correct, it nevertheless clearly seems they were scarcely prepared for so enormous an expansion of trade as that which has contemporaneously occurred and been fostered through the altered state of things. What seems in reality to have been forgotten in their craving to prove a depreciation in the value of gold is, that credit, in the first place, has been largely expanded; and that, in the second, the great activity engendered by competition—if even, as is daily witnessed, it is accompanied by ruin to the merchant, broker, or manufacturer—must prevent an appreciable enhancement of prices, or, at least, such a rise as would be attended with the danger predicted.

It is all very well to say that this would not have taken place had business been confined within its legitimate sources, the character of trading lately presented being purely exceptional, and not likely to recur. This explanation will, however, not suffice. Practical experience teaches that no legislation will properly restrain speculation or adventure; it will check, but not effectively prevent business being carried out on an unsound basis. The increase in the value of exports in 1848 from £60,000,000, to £122,000,000 in 1857, when the crisis took place, has presented sufficient proof of this difficulty, and notwithstanding the exposures which have ensued, a repetition will occur at future periods, in some shape or other, despite every precaution that may be attempted. To any one who will take the trouble to wade through the mass of evidence which can be presented in corroboration of this fact, in the course of the last half century, this conclusion will be abundantly plain, whether in the character of bank failures, mercantile breakage, or reckless speculation.

Through the enormous increase in the production of gold, which may be taken to have risen, including the estimated yield of the entire world, from £7,000,000 or £8,000,000 in 1849, to

* It is well known that scientific individuals have been consulted by the Emperor with the object of working supposed gold-fields, and that English adventurers have presented plans to his consideration for developing certain schemes.

£34,000,000 or £35,000,000 in 1858, the basis of credit may be fairly supposed to have been proportionably enlarged, and the buying and dealing power ranking in arithmetical ratio, the whole could not be conducted with profit, the counteractive effect of competition terminating in some branches with loss. With operations of this magnitude trebled, or quadrupled, as they might easily be, elements, however small, of doubtful trading could scarcely fail to intrude, and these, in their early stages, being attended with little inconvenience, have augmented as the tide of commerce swelled, till, as in late distressing cases, the bubble burst, and at once indicated the fragile foundation on which the huge overshadowing fabric had been reared.*

As showing to what an extent the produce of the gold-fields of Australia may have affected trade and its relations, the annexed figures are presented to trace the progressive rise that has taken place since the early date of the discoveries :—

YEAR.	VICTORIA.	NEW SOUTH WALES.	TOTAL.
1851	£480,000	£510,000	£990,000
1852	6,740,000	3,960,000	10,700,000
1853	9,530,000	1,950,000	11,480,000
1854	9,080,000	850,000	9,930,000
1855	12,430,000	230,000	12,660,000
1856	13,900,000	110,000	14,010,000
1857	......	......	15,000,000
1858	......	......	14,000,000

To imagine, therefore, that vast changes would arise out of this state of things, without entailing serious fluctuations and embarrassment, is simply to have reasoned on fallacious grounds, and the prophets who believed that they had exhausted the subject, and disposed of its principal characteristics, were sorely

* It is singular with what facility new firms are organized after old ones have broken down, and paid, probably, a few shillings in the pound. This was the case in 1847-48, and the same circumstance was again noticeable in 1851-52, while in 1857-58 it has been more than ever apparent. An apt illustration of the manner in which such a change is accomplished is furnished by the following anecdote: A house that stopped payment at the latter end of 1847 showed a miserable amount of assets, and with difficulty a dividend was declared. Some few months subsequently a heavy creditor was passing down a leading thoroughfare, and much to his astonishment, saw the name of the principal of the old firm, with that of a new partner on the door of well-appointed offices, even superior in situation to those previously occupied. Meeting, on his way home, the identical accountant who had been engaged in winding-up the estate, he mentioned the fact to him, and to inquiries concerning capital and prospects, received the curt response that he believed it was the exercise of the old principle, "*ship and draw.*"

chagrined when, in October, 1857, they found that the whole of their vaticinations had not been realized, and that the promised halcyon days were as far distant as ever. But if this were the case so far as regarded not only Australia but also California, what must have been the effects as concerned the trade of the entire world? The answer is readily afforded by the collapse visible in Germany and the north of Europe, and the other mercantile markets which suffered when England and America were convulsed. Some allowance in these considerations may be made for the drain of silver which has taken place, and it is evident, looking at the annexed approximate figures, that the yield of one kind of precious metal is out of all proportion with the other; and while, therefore, the process of absorption is carried on to the extent it is, through the requirements of India and China, any exceptional diversion to the Continent, such as may ensue from a state of war or other local disturbance, will rally create a scarcity and an advance in prices.

YEAR.	GOLD.	SILVER.
1800	£4,000,000	£6,000,000
1830	3,000,000	6,750,000
1840	5,000,000	6,750,000
1846	6,000,000	6,750,000
1848	7,000,000	6,750,000
1849	8,000,000	7,750,000
1850	11,000,000	8,500,000
1851	14,000,000	8,500,000
1852	22,000,000	8,500,000
1853	35,000,000	8,500,000
1854	34,000,000	8,500,000
1855	33,000,000	8,000,000
1856	34,000,000	8,000,000
1857	35,000,000	8,000,000
1858	34,000,000	8,000,000

Finally, the experience of the past shows that it is impossible, even through the attempted exercise of prophecy, to regulate the future. Investigation is, however, to be commended, since it lends assistance to the multiplication of facts and inferences, which otherwise would remain dormant and concealed; but these once eliminated and prominently displayed, will tell their own tale without the aid of embellishment, to give a particular colouring to events, or to raise a fanciful theory, the basis of which is frequently destroyed by the practical results of a few short weeks.

SECTION THE THIRD.

The Panic of 1857-58—The Previous State of Business, and the Apparent Indications of Prosperity—Sudden Change in the Aspect of Mercantile Relations—The Reaction in Credit in New York, its Influence on the Trade of the United Kingdom, and the immediate Revulsion succeeded by Failures in every Department of Commerce—The Report of the Select Committee—The Causes and Details of the Crisis—Its Spread and Effects in Germany and the North of Europe.

Before distinctly tracing the course of the panic of 1857-58, it may be permitted to make some reference to the previous condition of trade, and the state of relations which existed with foreign countries. It is necessary to do so, to bring the mind of the reader to the state of affairs as they stood at the date just prior to the commencement of 1857. It is true, and the fact is here referred to in order to elucidate the position of political influences, that although temporary differences with France and America had occasionally caused despondency, and given forewarning of what might be some day anticipated, it was not supposed that the revulsion which so speedily ensued was almost immediately to be encountered. Predictions had been, nevertheless, expressed that the trade of the country was making too rapid a progress, and that the extravagance of the day would lead to a collapse, either financial or commercial, which could not fail to entail serious consequences; but it was scarcely perceived that the deep-rooted system of fictitious credit had so thoroughly expanded through all branches of business, as to create that situation of things which was exposed by the subsequent events which occurred between September in that year, and the following February.

The period of excitement occasioned by the effects of the gold discoveries, and the belief that a reaction would have taken place after the expansion then experienced, had gradually diminished, and notwithstanding the loss of several millions in the various bubble gold companies, which proposed to develop, by the aid of machinery and imported practical mining-labour, the auriferous resources of California and Australia, still the sacrifices sustained by the merchants were paid out of profits realized from the first consignments made to those localities, in the days when 50 to 80, and as much as 150 per cent. were secured on special articles, such as spirits, beer, etc. Failures had in the interim occurred, and with the exception of one of those partial spasmodic changes which occasionally arise in the credit of the United States, there had been little to create alarm, or to arouse a suspicion that the business operations, not only of England, but of America, the north of Europe, and even the distant regions of Havannah and Porto Rico, would place them in a situation to be subjected to the vicissitudes of a great crisis, and to the consequences arising from such a severe shock.

It was, however, considered a marvel by most people, that the approach of the usual ten years had not been accompanied by some of those premonitory symptoms which indicated the presence of an unhealthy state of things, still the value of money continued to be, on the average, comparatively high, and the rates of discount, both in 1855 and 1856, had attained a point greater in proportion than in previous periods. This, it was nevertheless contended, only proved that the influence exercised by the gold discoveries had not been of the character which had been so strongly predicated, and it was at length ascertained that the increased supply of the precious metals led not only to the more rapid extension of commerce, but manifested its creative importance by its largely increased represented value on what ultimately became little better than worthless paper. The course of the crisis in England has been fully illustrated by the report published by the Select Committee appointed to inquire into the question under the authority of Parliament, and the influence exercised by the crash in the United States will be hereafter alluded to. But the turn which the foreign panic took was one of the most remarkable on record, extending, as its ravages did, through every market and in almost every conceivable direction.

The report from the Select Committee on the Bank Acts, which was appointed in the short session preceding Christmas, 1857, to inquire into the working of the Bank of England

Charter, and into the causes of the panic, then fresh in the public mind, presents a well-arranged survey of the circumstances that, during the preceding ten years, combined to bring about that alarming convulsion. These circumstances, according to the classification of the Committee, were three in number: an unprecedented extension of foreign trade, an excessive importation of precious metals, a monstrous development of banking system as an instrument for the distribution of capital.

The three kinds of evil thus enumerated were separately examined. With respect to the first, it was found that the total exports, which before 1848 had never exceeded £60,110,110, were set down for 1857 at the enormous sum of £122,155,000. As for the second circumstance, the increase of bullion, it appeared that during the same years, from 1851 to 1857, gold to the value of £107,500,000 was added to the European stock, while only £26,800,000 of silver was withdrawn, leaving on the side of increase a balance of £80,700,000. In the circulating medium there was naturally a corresponding increase, the total note circulation (unrepresented by bullion) being about £31,600,000, while the quantity of gold coin amounted to at least £50,000,000. As for the third cause of the panic, the augmented facility of obtaining credit, it arose naturally from the increase of trade and of bullion. The deposits in the joint-stock banks of London, alone, which in 1847 amounted to £8,850,774, had risen in 1857 to £43,100,724, and the increase in other quarters was proportional. These facts were not, in themselves, indicative of evil; on the contrary, the practice of opening accounts and depositing money with bankers having extended to numerous persons who never before thought of such an employment of their capital, was considered to show an advance of the middle classes in prosperity and habits of economy.

But when it was remembered that the aggregate of these deposits, which, flowing from all parts of the country, found their way to London, were employed either by the bankers themselves, who discount bills for their customers, or by bill-brokers who obtain it from the bankers, it was perceived that, at a time of excessive speculation, a practice in itself laudable might grievously heighten the prevailing malady. And this was the case in 1857. To give a notion of the facilities for obtaining accommodation from bill-brokers, shortly before the crisis, reference may be made to the statement of Mr. Neave, before the Committee, that to the knowledge of the directors of the Bank of England one broker had three millions and a-half, another four a third five, a fourth eight, and there was reason to believe that a fifth had between eight and ten millions.

Of these vast resources, which of themselves afforded temptation to the worst kind of commercial gambling, the mercantile community took every advantage. On the one side was an apparently boundless amount of capital waiting to be borrowed; on the other side, a throng of greedy speculators anxious to work the credit system to the utmost possible extent of which it was susceptible. In the lenders there was utter recklessness in making advances; in the borrowers unparalleled avidity in profiting by the occasion; and thus an unwieldy edifice of borrowed capital was erected ready to topple down on the first shock given to that confidence which was, in fact, its sole foundation. Such a shock was given by the American failures—this was a result of the same system, carried to a still more mischievous excess—and then the panic began. Bankers began to increase their reserve funds, and to limit their discounts to their own customers, and the bill-brokers, who had been carrying on enormous transactions without a cash reserve, fell back upon the Bank of England. On the 12th of November, the day on which the Treasury Letter, authorizing an extension of issues, was sent from Downing Street, a celebrated firm had asked and received of the Bank of England £700,000, bills to the amount of £2,250,000 had been discounted, and in the afternoon the reserved fund, which on the preceding day had stood at £1,462,000 (having been £2,706,036 on the 4th), had declined to £581,000, and the bullion in the vaults did not exceed £6,524,000. By the transfer of £2,000,000 of Government Securities from the banking to the issue department, the directors availed themselves of the discretionary power conferred upon them by the Government, and they sold stock to the amount of £1,000,000. The immediate pressure was thus relieved.*

Among the abuses brought to light by the exposure of mercantile doings, was a system of "open credits," by which certain English houses allowed persons abroad to draw upon them to an extent previously agreed upon. The drafts were negotiated on

* *Vide* Bank Act Report, 1857-58.

the foreign exchanges, and ultimately found their way to England, on the understanding that they were to be provided for when they fell due. However, the provision was made not by staple commodities, but by other bills that were sent to take up the drafts already issued. All that the English house gained by these transactions was a banker's commission, save in those cases where goods were consigned, and then, of course, a merchant's commission was charged. But such cases proved a rare exception to the general plan. In the report of the committee this system is exposed with great perspicuity, and it is stated that the chief business of a particular firm, which at the time of its suspension owed £900,000 upon a capital of £10,000, consisted in permitting itself to be drawn upon by foreign houses without any remittance previously or contemporaneously made, but with an engagement that it should be made before the acceptance arrived at maturity.

A series of failures had begun in America for several months before the commencement of the crisis, but did not attract much attention. Speculation is known to be one of the permanent maladies of the transatlantic republic; occasional embarrassments are merely regarded as the result of the peculiarity. The suspension of the Ohio Life and Trust Company, an old, respectable company, with a paid-up capital of £400,000 sterling, and deposits to the amount of £1,200,000, gave the first signal for a general alarm; and as other failures ensued, an immediate fall took place in all descriptions of stock, accompanied by a rise in the price of money. It is estimated that between the 25th and 29th of September, no less than 150 banks in Pennsylvania, Maryland, Virginia, and Rhode Island suspended specie payments. The New York banks at first weathered the storm, but the suddenness with which they had contracted their operations, by causing much damage to the country banks and the commercial interest, had awakened a feeling of jealousy in addition to the general alarm, and on the 13th a preconcerted run took place. The stoppage of eighteen banks was the immediate consequence of the pressure; the remaining establishments put themselves under the protection of the law. The number of failures that took place in the United States and Canada during the crisis is estimated at 5123, and the aggregate amount of liability at 291,801,000 dollars.

At first even the severe depreciation of American securities that took place during this state of things, caused no particular anxiety in England beyond the parties who were immediately interested in such investments. It was during the first fortnight in October that the prices of marketable commodities were generally affected, and several mercantile failures took place, while the Bank of England gradually raised the rate of discount from 5½ to 8 per cent. This period was marked by the failures of several Glasgow firms, whose debts to the Western Bank of Scotland amounted to nearly £2,000,000, and among whom the names of D. and J. Macdonald and Co., J. Monteith and Co., and J. and W. Wallace were conspicuous.

On the 27th of October, the Liverpool Borough Bank failed, after the directors had in vain applied for assistance to the Bank of England, who, however, announced their intention not to reject any good bill on account of the indorsement of the unfortunate establishment. At the commencement of November, the rate of discount went to 9 per cent., and on the 4th and 7th of that month occurred the suspension of two firms of the highest character—Naylor, Vickers, and Co., a Sheffield house, with property to the amount of £590,000, and Messrs. Dennistoun, Cross, and Co., a great American house, with branches in Liverpool, Glasgow, New York, and New Orleans.* The latter suspension brought with it the failure of the Western Bank of Scotland with ninety-eight branches, a paid-up capital of £1,500,000 and £5,000,000 of deposits, after a refusal of assistance on the part of the Bank of England, and of the other Scotch Banks, who had already advanced £500,000. This failure, which occurred on the 9th of November, was the first case of a suspended joint-stock bank in Scotland that had been known for many years, and the excitement created forthwith was immense; the circulation of one-pound notes, and the habit prevalent among Scotch operatives of depositing their savings in banks, causing the evil of a stoppage to be far more widely felt than in London. On the 11th of November, the rate of discount advanced to 10 per cent. on account of the demand for gold by the Banks of Scotland and Ireland, in which latter country, however, the effect of the panic was less severely felt than elsewhere. The large London bill-brokers, Sanderson, Sandeman, and Co., suspended payment, but the increased rate

* Both these establishments shortly paid in full.

of discount did not decrease the demand for gold made upon the Bank of England, the reserve fund of that institution, on the 12th of November, scarcely exceeding £500,000.

With the failure of the Wolverhampton and Staffordshire, on the 17th of November, commenced the extension of the panic into the iron districts. Several other failures, with liabilities little under £1,000,000, took place, and works were stopped that had given employment to 30,000 hands. On the 25th the Northumberland and Durham District Bank, which had received assistance from the Bank of England in 1847, was added to the list of suspensions—its subsequent management having involved it in greater difficulties than before. Nearly all the bank failures that occurred about this period carried with them a moral against a system which had existed for many years, but the unsoundness of which was thus forcibly revealed. Once and once again comes the story of reckless mismanagement on the part of directors, and of the sacrifice of helpless depositors. As for the sufferers by the failure of the Western, some of them were reduced to such a state of misery that an appeal on their behalf was made in the summer of last year for relief, while the wealth of the directors was a matter of notoriety. The banks of Northumberland, like those of Scotland, included the operative classes among their depositors, and could not suspend payment without a similar diffusion of misery. But even the fate of depositors was mild compared with that of shareholders, who having already lost their paid-up capital were subjected to heavy calls, not with the prospect of an ultimate profit, but simply to wind up the unfortunate speculations in which they had embarked.

Quitting the region of banks, their directors, shareholders, and depositors, another class of sufferers is presented, namely, the decent operatives, who were thrown out of work by the effect of the crisis in the manufacturing districts. In the cotton department alone, it was estimated that during the last three months of 1857 the aggregate amount of wages decreased to the extent of £1,064,700. Nor did the iron districts fare any better. Furnaces in Staffordshire, that had generally given employment to 28,000 persons, were not in work towards the latter end of the year; at the same time, in Scotland, about 16,000 men were out of employ under similar circumstances; in South Wales, wages were reduced one-fifth, and a fourth of the furnaces were closed. Indeed, every trade suffered more or less, with the single exception of the coal trade, and the records of town after town treat of wholesale cases of destitution, large meetings to petition poor-law guardians, charitable assistance in the shape of soup-kitchens, forced labour in public works—in short, of all the paraphernalia of distress. It is confidently asserted that one of the "sewed muslin" houses that suspended in Glasgow, had employed no less than 40,000 hands. The poor-law returns from the suffering districts were a safe index of the operation of the crisis on the industrious classes, and it was considered worth remarking that no material increase of rates took place in London or in the agricultural districts.

Scarcely less melancholy than this exhibition of physical distress was the aspect of a widely spread commercial immorality presented by the revelations of the crisis. Cases of mere over-speculation form the most agreeable part of the picture; the darker portion being filled with the records of fraud, and of a recklessness which was equivalent to fraud. A long succession of firms could be passed in review, in which assets and liabilities seem like so many figures, selected for no other object than that of illustrating a strong disproportion. And these "irregularities" reflected alike on creditor and debtor, who seemed, as it were, leagued together to keep up a rotten system of accommodation. For instance, when it appeared that a house which offered two shillings in the pound was a debtor to the Liverpool Borough Bank for £30,000, unsecured, who could say which was the more culpable party?

As a striking illustration of the manner in which the credit system was worked, the case of a daring genius was cited, who, about the middle of 1855, began to speculate in iron and cotton to a large extent, though he was not only without capital, but even a little in debt. He first shipped jointly with a friend, upon whom he drew for 25 per cent., while for the 75 per cent. he drew upon the firm to whom the goods were consigned. "Thus," he says with admirable *naïveté*, "I was enabled to pay for my goods without any capital." While performing the next operation, a shipment on his own and a joint account, he seemed to have been in a more solid condition, for he paid three-fourths in cash, and the agent drew for the remainder, and he extended his speculations to tea, shares, and "gray goods," thanks to a credit given by that most facile of institutions, the Liverpool Borough

Bank. The summer of 1857 found him somewhat in a dilemma; for he was dunned for money, and his whole property consisted of less than £800 in cash, and a shipment of cotton, then on the high seas, which, though mortgaged to a full extent, might yield a profit of nearly £3,000. The cotton must be held somehow, and to effect this desirable object, he bought jaconets to the extent of £5,000, which, of course, he did not pay, and divided them between his two most hungry assailants. In the social sphere the largeness of his mind was as brilliantly exhibited as in the world of commerce. His wine-bill for two years amounted to £638; the generous liquor was consumed to the extent of £388 in one of the four houses of which he was the munificent owner. His bill for jewellery during the same period amounted to £649, and of course his assets were small in proportion.

And just before all these revelations of vice and wretchedness took place, the country was supposed to be in a remarkably prosperous condition, because, forsooth, the export trade had greatly increased. No inference could possibly be more fallacious. The circular of a Manchester firm, which showed an increase of the exports in cotton yarn from £23,339,000 (in 1848), to £39,112,913 (in 1857), exhibited a corresponding decrease from £21,537,000 to £17,100,000 in the estimated home consumption of the same article. The fact was, that under an unwieldy credit system, allowing numberless persons to deal in what they please, at any price they please, without those restraints to enterprise that belong to a steady mode of conducting business, official returns of the extent of trade and manufacture convey no reliable information.*

By the countries in the North of Europe the crisis was felt with peculiar severity. During the few preceding years, the two kingdoms that are comprised under the common name Scandinavia had increased their trade to an enormous degree, and generally by the employment of borrowed capital. In some cases the English merchants sold to the Scandinavians on long credits; in others, a house in Hamburg would accept bills for their correspondents in Sweden, Denmark, and Norway, who thus effected their purchases by means of Hamburg credit.

* This circumstance is always strikingly illustrated on the occasion of every panic; statistics proving, in the most explicit manner, that over-trading is the cause.

Pending the continuance of prosperity, this system was profitable enough. The Scandinavians sold their goods in time to cover the acceptances of their Hamburg friends, who were therefore inclined to accommodate them anew, not hesitating for a moment to accept for houses who were not provided with even approximate means to meet their liabilities if a day of reckoning should ever come. With the explosion in America the day of reckoning did come. A ready sale of goods alone supplied the Scandinavians with the power of covering the Hamburg and other acceptances. Such a sale was rendered impossible by the American crisis, and equally unsaleable were the goods upon which advances had been made to Scandinavian houses. Sweden, Norway, and Denmark being thus wholly unable to meet their liabilities, the large firms in Hamburg and London, who had assisted the Scandinavians with their credit, were suddenly called upon to take up their acceptances. Hamburg felt the blow most severely. Disappointed in remittances not only from the North, but from London; destitute of available capital, and overstocked with unsaleable commodities; overwhelmed with mercantile paper, the merchants were reduced to a most hopeless condition. Soon discount became impossible, and so completely had the bill of exchange—from time immemorial the circulating medium in large commercial operations—lost its value in the estimation of the public, that a payment in hard silver was regarded as alone legitimate. It need hardly be said that a supply of precious metal sufficient to meet the enormous liabilities that had sprung up under an exaggerated credit system was not even to be hoped, much less to be obtained.

An attempt was made to raise the value of bills by guarantees of more than ordinary weight, but neither bills nor goods would be taken as an equivalent for silver, and the failure of the largest house connected with the Swedish trade bore awful testimony to the state of public opinion in this respect. The Government, after several vain attempts to remedy the evil, at last had recourse to foreign loans, and the Germans regard it as one of the worst features of the whole crisis that Austria, of all countries in the world, was the first to assist in extricating Northern Germany from a state of commercial difficulty. The first 10,000,000 marks banco were advanced by Vienna, and were immediately absorbed by the principal houses who at once were affected by the pressure. It is worth observing that the crisis was met in a different manner by different towns, and that of the three Hanseatic cities, Hamburg, Lübeck, and Bremen, the last-named came off with the highest degree of honour. The condition of Hamburg was one of blank despair; Lübeck attempted to stave off the evil by an alteration in the law of bills of exchange, that caused acceptances to fall less severely on her own citizens. The merchants of Bremen, on the other hand, though they exceeded those of Hamburg in the magnitude of their transactions with America, relying solely on their own exertions, and wishing neither a change in the laws nor assistance from their Government, weathered the storm so successfully that not a single failure of importance occurred in their city. It should, however, be mentioned that the credit system had not been carried to such an extent in Bremen as elsewhere, and that the proportion between speculation and capital was not so completely beyond all reasonable limits as at Hamburg. Moreover, Bremen sufficiently enjoyed the confidence of foreign states to be able to obtain precious metal on an emergency, although at an enormous price. The few failures that did occur at Bremen, and that might possibly be cited in contradiction to this statement, were, it is asserted, unconnected with the operation of the crisis.

Nowhere did speculation during the time immediately preceding the crisis assume so fantastic a form as at Paris, where the means of employing joint-stock capital promised to be even more various than in London before the year 1825. The so-called "Credit Mobilier," devised by M. Isaac Pereire merely invited the capitalist to bring in his money; the use to which it was to be put was left to the genius of the managers. Nothing could be more simple than the plan of the "Credit Mobilier." Employing either the money paid by the proprietary, or capital raised by credit, this enterprising company purchased shares in manufactories and ships, lines of omnibus traffic, paper-mills, plots of ground suitable for building; in short, anything in the market on which a profit could be raised. Nor was this enough to satisfy their desire for activity, they projected hotels, theatres, establishments, in fact, of the merest luxury, not so much meeting as creating a demand, to which they could respond with a supply. Wild, however, as the credit system appeared in Paris, and although the citizens of the United States are by far the largest purchasers of her fashionable commodities, France, nevertheless, felt the crisis less acutely than any other European state.* French commerce is not on so grand a scale as that of England and North Germany—it is less based on long credits; and it happened in the year 1857 that the shipments of French goods to America had almost ceased before the crisis broke out in New York. Moreover, there had been a particular French panic in the beginning of the year 1856, which had operated as a check on excessive enterprise in the subsequent period, and which caused an abatement of the former speculative activity, the decadence of the influence of the large associated undertakings having from that date been distinctly visible.

The effects of the change had also been rendered painfully apparent by the sacrifice of wealthy individuals, who, connected with the "Credit Mobilier," had operated upon the Bourse, and who, when unable to complete their engagements, sought a refuge in America, to avoid the responsibility of liabilities not reckoned by thousands, or tens of thousands, but by hundreds of thousands of pounds. High and respected financial names were sullied by these transactions, and though at the instant endeavours were made to conceal the magnitude of the transactions, and the discreditable circumstances by which they were attended, revelations speedily followed which established most conclusively the nature and character of the several defalcations. The special features of the Thurneyssen delinquency, the abstraction and appropriation of the Great Northern of France shares, and the irregularities indicated in the organization of the scheme for the construction of the Napoleon Docks, all more or less illustrated the evils resulting from excessive adventure, encouraged, stimulated, and fostered as it had been by the most exalted Government authorities, who were even alleged to have participated largely in the first-fruits, which, in the shape of profits, the various enterprises presented.

But if France suffered only in a second or third degree compared with England, America, or the North of Europe, she had, as it will be perceived, been convulsed at an earlier date, and, consequently, the lesson of experience inculcated no doubt produced its influence, and diminished the responsibility which an uninterrupted career of apparent prosperity would have entailed. Notwithstanding failures occurred in the American and Brazilian trade, and one or two weakly banking institutions

* This statement is generally believed, but it has been contradicted by parties who insist that the internal trade of France suffered severely.

were forced to succumb, the passage of France through the pressure was comparatively tranquil, and little calculated to excite apprehension.

Violent, nevertheless, as the storm was while it raged in the United States, and disastrous as were the effects of the collapse when its influence extended to Great Britain, succeeded in turn by the entire disruption of credit in the Hanseatic Towns and Denmark and Norway, its extreme fury was not of lengthened duration, considering the vast extent of interest compromised, and the important area of trade included. From the beginning of July to the middle of September, 1857, the American panic may be said to have commenced its short, sharp action, and to have concluded its great work of devastation, though the effects were subsequently traceable through continuous stoppages and banking crashes. In England the force of the pressure was felt from the beginning of September until the end of December, some weeks after the Government Letter appeared. This, compared with the course of events in 1847-48, was infinitely more protracted, because the crisis itself was of a doubly severe character, and involved larger interests.* The unstable foundation of business was general and more widely extended, and being diffused through almost every channel, the sacrifices occasioned were of the most onerous description. Meanwhile, the career of distress and embarrassment was proceeding abroad, contemporaneously exciting distrust, and producing failures of a serious character. Therefore, it was not until February, 1858, that the general panic diminished, although the Bank long previously attained a position to resume its ordinary functions, and to modify the unparalleled terms enforced for discounts.

The course of trade, subsequently, which had been so universally paralyzed that Venezuela, Porto Rico, Havana, and Dominica experienced in turn smart shocks, was by no means satisfactory; and eventually another phase of change was demonstrated, at a more advanced period of the year, by the break-up among the Honduras merchants, including several of the leading houses in London and Liverpool. With these events may properly be considered to have concluded the enormous

---

* *Vide* the celebrated speech of Mr. Disraeli on the crisis, and the accounts of the press at the same date.

catalogue of the disasters of 1857-58; and notwithstanding the latter suspensions could not fairly be said to be comprised in the actual history of the crisis, still they were in a measure accelerated by the weakening of credit and the prostration of prices which was previously experienced.* The last great mutation which afflicted capitalists and money-brokers, and disturbed one important centre of business, had not at this date been encountered, and it remained for April and May, 1859, to shadow forth a vast amount of mischief and suffering at the Stock Exchange, traceable, however, more directly to the effect of foreign political complications, than ordinary mercantile and financial embarrassments.

* Between February and June, 1858, it was generally noticed by mercantile observers, that there was only a slow recovery in business, the absence of confidence being very apparent. The writer, speaking to a high authority on 'Change on the subject, the latter remarked, "I, my dear sir, am not at all surprised, because you may depend that many firms who struggled through November and December, if they have surmounted their difficulties, have nearly had the breath squeezed out of their bodies." This eventually proved the case.

SECTION THE FOURTH.

Contrast of the Crises of 1825-26 and 1837-38 with those of 1847-48 and 1857-58 —Special Features of the Several Revulsions—The State of Distress and Disaster Occasioned—Duration of Panics, and the Consequences arising therefrom —Pernicious Results of Over-trading—The Liabilities and Dividends of the Failed Houses—The Management of the Provincial Joint-stock Banks—The Accommodation System and Inevitable Ruin to the Parties Engaged—The Termination of the Crisis of 1857-58—The Subsequent Suspensions, and the Lesson of Prudence Inculcated—Statistics of Suspended Firms, and Bank Acts Report.

The point in the "History of the Crisis" is now arrived at when it is essential to take a retrospective glance at its progress and effects, in order to ascertain what were its distinctive peculiarities, viewed in relation to antecedent events of a similar character. So far its course has been traced, not alone in connection with the trade of the United Kingdom, but also as identified with the revulsion that preceded the great and important change in the United States, and the later and more closely allied panic in the North of Europe.

A contrast of the several revulsions as they occurred shows, in the clearest possible manner, that, although brought about by almost similar antecedent circumstances, the consequences which followed were widely dissimilar, and varied according to the events which had inaugurated their commencement. History furnishes the details of the singular effect of the panic of 1825-26, and exhibits the distinctive features which accompanied the distress which ensued, principally exhibited in the explosion of bubble companies, the severe reaction which occurred in prices, and the breaking-up of a variety of private banks, which had been speculating and became involved through the unwarrantable inflation of their paper issues. Personal experience has given some insight into the results which attended the outbreak of the American crisis of 1836-37, with the several phases of its development. Although the course of events was marked by a number of important failures, and the situation of the Bank, through the extent of credit, raised on the false foundation of the paper circulated between New York, Liverpool, and London, was critical, still the dark shadow cast by that event was not of very lengthened duration. Of course, many months elapsed before anything like an approach to recovery was evident, and it was very apparent that the trade, both of England and the United States, suffered for a considerable period from the prostration experienced.

But the shock itself, notwithstanding it presented most alarming characteristics, was not to be compared to that which ensued from the extensive crisis of 1847-48, much less that of the later period of 1857-58. If the earlier may be characterized as one which exhibited a smart and effective blow to confidence, the second, which, through the enormous expansion produced by the railway mania, bound up, as it were, with the mercantile relations of the country, produced more wide-spread dismay and distrust, and, since the action of the bank directors told with more apparent force upon the course of the money-market, the consequence was very general disaster. It must, at the same time, however, be remembered that the position of the majority of the East Indian and Mauritius houses which succumbed, was anything but stable. Without imputing any improper motives to the heads of the several great establishments which, after maintaining a proud position for years, then collapsed, it is quite manifest that the system of business adopted facilitated their declension, for, instead of confining themselves to the legitimate operations of merchants and commission-agents, they entered deeply into various speculations, which involved their capital, and placed their immediate resources out of their own control.

Large advances upon estates, both in India and the Mauritius, swallowed up their capital; and the necessary drain to maintain the expenditure on the plantations, and the required outlay to keep them in condition, and develop the special manufacture of the staple in which they were engaged, whether cotton, indigo, sugar, etc., at once deprived them of the power of relieving themselves whenever a crisis arrived. The country, also, overweighted as it had been by the gigantic speculation, and losses produced by the collapse which followed the year 1845, was in a very impoverished situation, and the responsibilities, which had been assumed in the shape of railway calls, and which soon manifested themselves in the most startling manner, aided to create the pressure which was then so universally felt.

Meanwhile, although this fact was scarcely concurrent with the events which had shaken the commercial world to its foundation, the elements of strife appeared in France, and the revolution of 1848, so disastrous in its consequences, perpetuated that state of apprehension and distrust which was not for a lengthened period surmounted. With the close of 1848, however, the return to partial activity was exhibited, and although the effects were nevertheless protracted, the changes which affected the dynasty of Louis Philippe constituted almost a separate phase in the general course of affairs. The panic of 1847-48, properly so speaking, extended from the end of September to the middle of November, when the Government Letter was issued, and although, subsequently, a good deal of disturbance was caused by the continuance of failures, and the arrangements necessary to bring into working condition the state of business, it could not be said that it had in reality lasted longer than six or seven weeks.

But the peculiarities exhibited with respect to the crisis of 1857-58 were of a more marked and distinctive character, the antecedent causes having presented themselves in a new form. The first shock was certainly received by the disruption of mercantile credit in America; and the revulsion which followed so rapidly spread, that the attempt to prepare against the consequences which it was predicted would be experienced at once showed that fictitious credit had strangely compromised the position of those who were largely engaged in the export trade of the United Kingdom. Particularly was this the case in the provinces, and doubtful as were the examples manifested of London business, fostered through accommodation paper, many country houses presented statements who, with small means, had encouraged the open credit system to an extent that was positively alarming, which proved beyond question that trade had been maintained on a foundation of so slight a character that, whenever the first blow came, it could not hope to withstand its damaging impression.

More remarkable than all was the utter disregard paid to every sound commercial principle, and the very revelations which attended some of the failures in the Metropolis, as well as those in Liverpool, Leeds, Huddersfield, and the large manufacturing districts, gave evidence of the want of caution in granting facilities for adventures, or following out operations on a basis of capital in proportion to credit. If any testimony of this were required, reference need only be made to the revelations which ensued from the liquidation of the Western Bank of Scotland, the investigation into the affairs of the Liverpool Borough Bank, and the ultimate catastrophe connected with the Northumberland and Durham District Bank.* Exceptional cases were presented in which, fortunately, it was found that solvent houses, through the "lock up" of valuable assets were unable temporarily to meet their engagements; but these, when the current of adversity turned, were speedily realized, and those several establishments had the gratification of making a payment of twenty shillings in the pound, leaving them a large surplus to conduct their transactions, which had only experienced a temporary interference. Among these may be specially mentioned the London houses of Messrs. Dennistoun and Co.;† Messrs. Heine, Semon, and Co.; the Swedish firms of Messrs. Rew, Prescott, and Co.; Messrs. Albert, Pelly, and Co.; Messrs. Sewell and Neck; and the great Sheffield house of Messrs. Naylor, Vickers, and Co.,—the latter presenting a speedy and most encouraging liquidation.‡

* The details of these cases were positively frightful. Bolstered credit, facilities of the most extensive character, with false reports and manufactured dividends, furnished evidence of the extreme laxity of management.

† *Vide* some subsequent general remarks.

‡ "A circular has been issued to-day by Messrs. Naylor, Vickers, and Co., of Sheffield and Liverpool, whose suspension with large surplus assets took place during the height of the crisis last November, containing the satisfactory announcement of their resumption in full. It is understood that they are not only prepared to meet all claims with interest at 5 per cent., but even to discount every acceptance or other liability that may not yet have matured. At the time of the stoppage a balance-sheet was submitted showing property amounting to £590,000, to meet debts for £365,000, leaving a surplus of £225,000, besides private means to the estimated extent of £30,000 or £40,000. But as four-fifths of the assets were in America, where the firm had establishments both at New York and Boston, it was believed that under the best circumstances a long time would be required for realization, and the proposition made, therefore, was to liquidate by four quarterly instalments of 5*s*. each, the first maturing on the 15th of next July. The collections of the house, however, in all parts of the United States have since exceeded the best expectations, and, according to the general advices lately received, the losses not only of Naylor, Vickers, and Co., but of other firms of like repute, will, as compared with what was at one time apprehended, prove insignificant. The speedy restoration

With regard to the dividends on the general estates, there seems good reasons for believing, looking at the results of those which have at present been liquidated, that the distributions on the whole, since the recent crisis, have on an average exceeded those which were paid during 1847-48.* So large an amount of liability as that which was exhibited by the houses crippled through the late revulsion, must have been maintained by a more sufficient total of assets, otherwise the rates of payment would not have been so satisfactory. In the antecedent period the large establishments, with few exceptions, paid only fractional dividends, and the enormous engagements which they were under showed that their transactions had not been, except in very special instances, represented by real property. At the same time, it must be admitted that the estates those houses were interested in suffered very great depreciation, and in many cases were eaten up by their mortgage liabilities, which rendered them, when they came to be offered for sale, scarcely capable of any realization.

The case was different in 1857-8. Although the cycle of ten years had past, the houses which suspended were neither of so long a standing, nor so important a growth. They did not include, like those of 1847-8, firms which had been in existence some half or quarter of a century; but they were sufficiently prominent to include establishments carrying out operations on a most extensive scale, and whose facilities of credit had

---

of Naylor, Vickers, and Co. is the more remarkable from the fact, that the iron and steel trade in this country has as yet experienced but little revival, and under the circumstances it seems almost strange that they should have suspended at all. It must be remembered, however, that for the time the panic was so indiscriminate, especially in the case of all houses connected with America, that if the firm could have shown a probable surplus of two or three millions instead of £200,000 or £300,000, any attempt to obtain advances might have been equally futile. There are occasions on which it is a duty on the part of mercantile establishments, to the community as much as to themselves, to abstain from wild efforts, which can only add to the general confusion, and this was obviously a case of the kind."—*The Times*, May 19th, 1858.

* This was not the case with houses abroad. In Hamburg and in Stockholm the majority of the estates proved lamentably deficient, and the system of credit pursued appeared most unsatisfactory. The consequence was such an involvement that paper accumulated so rapidly as to absorb everything tangible in the shape of assets. Norway represented the most favourable appearances, and business was stated to have been conducted on a more legitimate basis, the accounts of the firms exhibiting no very discreditable features.

enabled them to enlarge their engagements to almost an indefinite extent. The system of business nevertheless presented most unsound symptoms, especially that which prevailed through open credits. But it is rather singular that at this juncture one or two of the houses which suspended in 1847-48 have, through the good management of their resources, and the arrangement of difficulties previously attending the adjustment of Mauritius claims, paid dividends in a greater ratio than was originally anticipated. These instances have, however, been rare, and can only be supposed to have occurred through exceptional circumstances, and do not form a contrast with the average of the distributions made subsequently to the panic of 1857-58.*

Notwithstanding the termination of the crisis, which, commencing at the end of September, lasted until the end of December, when the issue of £2,000,000 under the authority of the Government Letter was replaced (showing a duration of pressure of twelve or thirteen weeks), still the progress of business was far from favourable, and the paralysis which credit encountered was further affected by the failures which took place after that date.† From March to June, 1858, one or two important houses were brought to the ground, through the temporary "dead lock" in China, but the statements indicated that there was reason to expect—a fact since partially verified—that the assets would produce a liquidation in full, and leave a considerable surplus. The relations with Brazil were also in a

* At the very moment this sheet is passing through the press, the creditors of Messrs. Reid, Irving, and Co. have received notice of another dividend. This estate was wound-up in 1847-48, and it was thought would only pay about 1*s*. 6*d*. or 2*s*. 6*d*. in the pound. This arose through the inability to realize and convert estates. Ten years have elapsed, and through good management 6*s*. has already been received. Probably 10*s*. in the pound may eventually be distributed, and it seems certain that at least a total of 8*s*. or 9*s*. will be obtained.

† In the depth of the crisis, about the middle of November, a report became current that the affairs of a large American firm were in a critical position. For some days the rumour spread, but it could not be traced to any positive source. It then turned out that the wealthy house of Messrs. George Peabody and Co., the American bankers and merchants, had been compelled to apply to the Bank of England for assistance, and it was not until considerable negotiation had taken place that the arrangements were concluded. These were effected on the security of Mr. George Peabody's private property, and the capital of the firm, backed by the guarantees of several of the metropolitan joint-stock banks. The amount advanced was speedily repaid, and it was eventually stated that the total sacrifice sustained was only equal to about one year's profit.

measure compromised by the difficulties of some of the houses connected with the trade of Rio Janeiro and Buenos Ayres; but the more severe crash which afterwards ensued, involving, with few exceptions, the Honduras trade, comprised the firms of Hyde, Hodge, and Co., of London; Mr. J. Carmichael, of Liverpool; Archibald Montgomery and Co., of London, and a variety of smaller establishments.

In these cases, again, the disclosures were unsatisfactory, and from the out-turn of estates, it was manifest that the same irregularity had occurred as in former failures, the establishments having been involved in liabilities for years past. The smaller houses engaged in relationship with them were "under acceptances" for their accommodation to a great extent, and consequently found themselves compromised to an amount that included them in the general liquidation, and showed that the result was ruin to all who had mixed themselves up in these transactions. With the conclusion of 1858 came a termination of the distress and disaster occasioned by those extraordinary events, and, looking impartially at the results of the career of that particular epoch, it must be confessed that the crisis of that period was much more severe in its general aspect than that of 1847-48, and that though the results, as far as dividends are concerned, compare favourably with those of 1847-48, the distress and wide-spread effect following the revulsion of 1857-58 were much more extensive. To establish this circumstance, the evidence of the eminent public accountants, Mr. J. E. Coleman and Mr. John Ball, may be referred to in the Appendix to the Bank Acts Committee's Report.

Experience shows that the accommodation bill system as pursued in Scotland, the ramifications connected with which, as presented in the cases of Macdonald, Monteith, and Co., extended even to London, and brought down a number of small tradesmen, who had been made the convenience of these establishments, some of them receiving a commission for the use of their names, will, it is to be hoped, have inculcated a lesson which will not readily be forgotten, and the perils connected with which will act in future as a check to the recurrence of such operations.

The fate of the Western Bank of Scotland, the Liverpool Borough Bank, and the Northumberland and Durham District Bank, with the serious liabilities entailed upon the directors and managers, furnishes another page in the annals of financial misconduct; and although neither of these establishments has yet been brought to a final point of liquidation, it is painfully certain that the calls will be heavy, and absorb, except in cases of compromise, the whole of the assets of the unfortunate shareholders. The expansion of trade, which was encouraged through the facilities granted by these establishments, and which, spreading to London, crossed the Atlantic, and again in turn to the north of Europe, incontestably establishes the wild and reckless character of the operations engaged in, the whole of which being supported on a basis of credit out of all due proportion to the capital embarked, required a reaction only such as was inaugurated by the revulsion in America to create that state of affairs which so speedily engendered distrust, and brought with it the fearful consequences already described.

At the time of the general excitement, when the press was teeming with descriptions and statistics concerning the crisis, the following comparative summary was put forward in a leading daily journal:—"The papers, in commenting upon the recent failures which have taken place among the mercantile community, agree in what was stated in the House of Commons during the debate on the crisis, that the liabilities of the suspended firms reach at least £45,000,000, or double the amount of the total in 1847. Including in the estimate the debts of the five banks, taken at £23,000,000—viz., the Western of Scotland, the Northumberland and Durham District Bank, the Liverpool Borough, the City of Glasgow, and the Wolverhampton and Staffordshire—it is clear that the entire sum is probably nearer £50,000,000; and before the effects of the panic are exhausted, a large augmentation must be looked for. It is, nevertheless, pretty certain that the suspensions in London and the provinces in 1847—extending as they did to the termination of the year, and even into the commencement of 1848—represented a greater sum than £22,500,000, or £25,000,000; but the approximate calculation is nearly correct, and will, no doubt, be borne out in a great measure by results when the crisis shall have entirely passed. The bank suspensions on the present occasion, although not more numerous, have involved larger amounts, and have, consequently, increased the unfavourable proportion. In producing the list of the principal London failures, slightly amended and amplified by the description of the trade of the firm, it may be noticed that singularly enough, both in the crisis of 1847 and that of 1857, two leading bill-brokers have been compelled to succumb—Messrs. Sanderson and Co., who suspended in the former period, having again failed under the new firm of Messrs. Sanderson, Sandeman, and Co.; and Messrs. Bruce, Buxton, and Co., who liquidated under that pressure, again following out a similar process through the title of Messrs. Bruce, Wilkinson, and Co., the creditors, as before, in each case, being assured of 20*s.* in the pound. The appended failures occurred between October, the date of the actual commencement of the existing pressure, and the 19th inst., the Government Letter having been issued on the 12th of November, and the rate of discount fixed at 10 per cent.:—

Ross, Mitchell, and Co., Canadian trade	£396,282
W. H. Brand and Co., American trade	235,524
John Haly and Co., American trade	47,509
A. Hill, commission-agent and insurance-broker	61,268
Powles Brothers and Co., Spanish American trade	50,000
Dennistoun and Co., American bankers and exchange-brokers	2,143,701
Bennoch, Twentyman, and Rigg, silk dealers and manufacturers	257,694
Broadwood and Barclay, West India trade	212,020
Joseph Foot and Sons, silk manufacturers	27,640
Sanderson, Sandeman, and Co., bill-brokers*	5,298,997
Bruce, Wilkinson, and Co., bill-brokers, in liquidation	not stated
Wilson, Morgan, and Co., wholesale stationers	25,629
Fitch and Skeet, provision trade	55,000
Draper, Pietroni, and Co., Mediterranean trade	300,000
Jellicoe and Wix, drysaltery trade	not stated
José P. de Sà and Co., Brazilian trade	15,230
Bardgett and Picard, corn trade	85,142
Hoare, Buxton, and Co., North of Europe trade	466,601
Edwards and Matthie, East India and colonial trade	350,000
E. Sieveking and Son, North of Europe trade	400,000
Allen, Smith, and Co., North of Europe trade	20,306
Svensden and Johnson, North of Europe trade	not stated
Gorrissen, Huffel, and Co., American bankers and exchange-brokers	£125,310
Brocklesby and Wessels, corn trade	40,486
R. Bainbridge and Co., American trade	not stated
Hermann, Sillem, Son, and Co., Hamburg trade	not stated
Carr, Josling, and Co., North of Europe trade	300,000
A. Hintz and Co., North of Europe trade	101,439
Rehder and Boldemann, Hamburg trade	100,000

* Most of the creditors on this estate hold security in the shape of bills of exchange, which are rapidly running off. Hence the extent of the failure is more apparent than real.

Henry Hoffman and Co., Hamburg and German trade	£100,000
Herman, Cox, and Co., Hamburg and German trade	60,000
Bischoff, Beer, and Co., foreign trade	30,000
Mendes Da Costa and Co., West India trade	231,673
Kieser and Co., German trade	50,000
Barber, Rosenauer, and Co., foreign trade	32,488
Hirsch, Ströther, and Co., German trade	not stated
F. and A. Bovet, East India and China trade	not stated
C. A. Jonas and Co., foreign trade	100,000
Sewells and Neck, North of Europe trade	500,000
Albert Pelly and Co., North of Europe trade	170,000
Krell and Cohn, German trade	not stated
W. Caudery, general merchant	30,000
W. B. Filler, general merchant	140,000
Hadland and Co., Manchester trade	40,000
Lichtenstein and Co., German trade	80,000
J. H. Baird and Co., Australian trade	21,258
Heine, Semon, and Co., German bankers and exchange-brokers	700,000
Weinholt, Wehner, and Co., East India trade	300,000
T. H. Elmenhorst and Co., German trade	not stated
Montoya, Saenz, and Co., Spanish American trade	not stated
T. G. Ward, Smithfield, banker	not stated
H. and M. Toldorph, and Co., German trade	not stated
Rew, Prescott, and Co., North of Europe trade	150,000
R. Willey and Co., shawl manufacturers, etc.	50,000
G. H. T. Hicks, East India trade	151,900
Powell and Son, Manchester trade	60,000

"In the crisis of 1847 the bank failures numbered four of importance, viz., the Royal Bank of Liverpool, the Liverpool Banking Company, the North and South Wales Bank, and the Newcastle Joint-stock Bank; besides the Abingdon and Wantage, the Oldham Bank, and six other small private firms at St. Alban's, Salisbury, Shaftesbury, Shrewsbury, Honiton, and Bridport. The pressure at this period took place in August, when the suspensions in the corn trade were announced, and continued with unabated severity through September and October, the Government Letter having been issued on the 25th of the latter month, and withdrawn on the 23rd of November. During this interval the *minimum* rate of discount was 8 per cent. The extent of the liabilities of many of the firms which then suspended may be contrasted with the details previously given by reference to the annexed list:—

Alexander, L., and Co., corn trade	£573,188
Barclay Brothers and Co., Mauritius trade	389,504
Booker, Sons, and Co., corn trade	40,000

Bruce, Buxton, and Co., bill-brokers	£347,000
Bensusan and Co., Mogador trade	57,961
Boyds and Thomas, East India trade	38,684
Cockerell, Larpent, and Co., East India trade	619,393
Castellain, Sons, and Co., general merchants	69,651
Cotsworth, Powell, and Pryor, Brazilian and Spanish American trade	350,000
Cruikshank, Melville, and Co., East and West India trade	182,984
Coventry and Sheppard, corn trade	200,000
Douglass, C., and Son, corn trade	250,000
Fraser, W. T., East India trade	33,665
Fry, Griffiths, and Co., colonial brokers	90,979
Gower, A. A., and Nephews, Spanish and Mauritius trade	450,832
Giles, Son, and Co., corn trade	152,824
Hastie and Hutchison, corn trade	50,451
Johnson, Cole, and Co., East India trade	122,666
King, Melville, and Co., corn trade	200,000
Kingston, J. and Co., West India trade	25,245
Lackersteen, A. A., East India trade	185,529
Lackersteen and Crake, East India trade	133,091
Leaf, Barnett, Scotson, and Co., warehousemen	85,575
Lyall, Brothers, and Co., East India trade	340,387
Morley, J. and W., warehousemen	119,731
Nevins and Allen, corn trade	69,907
Perkins, Schlusser, and Mullens, East India and Baltic trade	127,327
Phillips, S., and Co., East India trade	101,474
Phillips, L., and Co., East India trade	18,368
Reid, Irving, and Co., West India and Mauritius trade	660,432
Rickards, Little, and Co., East India trade	144,626
Rougemont Brothers and Co., general merchants	109,450
Robinson, W. R., and Co., corn trade	94,362
Ryder, Wienholt, and Co., East India trade	34,587
Reay and Reay, wine trade	47,788
Scott, Bell, and Co., East India trade	99,629
Sargant, Gordon, and Co., colonial brokers	65,254
Sanderson and Co., bill-brokers	2,683,000
Thurburn and Co., East India trade	109,139
Thomas, Son, and Lefevre, Russia trade	401,760
Trueman and Cook, colonial brokers	379,104
Usborne, T., and Son, corn trade	59,457
Woodley, W. and J., corn trade	99,509."

Subsequently to the subsidence of the crisis, and when a sufficient period had elapsed to allow the principal estates estimated to produce 20*s.* in the pound, to be liquidated, the *Morning Herald,* in presenting some facts on the subject, thus alluded to the course of affairs :—

"In the midst of the fraud and deception daily practised in trade and business relations, it is highly gratifying to be in a position to refer to the speedy fulfilment of honourable engagements, and to the several favourable instances which have occurred through the progress of the late panic, in which houses of established credit and wealth did not hesitate, when it became their bounden duty to do so, to suspend, in order to protect the interests, not only of themselves, but also of those who were interested in their estates. Among the earliest of the large houses prostrated was that of Messrs. Dennistoun and Co., of Glasgow, Liverpool, and London, who, with connections in Australia and America, were under liabilities to the extent of upwards of £2,000,000. Notwithstanding the reputed position of the partners, and the knowledge of the existence of a surplus of at least £500,000, it was considered advisable, pending the fury of the storm, to arrest their business transactions, and to adopt measures for a liquidation. The certainty of 20*s*. in the pound, with interest at the rate of 5 per cent., was fully relied upon, since the accounts, when they were placed under investigation, exhibited the most encouraging results; but it was arranged to extend the instalments over a period running from January, 1858, to June, 1860. The assets were taken with due regard to ultimate realization, and although in the aggregate they exhibited an enormous excess, it was considered desirable to make allowance for depreciation, etc. Even including a large margin, they still represented a highly satisfactory total; and as the dates for the first instalments approached, not only were these paid, but succeeding ones also. In this manner have they proceeded, rapidly realizing their assets, until at length they have placed themselves in a situation to announce a further anticipation of payments, and to advise on the 30th of November their readiness to discharge the instalments due respectively the 31st of December, 1859, and the 30th of June, 1860. Circumstances may have in some degree influenced the value of the mass of securities they held, and a change in the prices of produce may have also assisted to bring about this agreeable conclusion; but at the same time the basis of business must have been very sound, and its conduct well regulated to have ensured so favourable an issue, surrounded as their concerns must have been by disaster and supposed loss. An event such as this in mercantile annals, especially after the experience of the last ten years, deserves something more than passing comment; but it is satisfactory to believe that in addition to the case of

Messrs. Dennistoun and Co., and those of the principal Swedish houses, who were, unfortunately, during the crisis, temporarily compelled to suspend, there are others in which the liquidation will not be less punctual, though the engagements and proportional assets may not prove so large. The favourable adjustment effected in the estates of Messrs. Sewells and Neck, Messrs. Albert Pelly and Co., and Messrs. Rew, Prescott, and Co., all engaged in the Swedish trade, with the prompt manner in which Messrs. Heine, Semon, and Co., the German bankers and exchange brokers, provided for their liabilities, have already been more than once referred to; but failures have occurred subsequently not to be comprised within the actual pale of the late panic, though probably some of their difficulties originated in the depression which followed that revulsion. The firms specially indicated are Messrs. Maitland, Ewing, and Co.,* and Messrs. Rawson, Sons, and Co.,† both engaged in the China trade, whose affairs are in steady course of arrangement, with assets accumulating, and every prospect of a full payment of 20*s*. in the pound. This is so far appreciable, as it shows that the commercial character of the country has not altogether descended to the depth of degradation supposed from the revelations made in other cases, though it is still difficult to close the eyes to the amount of fraud perpetrated under the guise of fair trading. Taking the number of houses which have paid in full, with the average of dividends declared under the various estates, it is thought that the results will contrast favourably with those apparent during and subsequent to the crisis of 1847-48. Only exceptional instances then appeared of 20*s*. in the pound, even among the best of the firms that suspended, and in this category could not be included Messrs. Cockerell, Larpent, and Co., Messrs. Barclay Brothers, and many others.

"The appended list makes a respectable exhibition for the several houses stated, and while the figures, through the process of liquidation, may have varied, they have not, like in the principal of those put forward in the preceding period, turned out altogether fallacious, and eventually exhibited serious

---

"* This house has liquidated in full, with interest.

"† Since this was penned an alteration has occurred in Messrs. Rawson's prospects, and it is now feared they will not liquidate in full.

deficiencies. In some of the cases the engagements were much larger at the date of suspension, but by the time the meetings took place they became considerably reduced.

			Liabilities.	Assets and Claims.	Surplus.
		1857.	£	£	£
Nov.	7.	J. and A. Dennistoun and Co. ...	2,142,701	2,935,992*	793,291
Dec.	5.	Sewells and Neck	150,675	188,257	37,582†
Dec.	7.	Albert Pelly and Co.	36,316	85,741	49,425‡
Dec.	10.	Heine, Semon, and Co.	93,084	130,916	37,832§
Dec.	14.	Rew, Prescott, and Co.	95,703	103,451	7,748
		1858.			
April	3.	Maitland, Ewing, and Co..........	164,392	193,049	28,650
June	2.	Rawson, Sons, and Co.	422,151	457,399	35,248‖

As supplementary matter to this Section, it has been thought desirable to give the Report of the Bank Acts Committee as corroborative of the general circumstances detailed.

THE SELECT COMMITTEE appointed to inquire into the OPERATION of the BANK ACT of 1844 (7 and 8 Vict. c. 32), and of the Bank Acts for Ireland and Scotland of 1845 (8 and 9 Vict. c. 37 and 38), and who were instructed to inquire into the Causes of the recent Commercial Distress, and to investigate how far it has been affected by the Laws for Regulating the Issue of Bank Notes payable on demand, and who were empowered to Report their Observations thereupon, together with the MINUTES of the EVIDENCE taken before them :— HAVE considered the matters to them referred, and have agreed to the following REPORT :—

1. The ten years which have elapsed since the last Committee sat under the same Order of Reference, viz., the Committee on Commercial Distress, which reported in 1848, have been marked by many circumstances of peculiar interest and importance. The foreign trade of the United Kingdom has in that period increased with a development unprecedented, perhaps, by any other instance in the history of the world.

"* This item included Liverpool Borough Bank shares held by partners, £208,873.

"† Private estates, estimated to realize £20,C00, not included.

"‡ Principal surplus, private property.

"§ Private property, £15,000, not included.

"‖ Separate estates of partners about £12,600, included in the account."

The exports, which before 1848 had never exceeded £60,110,000—the amount which they attained in 1845—have risen with little variation and with great rapidity; and in 1857, notwithstanding the severe commercial pressure which marked the latter portion of that year, they stood at £122,155,000.

2. In the year 1849, the newly-discovered mines of California began to add perceptibly to the arrivals of gold; and in 1851, the supply was increased by the still more fertile discoveries in Australia. The following figures, for which your Committee are indebted to the authorities of the Bank, will show how important an addition appears to have been made to the circulating medium of the world from these new sources of supply.

ESTIMATED INCREASE OF THE EUROPEAN STOCK OF BULLION IN SEVEN YEARS, 1851-1856.

	Imports from Producing Countries.		Exports to the East from Great Britain and the Mediterranean.	
	Gold.	Silver.	Gold.	Silver.
	£	£	£	£
1851	8,654,000	4,076,000	102,000	1,716,000
1852	15,194,000	4,712,000	922,000	2,630,000
1853	22,435,000	4,355,000	974,000	5,559,000
1854	22,077,000	4,199,000	1,222,000	4,583,000
1855	19,875,000	3,717,000	1,192,000	7,934,000
1856	21,275,000	4,761,000	479,000	14,108,000
1857	21,366,000	4,050,000	529,000	20,146,000
	130,876,000	29,870,000	5,420,000	56,676,000

GOLD.

The total import of gold in seven years has been, say	£130,000,000
The exports of gold bullion and British gold coin to India, China, Australia, the Cape, Brazils, the West Indies, United States, etc., may be taken at	22,500,000
Which would leave as the increase to the European stock of gold	£107,500,000

SILVER.

The exports of silver to India and China have been	£56,676,000	
The imports from the producing countries	29,870,000	
Making the amount of silver abstracted from the European stock		26,806,000
And the estimated increase in the European stock of bullion		£80,694,000

3. The remission of duties upon articles of necessity, and upon the raw materials of industry, and the great increase of trade to which your

Committee have referred, were naturally attended by a very remarkable improvement in the comforts and consuming power of the people, as exhibited in the imports; and especially in the vast increase in the clearances of those articles which enter most materially into the consumption of the working classes. It is probable that to this cause ought chiefly to be attributed the great increase which is believed to have taken place in the circulating medium of the United Kingdom. Mr. Weguelin, a Member of the Committee, and then Governor of the Bank, stated to the Committee of 1857, that this increase was estimated by those in whose judgment the Bank Directors placed the greatest reliance, at 30 per cent. in the six years then last elapsed. The total gold circulation is believed by him now to amount to nearly £50,000,000. The whole circulation of notes, which under the Acts of 1844 and 1845 are permitted to circulate, without being represented by bullion, retained for that purpose in the coffers of those who issue the notes, is £31,623,995, of which £14,475,000 are issued by the Bank of England; £7,707,292 by the English country bankers; £3,087,209 by the Scotch, and £6,354,494 by the Irish bankers.

4. With regard to bank notes, it is interesting here to observe, that in the smaller denominations, those, namely, which enter most into the retail transactions of the country, the number has considerably increased, concurrently with the increase of the gold circulation above referred to. The £5 and £10 notes of the Bank of England, which in 1851 were £9,362,000, had risen in 1856 to £10,680,000.

5. At the same time, for a reason which will presently be noticed, a great diminution has been observable in the use of notes from £200 and upwards.

6. The silver currency has in the same time increased as follows, viz. :—

SILVER COIN ISSUED TO THE PUBLIC IN EXCESS OF RECEIPTS FROM THE PUBLIC.

1851	£26,307
1852	420,418
1853	554,442
1854	36,803
1855	47,754
1856	289,142
1857	242,273

7. While this expansion of trade was in progress, and the precious metals received this remarkable addition, a new feature in the banking business of the country was observable. The joint-stock banks in London entered more and more into competition with the private banks, and by their practice of allowing interest on deposits, began to accumulate vast amounts. On the 8th June, 1854, the private bankers of London admitted the joint-stock banks to the arrangements of the clearing-house, and shortly afterwards the final clearing was adjusted in the Bank of England. The daily clearances are now effected by transfers in the accounts which the several banks keep in that establishment. In conse-

quence of the adoption of this system, the large notes which the bankers formerly employed for the purpose of adjusting their accounts are no longer necessary. The diminution in the use of these notes is shown by the following figures :—

Bank Notes of £200 to £1,000.	
1852	£5,856,000
1857	3,241,000

8. Meanwhile the joint-stock banks of London, now nine in number, have increased their deposits from £8,850,774 in 1847 to £43,100,724 in 1857, as shown in their published accounts. The evidence given to your Committee leads to the inference that of this vast amount, a large part has been derived from sources not heretofore made available for this purpose; and that the practice of opening accounts and depositing money with bankers has extended to numerous classes who did not formerly employ their capital in that way. It is stated by Mr. Rodwell, the Chairman of the Association of Private Country Bankers, and delegated by them to give evidence to your Committee, that in the neighbourhood of Ipswich this practice has lately increased fourfold among the farmers and shopkeepers of that district; that almost every farmer, even those paying only £50 per annum rent, now keep deposits with bankers. The aggregate of these deposits of course finds its way to the employments of trade, and especially gravitates to London, the centre of commercial activity, where it is employed first in the discount of bills, or in other advances to the customers of the London bankers. That large portion, however, for which the bankers themselves have no immediate demand passes into the hands of the bill-brokers, who give to the banker in return commercial bills already discounted by them for persons in London and in different parts of the country, as a security for the sum advanced by the banker. The bill-broker is responsible to the banker for payment of this money at call; and such is the magnitude of these transactions, that Mr. Neave, the present Governor of the Bank, stated in evidence, "We know that one broker had 5 millions, and we were led to believe that another had between 8 and 10 millions; there was one with 4, another with 3½, and a third above 8. I speak of deposits with the brokers."

9. It thus appears that since 1847 three most important circumstances have arisen, affecting the question referred to your Committee, viz.:—

1. An unprecedented extension of our foreign trade.
2. An importation of gold and silver on a scale unknown in history since the period which immediately succeeded the first discovery of America; and,
3. A most remarkable development of the economy afforded by the practice of banking for the use and distribution of capital.

10. In the years which immediately succeeded the great commercial crisis of 1847-48, the natural effect of such a crisis on the minds of persons engaged in trade was exhibited, and for a time prudence and caution were the marked characteristics of the commercial world. The bullion

in the Bank meanwhile accumulated, increasing, with little variation, until, in July, 1852, it amounted to £22,232,000. At this time the notes in the hands of the public ran to the unusually large amount of £23,380,300, yet scarcely exceeded the amount of bullion, while the reserve of notes in the banking department of the Bank of England was 12½ millions, and the minimum rate of interest two per cent.

11. The consequence of such a state of things was manifested in the year 1853, when the exports, which in 1852 had amounted to £78,076,000, rose to £98,933,000. The bullion at the same time declined, and was on the 22nd October of that year £14,358,000, while the reserve went down to £5,604,000, and the minimum rate of interest rose to five per cent.

12. In March, 1854, war was declared against Russia, and an expenditure of nearly 90 millions is estimated to have been incurred by England on this account. The foreign payments were largely made in specie, which to a great extent was hoarded in the East. Foreign loans were also contracted in London for the purposes of the war. The aggregate trade of the United Kingdom varied little. The Bank rate of discount was raised in May, 1854, from 5 to 5½ per cent., and continued at that rate till August 3, when it was again reduced to 5. On the 5th April, 1855, it was reduced to 4½, the bullion then standing at £15,079,000, and the reserve at £8,580,000. The bullion continued to rise, until in June it amounted to £18,169,000, and the reserve to £11,887,000. Before the end, however, of that year a great change occurred, and on the 27th December the bullion stood at £10,275,000, the reserve at £6,993,000, while the minimum rate of interest had been raised on 18th October to 6 per cent. for 60 days, and 7 per cent for 95 days, at which rate it stood till the following May. The changes in the rate of discount which took place from April, 1855, to March, 1857, are thus stated by Mr. Weguelin:—

"I have here a list of the various changes in the rates, beginning at April 5th, 1855, when the minimum rate of discount for bills having not more than 95 days to run was 4½ per cent. On May 3rd, it was reduced to 4 per cent. On June 14th, it was reduced again to 3½ per cent. On September 6th, it was raised to 4 per cent. On September 13th, to 4½ per cent. On September 27th, to 5 per cent. On October 4th, to 5½ per cent. The Committee will remark that very rapid rise in the rate of interest which was caused by the commercial demand for accommodation, and for the export of bullion, occurring at the same time with a considerable demand for bullion to supply the armies in the East. On the 18th of October the rate was 6 per cent. for bills having 60 days to run, and for bills having 95 days to run it was 7 per cent. In 1856, on the 22nd of May it was reduced to 6 per cent., and on the 29th of May to 5 per cent., and on the 26th of June to 4½ per cent., the minimum rate. There then occurred a great demand, and the rate was raised by order of the Governor, on October the 1st, to 5 per cent. That was not on the ordinary weekly court day, but in the interval of the court. On October the 6th (which was again not on a court day, but on a Monday) the rate was raised to 6 per cent. for 60 days' bills, and to 7 per cent. for bills not having more than 95 days to run. On November the 13th, the minimum rate for bills of all descriptions having not more than 95 days to run was raised to 7 per cent. On December the 4th, it was reduced to 6½, and on December the 18th to 6 per cent., at which it now stands. Here is also an account of the variations with regard to temporary advances upon stock. The first recent deviation from the practice that temporary advances on stock and Exchequer bills should be made at the Bank minimum rate ordinarily, and at a half per cent. below the minimum during the shuttings, seems to have occurred in July, 1854, when Exchequer bond scrip was in the market. The Bank minimum rate was then 5½ per cent., temporary advances were made at 5 per cent., and advances were made on Exchequer bond scrip at 4 per cent. I believe that was an especial arrangement at the time, which had not much reference to the state of the money-market. The term of those advances varied from 14 to 31 days. During the shutting for the dividends due in January, 1856, the allowance of a half per cent. on advances on stock, etc., was withdrawn, and no such advances have since been made at a rate below the Bank minimum. On the 8th of January, 1856, the demand for advances, chiefly on Turkish scrip and bonds continuing beyond the payment of the dividends, the term was contracted to 14 days. During the shutting for the April dividends this restriction was removed. After the April payment the general term was 14 days; but there does not appear to have been any restriction to that period. After the October payment the term was contracted to seven days; and on the 16th of October the Bank refused to advance on any Government securities except Exchequer bills. About the 11th of November the Bank declined to re-discount bills having more than 30 days to run; that is, bills which had been advanced upon by brokers. During the shutting for January, the usual course was resumed, without restriction as to stock or term. On the 9th of January, 1857, the rate for advances on Government stocks and Exchequer bills was raised to 6½ per cent., the rate on bills of exchange remaining at 6 per cent.; and this restriction remained in force till the present shutting. It is now 6 per cent. In addition to those restrictions, I may state that the Governors have placed certain restrictions upon the business conducted through the discount-brokers. In their business with them, when it suited the convenience of the Bank to have only short bills, they have limited their advances to the discount-brokers to 30 days, or have insisted upon their bringing in bills not having more than 30 days to run; the object being to obtain such a command of resources constantly returning to the Bank reserve as should keep the Bank safe in that respect."

13. Down, therefore, to the close of the inquiry of 1857, the Bank of England had continued, under the Act of 1844, to conduct its business without difficulty. The rate of discount had been raised, and *écheance* of bills shortened, as the drain for bullion appeared to the Directors to render these measures necessary from time to time. But neither the failure of the silk crop in Italy, with the bad harvests in France and other parts of Europe, and the commercial drain thence arising, nor the requirements of specie for the military service, nor both these causes combined, had occasioned any important derangement of our monetary system.

The course of trade may be collected from the exports of the years referred to, viz.:—

1852	£78,076,000
1853	98,933,000
1854	97,184,000
1855	95,688,000
1856	115,826,000
1857	122,155,000

These exports do not include shipments of stores in Government transports.

14. In the earlier part of the autumn of last year, the trade of the United Kingdom was generally considered to be in a sound and healthy state, and in the words of the Governor of the Bank, in reply to the following question :—

"Was there, in the month of August, any circumstance which caused you to be apprehensive of any reason for raising the rate of discount?—Not in the month of August; things were then pretty stationary; the prospects of harvest were very good; there was no apprehension that commerce at that time was otherwise than sound. There were certain more far-seeing persons who considered that the great stimulus given by the war expenditure, which had created a very large consumption of goods imported from the East and other places, must now occasion some collapse, and still more those who observed that the merchants, notwithstanding the enhanced prices of produce, were nevertheless importing, as they had done successfully in the previous years. But the public certainly viewed trade as sound, and were little aware that a crisis of any sort was impending, far less that it was so near at hand."

15. In this state of things, the bullion standing at £10,606,000, the reserve at £6,296,000, and the minimum rate of discount at 5½ per cent., the Bank, on the 17th of August, 1857, commenced a negotiation with the East India Company, which ended in a shipment of £1,000,000 in specie for the East. The general aspect of affairs continued without change until the 15th September, when the first tidings arrived of the great depreciation of railway securities in the United States, and immediately afterwards of the failure of a very important corporation, called the Ohio Life and Trust Company. Before 8th October the tidings from America had become very serious; news of the suspension of cash payments by the banks in Philadelphia and Baltimore was received; cotton bills were reduced to par, and bankers' drafts to 105; railroad securities were depreciated from 10 to 20 per cent.; the artisans were getting out of employment; and discounts ranged from 18 to 24 per cent. The transactions between America and England are so intimate, and so large, the declared value of British and Irish produce exported in 1856 to the United States having been £21,918,000, while the amount of securities held by English capitalists in America was by some persons estimated at £80,000,000, that this serious state of commercial disorder there could not but produce in this country great alarm.

16. In New York, 62 out of 63 banks suspended their cash payments.

In Boston, Philadelphia, and Baltimore, the banks generally did the same. The effect of the American calamity fell with the greatest weight upon the persons engaged in trade with that country, and Liverpool, Glasgow, and London naturally exhibited the first evidences of pressure. On the 27th October the Borough Bank of Liverpool closed its doors, and on the 7th November the great commercial house of Messrs. Dennistoun and Co. suspended payment. The Western Bank of Scotland failed on the 9th November, and on the 11th the City of Glasgow Bank suspended its payments, which it has since resumed. The Northumberland and Durham District Bank failed on the 26th, and on the 17th the Wolverhampton Bank for a time suspended payment.

17. Great alarm naturally prevailed in London, the centre of all the monetary transactions of the world. Vast sums deposited with the joint-stock banks, at interest, and employed directly by themselves, or by the bill-brokers, in addition to other moneys deposited by their other customers, were chiefly held at call; and the bill-brokers are stated to have carried on their enormous transactions without any cash reserve; relying on the run off of their bills falling due, or in extremity, on the power of obtaining advances from the Bank of England on the security of bills under discount. The inevitable result of this system, at a time of commercial pressure and alarm, was, that the banks limited their discounts almost exclusively to their own customers, and began to add to their reserves both in their own tills and at the Bank of England. It is well known that a periodical disturbance in the reserve of notes at the Bank of England regularly occurs at the time when the dividends upon the National Debt are paid. Interesting information will be found in the evidence of 1857 as to the effect of this disturbance in aggravating the panic of 1847. It had no such effect last year. By the 24th October that periodical disturbance was at an end. The public deposits also were in a satisfactory state, amounting to £4,862,000. It is interesting to observe, with regard to the private deposits, that the causes to which your Committee have above referred to, as affecting other bankers, tend to increase the balances of the Bank of England, the bank of last resort at a time of panic. Thus, for example, the deposits of the London bankers, which in ordinary times average about £3,000,000, continued to rise during the commercial pressure, and amounted on the 12th November to £5,458,000. The bill-brokers were compelled to resort to that establishment for assistance; and that to so great an extent, that the principal house went to the Bank to ask whether they could obtain discount to an indefinite amount, and actually received, on one day, the day on which the Treasury Letter was issued, no less a sum than £700,000. Two discount houses failed. Speaking of the general discount market, the Governor of the Bank stated: "Discounts almost entirely ceased in London, except at the Bank of England."

18. It is manifest, therefore, that in this emergency everything depended on the Bank of England; and it appears to your Committee that the proceedings of that establishment were not characterized by any want of foresight or of vigour. On the 16th July, however, before any

indications of the coming storm were visible in any quarter, the bullion read £11,242,000, the reserve £6,408,000, the discounts and advances £7,632,000, and the Directors reduced the rate of interest from 6 to 5½ per cent. On the 8th October, after the receipt of the American intelligence above referred to, the bullion was £9,751,000, the reserve £4,931,000, the discounts and advances 11,648,000, and the rate of interest was raised again to 6 per cent. Four days afterwards, the rate was raised to 7. The causes of this step are thus stated by the Governor :—

" Then four days afterwards there was another change?—Yes on the 12th. After having raised the rate to 6 per cent. we thought it necessary to give a guarded caution to our agents, showing that we began to be a little uneasy. The rate at Hamburg was 7¾; American discounts then were greatly higher. We also about that time were made aware that the East India Company would want £1,000,000 specie for shipment. The gold was then being taken for New York; we consequently raised the rate of interest under those circumstances to 7 per cent.

" The bullion which was wanted for the East being silver, was to be purchased by the export of gold; that gold to be exchanged for silver upon the continent of Europe, which silver was to be sent to the East?—That was the effect of it; the exports to India were very large each month; but as they were in silver, of course that silver had to be purchased on the Continent or imported from America.

" I think it was about the 12th of October that you were first apprehensive about the Western Bank of Scotland?—Yes; we had no direct application at that time, but there were rumours, and we had intimations which made us aware that they were in difficulties."

19. On the 19th October, the news from America continuing still more unfavourable, there were numerous failures in this country. The bullion had gone down to £8,991,000, and the reserve to £4,115,000, and the rate of interest was raised to 8 per cent. At this time the Bank of France, which in one week had lost a million sterling, raised the rate to 7½, Hamburg to 9. £300,000 in gold had left Liverpool for America.

20. At this juncture negotiations took place for sustaining the Borough Bank of Liverpool and the Western Bank of Scotland, which eventually failed, under the circumstances related by the Governor of the Bank.

21. There was great uneasiness out of doors (*i.e.*, in London), and the Bank had an application from the principal discount house for an assurance, that if it was necessary the Bank of England would give them any loans they might require. That application was made on the 28th October. There were also inquiries for assistance from other Scotch banks; and on the 30th October there was an express for 50,000 sovereigns for a bank in Scotland, part of £170,000, and £80,000 for Ireland. The first shipment of silver by the East India Company then took place. Under these circumstances the rate of discount was raised, on November 5th, to 9 per cent.

22. Between the 5th November and the 9th, an English bank received assistance from the Bank of England; the failure of Dennistoun's house for acceptances due upon nearly two millions occurred, and the Western

Bank failed on the 9th. Failures in London were on the increase. At this time (as was natural) the purchases and sales of stock in the funds were enormous. The transfers were much beyond what they had ever been before. The bullion had sunk to £7,719,000, and the reserve to 2,834,000. On the 9th the rate was raised to 10 per cent.

23. On the 10th November, a leading discount house applied to the Bank of England for £400,000. The Bank of France raised its rate to 8, 9, and 10 per cent. for the three different months. There was another English bank assisted. The City of Glasgow Bank suspended payment. The discounts for that day at the Bank of England rose to £1,126,000. The demand for Ireland was recommencing, and on the 10th and 11th alone the gold sent to Scotland was upwards of £1,000,000. On the 11th, Sanderson and Co., the large bill-brokers, stopped payment, their deposits were supposed to be £3,500,000. There was also an additional supply of gold required for the banks in Scotland. On the 12th the discounts at the Bank exceeded two millions. The following figures sufficiently exhibit the result of the foregoing operations, viz. :—

	Bullion.	Reserve.	Discounts and Advances.
	£ m.	*£ m.*	*£ m.*
10	7,411	2,420	14,803
11	6,666	1,462	15,947
12	6,524	581	18,044

24. The Government Letter was issued on the 12th, and was in the following terms :—

" Gentlemen, " Downing Street, 12th Nov. 1857.

" Her Majesty's Government have observed with great concern the serious consequences which have ensued from the recent failure of certain joint-stock banks in England and Scotland, as well as of certain large mercantile firms, chiefly connected with the American trade. The discredit and distrust which have resulted from these events, and the withdrawal of a large amount of the paper circulation authorized by the existing Bank Acts, appear to Her Majesty's Government to render it necessary for them to inform the Directors of the Bank of England, that if they should be unable in the present emergency to meet the demands for discounts and advances upon approved securities without exceeding the limits of their circulation prescribed by the Act of 1844, the Government will be prepared to propose to Parliament, upon its meeting, a Bill of Indemnity for any excess so issued.

" In order to prevent this temporary relaxation of the law being extended beyond the actual necessities of the occasion, Her Majesty's Government are of opinion that the Bank terms of discount should not be reduced below their present rate.

" Her Majesty's Government reserve for future consideration the appropriation of any profits which may arise upon issues in excess of the statutory amount.

" Her Majesty's Government are fully impressed with the importance of maintaining the letter of the law, even in a time of considerable mercantile difficulty; but they believe that, for the removal of apprehensions which have checked the course of monetary transactions, such a measure as is now contemplated has become necessary, and they rely upon the discretion and prudence of the Directors for confining its operation within the strict limits of the exigencies of the case.

"We have, etc.,
"(Signed) "PALMERSTON.
"G. C. LEWIS."

"To the Governor and Deputy-Governor
"of the Bank of England."

25. Whatever effect this letter may have had in other ways in calming the public mind, and so tending to mitigate the severity of the pressure, it did not immediately diminish the demand for discounts and advances. This continued to increase until 21st November, on which day the Bank had advanced in discounts £21,600,000, a sum exceeding the whole amount of their deposits, both public and private; a sum nearly three-fold the amount of their advances in July, when the rate was reduced to 5½ per cent., and more than double what they had advanced on the 27th October, when the first bank failed. Half of these loans were made to the bill-brokers, and were partly made upon securities which, under other circumstances, the Bank would have been unwilling to accept. They were made for the purpose of sustaining commercial credit in a period of extreme pressure.

26. The letter was issued on the 12th November, but whilst in 1847 it was not found necessary for the Bank Directors to avail themselves of the permission so given them to exceed the limits imposed by law, that necessity in this instance actually arose. An issue to the extent of £2,000,000 beyond the legal issue was made to the banking department. The following Account shows the sums actually issued from the Bank to the public:—

AN ACCOUNT, showing the Extent to which the Bank of England availed itself of its Power, under the Authority of Government, to issue Notes to the Public beyond the Limit allowed by the Act of 1844.

		Notes issued to the Public on Securities beyond the Statutory Limit of £14,475,000.
1857, November	13	£186,000
,,	14—15	622,000
,,	16	860,000
,,	17	836,000
,,	18	852,000
,,	19	896,000
,,	20	928,000
,,	21—22	617,000
,,	23	397,000
,,	24	317,000
,,	25	81,000
,,	26	243,000
,,	27	342,000
,,	28—29	184,000
,,	30	15,000

Average of 18 days ...... £488,830.

27. The causes which, in the judgment of the Bank Directors, immediately led to this result, are thus stated by them in their correspondence with the Treasury, laid before Parliament in December last:—

"On the 5th November the reserve was £2,944,000, the bullion in the Issue Department £7,919,000, and the deposits £17,265,000. The rate of discount was advanced to 9 per cent., and on the 9th November to 10 per cent.

"The continual drain for gold had ceased, the American demand had become unimportant, and there was at that time little apprehension that the Bank issues would be inadequate to meet the necessities of commerce within the legalized sphere of their circulation.

"Upon this state of things, however, supervened the failure of the Western Bank of Scotland, and the City of Glasgow Bank, and a renewed discredit in Ireland, causing an increased action upon the English circulation, by the abstraction in four weeks of upwards of two millions of gold to supply the wants of Scotland and Ireland; of which amounts more than one million was sent to Scotland and £280,000 to Ireland, between the 5th and 12th November.

"This drain was in its nature sudden and irresistible, and acted necessarily in diminution of the reserve, which on the 11th had decreased to £1,462,000, and the bullion to £6,666,000.

"The public became alarmed, large deposits accumulated in the Bank of England, money-dealers having vast sums lent to them upon call were themselves obliged to resort to the Bank of England for increased supplies, and for some days nearly the whole of the requirements of commerce were thrown on the Bank. Thus, on the 12th, it discounted and advanced to the amount of £2,373,000, which still left a reserve at night of £581,000.

"Such was the state of the Bank of England accounts on the 12th, the day of the publication of the Letter from the Treasury. The demand for discounts and advances continued to increase till the 21st, when they reached their maximum of £21,616,000.

"The public have also required a much larger quantity of notes than usual at this season, the amount in their hands having risen on the 21st to £21,554,000."

28. The Treasury Letter was the subject of discussion in the House, and an Act of Indemnity having passed, your Committee do not feel called upon to say more than that the evidence appears to them to show that the discretion of the Government was properly exercised.

29. Your Committee will now state to the House the general outline of commercial disaster, as it occurred in the United Kingdom.

30. The first occurrence in this country which caused alarm was the failure of the house of Macdonald and Co., of Glasgow and London, which took place in October, and was accompanied by the failures of Monteith and Co., and Wallace and Co., of Glasgow. The house of Macdonald employed a great many work-people in sewing muslin goods for the home trade and for the American market, and this they carried on to a very large extent. They had been in fair credit till very nearly the time of their failure, but shortly before that period they are described

as having given out that they had changed their mode of doing business for the purpose of embracing a wider field. This however is represented as having been a deception, intended to cover a system to which they had recourse of drawing fictitious bills, and to give to these bills the appearance of genuine business transactions. From the records of the public tribunals, it appears that a very considerable number of persons (one of the partners is said to have admitted as many as 75) in London and other places were employed by this firm, for a small commission, to put their names to fictitious bills, which were then discounted, a large proportion of them in Glasgow; and when the house of Macdonald failed it was found to be indebted to the Western Bank £422,000.

31. The house of Monteith and Co. was indebted to the same bank £537,000; that of Wallace and Co. £227,000.

32. The house of Messrs. Dennistoun and Co. stopped payment on November 7; it is expected to pay its liabilities in full, and its members bear the highest character. But it can occasion no surprise that, on the occurrence of such a crisis as that which took place in America last year, a house, with debts owing to it from that country of nearly two millions, losing at the same time £300,000 by the failure of the Borough Bank of Liverpool, of which the partners were shareholders, should, at a juncture when general alarm prevailed, have been obliged to suspend its payments.

33. During the month of October there was a very great gloom in Glasgow, occasioned by the commercial panic in America, Glasgow being very intimately connected in trade with America, with New York particularly. Towards the end of October that feeling was much increased, from its being well known that the Western Bank were in difficulties from their connection with the three houses which have been above referred to. The bank closed on the 9th of November, at two o'clock. The Western Bank and the City of Glasgow Bank had establishments open at night for the purpose of receiving the savings of small depositors. During the evening of the 9th, the Monday, there was a demand for gold by the savings bank depositors at the branches of the City Bank. On the Tuesday morning, when the doors of the banks were opened, a great number of parties appeared with deposit receipts, demanding gold; one witness, speaking of his own bank, says: "The office of our own establishment was quite filled with parties within a quarter of an hour of the opening of the doors; I think at half-past nine. This run or panic increased, and the continued refusal of the notes of the Western Bank added very much to the excitement. These people who came for money would not take the notes of any bank; it did not matter what bank it was; they refused everything but gold. Two of the banks sent a deputation of the directors to Edinburgh to confer with the managers of the Edinburgh banks on the subject, and to induce them to rescind a decision at which they had arrived, not to take the notes of the Western Bank. They failed in that; the notes of the Western Bank were refused the whole day on the Tuesday. The streets of Glasgow were in a very excited state; crowds were walking about going from one bank to another to see what was going on: there was an immense crowd of people. At the National Securities Savings Bank the run was very great indeed. The National Savings

Bank paid in notes, and then the depositors, having received their deposits in notes, went with those notes to the banks that had issued them to demand gold. The City of Glasgow Bank did not open on Wednesday the 11th. Troops were sent for by the authorities, who were afraid of some disturbance. The magistrates issued a proclamation either on the Tuesday night or on the Wednesday morning, and it was circulated very extensively, advising the people not to press upon the banks for payment, and to take the notes of all banks. The magistrates held a meeting on the Wednesday morning, and they issued an order to all the rate collectors over the city to take all notes presented to them; they did all they could to allay the excitement. In accordance with the provisions of the Act of 1845 the banks held a considerable quantity of gold, but they were under the necessity of having more gold from London; upon two occasions, on the Wednesday and the Thursday mornings, the 11th and 12th, large remittances of gold from London arrived about 10 o'clock in the forenoon; it was taken down in waggons to the banks, and escorted by a strong police force, and no doubt, seeing such immense quantities of gold come excited a great commotion in the town.

Mr. Robertson, the Manager of the Union Bank, is asked—

"What was the nature of that excitement; was it of a pleasurable character?—It was such a novelty; in the first place, a large bank stopping payment, and then such quantities of gold coming down from London; it was quite a new thing to the people altogether.

"Had it any effect in regard to the panic?—I should think it must have had an effect; the people saw there was gold there to pay them if they wanted it; but by the Thursday morning the panic was entirely allayed; it entirely ceased on the Wednesday afternoon about two o'clock; at half-past two I do not think there were half-a-dozen people in our establishment.

"To what do you attribute the cessation of the panic?—I cannot answer that question; whether the people thought better of it I cannot tell.

"When was it that it first became known that the other banks would take the Western Bank's notes?—I should like to speak of what I know positively; I understood that the Edinburgh banks on the Tuesday night, the 10th, had agreed to take the notes of the Western Bank amongst themselves. At the meeting it was announced to them that the City Bank had then failed; then there was an alteration again, and they agreed neither to take the notes of the Western Bank nor of the City Bank; and that was acted upon during the Wednesday by their agents in Glasgow, but not to the full extent after the Tuesday; they were partially taken.

"Had the notes of the Western Bank began to be taken in the course of the Wednesday?—Yes.

"And at two o'clock on the Wednesday afternoon you consider that the panic had come to an end?—Quite.

"And on the Thursday the Government Letter was issued?—Yes, I believe so."

34. It has been observed that the panic in Glasgow had ceased before

the Treasury Letter was issued, and that the demand at the Bank of England for advances and discounts did not cease with the publication of that letter; after which date it cannot of course be attributed to any fear that there was a limit to the quantity of bank notes. On the contrary, we have seen that the advances by discount kept rising continually, and though the rate of 10 per cent. was still maintained, they rose from £15,900,000, at which they stood on the day preceding the issue of that letter to £21,600,000 on the 21st November. It is obvious, therefore, that the principal causes of the commercial crisis of 1857 must be sought elsewhere. That calamity cannot be attributed exclusively or chiefly to panic occasioned by the operation of the Act of 1844. Since, too, the difficulties here experienced took their origin from America, where no such law is in force; and that crisis was felt in still greater severity than here, by countries in the north of Europe, whose currency is regulated by laws widely different from ours, it remains for your Committee to inquire whether any cause or causes, common to all those countries, and sufficient to account for the occurrence of commercial disasters in them all respectively, have been disclosed by the evidence.

35. For a general review of the failures which occurred in England your Committee have been indebted to Mr. Coleman, and to Mr. Ball, of the firm of Messrs. Quilter and Ball, both eminent accountants in London. These gentlemen do not profess to have studied abstruse questions of currency; they do not represent themselves as particularly conversant with the operation of the Act of 1844. They, however, assign what appears to your Committee an adequate cause for the recent commercial crisis. Availing themselves of their experience in 1847, the affairs of which have now been finally closed, to illustrate the transactions of 1857, which still appear in estimate, and are therefore liable to correction, they ascribe the calamities of both periods to the same principal cause, viz., the great abuse of credit, and consequent overtrading. They notice also this difference between the two periods: many of the houses which fell in 1847, had once been wealthy, but had long ceased to be so. Those of 1857 had, with few exceptions, never possessed adequate capital, but carried on extensive transactions by fictitious credit. In 1847, for example, one house, which had been originally wealthy, failed with liabilities amounting in the whole to upwards of £1,800,000, of which not quite £1,000,000 were to be paid by other parties, leaving more than £800,000 the direct liabilities of the house. The capital, as represented in their books at the time of suspension, was £215,000, and the assets, according to their own valuation, £800,000, or nearly sufficient to meet the whole of their liabilities. Very different, however, was the valuation of the accountant, who estimated their assets at £185,000, and even that was materially diminished in the result. The dividend ultimately paid was only 9*d.* in the pound. This firm, originally merchants, insensibly advanced their capital to planters in the East Indies, until it became necessary for them to be the planters themselves. They then were compelled to obtain advances from others, which they accomplished by the sale and circulation of bills in the East Indies upon the house to a great extent. Obtaining credit in that manner they postponed their fall many years, and ultimately

fell, paying only 9*d*. in the pound. In this case advances had been made on the credit of the next year's crop. This was an extreme case, and was connected with peculiar considerations at that time affecting the price of colonial produce, the principal property of the house. But Mr. Coleman, from whose evidence these particulars have been taken, says, that the estates which came under his notice as insolvent in that year paid generally very small dividends, not averaging more than 4*s*.

36. Another example of the same period is described by Mr. Ball as follows: It was that of a house which failed in 1847; they were engaged very largely as merchants in this country, and they were a house of very old standing. In the course of their business they came under advances to a house in one of the colonies, on the security of the crops to be sent forward from time to time. The parties to whom those advances were so made failed to repay them; that is to say, to recoup the London house for them; and eventually the London house was obliged to take upon themselves the business which was originally conducted by those whom they accommodated with advances; in other words, the merchant in London did practically become the planter and the owner of estates. After he had so become the planter, his position was changed from that of being a person who made advances, and he himself found it necessary to obtain advances. Most likely the course would be this, that the house on the other side, perhaps the correspondents themselves of the London house, and it might be identical with the London house, would draw upon the London house, or draw upon some third party and remit to the London house; which bill the London house would take to its banker and get discounted, and by that process would be placed in funds to provide from time to time for its own engagements. The result of which would be to sustain for some time the credit of the house, after the capital of the house had been exhausted. The effect would be to enable them to hold produce in expectation of better prices; the longer it was continued the heavier would be the ultimate loss. After an interval of ten years, this house has, within the last few months, paid a final dividend, making a total of 1*s*. 10*d*. in the pound.

Mr. Ball is asked—

"Looking back to the experience of the year 1847, were the dividends that were paid by the insolvent houses generally very small?—The average dividend would be small, as far as I recollect. Here and there there would be a house which would pay in full, or would pay a very large dividend; but the general result was, that a small dividend upon the whole was received by the creditors.

"Looking back now, with your experience, to the results of 1847, is it your opinion that if the law had afforded greater facilities for obtaining credit at that time for the purpose of sustaining these houses longer, the result would have been more advantageous to the houses themselves, or to the community at large?—Knowing what I do of the internal state of those houses when they did stop, I should say that had they been able to obtain further credit for a continued period of time, it would only have had a temporary effect upon their position, and that most of them (of course I have a reserve of some good cases in my mind) from their internal condition being worn out, and from the want of real capital in their concerns, must have failed ultimately, and that the longer the assistance was continued simply upon their credit, the greater the ultimate loss would be.

"Such is your view of the failures that took place in 1847, speaking generally?—That is my view."

37. Your Committee have thought it not irrelevant to place on record these instances, which it was not in the power of their predecessors in 1848 to give, because they furnish an instructive example how readily misfortunes are at the time attributed by the sufferers, and others sympathizing with them, to the operation of statutory enactments—which misfortunes, upon a full review of all the circumstances attending them, it is obvious that no wisdom of the Legislature, no regulation of the currency, could have prevented.

38. Your Committee have before them the particulars of 30 houses which failed in 1857. The aggregate liability of these houses is £9,080,000, of this sum the liabilities which other parties ought to provide for amount to £5,215,000, and the estimated assets £2,317,000. Besides the failures which arose from the suspension of American remittances, another class of failures is disclosed. The nature of these transactions was the system of open credits which were granted; that is, by granting to persons abroad liberty to draw upon the house in England to such an extent as had been agreed upon between them; those drafts were then negotiated upon the foreign exchanges, and found their way to England with the understanding that they were to be provided for at maturity. They were principally provided for, not by staple commodities, but by other bills that were sent to take them up. There was no real basis to the transaction, but the whole affair was a means of raising a temporary command of capital for the convenience of the individuals concerned, merely a bare commission hanging upon it; a banker's commission was all that the houses in England got upon those transactions, with the exception of receiving the consignments probably of goods from certain parties, which brought them a merchant's commission upon them; but they formed a very small amount in comparison with the amount of credits which were granted. One house at the time of its suspension was under obligation to the world to the extent of about £900,000, its capital at the last time of taking stock was under £10,000. Its business was chiefly the granting of open credits, *i.e.*, the house permitted itself to be drawn upon by foreign houses without any remittance previously or contemporaneously made, but with an engagement that it should be made before the acceptance arrived at maturity. In these cases the inducement to give the acceptance is a commission varying from ½ to 1½ per cent. The acceptances are rendered available by being discounted, as will appear hereafter, when the affairs of the banks which failed come under our notice.

39. The obvious effect of such a system is first unduly to enhance, and then, whilst it continues, to sustain the price of commodities. In 1857, that fall of prices which, according to Mr. Neave, far-seeing people had anticipated, actually occurred. Tables have been put in by more than one

of the witnesses, exhibiting an average fall of 20 or 30 per cent., in many instances much more, upon the comparison of July, 1857, with January, 1858. It needs no argument to prove what effect such a fall must have upon houses which had accepted bills, on the security of produce consigned, to the extent of one hundred times the amount of their own capital. The witness says—

"In the case which you are now describing to the Committee, these transactions had gone on to the extent of £900,000. The real guarantee was partly produce and partly bills of exchange; to whatever extent that produce was depreciated, of course the liability of the firm to meet such depreciation of produce was about one hundredth part of the whole of their liabilities?—That is so.

"Do you consider that case to be a fair illustration of the recent commercial disasters which have occurred?—I think it is, though I should mention that in some cases the proportion of capital possessed was larger than that which I have mentioned.

"In some cases, also, perhaps it might be smaller?—In some cases considerably smaller. In some cases I have known houses come under very large obligations, who had really no capital at all."

40. This practice appears to have grown up of late, and to be principally connected with the trade of Sweden, Denmark, and other countries in the north of Europe. One house at Newcastle is described as conducting before 1854 a regular trade in the Baltic. They were not great people, but were respectable people, and were doing a moderately profitable trade. They unfortunately entered upon this system of granting credits; and in the course of three years the following result ensued, viz., in 1854 their capital was between £2,000 and £3,000; in 1857 they failed for £100,000, with the prospect of paying about 2*s*. in the pound.

41. For other instances of this abuse of credit, your Committee refer to the evidence, concurring entirely in the opinions expressed by the witnesses, that the great abuse of credit is a feature common to the two years 1847 and 1857, and has been, in their judgment, the principal cause of the failures that took place in those years.

Mr. Coleman says—

"Speaking generally with regard to 1847, of which your experience is now complete, are you prepared to say that the failures which occurred in that year were owing to any imperfection of the law by which the facilities for obtaining credit were unduly curtailed?—No.

"With regard to the year 1857, what would your answer be to the same question?—That every house which applied and deserved assistance received it.

"From whence?—From the Bank of England, as far as I know; and more, that in the case of two houses which came under my personal control, I applied to know whether they could have assistance, and the answer was, Yes; guarantees were obtained to the amount required for one house, but I found that I could not advise their being used. The applications, when made by me, were immediately responded to by the Bank of England.

"The alteration of your opinion, I suppose arose from the fact of further investigation into the state of solvency of the concern ?—And the continued bad intelligence from the north of Europe with regard to failures.

"The failures of their correspondents in the north of Europe also being communicated ?—Yes."

42. The commercial crisis was very little felt in Ireland until the failure of some of the banks in England and Scotland. The trade of Ireland, with the exception of that of Belfast, being little connected with the United States, did not feel directly the effect of the failures there; but when failures began to take place at home there was an internal pressure consequent upon them, which, about the early part of the month of November, manifested itself severely in a demand for gold by depositors and holders of notes, and there was a run on the savings banks. The Bank of Ireland advanced to the Banks in Ireland requiring gold to the extent of about £250,000; and they were obliged to draw from the Bank of England from £1,000,000 to £1,200,000 besides. Belfast has a large trade with the United States, as well as a constant intercourse with Scotland, but there was no alarm until the time of the Scotch bank failures. There was then what had never been known before in Belfast since the institution of the joint-stock banks, a considerable run for gold in exchange for their notes. But the amount of gold which they held under the Act of 1845 was a source of strength. The banks appear to be well constituted, and no serious results ensued.

43. In London no bank failed. In Liverpool the Borough Bank, in Glasgow the Western Bank of Scotland, in Newcastle the Northumberland and Durham District Bank, failed in the months of October and November last. The City of Glasgow and Wolverhampton Banks suspended payment, but have since resumed.

44. Your committee have examined Mr. Joshua Dixon, who, in August, 1857, first assumed the post of managing director of the Borough Bank; Mr. Fleming, who has been since July, 1857, assistant manager, manager or liquidator of the Western Bank of Scotland; and Mr. Kirkman Hodgson, a member of the House, and director of the Bank of England, who, being well acquainted with the trade of Newcastle, went to that town in November, for the purpose of ascertaining how far it was right that the Bank of England should give assistance to the Northumberland Bank.

45. The state of these three banks at the time of their failure may be collected from the following summary, viz. :—

Mr. Joshua Dixon, for many years resident in the United States, and once a private banker at New Orleans, settled at Liverpool in 1852, and soon afterwards became a shareholder and director of the Borough Bank. This institution was originally a private Bank, that of Messrs. Hope, in whose hands it was prosperous, and they retired as wealthy men about the year 1834. In 1847, however, the Borough Bank was under the necessity of obtaining assistance from the Bank of England. When Mr. Dixon became connected with it, he found that the Board, which consisted of 12 directors, chose two managing directors and a chairman. The entire management of the Bank was amongst the managing directors and the manager. On the 1st of August, 1857, Mr. Dixon himself became a managing director, and thus describes the state in which he found the affairs of the bank:—"Its position," he says, "was, that of its available means being very much reduced, being far smaller than was at all consistent with the sound and safe position of any bank." Speaking irrespectively of any general commercial pressure, he tells your Committee, that—From the 1st of August, when his attendance at the bank was daily, as he became more and more thoroughly acquainted with the position of individual accounts, and with the whole circumstances of the bank in proportion as time lapsed, he became more and more convinced that the position of the bank was one of exceeding danger. When the commercial crisis showed itself, of course the danger to the Borough Bank became imminent, and they made an application to the Bank of England for assistance, some time between the 20th and the 23rd of October. The position, in general terms, of the bank was, that its assets were all locked up and unavailable, and that some £600,000 or £700,000 of its assets or claims on its debtors, which had, until a short time previously, been considered good, could not be relied upon, even for ultimate realization. About £3,500,000 bills were at that time in London under the indorsement of the Borough Bank of Liverpool; of which from £700,000 to £1,000,000 had no negotiable validity at all, except the indorsement of the Borough Bank of Liverpool.

46. Pending the negotiations with the Bank of England, there appeared in the *Times*, of October 27, an article, stating that arrangements had been made for giving assistance to the Borough Bank, in consequence of which a run took place, and the doors of the bank were closed. That run lasted only two or three hours, but the cash at their command was reduced to between £15,000 and £20,000; while their liabilities on deposits were in all £1,200,000, of which £800,000 were at call, and the remainder at periods varying from two to six months. The dividend of this bank, which had previously been seven per cent., had, at the last meeting, held on 10th July, 1857, been reduced to five; and a sum of £165,000 was, on the face of the report, acknowledged to have been lost. The total loss, so far as the witness could estimate it, amounted to £940,000, being the total capital of the bank. It is ascribed, not to advances improperly made to favoured persons, but to want of discretion in the management.

47. The Western Bank of Scotland was founded in 1832. In 1834 it was already in difficulties, and their correspondents in London dishonoured their bills. They applied to the other banks for assistance, and received it, upon certain conditions. In the year 1838 they applied to the Board of Trade for letters patent, which were refused. At this time the Bank of Scotland and other banks addressed a memorial to Mr. Poulett Thomson, alleging the breach of the conditions referred to. This memorial will be found in the Appendix. In 1847 the Western Bank was again in difficulties, and was assisted by the Bank of England, receiving an advance of £300,000. The then manager, Mr. Donald Smith, appears to have taken alarm from the occurrences of 1847; and,

in 1852, when he retired, the bank, though not in a satisfactory position, stood better than it had stood before, since 1847. When it failed on 9th November, 1857, it appeared that the four insolvent houses of Macdonald, Monteith, Wallace, and Pattison were indebted to it in the sum of £1,603,000; the whole capital of the bank being only £1,500,000. One of the conditions of the co-partnery was, "That if it shall at any time appear, on balancing the company's books, that a sum equal to £25 per centum on the advanced capital stock of the company has been lost in prosecution of the business of the company, such loss shall, *ipso facto*, and without the necessity of any further procedure, dissolve and put an end to the company."

48. Mr. Fleming became assistant manager in July, 1857, and at once examined the affairs. He estimated that even supposing the debts of these four houses (which had not yet become insolvent) were assumed to be good, there appeared on the face of the books as good assets £573,000 of bad debts; and deducting the rest and guarantee fund, which then amounted to £246,000, there remained an apparent deficiency or encroachment on the capital of the bank of £327,000. This of itself nearly approached the limit which dissolved the partnership and put an end to the existence of the board; and of this state of affairs Mr. Fleming believes that up to that time the directors were in a state of almost entire ignorance. In 1853, previously to the first meeting of the shareholders after Mr. Smith's departure, an examination was instituted preparatory to the annual balance. From a confidential paper, having marks upon it in the handwriting of the then manager, it appears that a sum of £260,000 was reported to him as irrecoverable on one branch of the assets, which, nevertheless, appeared as good assets in the published balance-sheet. The modes in which this kind of disguise can be accomplished will perhaps be best understood by stating the manner in which a debt called Scarth's debt, comprised in a different branch of the assets, was disposed of. That debt amounted to £120,000, and it ought to have appeared among the protested bills. It was, however, divided into four or five open credit accounts, bearing the names of the acceptors of Scarth's bills. These accounts were debited with the amount of their respective acceptances, and insurances were effected on the lives of the debtors to the extent of £75,000. On these insurances £33,000 have since been paid as premiums by the bank itself. These all now stand as assets in the books. Though this substitution took place in 1848, yet down to the time when Mr. Fleming's examinations began to bring to light the true state of affairs, the six directors appear to have regarded these sums as part of the available property of the shareholders. This being the actual state of the accounts, the dividend was raised in 1854 from 7 to 8 per cent., and in 1856 to 9. Nine per cent. was the dividend declared in June, 1857, at which date a very slight acquaintance with the books must have led to the strongest suspicion, not to say to the clear conviction, that for some time a considerable portion of the capital had been lost.

49. This bank had 101 branches throughout Scotland. It had connections in America, who were allowed to draw upon it for the mere

sake of the commission. At home it made advances upon "indents;" or, in other words, provided the manufacturer with the capital with which yet unmade cloth was thereafter to be produced. Its discounts, which in 1853 were £14,987,000, had been increased in 1857 (till 9th November) to £20,691,000. With what care this business was conducted may appear from the circumstance that Macdonald's bills were accepted by 124 different parties; that only thirty-seven had been inquired about, and in the case of twenty-one the reports received from the correspondents of the bank were unsatisfactory, or positively bad. Yet the credit given to Macdonald continued undiminished. The rediscounts of the bank in London, which in 1852 had been £407,000, rose in 1856 to £5,407,000. The exchanges of notes in Edinburgh have been always against the Western Bank, and for an average of the last six years, to an extent of not less than £3,000,000 a-year. This circumstance is accounted for by Mr. Fleming, chiefly by reference to the nature of the transactions with Macdonald's and other houses in accommodation bills; £988,000 were due to the bank from its own shareholders.

50. About the end of October the Northumberland and Durham bank applied for assistance to the Bank of England. It was declined, as they could not give any satisfactory explanation of their real position. They applied a second time, urging the great peril in which they were placed by the continued discredit, and by the constant drain of small deposits; they urged also the fear of disturbances and breach of the peace which might ensue if they were to fail, they being so largely connected with collieries and ironworks. Accordingly, on Tuesday, 24th November, Mr. Hodgson went down to Newcastle, and told the directors that he had been sent down by the Bank of England to examine into their books, and see whether it was possible to render them such assistance as would enable them to go on; but that the first condition of the Bank doing anything was that they should prove themselves solvent. The result was that Mr. Hodgson found the liabilities, as then stated, amounting to £2,600,000, of which there were £1,350,000 of deposits, £1,150,000 accounts current, and they had rediscounted £1,500,000, of which they expected that £100,000 would come back upon them, and for which they would ultimately be liable, making altogether £2,600,000. Their assets were of a very peculiar nature indeed, the early realization of which would be almost impossible. They held in securities about £1,000,000 of different kinds. They held in trade bills, that is to say, small bills on shopkeepers of Newcastle, about £250,000, bills which were probably good in themselves, but which were not available anywhere out of Newcastle; they were not bills which could have been discounted in any other part of the money-market. They had in overdrawn accounts £1,664,000, without any specific securities attached to them. Of these £1,664,000, there were £400,000 which one of the directors very candidly confessed must be considered as totally bad, and which ought to have been written off long before, but which still remained in the account as good debts. The capital of the concern was £656,000 nominally, but in reality it was considerably less than that; because in 1847 they had been in trouble, and in order to get out of that trouble they had made a call of £5 or

£10 a share, which was not paid upon some of the shares, which shares were forfeited, and taken by them into the stock of their bank, to be reissued should occasion warrant their doing so. The consequence was, that the subscribed capital of the bank was about £600,000. This statement at once showed that any attempt to help them, short of taking up the whole concern, and liquidating it for them, would be perfectly useless. It was evident that the whole capital was gone; and, looking at the character of the securities, Mr. Hodgson came to the conclusion not only that the capital was gone, but that the bank was totally insolvent. Being very much struck with the extraordinary loss which had taken place in the bank, which, when a private bank, he knew to have been a very flourishing one, he inquired whether there was not some old sore of which nothing had as yet been said. He was told that there was one; there was rather a disinclination to mention what it was, but he felt it his duty to press it, and they told him they had a very large debt with the Derwent Iron Company. He inquired the amount of this debt, and found, much to his astonishment, that it amounted to £750,000, the capital of the bank being £600,000. For that debt there was a kind of security, which consisted of £250,000 of what were called Derwent Iron Company's debentures, which were, however, in reality, nothing but the promissory notes of the directors, there being very few persons in this Derwent Iron Company. The bank had also £100,000 mortgage on the plant, and the remaining £400,000 was totally unsecured. In addition to this original debt then mentioned of £750,000, there is now another charge upon it of £197,000, resulting from bills which have not been paid, and which, in order that the Derwent Iron Company might get them discounted, the bank had endorsed or otherwise guaranteed. These have now come back, so that the total liability for which the Derwent Iron Company is indebted to the bank is about £947,000; very nearly £1,000,000. The Derwent Iron Company appears to have been, almost from the time of the conversion of the bank into a joint-stock bank, very intimately connected with it. Mr. Jonathan Richardson, who was the moving spring of the whole bank, in fact the person who managed everything, was, though not a partner in the Derwent Iron Company, very largely interested in it as holding the royalties upon the minerals which they worked. It appears that the concern has been worked extremely badly; that it has never made any profits at all, even in the very finest years for the ironmasters, and it has gone on absorbing the money of the bank unchecked by the directors. Mr. Hodgson says that £1,000,000 of securities were taken of the most extraordinary nature for any bank to hold that he ever saw;—that £1,000,000 of securities, which was the only tangible asset which they had against the £2,600,000 of liabilities, consisted of £350,000 of the Derwent Iron Company's obligations, £250,000 being debentures, and £100,000 mortgage on the plant. They had besides these, £100,000 on a building speculation at Elswick, near Newcastle, which however was not a primary mortgage, there being a mortgage of £20,000 on that land belonging to Mr. Hodgson Hinde. They had also another £100,000 on other building land and houses in the neighbourhood of Newcastle. They had about £350,000 in securities of works and manufactures of different sorts, and they had about £50,000 in navigation bonds guaranteed by the railway, but which railway was the only security to which they could look in any given time to realize any sum of money; that made about £1,000,000 altogether. The other securities were absolutely unmarketable. This bank had derived assistance from the Bank of England in the former crisis, that of 1847. Almost exactly the same circumstances arose then, which arose in 1857, and almost from the same cause. The bank, however, applied at that time to the agent of the Bank of England, at Newcastle, and he, on his own responsibility, made them a very large advance, which carried them through, he taking at the same time a very considerable security from them in various mortgages, pretty much of the character which has been above mentioned, but better in quality, although not any more banking securities than these; between £700,000 and £800,000 altogether.

"The whole of the advance made in 1847 was repaid to the Bank of England, was it not?—Yes. With regard to the late occasion I represented at the same time that, though the bank could not be assisted, yet the fact of its failing, which it would do the moment it was known that the Bank of England would not help it, would be at that moment a very serious thing for the district, because it was so much connected with the collieries and ironworks that it paid every week, either for persons who had balances with it, or for persons whose bills it discounted, and thus gave them the money, about £35,000, on which the wages of 30,000 were dependent; and as their pay-day was on the Friday, and the bank would stop on the Thursday, it was very desirable that something should be done to prevent the confusion which would arise if there was no prepation made for that conjuncture. In consequence of that the Bank of England requested me to go down again that night, with full powers to make arrangements with all persons who might have any tangible and good security, though, perhaps, not perfectly regular security, so as to provide them with the means of making their pays on the Friday. I went down accordingly, and arranged with almost everybody, or with everybody I may say, to make such advances as would enable them to meet the pays for that week and for the next, should it be necessary. I also advised the manager of the savings bank to open his bank on Saturday for payments, though it was not the usual day, and authorized him to draw upon the Bank of England for any sum of money which he might require for the purpose of making any payment; but owing to the fact of the Bank of England thus enabling the proprietors of the coal mines and the works to make their weekly payments, there was no run whatever upon the savings bank, and everything passed off quite quietly.

"Was there any limit to the authority which you had from the Bank of England to give assistance in Newcastle?—No, there was no limit; it was left to my discretion to do what might be necessary. We knew very well that it could not amount to a sum, under any circumstances, of much more than from £50,000 to £70,000.

"Are there any other particulars connected with the Newcastle Bank which you are able to lay before the Committee?—I will, if the Committee wish, give them the actual result of the accounts of the bank when it was finally wound-up in January this year, as compared with those in November, 1857; it will show a little difference. In November, 1857, the liabilities of the bank were £2,600,000; these consisted of deposits, £1,350,000; accounts current, £1,150,000; and estimated liabilities on rediscounts, £100,000; in January, when the bank was positively wound-up and the thing ascertained, it appeared that there were of deposits, £1,256,000; in accounts current, £766,000; and in liabilities on rediscounts, £231,000. The only great difference was in the accounts current, which were diminished about £400,000. That was principally, I believe, from the fact that many persons who had accounts current had deposit accounts also; they kept two accounts, one of which had a balance in its favour and the other was overdrawn; therefore, one account being set against the other, it diminished it by so much, and at the same time diminished the amount of overdrawn accounts; the assets, which were estimated in November at £2,500,000, had fallen in January to £2,000,000, and there was one peculiarity, which was, that while the debt of the Derwent Iron Company was taken as an asset in November at £750,000, in January it was taken as an asset at £947,000, and that is an asset of a very doubtful nature; the position of the bank is much worse in reality than is shown by the statement of the figures.

51. This disclosure was the result of an examination which lasted about two hours; yet the bank had declared, at the last half-yearly meeting, a dividend of seven per cent., making to the shareholders a statement the substance of which showed a very prosperous state of things. Mr. Hodgson mentions that he remarked on the fact of their having declared a dividend in June, when it was admitted that half the capital was lost, and he asked how they could have done so; it was stated, in reply, that there were so many persons who depended entirely for their livelihood on the dividends received, that they really could not bear to face them without paying any dividend.

52. Each of these three banks had been in peril in 1847, and though, by the assistance of the Bank of England, they were enabled to surmount it, they fell on the next occasion of severe commercial pressure, under circumstances still more injurious both to their own proprietors and to the public. Two bill-broking houses in London suspended payment in 1847; both afterwards resumed business. In 1857 both suspended again: —The liabilities of one house in 1847 were, in round numbers, £2,683,000, with a capital of £180,000; the liabilities of the same house, in 1857, were £5,300,000, the capital much smaller; probably not more than one-fourth of what it was in 1847. The liabilities of the other firm were between £3,000,000 and 4,000,000 at each period of stoppage, with a capital not exceeding £45,000.

53. These five houses contributed more than any others to the commercial disaster and discredit of 1857. It is impossible for your Committee to attribute the failure of such establishments to any other cause than to their own inherent unsoundness, the natural, the inevitable, result of their own misconduct.

54. Thus we have traced a system under which extensive fictitious credits have been created by means of accommodation bills, and open

credits, great facilities for which have been afforded by the practice of joint-stock country banks discounting such bills, and rediscounting them with the bill-brokers in the London market, upon the credit of the bank alone, without reference to the quality of the bills otherwise. The rediscounter relies on the belief that if the bank suspend and the bills are not met at maturity, he will obtain from the Bank of England such immediate assistance as will save him from the consequences. Thus, Mr. Dixon states: "In incidental conversation about the whole affair, one of the bill-brokers made the remark that if it had not been for Sir Robert Peel's Act, the Borough Bank need not have suspended. In reply to that, I said, that whatever might be the merits of Sir Robert Peel's Act, for my own part, I would not have been willing to lift a finger to assist the Borough Bank through its difficulties, if the so doing had involved the continuance of such a wretched system of business as had been practised; and I said, 'If I had only known half as much of the proceedings of the Borough Bank while I was a director' (referring to the time previous to the 1st of August, when I became a managing director), 'as you must have known, by seeing a great many of the bills of the Borough Bank discounted, you would never have caught me being a shareholder;' the rejoinder to which was, 'Nor would you have caught me being a shareholder; it was very well for me to discount the bills, but I would not have been a shareholder either.'"

55. It will be instructive now to turn to the North of Europe, to survey the condition of countries where, as in Hamburg, the currency is exclusively metallic, and to compare the state of things there with that which existed here under the laws which regulate the currency in this kingdom.

56. In Hamburg, on the 23d of November, commercial confidence is stated to have been entirely at an end; so that only the bills of three or four of the first houses were negotiable at the highest rate of discount. In the first instance, some of the leading houses and the banks originated a plan for relief, viz., the subscription of about £1,000,000, and the appointment of a committee to give, by indorsement, the credit of this fund to the current bills. At first it seemed that confidence was much restored, but in two days this hope vanished; and on the 25th the aspect of affairs was again very gloomy. On the 27th a meeting of the Bürgherschaft was held, and a new arrangement was proposed by the Senate for the issue of Government bonds on the deposit of goods, funds, and shares, to the amount of £1,125,000. On the following day the feeling of the Exchange was better in consequence of this Government measure, and of the arrival of considerable quantities of silver. Yet, on the 1st of December, our Consul writes, "The embarrassments of the mercantile community here still continue undiminished:" and on the 3rd, "There is no deficiency of silver in the Hamburg Bank; indeed the amount in the cellars of the bank is now much larger than it has been at any former period, but a total want of confidence prevents its holders from parting with it." The Government bonds could not be discounted. A loan was ultimately obtained from Vienna; but even the arrival of the amount in specie failed to produce the desired effect, until the Senate reluctantly proposed

that it should be intrusted to a secret committee, to be by them lent out on good security. On December 12, as soon as it was known that by the aid of the Government the leading houses would fulfil their engagements, the panic ceased. Money at once became abundant, and in about a fortnight the rate of discount for the best bills fell to two and three per cent.

57. The information on this subject, relating to the different countries in the North of Europe, which will be found in the Appendix, is most instructive. It shows the severity of the disaster there sustained, and also that the real origin of it was the undue expansion of commercial credit; and it confirms the proof that no system of currency can secure a commercial community against the consequences of its own improvidence.

58. In this place it may be convenient to notice two points on which considerable misapprehension appears to have prevailed. It is contended by some persons that the separation of the issue department by a local change, removing the office from the Bank premises, would have the beneficial effect of convincing the community that the law now regulates the issue of notes, and leaves to the discretion of the Bank Directors the purely banking business only. But strong evidence is given by the Bank Directors that much practical inconvenience would result from such a change; and your Committee think that repeated discussion and increasing knowledge will satisfy the public of a fact so obvious as this; viz., that without the interposition of the Executive Government, the Directors of the Bank of England have no power whatever to exceed in their issues the limit imposed by law. The duties which the bank discharges in this respect are purely ministerial, unaccompanied by any discretionary power. Whoever discharged these duties, it would be equally subservient to the general convenience of the public that the place of issue should be in the immediate neighbourhood of the place where the banking department is situate. Able papers by Lord Monteagle and by Mr. Arbuthnot on the subject of a state bank, will be found in the Appendix.

59. Another misconception has often perplexed those who have reasoned about the currency—that of supposing that by Act of Parliament the price of gold is fixed. If it had so happened that our sovereign, instead of being equal to $\frac{1}{4}$ oz. troy of gold of standard fineness, had been exactly equal to that weight, it would probably have been obvious to all that the word sovereign simply meant a quarter of an ounce of standard gold, with the Queen's head stamped upon it by the Mint; and the price of gold, as it is called, being thus exactly £4 an oz., anybody would have comprehended that the one was equivalent to the other. The use of the silver and copper coins as representatives or tokens of fractional parts of that gold, would probably have been intelligible, and this troublesome confusion would not have arisen. This topic has not much presented itself in the course of the present inquiry. But it is desirable that all persons who take an interest in this subject should understand how simple is the duty discharged by the Executive Government in relation to that money, viz., gold money, which alone is the standard of value in this country, so far as the transactions of our extended commerce are concerned. At the Mint a piece of standard gold, weighing 5 dwts.

3·274 grains troy, is verified by a stamp, and being then called a sovereign, is returned to its owner, and in this process no seignorage is charged. At the Bank five times the same quantity is received into the coffers for custody, and in return a paper, called a £5 note, is given to the owner of the gold. He is entitled at his pleasure to return the note, and demand for it sovereigns which contain an equal quantity of bullion. Upon every ounce of gold that thus passes in and out of the Bank an allowance for the double transaction of about 770465 grain troy weight of gold, or as it is expressed in our copper tokens, 1½*d.*, is retained by the Bank. This allowance is an equivalent for the loss of interest which it is comyuted the owner of the bullion saves by the transaction, inasmuch as he saves, by receiving notes from the Bank, the loss of time, and therefore of interest, which he would have incurred if he had taken his gold to the Mint to be coined into sovereigns. This allowance yields an annual profit, which is taken into account in the arrangements between the Bank and the Government. Your Committee have not entered into the question whether any charge should be made by the Mint for coinage. So intelligible and so simple is the relation between the Government and the issue of money; so entirely is the Bank of England excluded by statute from the exercise of any discretion whatever in this respect.

60. For the opinions of the most eminent writers on the subject of the currency, your Committee refer to the Evidence taken in 1857. It is interesting in the highest degree to all who make the scientific study of the most abstruse questions of political economy their pursuit. But a review of that Evidence would appear necessarily to involve subjects of controversy on which your Committee would not be able to arrive at any conclusion, without much difference of opinion, and they are therefore desirous of excluding these subjects from their Report. That the public welfare in times of commercial disaster requires the maintenance of an adequate supply of bullion at the Bank, is the opinion of Mr. Tooke, Mr. Newmarch, and Mr. Mill, as well as of Lord Overstone, Mr. Norman, and Mr. Hubbard. That the supply necessarily maintained in the coffers of that establishment, under the provisions of the Act of 1844, is greater than that which was ever maintained under circumstances of pressure in former times, is a fact beyond dispute. During the crisis of 1825, the bullion fell to £1,261,000; in 1837, £3,831,000; and in that of 1839, £2,406,000; while the lowest points to which it has fallen since 1844, have been, in 1847, £8,313,000; and in 1857, £6,080,000. That the opinion of the present Bank Directors is strongly in favour of maintaining the Act of 1844, appears in the Evidence. They say the assistance which they gave to the public, would not have been ventured on by them except for the Treasury Letter; nor would they have ventured to act on that letter if the bullion had been much lower than it was; for they must then have begun to think of the convertibility of the note which it would be their first duty to maintain; they attribute the maintenanee of that amount of bullion to the regulations provided by the Act; and while they affirm that the present Court of Directors, having had more experience, and having seen the gradual working of the Act of 1844, would probably, in their discretion, have adhered closely to the very regulations which the

Act required of them; yet, if they had not done so, but had been induced to issue more than the proportion which the law allowed, more gold would have gone out by the action of the foreign exchanges, and the consequences would have been that they would have been left with less gold as the panic came on; and then, even with the permission to issue more notes, they would not have felt warranted in hazarding the circulation by doing so. They further state that, for these reasons, it appears that the adoption of the policy which the Act now in force required, placed the Bank of England in such a position that it was enabled at the time of severest pressure to afford a larger aid to the commercial public than would otherwise have been in their power; that the true judgment of the Court would act in unison with the law; but yet it is not expedient to expose them to the influence of such a pressure as would inevitably be applied at such a time; and that, upon the whole, with a view to the operations of the Bank, including in that category their being able to afford aid to the commercial public, at the time of severest pressure, the Act of 1844 operated not as a fetter, but as a support, decidedly. They therefore recommend that no relaxation should be made in the provisions of that law.

61. In this opinion the Governor of the Bank of Ireland, the representatives both of the chartered and the unchartered banks of Scotland, the chairman of the association of private country bankers, and Mr. Alderman Salomons, of the London and Westminster Bank, concur.

62. Those who advocate what is called the theory of the Act of 1844 are guided by the following principles. They regard bank notes as being for every practical purpose, equally with the gold they represent, the money of the country—the measure of value—that which extinguishes debt—not as a mere form of paper credit, depending on the credit of the issuer, and constituting only the evidence and vehicle for transfer of a debt which still continues. If complete effect were given to their view, the result would be that for the whole United Kingdom there would be one description of note only, issued by the State, based on bullion in the custody of the State. This note, so secured by bullion, would be a legal tender everywhere, except at the place of issue. Experience having shown that even in the times when the paper circulation is most contracted, the sum in circulation with the public at large can never fall below a certain amount, and cannot therefore be presented to the Bank for payment in gold—they are satisfied that to this extent—so limited by experience—the actual deposit of bullion may safely be dispensed with, the notes in question resting on the security of the State. This is their justification for the permission accorded to the Bank of England to issue 14 millions of notes without the deposit of a corresponding amount of bullion. They consider any addition to the circulating medium of the country to be the act of the private individual who carries bullion to the Mint to be coined, or to the Department of Issue to be exchanged for notes; fixing the standard of money, and verifying the conformity of the pieces therewith by either of these processes to be the duty of the State; the use of money, and that only, they regard as the province of a bank, whether of a private person or incorporation, or of the banking department of the Bank of England.

63. These advocates of the theory, as it is called, of the Act of 1844, are far from contending that their theory is completely carried into effect by the provisions of the Act. The origin of that legislation is thus referred to by Lord Overstone:—"I had no connection, political or social, with Sir Robert Peel. I never exchanged one word upon the subject of this Act with Sir Robert Peel in my life, neither directly nor indirectly. I knew nothing whatever of the provisions of this Act until they were laid before the public, and I am happy to state that, because I believe that what little weight may attach to my unbiassed conviction of the high merits of this Act, and the service which it has rendered to the public, may be diminished by the impression that I have something of personal vanity in this matter. I have no feeling whatever of the kind. The Act is entirely, so far as I know, the Act of Sir Robert Peel, and the immortal gratitude of this country is due to him for the service rendered to it by the passing of that Act. He has never been properly appreciated; but year by year the character of that man upon this subject will be appreciated. By the Act of 1819, Sir Robert Peel placed the monetary system of this country upon an honest foundation, and he was exposed to great obloquy for having so done. By the Act of 1844 he has obtained ample and efficient security that that honest foundation of our monetary system shall be effectually and permanently maintained, and no inscription can be written upon his statue so honourable as that he restored our money to its just value in 1819, and secured for us the means of maintaining that just value in 1844. Honour be to his name."

64. But it does not appear by a reference to the speeches of Sir Robert Peel that he propounded the two measures of 1844 and 1845, as measures of theoretical perfection; on the contrary, they can only be regarded as having been designed to accomplish a great practical object by the least possible disturbance of existing interests. Thus Mr. Rodwell:—"Then the general result of those interviews was to leave upon your mind the impression that the measure was intended to be a great step in advance towards the establishment of one central issue, which was to be arrived at by voluntary arrangement?—I thought that the tendency of the views of Sir Robert Peel was, that that would be a natural result; but I thought that his view was, that whether that consequence ensued or not, the arrangement was a continuing arrangement with the country bankers, in order that that Act might pass without any opposition on their part."

65. And Sir George Clerk, the Deputy Governor of the Bank of Scotland, who was Secretary of the Treasury in Sir Robert Peel's Administration, in 1844, and Vice-President of the Board of Trade in 1845, and intimately acquainted with all that passed in reference to these measures:—"In the debate of the 25th of April, 1845, with reference to the £1 note circulation, Sir Robert Peel said, 'Whether or not the importance attached to the continuance of the privilege (of issuing £1 notes) can be perfectly justified by reason or argument, I know not. Whether there be not an undue value attached to them may be a fair question of doubt; still, in attempting to introduce principles which I believe to be good, I will not attempt to shock even the prejudices of the people, or to run the risk of encountering that opposition which I knew I should have to

encounter from Scotland almost universally. Without guaranteeing, therefore, the continuance of these notes, all I can say is, that we do not propose to prohibit them at present; I say nothing, however, as to the future. The discretion of Parliament must be left unfettered in respect to them. If the continuance of this privilege affects no interests, if it has no injurious effect upon the circulation either of Scotland or of other parts of the empire, there is no doubt whatever that a future Parliament will entertain the same forbearance, and will not disturb the settled habits of business of a whole country, or run counter to its feelings, for the mere purpose of carrying out some theoretical principle.' "

66. Your Committee have examined the operation of those statutes, not with a view to ascertain whether they constitute the most perfect system conceivable for regulating the paper circulation of an empire, but rather whether their operation has been such as to secure the main object for which they were designed. The main object of the legislation in question was undoubtedly to secure the variation of the paper currency of the kingdom according to the same laws by which a metallic circulation would vary. No one contends that this object has not been attained.

67. Mr. Rodwell says that before the Act of 1844 the country bankers were not all aware of the consequences of their issues; that if they had been such disasters would never have arisen, as arose in 1825; and he knows the practice to have been that it was considered as a part of the business of a country banker to get out as much of his issues as he could, which eventually turned back upon himself when he did not expect it, and was least prepared to meet it. He says that before 1844 they did not so fully understand the laws which ought to guide a banker in making his advances; but that now they look to the unemployed notes (in the Bank of England) as an infallible index of what it is necessary for the Bank of England to do, and for the country bankers to do also. In recent times the increased facilities of intercourse and of banking have increased the rapidity with which notes find their way back to the banker who issues them; while the restriction of bank notes in England and Wales to sums not less than £5 excludes them in a considerable degree from the retail transactions of the country. It may be laid down that in the opinion of every practical witness who is an advocate for the convertibility of notes, the amount of bullion retained in the coffers of the Bank under the operation of the existing law is not greater than a due regard to prudence would require, even if the law were altered. It appears that the present law ensures the maintenance in the coffers in the Bank of an adequate amount of bullion, whilst the history of past years proves that such an amount had not been maintained by the unassisted wisdom and firmness of the Bank Directors; and the present Court of Directors are unanimous in desiring that they should continue to be fortified by the provisions of the present Act.

68. No complaint against the Act of 1844 has been more popular, or more commonly employed out of doors, than one which may be expressed in the following words:—"That the trade of the country has increased, that a larger issue should be allowed, to supply the increased requirements of commerce; and that therefore a larger amount of notes, unrepre-

sented by bullion, should be issued." This question is thus disposed of by Mr. Weguelin in 1857 :—

"Do you consider that if the limit imposed by law of £14,000,000 were altered, for example, to £16,000,000, it would in truth add £2,000,000 to the active circulation?—By no means.

"Will you state what you think the effect really would be?—The effect would be either that those £2,000,000 would be held in the reserve of the Bank, or, in case it occurred that the increase took place at a time when there was an adverse exchange, those £2,000,000 would be exported from the country, and all the other figures would remain precisely the same.

"It would not add, under any circumstances, to the active circulation of the public?—It would not.

"You consider that the action would be, that either it would be added to the reserve of the Bank, or that the bullion held by the Bank would be *pro tanto* diminished?—That would be the action.

"Is there, in your opinion, any sufficient inducement, on the ground of public interest, to make an extension beyond the present limit of £14,000,000?—I see no advantage or particular object to be gained by it.

"The advantage of saving £2,000,000 of capital would not, in your opinion, be equal to the mischief that might result from the change?—I think it would be of an insignificant character, and it would diminish the amount of actual reserve of bullion in the country.

"Would not those £2,000,000 go out of the country at the first adverse exchange, and not come back; would not that be the ultimate effect?—That would be the ultimate effect."

69. It has been observed before, that while, on the one hand, the great increase of retail transactions has caused an increased demand for the smaller notes, concurrently with the increased demand for gold, yet, on the other hand, so great has been the effect of increasing facilities in banking, that a saving of a corresponding amount has been effected in the larger notes. The proportions are those represented in the following table :—

YEARLY AVERAGES OF NOTES WITH THE PUBLIC.

YEAR.	Notes of £5 and £10.	Per Cent. of Total Circulation.	Notes of £20 to £100.	Per Cent. of Total Circulation.	Notes of £200 to £1,000.	Per Cent. of Total Circulation.	TOTAL.
	£ *m.*		£ *m.*		£ *m.*		£ *m.*
1844	9,263	45·7	5,735	28·3	5,253	26·	20,241
1845	9,698	46·9	6,082	29·3	4,942	23·8	20,722
1846	9,918	48·9	5,778	28·5	4,590	22·6	20,286
1847	9,591	50·1	5,498	28·7	4,066	21·2	19,155
1848	8,732	48·3	5,046	27·9	4,307	23·8	18,085
1849	8,692	47·2	5,234	28·5	4,477	24·3	18,403
1850	9,164	47·2	5,587	28·8	4,646	24·	19,398
1851	9,362	48·1	5,554	28·5	4,557	23·4	19,473
1852	9,839	45·	6,161	28·2	5,856	26·8	21,856
1853	10,699	47·3	6,393	28·2	5,541	24·5	22,653
1854	10,565	51·	5,910	28·5	4,234	20·5	20,709
1855	10,628	53·6	5,706	28·9	3,459	17.5	19,793
1856	10,680	54·4	5,645	28·7	3,323	16·9	19,648
1857	10,659	54·7	5,567	28·6	3,241	16·7	19,467

70. The effect has been so great that, notwithstanding the great increase of trade, the whole amount of bank notes has actually diminished since 1844, and, under the present law, still continues gradually to decline. It must be taken, therefore, that in ordinary times, there is no cognizable advantage to be obtained by the commercial interest from the power of increasing the amount of notes which may be issued without the deposit of bullion.

71. It is here necessary for your Committee to advert to the question, whether the law should be left, subject only to that power which was contemplated by Sir Robert Peel and Mr. Huskisson, and was actually exercised by the two Governments of 1847 and 1857; or whether, on the other hand, provision should be made in advance for such contingencies, and the conditions expressly laid down on which the issue of an increased number of bank notes may in the time of pressure be allowed.

72. Your Committee think that such a provision could not be regarded as any violation of the principle of the Act of 1844. To have introduced such an express provision, when the law was itself first adopted by Parliament, or even when, as in 1848, it had only been a few years in operation and was comparatively little understood, was a far more serious question of policy and of prudence than it can in fairness be regarded at the present time. Yet the interference of Government in an extreme case must, in fact, be taken to have been contemplated by the framers of that Act. Mr. Cotton stated to the Committee of 1847-48, that this subject was considered when the Act was under preparation in 1844, and that Sir Robert Peel's opinion was thus expressed :—"If it be necessary to assume a grave responsibility, I dare say men will be found willing to assume such a responsibility." It scarcely therefore constitutes, of itself, a sufficient ground for bringing this important and difficult subject under the review of Parliament, and may properly await the decision of the Legislature when the other branches of the subject shall again be dealt with.

73. They would, however, here take occasion to observe, that if new provisions shall at any future time be made by Parliament, the great object of securing the maintenance at the time of severest pressure of an adequate supply of bullion should be guarded with the utmost caution.

74. In considering these new provisions, your Committee assume that no hazard will be incurred with regard to the foreign exchanges, but that the efficient action of the law in that respect will be firmly maintained. The mischief your committee are now considering is the domestic drain, occasioned by panic, and evidenced by hoarding, which in cases of commercial crisis supervenes upon a foreign drain, and creates an abrupt interference with the circulation, by withdrawing from it for a time, for the purpose of hoarding, a part of the ordinary circulating medium.

75. Your Committee have already touched upon other points in which the enactments of the Legislature in 1844 and 1845 fall short of the principles on which those enactments are founded, and desire to express their concurrence in the wisdom of adapting practical legislation in an important degree to the existing interests and wishes of the community. Of these questions, an important one is that of the small note circulation still existing in Scotland and Ireland. The advice of Adam Smith, that

no bank notes should be issued in any part of the kingdom for a smaller sum than £5, is enforced by the Bank Directors as a matter of principle, both in 1857, and again more strongly in 1858, after the experience of the autumn of last year. It is, however, still a question into which the same considerations enter, in a modified degree, by which all Governments and every succeeding Parliament have been influenced from 1826 to the present time, whether the application of this principle shall be extended to Scotland and Ireland. The failure of the Western Bank has now withdrawn £337,938, or about 1-10th part from the authorized circulation of Scotland, and the Act of 1845 operates with a greater proportionate effect, both upon Scotland and Ireland, as the population and trade of those countries increase, and the proportion of the retail or small bank-note circulation represented by bullion to the authorized or unrepresented part increases also.

76. On the other hand it appears from the evidence, that notwithstanding the expense which the requirements of the Act of 1845 impose upon Scotland, there has been a very remarkable increase in the number of branch banks established since the passing of the Act; so that the number of banks in Scotland, including branches, now actually exceeds the number of the whole of the banks and branches in England, or is very nearly equal to it; and that the amount withdrawn from circulation by the failure of the Western Bank has been supplied by an amount represented by bullion, without any more severe terms being imposed on the customers of the banks. Sir George Clerk says:—

"To the banks, I think you do not consider that so much an object as to the population?—I do not think the banks have so strong an opinion in favour of the retention of the £1 note circulation as they certainly expressed before the Committee which sat upon that very subject in 1826; but I believe that the general opinion of the banks would be strongly in favour of the continuance of the law as it at present stands.

"If it were to be altered, they would meet the case, as regarded themselves, by some slight increase of commission charged to their customers? —Probably in that way."

77. Without entering into any question respecting an issue of small notes on the credit of the State, the Committee desire it should be understood as their opinion that the subject of the issue of small notes in Scotland and Ireland, and of private issues generally in the United Kingdom, should be reserved, without prejudice, for the future consideration of Parliament.

78. The Bank of Ireland complained of a special prohibition laid on them with regard to mortgages; and Mr. Latouche attended on behalf of the private bankers of Ireland to complain of an Act passed by the Irish Parliament in 1759. There appears no very obvious reason for the continuance of antiquated restrictions peculiar to Ireland. But neither of these subjects properly belongs to the present inquiry.

79 Some smaller points connected with the wording and legal operation of the Acts of 1844, have been brought under the notice of your Committee, with which they do not think it necessary to encumber their Report. It will be desirable that these subjects, especially the question

whether a bank of issue which suspends its payments, even for the shortest time, should not lose the privilege of issue, should be carefully considered whenever the Executive Government shall next submit to the House a measure for the regulation of the relations which subsist between the Government and the Bank.

80. The pecuniary arrangements subsisting between the Executive Government and the Bank appear to your Committee to fall within the terms of their order of reference; and the Committee of 1857 took some evidence from the Governor on that subject. But your Committee understand from the Chancellor of the Exchequer that the subject is now under the consideration of the Treasury and the Bank. They think it doubtful also, whether, in case it shall be necessary to submit it to such a consideration, separate from the wider questions which have come under the notice of this Committee, it would not be expedient that a less numerous Committee should be appointed for that purpose. The appointment of a separate Committee may tend to produce the useful conviction, that it is not necessary to unsettle the great principles which regulate our monetary system, merely because the pecuniary relations between the Treasury and the Bank may require to be reconsidered.

81. It appears to your Committee that no mischief will result from at least a temporary continuance of the present state of things under which the Bank of England holds the powers given by the Act of 1844, subject to a notice of twelve months, which may at any time be given by the House of Commons through Mr. Speaker. They agree with the opinion expressed by Mr. Goulburn in 1844. The Bank Directors had suggested the propriety of renewing the arrangement for twenty years, with a power of giving notice at the expiration of ten, as has been done in 1833. Sir Robert Peel's government preferred the limit, which was actually adopted, of ten years; the Act, at the expiration of that period, to be terminable at any time upon a notice of twelve months; but, until such notice be given, to continue in force. Mr. Goulburn thus accounts for this decision. In making the proposal, he says, "The Government were mainly influenced by the consideration that it was not advisable unnecessarily to agitate questions affecting the banking interest and the currency of the country."

82. Your Committee have stated the reasons by which it is established, to their satisfaction, that the recent commercial crisis in this country, as well as in America and in the North of Europe, was mainly owing to excessive speculation and abuse of credit; and also, that in the time of pressure the houses which deserved assistance received it from the Bank of England in a manner in which that establishment would not have been able to give it, except for the bullion retained in their coffers; and your Committee are satisfied to leave in the discretion of the Executive Government, the time and prudent opportunity of giving further effect to those principles by which the convertibility of the Bank of England note has been kept above suspicion.

1*st July*, 1858.

SECTION THE FIFTH.*

The American Crisis of 1857 and its Results—General Course of Trading—The Banking System and its Defects—Inflation of Capital and Competing Powers of Speculation—Over-trading assisted by the Banks—Reckless Proceedings of Merchants and others in the Great Race of Business—The Natural Consequences—Universal Depression and Enormous Depreciation of Property—Railway Management Involving Fraud and Speculation—Recuperative Power of the States—National Characteristics—Resources and Means of Development.

THE inhabitant of one of our coasts who watches the operations of a spring-tide, views with surprise and pleasure, which no familiarity with the phenomenon can diminish, the rapid advance of the waters beyond their accustomed bounds; and as ridge after ridge of rock and wide bases of sand disappear beneath the tidal current, and life and buoyancy take the place of what had been stationary, if not positively monotonous, he not unnaturally experiences a feeling of exultation similar to that which arises from the display of unwonted energy. Again, as old ocean from afar calls the waters back in that mighty effort to establish a re-adjustment of the liquid element, the same spectator finds, exposed below the usual line of retrocession, a blank and barren waste, out of all proportion to the advance that had been made, composed, for the most part, of deposits only fit to be the resting-place of slimy monsters, and tainting the fresh air. Quite analogous in its way was that high state of apparent commercial prosperity, especially in our connection with the United States, which most persons were

* This Section has been principally prepared from the views of an English friend, long resident in the United States, and who, at my request, placed in order his opinions on the subject of the American crisis. Some portions have required modification, owing to changes which have occurred during the last few months, but on the whole the paper represents in an impartial spirit the causes and effect of that great revulsion. It will be found to be borne fully out by appended statistics.—D. M. E.

disposed to regard with satisfaction. Nearly everybody rejoiced at the display of new activities without caring to question their source, or to take into account their possible reaction. The great majority believed the apparent prosperity stable, and its conditions as permanent. Now, whilst a spring-tide can be anticipated, the period of its continuance accurately predicted, and the times of its recurrence, together with most of the leading phenomena of nature, brought under an ascertainable law, it is marvellous to notice how little the science of calculation has been brought to bear on those mercantile crises which from time to time shake nations. There are, indeed, persons who confidently augur stated periods for their appearance, though on doubtful data; but most persons regard them as one would the occurrence of an earthquake, neither to be anticipated nor prevented, but passively submitted to. It is certain, however, that if the phases these phenomena present were more accurately examined, and the conditions under which they transpire searchingly analyzed, the community would soon be put in possession of an accurate diagnosis, and be better prepared to deal with premonitory symptoms. There are many people, it is admitted, who would be disposed to attach but little credit to the deductions thus obtained, even after experience had confirmed their correctness; but the general result would no doubt be, that greater caution would characterize dealings in the direction indicated, and this without any prejudice to the spirit of legitimate enterprise. As thinking individuals are far from believing that a commercial crisis, commence when or where it will, is out of all law, perhaps the most direct line of investigation as to the origin, nature, and results of such an occurrence, is to look into the character and actions of those who are to be viewed as the chief parties to it, and, together with the results, to examine the organization, financial or otherwise, by which it was favoured, and the circumstances by which it was attended. Were this course more generally pursued the public would be better prepared for the evil day, and although unable personally to influence such catastrophes, the damage that would otherwise ensue might be diminished by anticipation, leaving the distress to fall chiefly among those firms whose operations had not been conducted with the prudence which ought to be observed in all mercantile relations.

Had more attention been previously paid by merchants and

others in this country to the various banking systems of the United States, and their relations to trade and commerce, sufficient prognostications of succeeding events might have been discerned, to admit of some forewarning. This neglect is the less pardonable from the fact that banking operations in America are far from being shrouded in mystery; their periodical published statements furnishing all the information that is desirable. Even now, it is only with a full knowledge of the monetary operations of these institutions that the commercial interests can hope to escape a renewal of the same dangers as those already encountered.

The public are familiar with the collapse which overtook a large portion of the trade of this country upon the news of the late American crisis, with the suspensions which were the result of the stoppage of remittances from the States, with the rumours circulated in regard to the stability of various mercantile houses, the high rates of interest, the depression in produce and other articles, the diminution of bullion in the Bank, the shipments of specie to New York, the decline in the funds consequent on the war with India; to which might fairly be added, abstinence from large undertakings, gratuitous alarms, and no end of prognostications of further calamities.

The disasters that ensued in the United States may be traced, primarily, to the relative position and action of the banks in connection with the enterprises which capitalists had embarked in far and wide—the facilities then afforded preventing the exercise of that caution which should have governed their operations. In fact, the crisis was a foregone conclusion, and in proportion as the perils of the future loomed up, the spirit of recklessness which these banks did their best to feed counteracted every influence that made its appeal to timidity and apprehension. Lands were sold and stock bought up. Those who took only a quarter of a million dollars' worth of goods from England, under favour of the banks, advised one-half, though it was well known that the major portion of these purchases would have to go to auction. These operations on the part of thousands merely represented a race against time, on the strength of bank favours and foreign credit. The current notes of the banks throughout the States had, for a long period previous to the crisis, by their immense expansion and credited guarantees, led to that depreciation of the general currency which is to be estimated by the increase of the average prices of commodities generally. This, instead of lessening, gave new stimulus for a period to all operations; labour and industry gained new life. The tradesman dealing with the merchant in New York, Boston, or Philadelphia, after laying in his ordinary stock of goods, had met with a ready and profitable sale; but when about to replenish his store, he found that he had to purchase at an advanced price. Again he would meet with a sure demand, and again, when he replaced his stock, the price was raised. He would now take a larger amount of goods to provide against the constant augmentation of price; and the success of this operation would induce him to repeat it on a larger scale. In such a state of things credit became extended, and though a collapse was threatened, yet to those who took their observations from accustomed points of view, a general and well founded confidence in pecuniary engagements appeared to prevail. Labour and industry, that had been the first to profit by this virtual depreciation of the currency, through an almost limitless issue of paper, were the first also to feel its ruinous effects. With the continuous and rapid rise in prices, the demand for their products fell off; they could not cope with it. The dealer came next; the money he received was not available to discharge his engagements; the merchant looked for accommodation to the banks to enable him to support increasing burdens, applying himself to enlarge at any hazard, in the hope of gaining time, his indebtedness to this and other countries, and applied his undivided efforts to obtain the extension of his bills, and to reduce the more pressing and immediate obligations. This was nothing more nor less than taking a long lease of credit, but as for any hope of speedy relief, none, it is believed, was entertained by even the most sanguine American merchants. Those who had confined themselves to safe transactions anticipated silently and confidently—for it was not their interest to expose it—the coming catastrophe. It is certain that the most stable commercial houses failed to communicate to houses in this country, as they might have done, the knowledge in their possession as to the true state of affairs in the United States, the weakness of the tenure of so much seeming show and prosperity. A feeling hardly to be called immoral, a hankering after individual and national prosperity (terms almost exchangeable in the American mind), at whatever risk and cost to other nations, kept their mouths closed and their pens fettered. A "mighty smart" thing was about to be done. It would have been thought a pity to spoil the joke. The banks were taking no effective measures to reduce their circulation in quantity, so that the result was inevitable. The next phase to note in the progress of those events which ripened into the American crisis, is the activity of the commerce carried on between the United States and this and other countries the preceding season—an activity with no determined basis except that which lay in the means and capacities of France, Germany, and England to sustain the losses which were waiting them the opposite side of the Atlantic.

It is in the extreme and unwarrantable extension of their note circulation that the banks of New York, Boston, and Philadelphia are chargeable with promoting the late crisis. The notes of these banks are far from being absolutely secured by their being based on State stocks, as late transactions have shown. The evil is, that in addition to contributing an undue share to the general currency, these banks, in tendering their services to transact, as far as may be necessary, the pecuniary business of this community itself (including all sorts of speculators), put an equitable limit on such transactions. Whilst recklessly dealing out, these banks were equally arbitrary in contracting or extending their accommodation. The bills of exchange in circulation in the cities and towns of the United States, just previous to the crisis, depended almost wholly for existence on the banks, a very considerable portion being of their own drawing. The banks had made the bills currency by accepting them in account, discounting them, and promising to to take them up when due. Speaking generally, the acceptors of them could not make them payable out of the United States, and the drawers could make no use of them without the banks. The destruction of the latter would have proved the destruction of the bills. These banks thus enabled the smaller traders to double the business they would otherwise have been able to transact, and men in business, to transact their payments regularly, must have facilities, or there will be eventual bankruptcy. Both these parties, therefore, not to speak of brokers and various grades of operators, were interested in upholding the circulation. There must, undoubtedly, exist a circulating medium, the representative of property, to answer to a sufficient extent the purposes of the exchange of property. But where, by the importation of foreign manufactures, a debt is contracted abroad to a greater amount than the surplus of the raw material will pay for, the difference must be paid in specie. This will occasion annually a diminution of the solid circulating medium, and with an increase of paper credit, as extensive, and for as long a period, as the folly of the borrower and the capital of the banks will permit. This abundance of paper currency in the United States depreciated the means of support of all who lived upon a specified money income, tempted to adventurous speculations in trade and to indiscreet expenses, while by the smiling aspect of seeming prosperity it hid from the thoughtless multitude the day of retribution. The banks, at length alarmed at the prospect of the disappearance of their specie, consequent on an adverse balance of trade, and at the extent of credit to which the desire of gain had tempted them, retrenched at once their discounts, and called upon their customers to pay their debts. These customers called upon the consumers. But now the paper medium was retrenched, the solid medium of trade was gone, the payment in money could not be made, and lands and other kinds of property had to be sold at a sacrifice not only of the adventitious value they had acquired, but at one-half and two-thirds of their real value. Then commenced a scene of failure and fraud, and sacrifice of property, of blasted hopes and family distress, of national embarrassment and stagnation of business, which almost defies description. The evil was radical in the system of basing prosperity on artificial credit. So far as respected the outstanding circulation of unredeemed and unredeemable bank notes at the time of the crisis, Philadelphia and Pennsylvania generally had a full share of suffering; and as the city largely depends for prosperity on the transit of business—the forwarding of goods and manufactures into the interior, and again concentrating within herself the yield of the neighbouring coal regions, and of the western mineral districts by means of costly and gigantic schemes of railway, together with the agricultural produce of the north-western, western, and southern regions, it could not be otherwise than that the effects of the collapse as it affected her were far and widely felt. It was in Philadelphia that the way in which the inducements of those banks were the means of increasing disasters and hastening the crisis was most apparent. Competition in banking is in Philadelphia limited by law; but this very restraint, as to discounting privileges, leads those that are established to push their business to the extreme verge of their capital. The amount of discount and loans in proportion to capital has always been larger in Philadelphia than in New York. The proportion of borrowers being always in excess of the lenders, these banks had it in their power, and exercised that power for some period before the crisis of November, to impose the very hardest conditions. Illegal interest was not directly taken, but exactions were made of large balances given to remain on deposit. Particular cliques and special business interests were accommodated, and those less favoured had to pay the accustomed tribute to usury. With dealings in the discounting of paper conducted on such a partial and unsound policy, it was no wonder that the majority of the enterprises in which these banks trusted "went by the board." The banking system of Pennsylvania has been regarded by astute Pennsylvanians themselves as less secure than that of any other State of the Union. The restriction on the competition in banking gives to them all the character of a monopoly. With the prospect of usurious tribute, they are liable at any time to be tempted into too great an expansion of their operations.

The whole of the distress is not to be ascribed to the conduct of the banks, whether, first, by extravagant emissions, and then by pressing on their debtors. The support and stay of banks is specie, and the extravagance of commercial operations having led to that alarm which caused its withdrawal by individuals from circulation, and especially by foreign banking houses, such material was wanting. It was not, however, to the interest either of the British or the American community that the banks should be further supported, unless indeed to postpone what must have eventually occurred. Little, nevertheless, can be said in defence of these institutions. There are in many of the cities of the United States three or four times the number that are necessary; a great portion of them are very ill managed, and have done much mischief. But when the great mass of distress existing in the country is charged to the account of these establishments, the effect is mistaken for the cause.

It may here be remarked that the Americans, beyond any other people on the face of the globe, "calculate" on the value of time—and time is an element of strength on which they largely depend for recovery from the late disasters. In fact, knowing the resources of the country, and looking to their future development, they are in a manner, and from a European point of view, reckless of the present. This, in fact, is the key to the great reactive effects of those principles which attach to the commercial and trading system of the United States—principles which have really no limitation but those resources without which, in any practical sense, credit must prove utterly worthless. It could be readily shown that the credit which America has opened up with this country (not the credit system, which is always the same) must go on indefinitely increasing, despite the unsound schemes which are borne along with it. Whatever the sum of the phrases "prosperity misused," "stocks falling in price," "real estate declining," "ships becoming a drug," "railroads suspending their dividends," with the declared absence of any guarantee for the payment of claims—notwithstanding these expressions, the public may be satisfied that the United States is not fated to perpetual and overwhelming disasters. Time alone, not counting on anything else, will save it. This is the element of calculation which is left out in all those gloomy conclusions to which the English community yield. Even the telegraph, with its wondrous aids, and the mechanical improvements so subsidiary to the operations of trade and commerce, constitute securities for the continued and increasing prosperity of the land of "stars and stripes." The true relation of time to credit, in connection with these scientific aids, has found as yet no exposition in the published theory of any system of finance. Financial theories, in fact, are of old date, concocted at a time when commerce did not embrace such extensive interests as at the present time, when the electric telegraph—an agency causing millions of dollars' worth of goods to change hands in a moment, in cases in which money and weeks of time to connect remote distances were formerly required, linking places thousands of miles apart, and answering all the purposes of a brisk monetary circulation, so far as it is the medium of exchange—was unknown, when steam and clipper ships had not begun to stimulate exchanges, or an improved system of credit had affected materially the old routine of practice, with its standard regulations, and measured allowances of weeks and months for results that are now accomplished in a day or an hour. What is chiefly to be complained of is, that the credit of the United States has been upheld on false pretences; not that it has had a credit which was to some extent a necessity. The system has been sustained with them on a false principle, that of inflation; and the people have yet to learn that to check this expansion is not to diminish, but to secure their credit. It would take tens of millions of dollars to accurately represent the value of the decline in United States railroad stocks and bonds during the twelve months ending December, 1857. Not that the country is so much the poorer, but that this amount has been abstracted by mere speculators from the pockets of shareholders. Such a result is injurious to established relations between man and man. To order more than can be paid for to cover the errors of borrowed money, has been the crime of railroads, corporations, and private individuals, and largely indebted as they have been to these banks for the opportunity of enriching themselves by fortunate speculation, it is natural they should associate them indissolubly with every degree of prosperity. In the temptation which these institutions offer to an undue extension of individual credit, and in that want of probity which is the last result of temptation, most persons are disposed to trace, more than in anything else, the concatenation of evils known as the American crisis. The American mind, it is nevertheless thought, has sufficient acuteness to discern the tendency and relative value of any given principles of action, sufficient practical insight into the conditions by which a harmony of interests is to be maintained, and such an amount of what may be termed natural conscientiousness, when this is not swerved by an avaricious impatience of results, to see its own interest in adhering to some higher principle than that which, hitherto, has practically regulated it. On this thorough purgation and self-reform must, or should, depend the renewal of English confidence. Indeed, prosperity has been so greatly misused in the United States as to tend to general bankruptcy, to a practical suspension of all the laws and dealings which govern right. As a result, foreign merchants and bankers demand of the financier merchants of the United States more substantial security in making their advances. Inflations with corresponding reactions, casual or periodic, are felt to be grievous evils, and as far as general trade is concerned, the sooner the late existing state of things vanished, the better. At the same time, individuals should be on their guard against contracting within the too narrow limits of selfish suspicion, in the absence of fair explanatory statements. Seeing how largely the financial trading and commercial operations of the United States partake of the character of the people who enter into them, and to what extent they are removed from those systems of combination which have grown up amidst ourselves, it may be hoped that the correction of personal character through the lesson which has been lately taught, may result in the abandonment of false and adventitious grounds of reliance. The country, in its condition of rapid development, has certainly little need for such dependence.

It is a remarkable fact that whilst as much as fifteen months before the American crisis, a general distrust of the railway system prevailed through the United States, public opinion in England was unaffected up to the verge of August and September, 1857. The character of the widely diffused class of securities was here stationary, and there was no reluctance on the part of capitalists to embark in new enterprises. The distrust prevailing in the United States had sprung mainly from a prevalent belief that the chief lines were corruptly managed. English manufacturers, too, continued to make advances, though industry in the States was slack and work was scarce, and people there were looking to the future with timidity and apprehension. The fearful indiscretion in the financial management of American railways has been among the prominent results of the late crisis. It had been noticed beforehand how uniformly the "total cost of construction" account was increased from year to year. The reasons, or rather the excuses, which were given for this unpromising feature, showed a good deal of ingenuity. In one report, it was the difference in the cost of a new bridge, and the original cost of an old one, difference in the cost of new and old iron, difference in the cost of heavy and light locomotives, enlargement of depots, raising the lines where a deficiency of gravel or ballast had occurred, and, in fact, almost any change in any department of the road or machinery was taken advantage of with the view of charging something to the "construction account," so that the net earnings each year might make a favourable show. It was very pleasant for the owners of a railway to hear that it earned eight, ten, or twelve per cent., but these eventually experienced to their regret that the items charged had to come out of their pockets, that the net earnings must cover a much larger surface, and that there was almost an entire cessation of payment of dividends. In many roads, however, the "open construction account," which assisted "to make things look pleasant," has been closed, but not on all of them, and therefore shareholders should look to this as well as to the floating debt system, where money is borrowed on shares at from ten to thirty per cent., the usury finally finding its way into the capital, upon which a dividend has to be earned just as well as upon the *bona fide* subscribed and paid amount. The "open construction account" was a delusion. It was easy to leave the increase of the account at the option of a few scheming and cheating speculators, who thus contrived to feather their own nests, to cock up an annual report at the expense ultimately of a cessation of dividends. What a declination has been witnessed of splendid dividends of ten and twenty per cent., payable in stock or bond!

The extraordinary and undue expectations entertained not only in the United States but in this country as to the capability of California, unquestionably aided in multiplying and extending the disasters consequent on the American crisis. When it was again and again stated, both in London and Boston, in regard to shipments to San Francisco, that six, or at most eight, moderately sized or assorted cargoes per month were all that were required or that could be consumed; instead of that quantity eastern shippers despatched twelve to fifteen first-class ships per month, fully laden with assorted merchandise. Some of these ships would have on board more than three thousand tons of goods. It is not too much to say that twelve ships now are equivalent to twenty-four, five or six years back, for then at least one-half the tonnage going to California was taken up by lumber, flour, wheat, oats, corn, and other bulky articles, of which California does not now require one dollar's worth—being itself an exporter. The cargoes sent were out of all proportion to the wants of the State. Parties in the East interested in ventures were slow to learn there was a glutted market, notwithstanding the accounts of sales of goods and the remittances for them were less and less satisfactory. On several occasions considerable excitement was created in New York among those who had made shipments of merchandise to California, by the receipt of letters from commission houses in

San Francisco containing accounts of sales. The charges swallowed up the proceeds of the sales. It was once stated in dry-goods circles, that one of the largest auction houses sent over two hundred thousand dollars' worth of dry goods, for which it never received one cent.

It is evident that in proportion as the commerce of the United States depends on agriculture as its basis, and "trading" conducted on the capital of the country, will both become independent of fluctuations; and that late events tend to this result, there can be no doubt with those who have watched the order of things from the commencement of the new epoch, each day serving as it does to mitigate its calamitous consequences, and the experience of the past having already led to a practical application of a remedy. The effects of the disasters that have overtaken the United States can be but temporary. The States, indeed, must chiefly look for the liquidating of their indebtedness to the fruits of agricultural industry. The late crisis showed that American merchants disregarded the valuable lesson of Æsop's fable of the goose that laid golden eggs. They were in full course of killing the goose by their determination to enjoy all the benefits of extended trade at once, without waiting for the gradual improvement of those resources which time would have accomplished. Thousands in this country, with shattered fortunes, lament the infatuation that led them to inundate the United States with merchandize, but the responsibility rests on the other side of the Atlantic, where was manifested, with a like infatuation, an unpardonable cupidity. That distress and embarrassment still pervade the United States, notwithstanding late favourable financial statements, to an extent probably never felt before there, except during the period that elapsed between the close of the revolutionary war and the adoption of the federal constitution, cannot be denied. A large proportion of the manufacturing establishments are suspended, and nine-tenths of those that are in operation have greatly curtailed their business. Of the proprietors many are ruined, and those whom strength of capital and other advantages have enabled to maintain the struggle, are but weakly encouraged to persevere. Numerous emigrants, who have gone to the United States in the flattering expectation of having full employment in the various arts and trades, have realized disappointed hopes and broken spirits.

It is to the agricultural portions of the United States, particularly to the resources of the great West, every one looks for the recuperative power which in time will diffuse a new energy over the country. Enterprising men are hastening thither, towns are springing into life, valuable roads that pass through rich and fertile countries are being constructed, the United States Government is quickly disposing of its waste lands, and the produce of copper-mining districts, and of new coal regions recently discovered, add important elements of profit. Attention may also be directed to one further element of strength, and that is—all other appearances to the contrary—the indissoluble character of the Federal Union, by which the States are linked to one another, and this without the remotest prospect of open collision. The people in the North regard slavery in the Southern States as a fact with which they have no right to interfere, and which they feel bound by the letter and spirit of the constitution to leave to the control of those on whom it is fixed. At the formation of the Union there were twelve slave States out of the thirteen, and at this date there are but thirteen slave States out of thirty-two of the Union. Slavery has shifted its location, and is now dying out of Virginia through the large introduction of free labour, and the fact that it is becoming unprofitable. It will flicker and pass away in the South, as it did in the North, as indeed it has done in all nations by the same process, that is, when it ceases to pay and becomes an insupportable evil to those who have to endure it. It will undoubtedly have to be left to what may be termed self-legislation as respects the States in which it is located, for intermeddling with it has been found to contribute to its perpetuation, because enlisting the spirit, pride, independence, and sectional feeling of the localities interested. If Kansas had been left to the natural course of events, or settlement and colonization, it would have escaped even the chance of becoming a slave territory. Such was the general sentiment and expectation in the South; but this slow, quiet, and certain process would not content the agitators. They could not be satisfied without threatening the South, and thus aroused that section to active and self-sacrificing efforts to extend its influence into this new region. This may serve to explain ebullitions of feeling that have occurred, and which have been widely misinterpreted in this country, though in no way affecting the stability of the Union. The entire abolition of slavery in the United States will probably be brought about as a question of profit and loss long before the terms on which one race should be subjected to another are fairly settled by casuists.

The correctness of the description thus far given of the great American crisis will scarcely be disputed, and unless this statement be impugned, it would seem difficult to forbear omitting the necessary conclusions to be deduced from it. The direct effect of this crisis has been to shake the attachment of the people of the Continent to the United States, and to produce a conviction, however vague or ill-grounded, that there is something inherently defective in the principles dominating throughout transatlantic society. There is no question that the character of the business portions of the communities in the commercial cities of America has been deeply compromised.

In addition to a thorough change in general principles of action, under the great law of imposed necessity, so far as regards the future dealings with this country, it should be urged as an essential condition, the adoption of sounder policy in the working of American banks. As to the advantages which these banks are supposed to be capable of yielding to the community, a deplorable and portentous public delusion appears to prevail throughout the United States. As the leading mercantile, manufacturing, and trading men throughout the States, in most cases began life with nothing, and owe what they possess in a very large degree to the aid they received from the banks when they commenced, let them learn to profit by their experience, and with less eagerness and less unscrupulousness to turn their intelligence, their skill, and the advantages by which they are surrounded to better account. Let them not endeavour to force a prosperity beyond that provided by the natural resources of the country; to accelerate national growth, and make individual fortunes by dint of mere scheming and on purely surreptitious grounds; lastly, in a country where labour is more respected than in any other, let them look to it as the only assured means they possess of permanent wealth and advancement. With the exercise of due vigilance on the part of the English, they can never again have an opportunity of making such breaches in the credit system as those respecting which such grave complaints are made.

Looking to the future, a regeneration in trading morality may be hoped for. No people ever possessed in a higher degree the means of national prosperity. An ample territory, fertility of soil, variety of climate and produce, a sea-board of five thousand miles facilitating foreign commerce, fisheries, and the coasting trade, and the mountainous divisions of the interior intersected by rivers, canals, and steam navigation, which create unparalleled conveniences for intercourse. The greater proportion of this territory is tilled by men as hardy, intelligent, and enterprising as ever turned the soil. A market abroad is always to be obtained for the surplus of all the raw material and the surplus of all the labour. Enterprise is everywhere the guiding motive. It has, undoubtedly, been carried to undue bounds, but is a good motive still. In the meantime, there are few artificial combinations, such as those which have grown up amidst us from the feudal ages, to embarrass advancement. The independent existence of the people was commenced in a state of civilization. The whole land was before them, to frame their laws and fashion their institutions as experience and an enlightened intellect should dictate. Nor is any compromise of local temporary interest demanded. The effect of the late crisis will no doubt be to throw the people more completely on their moral resources, to teach them that they must be more faithful to principle; that to be great, wealthy, and happy, it is not only necessary that manufactures should be extended, that commerce and trade should be enlarged, that capital should be augmented, that industry should exhibit new energy, that roads should be improved, arts and sciences patronized, schools be built, and universality be given to institutions of religion; but that the morality, prudence, and integrity which enable them to look along extended lines of possible action, should be combined with that remarkable tact which undoubtedly characterizes the Americans.

AMERICAN FINANCIAL MOVEMENT FROM 1811 TO 1852.

(From the New York Herald.)

THE annexed statement exhibits the discounts, deposits, circulation, and specie, of the banks, at different periods, with the value of imports, and the consumption of foreign imports, per head of the total population, at corresponding dates, showing also the state of financial matters each year, and the effect of the currency upon commercial affairs:—

BANK MOVEMENTS AND COMMERCE OF THE UNITED STATES.

January.	Banks.	Deposits.	Discounts.	Specie.	Circulation.	Imports.	Consumption per head.	
		Mil. $	Mil. $	Mil. $	Mil. $	Mil. $	$	
1811	89	—	—	15	28	—	—	Steady expansion.
1815	208	—	—	17	45	—	—	Great speculation.
1816	246	—	—	19	68	—	—	Revulsion.
1820	308	36	—	19	45	41	4·14	Contraction.
1830	330	55	200	22	61	56	4·39	Expansion.
1834	506	75	324	—	95	103	7·09	Expansion.
1835	704	83	365	44	103	129	8·64	Speculation.
1836	713	115	457	40	140	168	10·93	Speculations & shin-plasters.
1837	788	127	525	37	149	119	7·53	Revulsion.
1838	829	84	485	35	116	101	6·23	Contraction.
1839	840	90	492	45	135	144	8·68	Slight revival.
1840	901	75	462	33	106	89	5·21	Contraction.
1841	784	65	386	34	107	112	6·38	General bankrupt law.
1842	692	62	323	28	83	88	4·87	Contraction.
1843	691	56	254	33	58	58	3·11	Contraction.
1844	696	84	264	50	75	97	5·03	Expansion commenced.
1845	707	88	288	44	89	102	5·15	Expansion.
1846	707	97	312	42	105	110	5·42	Do.
1847	715	92	310	35	105	138	6·60	Do.
1848	751	103	344	46	128	134	6·25	Do.
1849	782	91	332	43	113	134	6·13	Do.
1850	824	109	364	45	131	163	7·25	Do.
1851	910	130	440	48	150	194	9·25	Speculation.
1852	930	150	520	52	175	220	10·15	Do.

In giving the banking movement and value of imports, we have left off the hundreds and thousands, putting down only the millions. The above table gives the effect of all the different movements in the currency on the commerce of the country, and it is easy to trace from the cause to the effect. A steady expansion in the currency usually leads to an increase in the value of imports, to an increase in the average consumption per head of total population, to individual extravagance and expenditure, and of course to an extension of private credit. It appears by this that up to 1836 the consumption increased to nearly eleven dollars per head, or more than doubled in six years. In 1837 the revulsion spread over the country, and the consumption of foreign manufactures fell from $1093 to $753 per head. In 1839 a slight reaction was realized, and the consumption per head slightly increased. In 1840 we had a second edition of the revulsion of 1836, and everything throughout the Union, connected with trade and finance, was completely prostrated—so much so that Congress passed a general bankrupt law; and in 1843 the banking movement had reached the lowest point, and the importations had become so much reduced that the average consumption per head amounted to only $3 11—a smaller sum than had been known within the previous twenty-five years. That was really the deepest depth. From that time up to 1849 there was a gradual but speedy recovery. Our progress was slow, but apparently sure. In 1848 the banking and commercial operations of the country received an impetus from the discovery of gold dust in California; but the effect was not general until about the commencement of 1850, when the movement exhibited itself decidedly, and with unmistakable evidence of becoming strong. In 1851 it had spread over the country, and the effect was everywhere visible—petitions came crowding into every legislature in session for the adoption of a more general system of banking. The old plan of applying for special acts of incorporation was considered too tedious—too slow for this fast age—and a general law, like the free banking law of the State, was considered just the thing to meet the wants of the speculating financiers of the day. State after State adopted the general law, and thus threw open the doors to all. The effect was seen at once in the increase of banks. Illinois, Indiana, and Wisconsin adopted the new law, and in fact very few States in the Union still adhere to the old restricted system. The facilities for establishing banks are daily increasing. The largest liberty is given by the general law, and there is very little danger of any scarcity of securities for deposit with the State authorities, as a basis for circulation. Within the past three years the manufacture of such securities has proceeded with frightful rapidity. Millions upon millions have been turned out, apparently without the first thought of where the means of payment were coming from. At present a moderate amount of discrimination is exercised in the selection of securities; but the supply of such as are issued upon a proper basis is likely, sooner or later, to fall short, and then recourse must be had to the next best in the market. This will be the course of the movement, and it is not difficult to foresee the ultimate result. It will come in a familiar shape when it does come. It will bring us back again to the times of 1817 and 1837, but we fear with ten times the force, and with an effect ten times more disastrous. The past has been too full of dear experience not to be heeded. We have so many precedents that it will be our own fault if we do not take timely warning. We have, however, very little faith in the lessons of the past. They are forgotten or unheeded in the excitement of the moment, and all press onward with apparently one object in view—the accumulation of the greatest fortune in the shortest possible time—entirely regardless of the means used in the accomplishment of the purpose. The end sanctifies the means. This is the secret of the ruinous fluctuations in this country in all things connected with commercial affairs; and so long as this spirit governs the financial and mercantile classes, so long will the same uncertainty attend all their operations, and so long will similar results follow all expansions.

We are subject, from ordinary causes, to periodical expansions, speculations, sudden contractions, and revulsions. Within the past fifty years, four or five severe collapses have prostrated every important interest, and swept away nearly every vestige of our previous prosperity. Heretofore the basis of all extensions of credit has been artificial. Upon the weakest and most defective foundations we have built enormous commercial systems, which have, in time of need, proved deceptive and disastrous. Notwithstanding all this we are just as ready again to raise confidence upon the same material, and risk every dollar upon the result. The present movement differs somewhat from those of former years. The basis is far more substantial, but it will only induce us to place more confidence in it, and carry the expansion to a much greater extent. The large receipts of gold from California gave the first important impetus to the present system of credits, and so far it has not progressed beyond a sound and healthy proportion. The banks have not, it is true, been the recipients of much of this gold, but most of it has gone into active circulation in the channels of commerce, where it exercises rather a conservative influence upon the currency, by restraining, to a certain extent, the issues of the banking institutions of the country at large. The receipts of gold-dust at the United States Mint from California, up to the present period, have not been so large as the extravagant statements which have been made, from time to time, induced many to expect. They have, however, been large enough to form a legitimate basis for the expansion of public and private credits to the extent already realized. Whether the supply will continue at this rate or not is a most important question, as upon that point alone depends the continuance of our present prosperity, or freedom from one of those revulsions which capitalists fear so much.

California Gold Received at United States Mint and Branches.

For the year ending December 31,		1848	$45,301
”	”	1849	6,151,360
”	”	1850	36,273,097
”	”	1851	55,938,232
To July 17		1852	30,000,000
Total receipts			$128,407,990

About one-half of this amount has been exported, leaving upwards of $60,000,000 in the country, only about $6,000,000 of which have gone into the bank. This is a very fortunate circumstance, for it will be seen by the table given above, that on the $6,000,000 increase in specie, the banks expanded their circulation $47,000,000, and in the same period augmented their discounts $180,000,000. This expansion upon such a small addition to the previous supply of precious metals in hand, gives a pretty good idea of what would have been the result had the bulk of our receipts from California gone into the vaults of our banking institutions. The free banking system permits an issue of bills for circulation upon public securities, instead of the precious metals, which accounts, in a measure, for the absence of coin in the banks. This system gives abundant facilities for increasing the paper currency of the country, and there is very little doubt but that the banks in existence, and about being established, will use them to the utmost.

We cannot expect that even the present proportion of paper to specie will be long maintained, or that the expansion of credits of all kinds will not become more rapid as the volume enlarges. Every day something new springs up—some new enterprise presents itself for aid and support, and they are soon forthcoming. Every week gives a greater impetus to the various movements; and where credits formerly increased by thousands, they will soon increase by millions. We have never yet continued long in the rear of our resources, and there is very little probability of our doing so in this railroad locomotive age. If we go on at the rate realized during the past three years, for any length of time, we shall most assuredly bring up where we have heretofore. We shall, before the lapse of many years, be as far in advance in our paper credits of the supply of gold and silver, as we ever were before the discovery of gold-dust in California and Australia; and as it will be necessary, before reaching that point, to more than quadruple at least the present inflations, we can form some idea of a revulsion coming with such an expansion. It would be almost fatal to every class, to every interest. There is probably no nation on the face of the earth which will be affected, both favourably and unfavourably, to the same extent, by the immense accession to the supply of gold and silver, as this. It will accelerate the growth of the country, build up great cities, every kind of internal improvement; extend the cultivation of the soil, and increase the products enormously; it will build palaces and fill them with extravagance; but it will ruin most of those who have participated in all these things. It will deprive them of all their luxuries, and plunge them into the lowest depths of bankruptcy. Property which has cost millions will change hands for thousands. Disaster and despair will mark the course of the revulsion, and years of economy, contraction, and inactivity be required to recuperate the energies of those who over-taxed them in their prosperity.

THE REVULSION OF 1857—ITS CAUSES AND RESULTS.

(*From the New York Herald.*)

The depreciation in leading or staple articles of produce the past year was remarkable. There was scarcely an article of trade known to the commercial world which was not lower at the close of the year 1857 than it was at its commencement.

We have not time nor space to go into details, but we have based generally our estimates of losses upon leading articles of domestic produce which make up the bulk of our exports, and also upon the prominent articles of importation.

Taking the first class of articles, we are led to believe that the losses from a decline in value during the past year amounted on an average of from 25 to 33 per cent. Estimated at the lower figure the result will give an aggregate of about $77,646,582
To which may be added the decline of values in manufactured products, shipping, etc., of about 30,000,000

Total $107,646,582

If we estimate the average loss on articles of foreign importation in 1857, compared with 1856, at 25 per cent., it will sum up about $35,000,000.

The shipping interest has also greatly suffered, and tonnage is now, and was for the greater part of 1857, in excess of demand. No new merchant vessels are at present on the stocks, and if things remain as they are, none will be required for a year or two to come. This depression in the shipping interest has acted very unfavourably upon the labour of our commercial cities. Large numbers of mechanics and labourers have been thrown out of employment. A heavy loss to the country has been sustained by the suspension of labour, the source of wealth and progress. Every able-bodied labourer or mechanic who is compelled to stand idle for the want of work is a positive loss to production, on the average of not less than one dollar and twenty-five cents per day. The seaboard cities and manufacturing districts of the interior probably at this time contain not less than one hundred thousand unemployed adult persons, which is equivalent to a loss of productive labour of about $125,000 per day, or about $875,000 per week, making an aggregate of about $2,800,000 per month; and, estimating the suspension of labour for four months at the same ratio, will give us a total of about $11,200,000. Summing up the total losses during the year 1857, we have the following grand result:—

Losses on domestic exportable produce	$77,646,582
Ditto on mining and manufacturing products, shipping, etc.	30,000,000
Ditto on articles of importation	35,000,000
Ditto on labour	11,200,000
Ditto on railroad and other corporated securities in 1857 alone, about	50,000,000
Total	$203,846,582

It is estimated that the railroads in the United States cost about one thousand millions of dollars, and that by the depreciation in their bonds and stock in the past six to eight years, that about $300,000,000 of that nominal amount of capital has disappeared, about $50,000,000 of which, with other corporate securities, we estimate, was sunk during the year of 1857.

This general view of the subject is liable to some variation on account of the fluctuations in the prices of articles within the last month or two. Thus cotton has recovered nearly two cents per pound from the lowest point current about the close of the year, or nearly $7 50 per bale, which, if applied to about 2,600,000 bales of the estimated crop of 1856-57 remaining in the country, makes a total increase of value equal to about $18,750,000, leaving a decrease in value, if no other change occurs, compared with the previous year, of only about $33,250,000.

Cotton and tobacco form almost the only exceptions to the continued decline which pervades the markets of the world for all other articles of commerce.

By scrutinizing the chief causes of failures and bankruptcies we shall find that they were brought about by the extraordinary expansion of the credit system and inflation of prices, induced by the action of the banks, bankers and moneyed corporations of the Old and New World. When a barrel of pork was pushed up to $24 per bbl., flour to $10 per bbl., wheat to $2 per bushel, sugar to 13*c*., cotton to 15*c*. to 16*c*., and coffee to 12*c*. to 16*c*. per pound; and the revulsion came, the speculators had to let go, because the banks found themselves unable any longer to sustain them.

The trading public had for some time based its credits upon the prevailing high prices of produce and merchandise. Bills had been made and notes given at ninety days to six months, predicated upon the prevailing high prices of produce noticed above. Hence, when the breakdown in the artificially inflated prices came—made more stringent and severe by its suddenness—bankruptcies followed on every side. People who expected to pay in pork at $24 per barrel find its value reduced to $15. Their flour, instead of selling at $10, has to be sold at $5 to $6; their wheat at $1 25 to $1 35, instead of $2; sugars at 5½*c*. to 7*c*. instead of 13*c*.; cotton at 10½*c*. instead of 15*c*. to 16*c*.; coffee at 8½*c*. to 10½*c*. instead of 12*c*. to 13*c*. for Rio; and other articles in about the same proportion, and railroad bonds and stocks, with bank and other corporation securities, instead of at par, at 50*c*., on the dollar's worth. The loss on leading articles of importation may be judged of by the comparison of prices which they bore at the close of 1856, with those current at the close of 1857. The articles on which the largest losses were sustained were on East India goods, and especially flax-seed and spices; on the importation of hides from Buenos Ayres, and on European manufactured goods, including silks and other articles from France.

The losses sustained on imported goods and merchandise can never be correctly ascertained, and can only be approximately estimated, with articles of domestic produce, at an average of not less than about 25 to 33 per cent., chiefly realized within the last six months of 1857. The whole amount of exports for the financial year ending on the 30th of June, 1857, was $310,586,330. Allowing the exports for the six months from the 30th June to the 31st December, 1857, to be about half in nominal value, or $155,193,165, a decline of 25 per cent. on this sum would give a total falling off in value of about $38,823,291. The same ratio applied to the produce remaining in the country to go forward in the next six months, or by the 30th June, 1858, will give about an equal amount of decline in value of about $38,823,582—making a grand total of about $77,646,582. But as some articles, such as cotton, may regain a good portion of the decline before the products reach market, it will be safe to fix the average loss at 25 per cent. The imports for the year ending June 30, 1857, amounted to $314,639,942, which would give the imports for the half-year ending 31st December, 1857, at the same ratio, half that amount, or about $157,318,871. But as the importations rapidly fell off towards the close of the year, and prices also in many cases were lessened by the operations of the late Tariff Act, we may reduce the estimate to about $140,000,000—which, at an average loss of about 25 per cent., would give $35,000,000. The loss on raw wool and on the products of the mines, of the forests, and of the fisheries, and with the depreciation also in the value of shipping and various other kinds of property and produce, may be set down at about $30,000,000 more, making up the grand total as stated in the foregoing table.

These aggregate losses exceed the amount of the total losses by failures in the United States in 1857, which were estimated at $143,780,000; yet they bear some proportion to them. It must also be recollected that many large houses, who have neither failed nor suspended, have yet suffered immense losses by the revulsion, and which, if taken into the account, would probably swell the amount of losses to the sum we have embraced in the foregoing table. If we apply a similar test to the losses sustained by failures in Europe, we shall find that similar causes have operated to bring them about, and that the fall in produce, merchandise, and securities has borne about the same ratio to the total losses sustained by failures and suspensions, and in houses which have withstood the storm. The failures in Europe involved liabilities, it was estimated, to the extent of about $700,000,000, of which about $140,000,000 were set down as a total loss, but we imagine the total to have been not less than about $200,000,000, while the loss by the fall in produce, merchandise, manufactures, stocks and bonds, was not less than about $300,000,000—thus making the grand total decline for Europe and America, including Canada, as follows:—

Losses by a decline in values in the United States and Canada	$203,846,582
Do. do. in Europe	300,000,000
	$503,846,582

Each succeeding panic but exhibits the recuperative energies of the United States. The late revulsion was inevitable. The inflation of prices had been forced to a point at which they could not be sustained without ruining consumers. The reaction was necessary to restore trade and commerce to their healthy and legitimate channels—to bring about a greater equality between the value of labour and the value of products to be given in exchange for it, and free them from the disturbing influence of the unnatural expansion of credit, fostered by banks and other corporations, to which the guarantees of states, counties, and towns, had been lent.

Each panic has resulted in making the city of New York the centre of finance and of trade for this continent. In 1837 it stood on a sort of struggling emulation with Philadelphia and Boston. The revulsion of that period decided its position and gave it an advance over them, which it has not only maintained, but has increased ever since. The rivalry between New York and other cities on this continent has ceased. The late struggle of 1857 was in a great degree between New York and London, and has terminated to the advantage of the former city. And the time must ere long arrive, when New York, and not London, will become the financial centre, not only of the New World, but also to a great extent, of the Old World.

We conclude by giving a comparative statement of prices on the first of January, 1857 and 1858:—

Ashes form a considerable article of trade in this market, and we proceed to give the prices at respective periods as follows:—

	January 1, 1857.			1858.
Pots per 100 lbs.	$7 62½	to	$7 75	$6 60
Pearls	8 0	„	0 0	5 75

—Prices of both sorts have, since the 1st instant, receded to 5*c*. to 5¼*c*.

Coffee reached its highest figures during the early part of the past summer. We give the comparison of prices, as follows, on the first of January of each year since 1852:—

	Brazil.			St. Domingo.		
1852 per lb.	8*c*.	to	9¼*c*.	7¾*c*.	to	8*c*.
1853	8½	„	9¾	7¾	„	8
1854	11	„	12½	10½	„	11
1855	8½	„	10½	8¼	„	9¾
1856	10	„	12¼	0	„	10⅜
1857	10	„	11½	0	„	10¾
1858	8½	„	10¾	7	„	8

—The stock of Rio on the 1st of January, 1858, was about 99,000 bags, against about 73,500 at the same date in 1857.

The highest prices obtained for cotton were in July and August, 1857, when middling uplands rose to about 15*c*. to 16*c*. per pound. The following is a comparison of the average of prices in New York on the first of January for each year, as follows:—

	Uplands and Flor. Fair.			N. O. and Mobile. Fair.		
1851 per lb.	14⅛*c*.	to	0*c*.	14¾*c*.	to	0*c*.
1852	8¾	„	0	9½	„	0
1853	10	„	0	11¼	„	0
1854	11¼	„	11⅜	12	„	12½
1855	0	„	9⅜	9¾	„	10⅛
1856	10	„	0	10½	„	11
1857	13¾	„	0	14	„	0
1858	10	„	0	10¼	„	0

During the panic in October, 1857, cotton rapidly declined, and continued to recede until the first days of January, 1858, when middling uplands sold as low as 8⅝*c*. to 8¾*c*. On the 7th of January the Atlantic's news was published, which produced a reaction, and prices have since advanced (January 20) to 10½*c.* per pound, or about 1¾*c.* to 1⅞*c.* per pound—equivalent to about $7 and $7 50 per bale.

EAST INDIA GOODS.—Among the most disastrous losses of 1857 were those sustained by houses in the East India trade. Gunny cloth, which sold in January, 1857, at 11*c.*, sold at the same period in 1858 at 9*c.* Gunny bags, which were sold as high last summer as 14*c.*, declined with the close of the year to 11½*c.* Saltpetre, which sold in January, 1857, at 8*c.* to 8½*c.* for crude, sold at the same period in 1858 at 6½*c.*, duty paid. The heaviest decline, however, was realized in flax-seed, which cost in Calcutta $2 and $2 20 per bushel, and receded in the United States to $1 20 and $1 25.

Manilla hemp declined from 9*c.* to 7*c.* per pound. Spices, including cassia, nutmegs, pepper, etc., have also largely fallen. We give the annexed table of the highest prices of last year, compared with those occurring in January, 1858 :—

	January, 1857.			January, 1858.		
Nutmegs	90*c.*	to	$1 0*c.*	50*c.*	to	52*c.*
Mace	80	„	0 85	40	„	45
Cassia	42½	„	0 0	26	„	0
Pepper	12	„	0 0	8½	„	0
Cloves	12	„	0 0	7	„	0

The duty on these articles was stricken off by the late tariff law of Congress, which threw them into the free list, and has no doubt contributed to reduce prices. Yet the decline exceeds the amount of impost taken off.

The comparison in the prices of flour and grain will be seen from the following tables. Owing to the short crop of 1856 prices were fully maintained during the summer of 1857, or until the crop of that year began to reach market. The favourable harvest in Europe, combined with the late panic, at once sent down prices; and 1858 opened with a fall in flour of fully $2 per barrel, compared with the same period of 1857. We give the following average of prices in New York, on the 1st of January for each year, as follows :—

		State.			Western.		
1852	per bbl.	$4 37½*c.*	to	$4 50*c.*	$4 50*c.*	to	$4 62½*c.*
1853		5 56¼	„	5 62½	5 62½	„	5 75
1854		7 75	„	7 87½	7 75	„	7 87½
1855		9 0	„	9 37½	9 12½	„	9 50
1856		8 25	„	8 50	8 25	„	8 50
1857		6 10	„	6 75	6 40	„	7 0
1858		4 20	„	4 70	4 25	„	5 50

And the stocks at the same periods were as follows :—

Flour.		1855.	1856.	1857.	1858.
Western Canal	bbls.	120,000	439,600	255,000	476,900
Canadian		20,700	27,300	18,600	7,800
Southern		43,000	130,100	119,500	118,450

The same causes which operated on the prices of flour also operated on those of grain. We give the following average of prices of grain in New York on the 1st of January, in the years named below:—

	Wheat.			Corn.			Rye.		
1852 ... per bushel	$1 0	to	$1 15	$0 77	to	$0 0	$0 66	to	$0 67
1853	1 25	„	1 30	0 93	„	0 0	0 68	„	0 70
1854	1 90	„	2 08	1 22	„	1 24	0 77	„	0 83
1855	1 80	„	2 50	1 38	„	1 40	0 97	„	1 00
1856	1 90	„	2 20	0 80	„	0 93	1 30	„	1 31
1857	1 30	„	1 80	0 68	„	0 75	0 88	„	0 90
1858	0 90	„	1 35	0 53	„	0 66	0 72	„	0 74

And the stocks at seveal periods were as follows:—

	1855.	1856.	1857.	1858.
Wheat......... per bushel	74,000	789,796	531,650	389,000
Corn	500,000	645,962	1,967,500	97,000

Hemp also fell off in prices as follows:—

		January 1, 1857.				1858.		
Russia.....................	per ton.	$275 0	to	$280 0	...	$210 0	to	$0 0
Manilla	per lb.	0 9	„	0 0	...	0 7	„	0 0
American dew rotted...	per ton.	210 0	„	215 0	...	100 0	„	110 0
„ dressed	„ ...	240 0	„	260 0	...	140 0	„	160 0

Hides were imported in 1857 largely in excess of demand, chiefly from Buenos Ayres, and immense losses have been sustained by their fall in prices. Several houses in the trade have been swept entirely away by the ruinous decline.

The total imports and average prices for a series of years may be seen from the following tables:—

TOTAL IMPORTS FROM JANUARY 1 TO DECEMBER 31.

	1854.	1855.	1856.	1857.
Foreign, No......................	1,360,429	1,210,185	1,310,815	1,259,271
Coastwise	310,869	344,958	444,105	551,943
Total........................	1,671,298	1,555,143	1,754,920	1,811,214

PRICES IN NEW YORK, JANUARY 1.

	B. Ayres.				Orinoco.		
1852 ....per lb.	11½c.	to	13c.	....	11c.	to	—c.
1853	15¼	„	16	....	0	„	14
1854	22	„	23	....	21½	„	22
1855	19	„	20	....	0	„	17
1856	0	„	26	....	0	„	24
1857	0	„	31	....	0	„	29
1858	0	„	20	....	0	„	17

Leather also declined something in the same ratio, as may be seen from the following tables:—

PRICES IN NEW YORK, JANUARY 1.

	R. G. and B. Ayres. Light and Middle.			Orinoco, etc. Light and Middle.		
1852 ... per lb.	14c.	to	14½c.	13c.	to	13½c.
1853	17	„	17¾	16	„	16¾
1854	0	„	23	0	„	22
1855	19	„	20	17½	„	18½
1856	23	„	25	22	„	23½
1857	30½	„	31½	29½	„	30½
1858	22	„	23	19	„	21

The receipts were as follows:—

	Receipts.	Sales.
1856	2,570,450	2,609,350
1857	2,683,737	2,395,600

	Hemlock.		Oak.	
	1856.	1857.	1856.	1857.
Receipts of sole the past week, sides	54,200	41,500	8,800	8,900
Sales	52,500	38,400	9,100	3,100
Stock	21,500	315,100	5,700	42,500

Same time.	1855.	1856.
Receipts	37,800	8,200
Sales	34,800	4,400
Stock	53,300	11,900

Metals have not escaped the general depression, and especially iron and lead. Scotch pig dropped down from $30 to $31 per ton in January, 1857, to $24 to $25 in January, 1858, having at one time, last summer, rose to $35 to $38 per ton. American pig and English bars also suffered a decline. The termination of the Russian war sent down the prices of lead, from which the article has not recovered, while foreign imports have declined. The article sold at 6c. to 6¾c. in January, 1857, while in January, 1858, it fell to 4¾c. to 5½c. for Spanish and English. Copper and tin have stood their ground better than other metals.

The product of the Lake Superior mines in copper during 1857 amounted to about 3,800 to 4,000 tons, showing an increase of about 500 tons over the previous year. Sales of Lake Superior copper were mostly made for export to France at 19½c. to 20c. per lb. Taking the inside yield at 20c. per lb. will give the value of Lake Superior copper for the last year at $1,520,000.

Naval stores also suffered a depreciation, as will be seen by the following table of

PRICES IN NEW YORK, JANUARY 1.

	Turpentine. (Wil. and N. C.)			Spirits of Turpentine.			Common Rosin.		
1852	$0 0c.	to	$3 6¼c.	34½c.	to	37c.	$1 20c.	to	$1 30c.
1853	4 0	„	4 25	62	„	63	1 40	„	1 55
1854	4 75	„	5 0	60	„	62	1 70	„	1 90
1855	4 0	„	4 37½	45	„	46	1 80	„	1 87½
1856	3 12½	„	3 37½	0	„	42	0 0	„	1 60
1857	3 75	„	4 0	48	„	49	1 50½	„	1 62½
1858	2 75	„	3 0	0	„	38	0 0	„	1 30

Oils of all kinds closed lower in January, 1858, than they were at the same time in 1857—the heaviest fall, as previously noticed, having been in linseed. Provisions, also, with the commencement of new supplies in the autumn of 1857, underwent a marked decline, which continued to droop until the present time, and especially pork and beef. The following table shows:—

PRICES IN NEW YORK, JANUARY 1.

	Mess Pork. Per bbl.	Mess Beef. Per bbl.	Lard. Per lb. (bbl.)
1852	$14 75 to $15 12½	$8 00 to 11 00	$8¾ to 9
1853	19 00 „ 19 50	9 25 „ 10 25	12 „ 12¼
1854	13 50 „ 15 75	8 50 „ 14 00	9⅜ „ 10
1855	12 37½ „ 13 00	8 25 „ 11 25	10 „ 10½
1856	17 25 „ — —	10 50 „ 14 00	12 „ 12¼
1857	19 37½ „ 20 00	10 50 „ 14 75	12½ „ 12¾
1858	15 50 „ 16 00	9 00 „ 10 00	8¾ „ 9½

The stock of pork, 1st January, 1858, was 10,558 bbls. against 13,046 ditto, 1st January, 1857, while the stock of beef on 1st January, 1858, was 39,144 bbls. against 10,488 ditto, 1st January, 1857.

The market for pork reached its highest point during the year in the summer months of 1857, when mess sold at about $24 per barrel. Beef also attained prices in about the same ratio; prime mess having sold as high as $28 and $31 per barrel. Lard also went up to 14*c.* to 15*c.* per lb.

Rice at the opening of 1857 sold at 4*c.* to 4⅜*c.* per lb., and on the 1st January, 1858, at 2¾*c.* to 3½*c.*

Sugars and molasses underwent the greatest fluctuations. The following table will give the—

PRICES IN NEW YORK, 1ST JANUARY.

	New Orleans.		Clayed Cuba.
1852	per gallon $28 to $—		$18 to $19
1853	30 „ 31		20 „ 21
1854	27 „ 28		23 „ —
1855	23 „ 26		22 „ 25
1856	48 „ 49	New	40 „ 42
1857	— „ 80	Old	38 „ 40
1858	— „ 35	Old	17 „ 20

STOCK IN NEW YORK, JANUARY 1.

	1855.	1856.	1857.	1858.
Cuba, mus., hhds.	$ —	$—	$1,161	$1,518
Cuba	276	—	—	—
Cuba, bbls.	—	—	—	4,041
Cuba, clayed, hhds.	—	281	370	2,820
Cuba, clayed, tcs.	—	14	—	—
Cuba, clayed, bbls.	—	17	—	—
Porto Rico, hhds.	—	—	267	275
New Orleans, bbls.	3,730	200	100	471

The crop of sugar in Louisiana having only reached about 73,900 hhds. in 1856, showed a decrease of about 157,000 hhds., compared with that of 1855. The year 1857 opened with a hardening tendency in prices, but it was late in spring before they assumed the highest rates—which ranged from 12*c.* to 14*c.* With the close of the summer and the advent of the panic in the autumn, they commenced receding, and closed with the year at a much lower range of figures. We annex tables of quotations and stocks at the periods named:—

PRICES IN NEW YORK, JANUARY 1.

		New Orleans.	Cuba.
1851	per lb.	$5\frac{1}{4}$*c.* to $6\frac{3}{4}$*c.*	$4\frac{1}{2}$*c.* to $6\frac{3}{4}$*c.*
1852		$4\frac{1}{2}$ „ 6	4 „ 6
1853		$4\frac{1}{2}$ „ 6	$4\frac{1}{4}$ „ $5\frac{3}{4}$
1854		4 „ $5\frac{1}{2}$	$4\frac{1}{4}$ „ $5\frac{1}{4}$
1855		$4\frac{1}{4}$ „ $5\frac{3}{4}$	4 „ $5\frac{1}{2}$
1856		8 „ 9	7 „ 8
1857		9 „ 11	8 „ 10
1858		$5\frac{1}{2}$ „ 8	$5\frac{1}{2}$ „ $7\frac{3}{4}$

STOCK IN NEW YORK, JANUARY 1.

	1855.	1856.	1857.	1858.
Cuba, hhds.	2,020	2,314	7,595	6,959
Porto Rico	99	1,555	2,819	1,306
New Orleans	4,592	1,547	62	1,442
St. Croix	—	—	—	250
Texas	24	—	—	—
Jamaica	—	—	520	—
Total	6,745	5,146	10,976	9,957
Molado	—	—	—	6,079
Cuba, boxes	—	10,788	19,931	8,731
Manilla, bags	14,201	—	—	—
Brazil	—	—	—	1,000
Cuba	—	—	—	400

The market for foreign spirits has also undergone a decided depreciation in prices, induced partly by the reduction of the duties, but mainly by the general causes which have operated to reduce the prices of other articles.

Tobacco, which had ruled high for the two or three past years, and maintained the advanced rates at the opening of 1857, unlike other articles, owing to unusually light stocks, as far as prices were concerned, resisted the influence of the panic of last fall, though the market was rendered dull and inactive by it. This inactivity remained at the close of the year, though had stocks and assortments been adequate to the wants of the trade, greater animation would have been witnessed in January, 1858,

The following gives the total inspections of leaf tobacco in New York for each of the years named below :—

		Ky.	Va. & N. C.	Ohio.	Md.	Total.
1849	hhds.	10,753	2,254	29	100	13,136
1850		12,207	1,437	28	122	13,794
1851		12,285	655	6	100	13,046
1852		20,107	361	1	3	20,471
1853		11,295	154	4	4	11,457
1854		9,295	295	21	—	9,611
1855		8,679	1,720	2	—	10,401
1856		12,683	2,009	9	—	14,701
1857		8,963	1,376	—	—	10,339

The monthly returns of stocks averaged from 5,000 to 6,000 hhds., leaving in the inspection warehouses on the 1st January, 1858, 4,644 hhds. against 5,746 hhds. 1st January, 1857. The stock of Spanish was much larger than usual, which was as follows :—

	1st Jan. 1856.	1857.	1858.
Havana, Cuba, Sagua, Yara and Cienfuegos, bales......	5,921	3,640	10,767

The prices were as follows :—

	1857.			1858.		
Virginia and Kentucky leaf...	*7c.*	to	*16c.*	*7c.*	to	*18c.*
Mason county	10	„	16	8	„	18
Havana fillers	24	„	$1 25	28	„	$1 00
Cuba	25	„	28	16	„	22
Yara	35	„	36	30	„	35

The fluctuations in the prices of tea, induced by the war with China and consequent speculation, we have noticed in a previous article headed the "Commercial Retrospect for 1857."

The heaviest decline in staple articles has been that in wool. The years 1855 and 1856 were years of high prices and speculation in wool, while the year 1857 may be recorded as one of the most disastrous known to the trade for a period of twenty years. Both producers and manufacturers, with speculators, were alike in many cases made bankrupt by the fall. With the suspension of woollen factories last autumn the article became nominal, and for a while unsaleable.

The stock of wool held in the seaports at the commencement of 1857 was estimated at 6,000,000 lbs., and in the interior about 3,000,000 lbs.—giving about 9,000,000 lbs. for the whole country. The annual consumption of wool in the United States, prior to 1857, was estimated at 72,000,000 lbs., or nearly 3 lbs. per head to the entire population. Of this 72,000,000 lbs., the United States supplied about 52,000,000 lbs.—leaving about 20,000,000 lbs. to be supplied from foreign countries. The reduction in the duty on coarse wools will, should trade and manufactures revive, lead to a large increase in the importation of cheap wools. The following table gives the prices of wool in New York on the 1st of January :—

	1857.				1858.		
American, Saxony fleece ....per lb.	55c. to	60c.	......	40c.	to	45c.
American, full blood merino	50	,, 54	......	35	,,	40
American, ½ and ¾ merino	44	,, 48	......	30	,,	35
American, native and ¼ merino	33	,, 37	......	27	,,	32
Extra pulled	45	,, 48	......	26	,,	28
Superfine pulled	38	,, 41	......	22	,,	24
No. 1 pulled	33	,, 36	......	20	,,	21
California fine, unwashed	23	,, 26	......	22	,,	25
California common, unwashed	12	,, 17	......	10	,,	15
Peruvian, washed..............................	30	,, 36	......	—	,,	—
Valparaiso, unwashed	14	,, 15	......	10	,,	13
South American common, washed	13	,, 15	......	10	,,	13
South American, Entre Rios, washed	16	,, 20	......	15	,,	18
South American, unwashed	8	,, 12	......	10	,,	14
South American, Cordova, washed	25	,, 27	......	20	,,	25
East India, washed	26	,, 28	......	18	,,	20
African, unwashed	9	,, 18	......	9	,,	18
African, washed	18	,, 30	......	16	,,	28
Smyrna, unwashed	14	,, 20	......	14	,,	18
Smyrna, washed	24	,, 28	......	23	,,	28
Mexican, unwashed............................	15	,, 18	......	12	,,	16

Freights also participated in the general depression of 1857. We give the—

RATES TO LIVERPOOL AND LONDON, JANUARY 1.

Liverpool.	1857.				1858.		
Grain	0s. 7d.	to	0s. 8½d.	....	0s. 5½d.	to	0s. 6½d.
Flour.....................	2 3	,,	2 6	....	—	,,	1 10½
Cotton	7-32d.	,,	¼d.	....	3-16d.	,,	—
Heavy goods...........	27 6	,,	—	....	20 0	,,	—
London.							
Flour.....................	2 10½	,,	3 0	....	2 6	,,	—
Heavy goods...........	30 0	,,	32 6	....	25 0	,,	26 0

NUMBER OF VESSELS IN PORT, JANUARY 1.

	1857.		1858.
Steamers	35	....	33
Ships	111	....	145
Barks	87	....	92
Brigs	96	....	107
Schooners	270	....	314
Total	599	....	691
Sterling exchange	8¼ to 8¾	....	8¾ to 9½

We have thus given, at some length, statistics, or actual facts, on which we base the primary causes of the extraordinary revulsion of 1857.

This remarkable reaction was as necessary as a thunder-storm in a mephitic and unhealthy tropical atmosphere. It purified the commercial and financial elements, and tended to restore vitality and health, alike conducive to regular trade, sound progress, and permanent prosperity.

THE FAILURES IN AMERICA.

OFFICE OF THE MERCANTILE AGENCY,
314 and 316, Broadway, New York, Jan., 1858.

TO OUR SUBSCRIBERS.

THE entrance of the new year induces us to present ourselves again before you, glancing back at the principal events of the past year in the mercantile world, and observing a few suggestions dictated by the interest which we feel in your welfare.

The financial revulsion through which our whole country has just passed—the disastrous effects of which will be felt for a long time to come—exceeded, in intensity, the apprehensions of the most timid and cautious in our midst. A stringent money-market was looked for. There had been over-trading, too much competition in business, too heavy expenses, too long and large credits, and there was still quite a land speculation in the West. Some failures were of course expected. The tendency to overdo, however, you will remember, had been very much restrained by causes which have operated every year or so for the last six years. During that time, drought and failure of crops both at the South and West, and two or three serious financial revulsions, contributed to put jobbing merchants on their guard against excessive over-trading, in which we do not believe they indulged to any serious extent. After the hard years referred to, great anxiety was felt in regard to the crops for 1857. A kind Providence gave us an abundant harvest, on the basis of which a sound and healthy trade was anticipated. Our importations were heavier, it is true, than they ought to have been; but our jobbers stocked up moderately and were looking for their usual fall trade when the contraction commenced, which was followed by panic, and the crash came, destroying all demand for goods, depreciating stocks, deranging our interior exchanges, and stopping collections and the forwarding of the crops.

Was there sufficient disease in the body commercial to cause such an arrest of business and a general liquidation? The number of houses that have resumed or will pay in full; the large number that have already compromised as high as sixty to eighty cents; the small losses our banks have met with, and the quickness with which they were able to collect in and place themselves in a strong position—not to speak of the value of the crops on hand and the many millions of specie hoarded throughout the country—justify us in believing that the over-issues of paper money, over-trading, speculation, etc., would not have warranted, at least after the export demand for specie ceased, more than a moderate contraction, which should have been even then very gradual and discriminating, and not a ruthless assault upon private credit.

We say that the larger part of the contraction which took place was caused by panic, by which we mean, "terror inspired by a trifling cause or misapprehension of danger." Death from fright is as serious a calamity as from any ordinary cause. We have reason to be thankful that this panic came in a year of plenty; and with smaller indebtedness, we believe, from the country than usual. Had it occurred at any time within the last three or four years—the year of drought and failure of crops, for instance—there would have been more apology for it; but the losses would then have been much greater. As it is, we may reasonably hope that the houses which have been bankrupted by this crisis will make better dividends than were ever made before under such disastrous circumstances.

We entertain the common opinion that the action of the officers of four or five of our strongest banks was the chief cause of the great disasters of the season. They concerted together and forced a rapid and merciless contraction upon all our city banks, carrying along with them those of the whole country. If the banks of New York city are to control the action of the banks of the United States, it is to be regretted that so small a number of them should be able by combining to determine the line of policy to be pursued by the whole body. The whole country is, therefore, interested that New York should have a steady and safe banking system; and if our Legislature should impose some such restriction as the press generally have recommended, like the law of Louisiana, requiring them to have one dollar in specie for every three of liability, we should have as perfect a system, probably, of paper currency, as has ever been devised. If the control of our banks is trusted to unskilful hands, the system ought not to be blamed for the difficulties which their bad management has caused. A wise action on the part of the banks would have driven into insolvency all that class whose condition warranted such treatment, without putting every one on the rack in order to discover the weakness of the few.

The tendency to excessive importation, which has affected all classes of merchants, may be largely ascribed to the fact that European letters of credit have been so easily obtained, and that they have afforded such fine openings for "kiting" operations. In this regard the times are like those which preceded 1837. There has not been so much over-trading on the part of the jobbers with the interior as has been supposed. As a class they manifest more caution in giving credits than in former years. Where the organization of the house is good, where it combines other talent and experience than that of the mere salesman, with the judicious use of the mercantile agency, they can and do show good ledgers and a prosperous state of things.

It is considered safe for a merchant doing an ordinary jobbing business, buying on eight months and selling on six, to sell three or four or even five times the amount of his capital; and if his business is well attended to, and his goods placed in safe hands, he may be, always, easy in his finances, anticipating his paper, and entirely independent of banks. We can point to such houses among our subscribers in all kinds of jobbing trade, even that most slandered of all, the dry goods trade. We have in our mind the case of one such house, a leading one, too, which, from the commencement of its business, some years since, has never borrowed a dollar nor asked a discount till within a few weeks. We think of a silk house, which, for seventeen years, has not, till this crisis, asked a discount nor borrowed a dollar.

We have also in our mind the case of one of our largest umbrella houses, which, for ten years, has asked for no discount nor borrowed a dollar; and we could mention others. We admit that these instances are rare, but it shows what can be done and what ought to be done more commonly than it is. The risky houses are those which sell ten, twelve, or fifteen times the amount of their capital. In easy times they are able to do this by using all the credit they can get at bank; as well as that they get from the importer and manufacturer.

The jobbing merchant's embarrassments arise mostly from the want of promptness on the part of his country debtor; no definite time being fixed when the debt must be paid. The country storekeeper cannot be prompt when selling to farmers on a credit of twelve months or longer. He should sell for cash or short credit. Then he would be prompt to his own payments, and jobbers could begin to make calculations on the certain and prompt payment of their bills receivable. This would be striking at the root of the evil. The prosperity of the manufacturer, importer, banker, and jobber is largely dependent on the manner in which the country merchant does his business. If his course tends to make or break all those above him, how important that he should do business on sound principles? He should be required to give notes for all his purchases, and to meet them at maturity. He would then, of necessity, buy carefully, sell prudently, collect closely, and thus enhance his own prosperity as well as that of trade generally. The farmers, it is conceded, are the richest body of men in our country. They can now do, with convenience, that which twenty or twenty-five years they could not. If, however, they cannot pay cash, they ought to be brought to quarterly settlements at all events. There are 157,394 village and country stores in the United States; and they are all the while indebted to jobbers of the cities, on the average of say $14,500 each;* or, in the aggregate, $2,282,000,000. They are, season after season, paying off and incurring this debt afresh. These merchants, we are proud to say, embody as much industry, business talent, and integrity as are to found in any of the walks of life, and the per centage of loss by them, including that which is made by adventurers and innovaters is, on the amount of business done, small indeed.

Another great cause of embarrassment, admitted to be such by a wealthy class of jobbers in some of our cities, has been the fact, that when they have sold to men whom they knew to be good, they have been indifferent as to whether the debt was paid them when it was due or not, preferring to make it secure at large interest, and in this way, locking up their capital, they have relied upon the banks to supply them in time of need with accommodation facilities. This accounts, in part, for so many houses of this description having been obliged to ask for some extension of their paper.

---

* This seems large, but the table shows it to be below the average of the indebtedness of 2771 country merchants.

We beg your reference to the following table, giving such particulars as may be of some interest. It has been gotten up in great haste, necessarily, but we have avoided no pains nor expense to make it accurate. In the large cities we have, in three-fourths of the cases, the statements of the parties themselves. In some recent cases of failure no exhibits have as yet been prepared. In all instances where these statements were not to be had, we have put a careful estimate upon each case, our own records furnishing us data by which we could more nearly approximate it than by any other method.

We have not included such houses as suspended for a short time, and resumed payment; nor such as, during the panic, had partial or even general extension, where it has been upon short time; nor have we included the losses by the failures of banks or railroads.

These returns omit California, our arrangements for that State not yet being as perfect as elsewhere. They embrace, however, the Canadas, New Brunswick, and Nova Scotia.

It will be seen that the number of firms in the United States (California excepted), by our records—and they embrace all but a class of small retailers in the larger cities—is 204,061, or, estimating the population at 25,000,000, that there is a store to every 123 of our inhabitants, or to every 25 families.

There has been lost by 337 swindling and absconding debtors, $5,222,500, and by 512 firms which will pay nothing, their losses and confidential debts absorbing everything, $20,309,000. There are 3,839 concerns owing $197,080,500, and they are such cases as usually average 40 to 50 cents; and there are 435 houses owing $77,189,000, which will pay in full if the times ahead prove to be ordinarily prosperous.

The total amount of the liabilities of the 5,123 failures is put down at		$299,801,000
But there will be realized from those who will pay in full	$77,189,000	
And on the amount of "ordinary" failures $196,080,000 at 40 cents	78,832,000	
		156,021,000
Leaving a final loss of		$143,780,000

The columns exhibiting the number of counties in each State, with the number in which, during the whole year, no failures have occurred, will surprise you with their showings. No other conclusion can be formed than that the trade of the country was in good hands, and in a prosperous condition prior to the panic.

If the country storekeepers' general debt is, as herein before estimated, $2,282,000,000, it would be fair to infer that ordinarily the year's business would amount to all of double that sum, or $4,564,000,000. From this we would deduct for light trade last spring, and the injury to sales this fall, 40 per cent., which would make the business for 1857, with the country merchant, amount to $2,738,400,000, and as the losses for the year by this trade were $41,838,000, the per centage is about one and a half.

For the convenience of those of you desiring to contrast the city with the country trade, we have separated the one from the other, as will be readily perceived by reference to the table. Twenty-six cities are enumerated. Their failures have been 2,352 in number, and the liabilities of the same amount to $257,963,000—or an average of about $109,000.

Those of you who preserved our last year's circular, will be able to compare the statistics therein given with those contained in this.

STATISTICS AS TO FAILURES

FROM DECEMBER 26, 1855, TO DECEMBER 25, 1857.

Places.	Present Number of Stores.	Failures.	
		Number.	Liabilities.
NEW YORK.			
New York city*	13,854	915	$135,129,000
Albany	721	35	838,000
Buffalo	793	72	4,224,000
Oswego	204	13	161,000
Rochester	408	31	850,000
Syracuse	305	29	436,000
Troy	391	24	1,607,000
Utica	298	20	585,000
Balance of the State	15,875	447	6,789,000
MASSACHUSETTS.			
Boston	4,374	253	41,010,000
Balance of the State	10,257	230	2,611,000
PENNSYLVANIA.			
Philadelphia	7,404	280	32,954,000
Pittsburg	1,374	28	1,183,000
Balance of the State	13,526	226	2,283,000
ILLINOIS.			
Chicago	1,350	117	6,572,000
Balance of the State	11,459	199	2,766,000
OHIO.			
Cincinnati	2,513	96	3,898,000
Cleveland	550	30	612,000
Balance of the State	15,746	220	2,357,000
LOUISIANA.			
New Orleans	2,230	58	6,285,000
Balance of the State	1,667	5	246,000
MISSOURI.			
St. Louis	1,580	49	5,522,000
Balance of the State	4,851	29	433,000
RHODE ISLAND.			
Providence	1,100	35	4,564,000
Balance of the State	566	4	105,000
MARYLAND.			
Baltimore	1,970	58	3,206,000
Balance of the State	3,368	41	725,000
Carried forward	118,734	3,544	$267,951,000

* Includes Brooklyn and Williamsburg.

Places.	Present Number of Stores.	Failures.	
		Number.	Liabilities.
Brought forward	118,734	3,544	$267,951,000
MICHIGAN.			
Detroit	649	34	1,514,000
Balance of the State	3,706	98	1,004,000
IOWA.			
Dubuque	403	36	735,000
Balance of the State	4,308	108	1,333,000
KENTUCKY.			
Louisville	1,080	19	757,000
Balance of the State	5,715	31	1,007,000
SOUTH CAROLINA.			
Charleston	900	31	922,000
Balance of the State	2,538	24	305,000
Territories	1,697	63	1,705,000
Indiana	7,337	139	1,636,000
VIRGINIA.			
Richmond	1,583	30	781,000
Balance of the State	7,781	90	982,000
WISCONSIN.			
Milwaukie	633	19	380,000
Balance of the State	3,757	101	1,244,000
North Carolina	3,233	62	1,171,000
New Jersey	4,433	86	1,142,000
Connecticut	4,209	61	1,129,000
Maine	4,912	81	1,060,000
New Hampshire	2,700	70	928,000
Vermont	1,962	57	473,000
Georgia	5,339	32	925,000
Delaware and District of Columbia	2,727	20	261,000
Arkansas	1,179	7	309,000
Alabama	2,694	16	295,000
Mississippi	2,235	11	445,000
Tennessee	4,387	40	712,000
Florida	783	7	250,000
Texas	2,447	15	393,000
Total United States	204,061	4,937	$291,750,000
CANADA WEST.			
Toronto	389	25	$2,714,000
Balance of Canada West	3,444	109	2,172,000
CANADA EAST.			
Montreal	909	15	523,000
Balance of Canada East	1,764	15	1,267,000
Nova Scotia and New Brunswick	1,797	22	1,375,000
Total British provinces	8,303	186	$8,051,000
Total United States and British Provs.	212,364	5,123	$299,801,000

Places.	Ordinary failures.		How many have arranged with creditors, and at what average.
	No.	Liabilities.	
NEW YORK.			
New York city*	600	$83,951,000	218 av. 51*c.*
Albany	18	480,000	10 „ 42
Buffalo	53	2,795,000	13 „ 43
Oswego	12	156,000	...
Rochester	27	707,000	8 „ 48
Syracuse	22	268,000	4 „ 31
Troy	12	682,000	3 „ 48
Utica	9	376,000	5 „ 47
Balance of the State	378	5,565,000	...
MASSACHUSETTS.			
Boston	212	32,255,000	182 „ 48
Balance of the State	202	1,711,000	...
PENNSYLVANIA.			
Philadelphia	155	16,995,000	63 „ 54
Pittsburg	22	918,000	23 „ 47
Balance of the State	204	2,005,000	...
ILLINOIS.			
Chicago	82	4,571,000	11 „ 54
Balance of the State	149	2,093,000	...
OHIO.			
Cincinnati	69	2,387,000	54 „ 48
Cleveland	24	390,000	10 „ 47
Balance of the State	178	1,742,000	...
LOUISIANA.			
New Orleans	36	4,388,000	8 „ 55
Balance of the State	2	26,000	1 amt. 18,000 pay 50
MISSOURI.			
St. Louis	25	3,585,000	4 av. 50
Balance of the State	17	247,000	...
RHODE ISLAND.			
Providence	22	2,136,000	12 „ 40
Balance of the State	3	60,000	...
MARYLAND.			
Baltimore	39	2,472,000	17 „ 44
Balance of the State	37	708,500	...
MICHIGAN.			
Detroit	24	1,199,000	13 „ 41
Balance of the State	71	722,000	...
IOWA.			
Dubuque	21	463,000	4 „ 44
Balance of the State	79	1,059,000	...
KENTUCKY.			
Louisville	12	412,000	...
Balance of the State	24	496,000	...
Carried forward	2,840	$178,020,500	

* Includes Brooklyn and Williamsburg.

Places.	Ordinary failures.		How many have arranged with creditors, and at what average.
	No.	Liabilities.	
Brought forward	2,840	$178,020,500	
SOUTH CAROLINA.			
Charleston	23	812,000	8 av. 42
Balance of the State	20	245,000	...
Territories	46	1,302,000	...
Indiana	114	1,411,000	15 ,, 49
VIRGINIA.			
Richmond	22	694,000	3 ,, 58
Balance of the State	70	749,000	...
WISCONSIN.			
Milwaukie	14	312,000	3 ,, 73
Balance of the State	92	1,150,000	...
North Carolina	42	668,000	...
New Jersey	72	836,000	...
Connecticut	50	995,000	...
Maine	71	832,000	...
New Hampshire	60	775,000	...
Vermont	49	382,000	...
Georgia	21	681,000	...
Delaware and District of Columbia	18	258,000	...
Arkansas	5	285,000	1 am'g to over $100,000 will pay nearly all
Alabama	14	265,000	...
Mississippi	10	435,000	2 av. 50
Tennessee	28	618,000	...
Florida	5	228,000	2 ,, 52
Texas	12	353,000	...
Total United States	3,703	$192,305,500	
CANADA WEST.			
Toronto	17	$1,270,000	3 ,, 58
Balance of Canada West	73	1,631,000	8 ,, 45
CANADA EAST.			
Montreal	12	445,000	4 ,, 52
Balance of Canada East	13	66,000	3 ,, 35
Nova Scotia and New Brunswick	21	1,363,000	3 ,, 50
Total British Provinces	136	$4,775,000	
Total United States and British Provs.	3,839	$197,080,500	

Places.	Swindling and absconding debtors.		Not classed dishonest, but will pay little or nothing.	
	No.	Liabilities.	No.	Liabilities.
NEW YORK.				
New York city *	37	$988,000	111	$8,033,000
Albany	1	5,000	10	144,000
Buffalo	6	74,000	9	534,000
Oswego	1	5,000	...	...
Rochester	...	...	3	123,000
Syracuse	3	85,000	3	53,000
Troy	1	20,000	7	230,000
Utica	1	12,000	9	147,000
Balance of the State	25	220,000	27	255,000
MASSACHUSETTS.				
Boston	2	109,000	20	636,000
Balance of the State	14	62,000	8	75,000
PENNSYLVANIA.				
Philadelphia	13	481,000	60	3,704,000
Pitsburg	1	15,000	3	40,000
Balance of the State	4	42,000	14	128,000
ILLINOIS.				
Chicago	11	203,000	12	407,000
Balance of the State	25	210,000	17	291,000
OHIO.				
Cincinnati	7	99,000	5	100,000
Cleveland	2	5,000	2	38,000
Balance of the State	26	380,000	16	235,000
LOUISIANA.				
New Orleans	4	213,000	6	342,000
Balance of the State	2	70,000	1	150,000
MISSOURI.				
St. Louis	4	205,000	8	498,000
Balance of the State	5	51,000	3	35,000
RHODE ISLAND.				
Providence	3	18,000	5	1,456,000
Balance of the State	...	...	...	...
MARYLAND.				
Baltimore	3	30,000	9	352,000
Balance of the State	4	16,500	...	...
MICHIGAN.				
Detroit	4	23,000	4	142,000
Balance of the State	13	124,000	6	48,000
IOWA.				
Dubuque	2	26,000	8	76,000
Balance of the State	10	81,000	11	61,000
Carried forward	234	$3,872,500	397	$18,333,000

* Includes Brooklyn and Williamsburg.

Places.	Swindling and absconding debtors.		Not classed dishonest, but will pay little or nothing.	
	No.	Liabilities.	No.	Liabilities.
Brought forward	234	$3,872,500	397	$18,333,000
KENTUCKY.				
Louisville	2	7,000	3	78,000
Balance of the State	3	310,000	3	107,000
SOUTH CAROLINA.				
Charleston	2	31,000	5	67,000
Balance of the State	2	35,000	2	25,000
Territories	11	154,000	4	88,000
Indiana	9	78,000	11	107,000
VIRGINIA.				
Richmond	2	3,000	5	44,000
Balance of the State	9	73,000	7	103,000
WISCONSIN.				
Milwaukie	1	10,000	3	50,000
Balance of the State	...	...	7	61,000
North Carolina	5	57,000	13	428,000
New Jersey	4	39,000	6	122,000
Connecticut	5	35,000	6	99,000
Maine	2	58,000	8	170,000
New Hampshire	4	20,000	4	23,000
Vermont	...	...	3	15,000
Georgia	7	45,000	3	180,000
Delaware and District of Columbia	2	3,000	...	...
Arkansas	2	24,000	...	...
Alabama	1	14,000	1	16,000
Mississippi	...	...	1	10,000
Tennessee	7	82,000	5	13,000
Florida	1	10,000	1	12,000
Texas	2	25,000	1	15,000
Total United States	317	$4,985,500	499	$20,166,000
CANADA WEST.				
Toronto	2	153,000	2	17,000
Balance of Canada West	16	76,000	9	113,000
CANADA EAST.				
Montreal	2	8,000	...	...
Balance of Canada East	...	...	1	1,000
Nova Scotia and New Brunswick	...	...	1	12,000
Total British Provinces	20	237,000	13	143,000
Total United States and British Provs.	337	$5,222,500	512	$20,309,000

Places.	Likely to Pay in Full.		No. of Counties in each State.	No. of Counties without Failures in 1857.
	No.	Liabilities.		
NEW YORK.				
New York city*	167	$42,157,000	...	...
Albany	6	209,000	...	...
Buffalo	4	821,000	...	...
Oswsgo	...	...	...	...
Rochester	1	20,000	...	...
Syracuse	1	30,000	...	...
Troy	4	675,000	...	...
Utica	1	50,000	...	...
Balance of the State	17	749,000	60	...
MASSACHUSETTS.				
Boston	19	8,010,000	...	...
Balance of the State	6	763,000	14	...
PENNSYLVANIA.				
Philadelphia	52	11,774,000	...	...
Pittsburg	2	210,000	...	...
Balance of the State	4	108,000	64	19
ILLINOIS.				
Chicago	12	1,391,000	...	...
Balance of the State	8	172,000	101	51
OHIO.				
Cincinnati	15	1,312,000	...	...
Cleveland	2	180,000	...	...
Balance of the State	...	...	88	17
LOUISIANA.				
New Orleans	12	1,342,000	...	...
Balance of the State	...	...	48	43
MISSOURI.				
St. Louis	12	1,234,000	...	...
Balance of the State	4	100,000	111	90
RHODE ISLAND.				
Providence	5	954,000	...	...
Balance of the State	1	45,000	5	2
MARYLAND.				
Baltimore	7	352,000	...	...
Balance of the State	...	...	21	6
MICHIGAN.				
Detroit	2	150,000	...	...
Balance of the State	8	110,000	46	20
Carried forward	372	$72,918,000	...	...

* Includes Brooklyn and Williamsburg.

Places.	Likely to Pay in Full.		No. of Counties in each State.	No. of Counties without Failures in 1857.
	No.	Liabilities.		
Brought forward	372	$72,918,000	...	...
IOWA.				
Dubuque	5	170,000	...	...
Balance of the State	8	132,000	100	73
KENTUCKY.				
Louisville	2	260,000	...	...
Balance of the State	1	94,000	100	78
SOUTH CAROLINA.				
Charleston	1	12,000	...	
Balance of the State	...	...	29	16
Territories	2	161,000	...	...
Indiana	5	40,000	92	37
VIRGINIA.				
Richmond	1	40,000	...	...
Balance of the State	4	57,000	140	94
WISCONSIN.				
Milwaukie	1	8,000	...	...
Balance of the State	2	33,000	47	15
North Carolina	2	18,000	82	51
New Jersey	4	145,000	20	5
Connecticut	...	...	8	...
Maine	...	...	13	...
New Hampshire	2	110,000	10	...
Vermont	5	76,000	14	...
Georgia	1	19,000	105	89
Delaware and District of Columbia	...	...	...	...
Arkansas	...	...	54	48
Alabama	...	...	52	44
Mississippi	...	...	60	51
Tennessee	...	...	79	62
Florida	...	...	30	23
Texas	...	...	91	79
Total United States	418	$74,293,000	...	...
CANADA WEST.				
Toronto	4	1,274,000	...	...
Balance of Canada West	11	352,000	42	10
CANADA EAST.				
Montreal	1	70,000	59	55
Balance of Canada East	1	1,200,000	26	22
Nova Scotia and New Brunswick	...	...	...	...
Total British Provinces	17	$2,896,000	...	...
Total United States and British Provs.	435	$77,189,000	...	...

You who have survived the panic have a cheering prospect before you for doing business to advantage. Trade, as it springs up, will be more healthy. We shall rejoice in your prosperity, and hope you may soon retrieve any losses you have sustained through the misfortunes of others.

We shall be thankful to have you speak well of and recommend the Agency to the favour of those of your friends who may chance to be doing business without the benefits it affords. The *Bankers' Magazine* for this month contains a short article showing the object and benefits of the Mercantile Agency, to which we would refer for all needed information. Since 1841 we have laboured to make it an indispensable auxiliary to every man's business. We know it is instrumental of saving many millions yearly to our city, and that its branches are equally valuable to their several localities. Who can tell what the financial difficulties of 1857 would have aggregated, if it had not been for the conservative influence which the Agency has exerted in time past, and even during the whole period of the pressure? The exigencies of those troublous times, 1837 and 1841, originated it; and we hope it will ever be regarded as the strong bulwark and reliance of our merchants. Wishing you the compliments of the season, we remain, yours faithfully,

B. Douglass & Co.

THE AMERICAN COMMERCIAL AGENCY.

TAPPAN AND M'KILLOP,

5, Beekman Street, New York,

Established 1842.

Associate Offices.

Boston. Philadelphia. Baltimore. Chicago. Cincinnati. St. Louis. Detroit.

With upwards of 3000 corresponding Agents in the United States and Canada.

Gentlemen,—Annexed, please find a statement, which has been compiled with much care from the books of this Agency. I believe it to be as accurate as it can be made. In many cases Messrs. Tappan and M'Killop have been compelled to estimate the probable amount per cent. that will be paid, but they think the data on which they have founded their estimate sufficient to justify the belief that it is very nearly correct.

In the number of failures they have not included any extensions, no matter how long, where time only was asked. Some of these have proved failures since January 1st, the date to which this statement is made up. Neither have they included in the statement banks, railroads, etc. The figures show simply the number of commercial failures during the year 1837.

Purchases made in the summer and fall of 1857 are now maturing, and further loss may be anticipated on them; but the statement includes the losses on the fall purchases of 1856. The balance will, no doubt, be against the sales made in 1857, but will not greatly increase the aggregate loss. They state, however, that during January, 1858, 640 failures have been reported to them, while in January, 1857, only 310.

By referring to the annexed statement it will be seen, that of 227,048 firms reported on in the books of the Agency, over 6,000 have failed during the year; and that of these 741 have been total or fraudulent. By total, they mean where no dividend will be paid to general creditors, and confidential creditors will not be paid in full. The aggregate loss by these is nearly 20 millions of dollars. From the 6,022 reported, deduct 741, and it leaves 5,281 which will pay a dividend on an indebtedness of over 280 millions—the dividend will not exceed 40 cents, and the loss will consequently exceed 150 millions, making a total loss for the year of about 170 millions of dollars. If to this is added losses by railroads, banks, etc., the aggregate will be very great. Without doubt there has been undue expansion, too much has been invested in railroads, unproductive lands, etc., but it is not the business of this office to attempt the task of setting forth fully the causes of the great commercial distress which has prevailed. Yet no country is in a position to recover more rapidly from such distress than the United States.

Respecting the Commercial Agency, Tappan and M'Killop state that it has been in operation over fifteen years in America, and has now extended its sphere to Great Britain and the Continent of Europe, they ask special attention to the list of Associate Offices already named, by which it will be seen that they have facilities beyond those of any similar establishment. More than once they have been favoured by the press with voluntary testimonials in their favour. These came unsought on their part, and were not the product of their own pen, nor paid for by them; they can, therefore, refer to them with satisfaction. They conduct the business in a way to serve commerce, without injuring the individual. They have steadily kept this in view.

That great care is necessary to avoid suits, will be seen by reading the following copy of a report taken from their books, the names being of course suppressed:—

"August, 1847. ———, styling himself general commission merchant and dealer in foreign wines, liquors, and cigars, in ——, is very anxious to purchase goods in New York. He will refer to —— and —— and ——. Some of these do not know him, and others are of the same stamp as himself. They represent him as honourable and respectably connected, and are very anxious that a good report of him should be given by the Agency. He bought a store in ——, stock about $10,000, for which he gave his notes; then pledged the goods to a broker for an advance of $2000; raised $1000 cash additional somehow, gave a mortgage on the goods, and agreed to mortgage his New York purchases for aid to make them on credit. He was at one time employed to solicit trade for ——, in ——. He introduced and recommended to them the most finished set of scamps that ever visited that or any other city. Let no man sell to him unless he brings the gold to pay for the goods, and, before delivery, send the gold to the assayers to be sure of its genuineness."

From the foregoing date to 1853 sundry operations are recorded, and in 1853 his last is obtaining $5,000 from —— bank by means of a fraudulent cheque. In October, 1861, he will be again free, but just at present he is not, having been, in October, 1854, sentenced to seven years' imprisonment for the bank affair. Such reports are communicated only to principals and never in writing.

Subscribers are requested to hand in a list of parties in whose position they are interested. By so doing their names will be put opposite these firms in the books of the Agency, and all changes which take place from time to time affecting their credit will be advised without inquiries.

The accuracy of their reports has become well known, and the care which is exercised in making investigations has secured for them the confidence of the mercantile community in America. Being assured of their determination and ability to do justice to their constituents in this country, I invite a free use of the Agency, and a full investigation of their ability to serve your interests. PATRICK ROBERTSON.

London Agency, 1, Sun Court, Cornhill, E.C.
March 1, 1858.

STATEMENT OF FAILURES FOR 1857.

	No. of firms on our books.	No. of failures reported.	No. of frauds and fraudulent failures.	Amount of failures.	Amount of total and fraudulent failures.	Average probable payments.	
						Failures.	Frauds.
				$	$	c.	
New York city	14,136	869	87	96,454,000	3,711,000	37	Nothing.
Do. state	18,984	777	43	21,334,000	1,317,000	41	
Philadelphia	7,203	317	74	35,162,000	3,113,000	28	10 cents.
Pennsylvania state	15,202	316	27	5,213,090	1,217,000	34	5 "
Boston	5,420	304	31	52,231,000	827,000	46	
Massachusetts state	14,198	224	19	2,433,000	143,000	52	
Baltimore	2,130	72	18	4,119,000	415,000	29	
Maryland state	3,502	39	6	689,000	18,000	25	
Alabama	2,504	22	3	362,000	38,000	48	
Arkansas	1,190	18	6	423,000	42,000	50	
Connecticut	5,123	86	14	1,415,000	172,000	48	
Delaware and District of Columbia	3,513	25	4	324,000	4,500	47	
Florida	792	7	2	250,000	22,000	50	
Georgia	5,518	56	15	1,013,000	342,000	33	
Illinois	12,957	363	43	6,713,000	422,000	40	5 cents.
Iowa	4,654	262	23	1,610,000	213,000	41	Nothing.
Indiana	7,614	191	12	1,476,000	110,000	36	6 cents.
Kentucky	6,580	96	15	2,327,000	617,000	34	
Louisiana	3,813	74	17	7,213,000	809,050	42	
Maine	7,196	92	9	1,712,000	232,000	54	
Michigan	4,420	168	30	3,113,000	423,000	28	
Mississippi	2,412	14	2	463,000	15,000	38	
Carried forward	149,061	4,392	500	$246,049,090	$13,222,550		

	No. of firms on our books.	No. of failures reported.	No. of frauds and fraudulent failures.	Amount of failures.	Amount of total and fraudulent failures.	Average probable payments.	
						Failures.	Frauds.
				$	$	c.	
Brought forward	149,061	4,392	500	246,049,090	13,222,550		
Missouri . . .	6,727	92	24	6,319,000	821,000	37	
New Hampshire	3,256	64	7	897,000	37,000	54	
New Jersey . .	4,398	108	12	1,228,000	178,000	51	
North Carolina .	3,122	65	17	1,192,000	415,000	46	
Ohio	18,392	467	48	5,475,000	653,000	35	5*c*. to 10*c*.
Rhode Island .	2,213	41	7	4,737,000	1,403,000	47	
South Carolina .	3,313	65	12	1,413,000	162,000	43	
Tennessee . .	4,294	59	12	818,000	93,500	46	
Texas	2,616	13	2	377,000	27,700	30	
Vermont . . .	2,805	64	5	617,000	27,000	52	
Virginia . . .	9,284	123	27	1,927,000	236,000	45	
Wisconsin . .	4,628	209	13	1,454,000	15,650	50	7*c*.
Territories . .	2,727	70	21	1,714,000	327,000	25	
British Provinces	10,112	197	34	8,118,000	393,000		
Totals . . .	227,048	6,022	741	$282,335,000	$19,110,400		

I invite all parties either directly or indirectly connected with American trade to subscribe at once to this office, and, by handing in a list of the firms they are interested in (which will be entered in the books of the Agency), they will be kept advised of any changes which may take place affecting the credit of their correspondents, without injury on their part. The value of this must be apparent.

I remain yours, faithfully,

PATRICK ROBERTSON.

London Agency, 1 Sun Court, Cornhill, E.C.
14th *April*, 1858.

COMMERCIAL FAILURES SINCE THE FIRST OF JANUARY—A BANKRUPT LAW WANTED.

(*From the New York Herald.*)

WE publish to-day a list of the failures which have occurred in the United States from the 1st of January, 1858, to the 26th of March, of the present year, and the total will be found somewhat alarming, though by no means unexpected, after the late fearful financial panic. The number of failures for the first eighty-five days of the year was 1,495; and the total amount of liabilities is set down at $30,639,000. If we add to this 45 failures in the British Provinces, figuring up $1,094,000 more, we have the grand total for the United States and Canada of 1,540 failures, and $31,733,000 of liabilities.

The following is the list:—

	No. of Failures.	Amount of Debts.
New York state	183	$1,632,000
New York city	74	3,784,000
Massachusetts state	25	239,000
Boston city	34	1,731,000
Pennsylvania state	104	1,325,000
Philadelphia city	27	845,000
Maryland state	10	164,000
Baltimore city	23	1,535,000
Alabama	25	284,000
Arkansas	3	70,000
Connecticut	29	400,000
Florida	2	30,000
Georgia	19	336,000
Illinois	195	3,743,000
Iowa	93	504,000
Indiana	81	638,000
Kentucky	41	916,000
Louisiana	39	2,884,000
Maine	9	171,000
Michigan	74	1,490,000
Mississippi	6	276,000
Missouri	31	2,100,000
New Hampshire	13	140,000
New Jersey	23	230,000
North Carolina	33	594,000
Ohio	174	1,321,000
Rhode Island	25	943,000
South Carolina	11	211,000
Tennessee	32	464,000
Texas	5	63,000
Vermont	19	189,000
Virginia	41	577,000
Wisconsin	68	422,000
All the Territories	19	368,000
Total in United States	1,495	$30,639,000
Total in British Provinces	45	1,094,000
Total for U. States and British Provinces	1,540	$31,733,000

Last year there were over 6,000 failures, with debts to the amount in all of nearly 300,000,000. While the number of failures for the first quarter of this year is in proportion with last, it will be seen that the amount is disproportionately small, showing that it is small concerns which are breaking down now.

One singularity about this statement is, that it shows that the Central, the Western, and the sugar-growing States are in a woful condition, while the cotton-growing States of the South, and all the New England States, are sound.

All the failures which we record to-day should legitimately have occurred last year. They have only been protracted by the parties obtaining extension That grace is exhausted now, and they fall by sheer inanity. This was inevitable; for these houses based their engagements on the prices at which goods and produce rated before the revulsion, and, consequently, they are unable to meet them now that the rates have fallen from 35 to 50 per cent. In 1837 the result was the same; but then the smashing continued for almost three years, and in all probability we shall see the same thing now. In the first few weeks of 1837, over 500 merchants in New York alone broke down. Others held on for a time, propping themselves by extensions; but, until 1840, they continued to drop, when the number of failures in the Union amounted to nearly 40,000, and the Bankrupt Law which went into operation in 1842 wiped out $450,000,000 of debts, and released over 38,000 bankrupts. In a panic, large houses are the first to go; then follow the banks, and little concerns follow them. The great drag down the small. The banks recovered themselves quicker in this revulsion than in 1837, because they held more specie, and their bills were issued on a surer basis; but we see that the failures of commercial houses are still going on. There is but one remedy for this, and that is a good bankrupt law, applicable to individuals, banks, railroads, and all corporations. If the houses that are breaking now could have settled with their creditors long ago, by the operation of a bankrupt law, and started afresh, basing their business on the new state of affairs which panics always leave behind them, many of them would be now in a safer and more thriving condition; but by the present system commercial disasters are protracted, but not avoided.

We are indebted to the commercial agency of Tappan, M'Killop, and Co., No. 5, Beekman Street, for the list which we publish this morning. We gave, at the beginning of the year, the list of failures for 1857, furnished by the commercial agencies of B. Douglass and Co., and Tappan, M'Killop, and Co.; but, for some reason or other, we have been able to obtain the list for the first quarter of this year from the latter house alone. Owing, most probably, to the superior manner in which the books of Tappan, M'Killop, and Co. are kept, they were able to furnish us with the list in a few days. The others, we suppose, keep their books on a different system, like some of the Wall Street Banks, perhaps, who only know how they stand at the end of the year, and many of them not even then. Every well-conducted establishment, whether bank, commercial house, or newspaper, is able to tell exactly how it stands, at least after a few days' notice. In the *Herald* office, we know precisely how we are every week, but there are newspaper concerns in Wall Street, which, like the banks, are in happy ignorance of their condition until the end of the year, and even then they often find themselves in a fog.

It is clear, from this statement of the failures still going on, that the effects of the panic continue to be felt throughout the country; but if we had a bankrupt law things would have been reduced to a level long ago.

THE NEW PHASE OF THINGS, 1859.

WHAT THE COMMERCIAL WORLD IS DOING.—NEW YORK BANKS.—IMPORTS AND PRICES.

(*From the New York Herald, August,* 1859.)

As the season approaches for the heavy fall payments, the price for money steadily advances. Paper which was in demand at $4\frac{3}{4}$ to $5\frac{1}{2}$ per cent. in April, is now slow at from 8 to 9 per cent. This advance, great as it is, does not, however, give a full idea of the extent of the change that has taken place in the views of capitalists. Amongst all capitalists there is a strong and growing indisposition to do anything except very prime, short paper. There is an absence of demand for long dates and second-class paper, which betokens in the minds of our capitalists a renewal of that want of confidence which arose from the recollections of 1857. Various causes conspire to produce and increase this want of confidence in the future prices of money during the fall months. The advance in the price of money to 8 or 9 per cent., were it from a healthy, general increase in trade, would be cause for congratulation, and not regret; but there are doubts as to this being the case. The domestic trade of the country is still moderate and prudent, and, as far as we can learn, free from expansion or unnatural activity. The foreign trade, however, is not so. With money cheap in Europe and New York, with an exceedingly profitable year's business in 1858 and the first three months of 1859, the importers were tempted to try and make a double season; hence the enormous arrivals of foreign goods in April, May, and June. These sold to a loss.

In January and February, the jobbers, finding goods brisk and profitable in the importers' hands, grudged the importers their profits. The jobbers, with their notes drawn to their own order at eight months in demand, and easily cashed at 6 to 7 per cent., determined to save the importers' profit for the fall trade. The commission houses in Manchester, Paris, etc., were energetic in showing the jobbers what profits the importers were making in January and February. They offered the jobbers six and eight months' credit, and at the end of that period to take a four months' draft of the Anglo-American banking firms if necessary.

The consequences were natural. The jobbers ordered from Europe a liberal import for the fall; the importers did the same. Thus the double import of the spring was produced by the greediness for speedy gains by the importers, stimulated thereto by the greediness of the banks for paper. The losses on the late spring imports were not ascertained sufficiently early to enable importers to curtail. Every one, also, until the peace, was making great calculations on the demand for goods from the West. The importers, no doubt, made a liberal import for the fall, basing their estimates on a large Western demand. But the double import for the fall arises from the jobbers doing the same thing. Every jobber, and many retailers, have been drummed by the commission houses in Manchester, Paris, etc., and have given liberal orders, tempted thereto by the six and eight months' credit offered to them, and the extra profit they hoped to make. The importers thus find themselves forestalled with their best customers by the European commission houses. They are in a very unpleasant and embarrassing position. The backwardness of their sales is owing to this cause. The jobbing trade, as a body, have enough goods from their own importation to fill their early sales.

This makes the jobbers comparatively independent of the importers until about September. If the imports should, however, prove not to be more than the legitimate wants of the country, the importers have only to hold their stocks firmly until the jobbers dispose of their own importations. The importers will then have a good season, and as the bulk of importers date all goods bought previous to September 1 as eight months, and to A1 Southern and Western houses nine months from September 1, their notes will mature at nearly the same dates as if their sales had been early. The danger to prices arises from weak or nervous importers crowding goods at private sale, or rushing them into the auction-room, to raise money to remit to Europe. It is the fear of specie shipments that is operating on the minds of capitalists; the uncertainty about what per centage of the amount of deposits held by the New York banks belongs to European banking firms, importers, auctioneers, and other parties, who may at any moment draw down their deposits in specie, and remit it to Europe. There is also great uncertainty felt as to what probable amount the New York banks will discount of notes belonging to importers, jobbers who have imported, and auctioneers, all of whom, of course, want the discounts for remittances to Europe.

The aggregate excess of the spring import was $67,000,000. The payments for importers, who take about four months' credit from date of purchase or shipment, with the sixty day bills of exchange as remittance, give America about five months credit, so that January imports bought or shipped in December, would be remitted for in produce bills or specie in May, and so on for the succeeding months. We are, therefore, now in process of paying for April imports bought or shipped in March. This real state of our payments to Europe will tend to allay the uneasiness and want of confidence felt by capitalists from the large imports we are now receiving. We give below a table of the imports at New York, with the month they would probably be remitted, if bought by the regular importers on four months' credit. As the increase on the spring import was chiefly by the regular importers, the remittances would be made in about five months, and mature much as follows:—

IMPORTS OF FOREIGN GOODS AT NEW YORK.

January, 1859	$19,377,000,	remitted for	shipment	May.
February	18,756,000,	,,	,,	June.
March	20,739,000,	,,	,,	July.
April	22,153,000,	,,	,,	August.

May, 1859	23,430,000, remitted for shipment		September.
June	23,584,000,	,, ,,	October.
July	27,111,000,	,, ,,	November.

For the regular importers' part, and for the jobbers' part, about January to March, 1860.

Consignments sent to this country are generally advanced upon to 75 per cent. of the amount of invoice in a four months' bill from date of shipment. Others are sent out, and are realized on at auction, and remitted for ten days after sales.

It is quite a fallacy to suppose that consigners do not require prompt remittances. As a rule, they are very needy, and often consign to raise cash. The sacrifice made here is not so likely to be found out as in their own market, and their credit does not suffer.

The importations of the jobbers are those which give America the largest credit. Although this is gratifying to the banks and capitalists, who look only to an immediate shipment of gold, yet it is questionable whether America is not more injured by the enhanced prices jobbers are charged by the commission houses to cover the heavy expenses and risks attending long credits, than it would be by gold shipments. It certainly conceals over-trading for a longer period. Had the jobbers' part of the spring import been as large as that of the fall, we should not have had so heavy a drain of gold. Their part of the payments would have been deferred two to four months later. The spring import being chiefly by importers, our remittances are more prompt than they will be for the goods arriving in July and forwards. This demand for gold unfortunately happens at a time when we have no produce to ship. It necessitates gold shipments, or relief by the Anglo-American banking firms lending their sixty days' and four months' kites to the importers.

These bills can be readily discounted in London. British capital can thus be made to step in and bear the load of our heavy imports. Money is so easy and abundant in London that what would embarrass New York, will be a positive benefit to London. To force this, the New York banks must refuse to discount importers', auctioneers', and all paper that can be drawn in gold and shipped to Europe. With twenty millions specie in their vaults they will have all they can do to supply the interior with specie to move the crops. They cannot prevent some gold going to Europe. Last year, with less than half the imports, we shipped to Europe all the specie we received from California from July to December.

Were capitalists to be assured that the New York banks would pursue this course as to importers, auctioneers, etc., it would remove their dread of a specie drain to Europe. The effect of a specie drain is to cause the banks to contract suddenly. The disastrous result of a sudden or great contraction by the New York banks is to reduce the prices of everything that must be sold during that contraction. Those who can afford to hold till the banks expand again and trade is restored to its former activity are not affected. Take, for example, the prices of the following bank stocks, etc. :—

PRICE OF BANK STOCKS, ETC., OCTOBER, 1857.

Bank	Price	Bank	Price
Bank of Commerce	83	Commonwealth Bank	65
National Bank	74	North River Bank	50
Mechanics' Bank	90	Market Bank	65
American Exchange Bank	75	Continental Bank	65
Hanover Bank	50	Shoe and Leather Bank	70
Metropolitan Bank	75	Importers and Traders' Bank	70
Bank of New York	80		
Phœnix Bank	60	STATE STOCKS.	
Union Bank	85	Missouri 6's	66
Bank of America	80	Tennessee 6's	72
Merchants' Bank	75	Virginia 6's	82
Bank of State of New York	50	New York State Loan	85

All these, previous to the bank contraction, were at prices from 105 to 125.

It is clear from the above statement that if any holder of these stocks had foreseen that the drain of specie to pay for the large spring import of 1857 was going to compel the New York banks to curtail their loans as they did, he could have doubled his money by selling out before the bank contraction, and buying again precisely the same thing he had sold after the bank contraction. Those who held and continued to hold till the banks expanded again were neither much better nor much worse.

All property, dry goods, real estate, state stocks, that holders were compelled to sell during the contraction of the banks, was realized on at a like enormous sacrifice, thereby greatly enriching those who were fortunate or far-seeing enough to get and hold all the cash they could, ready for the reduction in prices consequent on such a contraction of bank loans.

The New York banks commenced their contraction in 1857 as follows:—

Date	Loans	Date	Loans
August 22, 1857, their loans were	$120,000,000	September 26	$107,000,000
August 29	116,000,000	October 3	105,000,000
September 5	112,000,000	„ 10	101,000,000
„ 12	109,000,000	„ 17	97,000,000
„ 19	108,000,000	„ 24	75,000,000

There are many shrewd capitalists waiting for a similar contraction and fall in prices. If the banks adopt a rigid refusal of all paper from importers, auctioneers, etc., the drain of specie to Europe will be prevented. It will postpone payment for our spring imports until we can send cotton and other produce instead of gold ; it will prevent the fall in prices that these capitalists are keeping their money for; it will prevent the price of money rising to an exorbitant rate ; it will confine the loss or embarrassment arising from the successive importations to a few importers in the city of New York, and compel their European friends to help them.

The banks will be free to give our domestic and internal trade and produce all the accommodation they want, liberally. The internal trade demands little specie, and it soon comes back again. The banks will be able to get all the domestic paper to discount they ought to discount. The position of the banks will be safe, because a foreign drain for specie to pay for excessive imports is our only trouble. Why should over-trading on the part of a few importers and foreigners be permitted by the New York banks to tax the whole trade of the country, when Secretary the Hon. James Guthrie, of the Treasury, published the following carefully prepared estimates of our internal trade in 1856:—

Capital in manufactures	$843,000,000
Products in manufactories	1,688,000,000
Exports	281,000,000

If an excess of foreign imports of $67,000,000 on the half-year enforces such action on the New York banks as will raise the price of money to 15 per cent. instead of 7 per cent., it is clear the 8 per cent. additional is a tax caused by the over-imports, to be levied on about $2,000,000,000 of domestic trade for such time or portion as may want discounts or loans.

If the foreign trade is thrown on Europe this vast amount of $2,000,000,000 domestic trade will remain undisturbed, with no immoderate price to pay for money.

The responsibility of all this rests with the New York banks. Their action guides the rest of the Union. They occupy now much the same position the Old United States Bank did—they are the regulator of the banks of the Union.

The principles which they call sound, legitimate, healthy banking, are all very good when we have not importers, auctioneers, etc., ready to ship in gold whatever the banks discount for them. To meet a demand for specie, the more specie they have the stronger they are. If they were to discount during August and September $15,000,000 for importers, auctioneers, etc., how much would be left of their $20,000,000, even though they were relatively stronger, as the banks reckon strength? This is the true way to look at their position. The present rule the banks use to test their strength will not bear the test of practice; it neither keeps their specie in their vaults nor their loans from unhealthy fluctuations.

We subjoin a statement of the bank mode of estimating their relative strength. We give the same dates, chosen by one of themselves, signed "Banker," and published in the *Herald's* money article, August 3. The figures are official—published by the Clearing House, and signed George D. Lyman. Copy appeared in the *Herald's* money article, August 10:—

	Loans.	Specie.	Net Deposits.	Circulation.
August 1, 1857	$120,597,651	$12,918,014	$68,682,088	$8,965,422
August 7, 1858	120,892,857	35,154,844	90,339,678	7,784,415
July 30, 1859	119,347,412	20,764,464	74,474,895	8,214,259

Liabilities.	1857.	1858.	1859.
Net Deposits	$68,682,688	$90,339,678	$74,474,895
Circulation	8,965,422	7,784,415	8,214,959
Total	$77,647,510	$98,124,093	$82,689,854

Assets.					
Loans, Specie, and Stocks	133,515,065	...	156,047,701	...	140,111,976
Surplus strength............	$55,867,555	...	$57,923,608	...	$57,422,122
Banker's wrong surplus strength caused by taking gross deposits for 1857 and 1858 instead of net	$38,520,126	...	$49,253,986	...	$65,423,081

"Banker" fell into the above errors—inadvertently, no doubt—by quoting the gross deposits, instead of net deposits, for 1857 and 1858; and by comparing the gross deposits of those years with the net deposits which he had in his statement of 1859, made out that the banks were, July 30, 1859, $16,170,000 stronger than they were same date last year, and $26,900,000 than they were at same date in 1857.

The actual position of the banks at these three dates, as the preceding statement shows, did not differ at all.

This statement of "Banker's" has been extensively used by the partisan press of the importers and banks to cavil at an article that appeared in the *Herald*, headed "Banks, Banking-houses, and Importers."

To show how fallacious the whole thing is to throw any light at all upon the strength of the banks to meet a specie drain to Europe, we give similar comparisons for October 10, 1857—three days before the banks stopped payment, and for October 17—four days after they stopped payment.

The gross deposits not being published then, we deduct $16,000,000 and $17,000,000 from the gross, which is probably an under estimate. We wish to give the banks the advantage. With these exceptions the other figures are official:—

New York City Banks.

	Loans.	Specie.	Net Deposits.	Circulation.
October 10, 1857...	$101,917,569 ...	$11,476,294 ...	$46,000,000 ...	$7,523,599
October 17, 1857...	97,245,826 ...	7,843,234 ...	36,000,000 ...	8,087,441

Liabilities.	October 10, 1857.		October 17, 1857.
Net deposits............................	$46,000,000	....	$36,000,000
Circulation	7,523,599	....	8,087,441
Total	$53,523,599	....	$44,087,441
Assets.			
Loans, specie, and stocks	$113,393,863	....	$105,089,060
Surplus strength..............	$59,870,264	....	$61,001,619

We also give the following dates, as they have been cited to prove the banks stronger July 23 than January 1859. It will be seen their so-called relative strength is much the same.—

	Loans.	Specie.	Net Deposits.	Circulation.
January 22, 1859...	$129,540,050 ...	$29,472,056 ...	$95,066,400 ...	$7,457,245
July 23, 1859...	119,934,160 ...	21,196,912 ...	75,301,943 ...	8,170,626

Liabilities.	Jan. 22, 1859.	July 23, 1859.
Net deposits	$95,066,400	$75,301,943
Circulation	7,457,245	8,170,626
Total	$102,523,645	$83,472,569
Assets.		
Loans, specie, and stocks	$159,012,106	$141,131,072
Surplus strength	$56,488,461	$57,658,503

We have briefly glanced at some of the causes that produce our tightening money-market and want of confidence amongst capitalists. We have tested the accuracy and soundness of the rule by which our New York bank managers regulate their loans and judge of the strength of their position. That there is something radically wrong in the practical working of that banking rule that guides their action, is obvious to any one that chooses to examine the few illustrations we have given; or, what will be still better, and not unprofitable, let business men take the bank statements and work out more illustrations for themselves, taking care to make the distinction between gross and net deposits. Our illustrations of the banking principle they profess to evolve prove the banks had the most strength just before and after they failed, October 10 and 17, 1857. To express that strength in millions, according to the bank managers' method, the banks were from two to six millions stronger, October 10 and 17, 1857, than they were August 1, 1857, August 7, 1858, and July 30, 1859.

This is not the strength required to meet our present trouble—excessive exports of gold to Europe, caused by excessive imports. We see that when the banks were, by their rule, strongest, they had to stop paying specie, October 13, 1857. We are very much afraid that a large specie deposit is what is required to meet a large foreign export of specie. We give the following statistics for business men to study. We include 1857, not that we think the country has ever yet been in a condition to consume and pay for, conveniently, an import as large as 1857. We, of course, know that nobody wants to see in 1859 either import, bank expansion, or anything else that took place in 1857:—

First week in August.	1857.	1858.	1859.
Foreign imports for the year to date August 8	$151,628,551	$82,536,928	$158,614,860
Dry goods imports for the year to date August 8	66,716,293	33,750,174	75,623,412
Bank loans	122,077,252	120,892,857	118,938,059
Specie in bank	11,737,367	35,145,844	20,083,877
Specie in sub-treasury	...	5,553,400	5,330,508
Exports of specie for the year to date August 8	28,216,619	15,775,719	44,396,190
Exports produce (exclusive of specie) from New York from January 1 to August 3	41,780,965	37,982,292	38,861,220

First week in August.	1857.		1858.		1859.
Imports of specie from California for the year to date August 8	$23,106,411	...	$20,884,302	...	$21,116,448
Gold received from California from August to December 31	15,815,613	...	14,139,685	...	...
Specie exported from August to December 31	14,881,098	...	10,933,299	...	...

The excess of foreign imports reduced fifty millions in silks, laces, and other gewgaws, would injure only foreign manufacturers. Importers lose from a fall in prices from a glutted market. These importations pamper the female vice of love for dress. American women would be better physically and morally by wearing less of them. American men would be better temporally and spiritually by having to pay for less of them. Many an unhappy home is caused by the extravagance of wives and daughters. Much vice and misery are the consequences. The evil of extravagant foreign imports, like all evil, ends not with itself. Evil breeds evil.

The subject is not merely a financial one—it is spreading a social cancer, making our women lazy and extravagant, our men, worn out with hard work, to pay for that extravagance. Our men, finding their homes, no homes for them, but hotels for their wives and daughters to make a display of their charms, and their dress, and their furniture, and their curtains, to groups of flattering friends, seek solace and comfort elsewhere. The financial view is, however, that which business men are considering.

Till it is shown what means the New York banks will adopt to keep their specie safe from European banking firms, importers, auctioneers, etc., the public mind will remain in its present anxious state. In any event, prudent men, who are solvent and mean to pay their debts, will buy as little as they can and shorten credits. They will not contract debts for merchandise which a glut in importations and contraction by the banks may cause to fall in price on their hands.

SECTION THE SIXTH.

The Stock Exchange Panic of 1859—Heavy Appearance of Prices from the Commencement of the Year—Inauguration of the Italian Difficulty—Progress of Embarrassment and Outbreak of the War—Early Symptoms of the Depreciation of Prices in March—Increased Depression at the Beginning of April, and Final Collapse during the Remainder of that Month and May—Serious Revulsion in Values through the Alleged Russo-French Alliance—General Prostration of Credit—Disastrous Failures and Comparative Tables Exhibiting the Actual Decline in Securities—Measures of Amelioration Adopted by the Committee of the Stock Exchange—Slow Return of Confidence and Eventual Recovery.

It could scarcely have been supposed, if subsequent events had not rendered the truth apparent, that the Stock Exchange could have been so deeply involved as it was by the events of the early part of the present year. If the assertion had been made, on the announcement of the probability of war in Italy, that the credit of the house would have been so severely shaken, few would have been prepared to believe it; and when the state of business is taken into consideration, it must be confessed that the influence exercised by the progress of affairs was more startling than could have been reasonably imagined. Now that the crisis which affected that establishment has passed, and the weight of responsibility experienced in connection with the engagements incurred has been adjusted, persons can look back with calmness and view the circumstances which then produced that collapse, and exercise a sound and impartial judgment. Various as have been the opinions entertained on the question, and serious as have proved the allegations respecting the motives which prompted influential individuals in exaggerating the importance of the rumoured alliance between France and Russia, it must be admitted that the basis of operations was too widely extended in particular markets, and that the disposition to hold for a recovery caused much of the disaster then experienced in the several departments. So completely was this state of things manifested and established, by the rapid fluctuations which occurred in some descriptions of securities, that for months after the sudden drop which affected all classes of securities, symptoms of recovery were long wanting. Nevertheless the reaction at the fatal moment, when the important change took place, so completely paralyzed the members, that the fright, probably more than the advices themselves, created that crash which, when it once commenced, produced such terrific havoc on the credit of the dealers and operators, that at one moment the bankruptcy of the house, it was thought, must inevitably follow. This painful position of affairs may be well gathered from the circumstance that twenty and thirty failures occurred in a day, and that the link of connection between the respective markets was so closely associated that the suspension or embarrassment of one member frequently jeopardized the position of seven or eight. The occurrence, therefore, of difficulties of this character not only affected the position of the individual himself, but also those to whom he had to pay balances; and even in many instances where operators had profits on their several accounts, they had to pay large amounts to maintain their own status, or to support the credit of their principals, who were not immediately in a situation to provide payment for the differences which their numerous accounts exhibited. Through these consequences the fearful amount of embarrassment increased, and as the extent of liability was enormous, it was only by measures of forbearance and partial indulgence that order was re-established, and that the value of the various classes of negotiable property once more eventually returned to anything approaching its original level.

The Easter holidays of 1859 will long be remembered by those connected with financial affairs as the date of the dark crisis in the position of the prices of public securities, and the inauguration of the conflict between France, Sardinia, and Austria, which temporarily so disastrously affected the peace of Europe. Although, since the 1st of January, quotations, not only of Consols but of all descriptions of securities, had been drooping through the effect of the shock of the warning given to M. Hubner, it was not fully apparent until it was announced that the Austrians had determined to cross the Ticino. This, as in the similar case of the passage of the Pruth by the Russians, seemed to leave no hope of a settlement of the question, and prepared most people for the results that were subsequently witnessed. Good Friday was as usual observed as a holiday at the Stock Exchange; but after hours on Thursday, April 21, and when it was determined that the house should be closed on Saturday, a sudden fall took place in prices, which spread terror and dismay throughout all circles. Intelligence, it turned out, had been received that Austria had forwarded to Sardinia an ultimatum calling upon her to disarm, and had refused the last proposition made by England to accept mediation. The Indian loan had just been completed at the minimum price of 95, and the greater part of the seven millions was taken, which was considered a favourable circumstance, looking at the doubtful prospects which attended the Italian question. Notwithstanding the Stock Exchange was closed on the Saturday by order of the committee, as settled on the Thursday, arrangements were made for admitting the members, and prices were affected by the few non-official transactions, the whole of which nevertheless evinced the disposition of the speculators to sell for the fall.

From that date forward to the subsequent progressive stages in the contest, from the announcement of the French loan of twenty millions to the departure of the Emperor Napoleon for the seat of war, panic reigned throughout the markets for the various public securities. On some occasions the decline was as much as 1½ per cent. in Consols; Turkish went down 10 to 12 per cent. in a day; Sardinian 3 to 9 per cent.; and English, French, and Italian 2 to 3 per cent. So great and indiscriminate a reaction was scarcely ever before known, and the fluctuations on the foreign bourses were equally extensive, a circumstance indicating that confidence throughout Europe was completely prostrated. About the 27th of April, when the preparations for the foreign and share settlements were in course of progress, early evidence of the unfavourable situation of the jobbers and brokers speedily became apparent. The first intimation of an alleged treaty between Russia and France, offensive and defensive, was then made public, which tended to increase the excitement prevalent; but although it was contradicted on the most confident authority, the influence was so paralyzing, that seven failures immediately took place. It was now felt that the blow had been so severe that no recovery could be immediately looked for, and as it was not possible to obtain advances upon the different classes of securities on any terms, a further fall speedily followed. Again, Consols, Turkish, and the other securities were extensively offered, and the despondency was almost universal; every telegram received with prices from Paris occasioning a variation which was immediately responded to by a change in values in proportion to the rise or fall exhibited on the other side of the Channel. Indeed, it was quite evident that the quotations from the French Bourse ruled the general course of prices.

At this date it was confidently assumed that the Bank of England would be compelled to advance the rate of discount, and on the 28th of April they at once put the terms up to 3½ per cent. The adjustment of the half-monthly foreign stock and share accounts exhibited the most painful results of the sudden depreciation, and whereas seven failures had previously occurred, these were now increased to between twenty-five and thirty. Immediately the first movement upwards in the rate of discount occurred, the anxiety was most intense, and notwithstanding little faith was placed in the reported alliance between France and Russia, as engaged in with a view to jeopardize England, the fall in prices continued. On the 28th of April Consols touched the low rate of 88¼, and the continuance of failures, in addition to the unsatisfactory news from the Continent, produced such weakness that there was no corresponding rally. The extent of the mischief was apparent when it transpired that additional suspensions had taken place, and although several represented liabilities for large amounts, their default was principally occasioned through the position of other members. Business was now prolonged to very late hours in the evening; the anxious appearance of the brokers and jobbers betrayed the deep-seated alarm which pervaded the markets, and every day brought new and unexpected failures. In consequence of the almost universal embarrassment some of the first houses were talked of, and it was asserted that had they not secured assistance they would have been compelled to succumb. The fact subsequently transpired that this was actually the case in two notable instances; Messrs. Rothschild having supported one firm through its difficulties, and Messrs. Overend, Gurney, and Co. the other.

There now came a slight turning point in the value of the various classes of securities, and as it was supposed that the exaggerated notions formed with respect to the alleged treaty were premature, the recovery continued to make progress. The public, finding securities at depreciated prices, came in as purchasers, and from 88½ Consols returned to 91 in the course of two days, notwithstanding the total number of failures meanwhile had risen to fifty-two. The foreign stock and share settlements were completed in the midst of much excitement, and then the issue of the Consol account was awaited with serious apprehension. It was well known that many parties must be compromised, not through individual liability, but through their inability to realize their securities, and hence every disposition was shown to grant an extension of time for the adjustment of balances. After the great fluctuation had terminated, and the difficulty of the situation was viewed with less apprehension, it was discovered that many of the members who had considered it advisable to suspend, might, by the exercise of a little discretion, have been enabled to "pull through." Still, the fall was found to be very embarrassing, and as is usually the case in panic periods, the endeavours made by those alarmed for their security to provide means to meet engagements led to the accumulation of large balances in the hands of the bankers, and created a very pressing inquiry for notes. The progress of the general elections, and the unfavourable views advanced with respect to the intentions of Louis Napoleon, continued to create anxiety, but the pressure was limited by the disposition evinced on the part of the committee of the Stock Exchange to alleviate the difficulties of the members in regulating the payment of dividends and supporting the credit of the establishment. Meanwhile the new French loan of twenty millions, and the withdrawal of the announcement of the Russian loan of twelve millions, caused a tendency to stimulate fear, and the further adverse movement on the 5th of May by the Bank directors, in carrying the rate to 4½ per cent., indicated the policy that was likely to be adopted if affairs continued to exhibit a discouraging appearance. The influence of this advance, however, was not so great as that which distinguished the previous rise; and after a fortnight or three weeks of the most "troublous times" known at the Stock Exchange since the unparalleled fluctuations of 1825, the markets began to assume a better and more steady appearance. On the termination of the Consol account it was asserted that the entire number of defalcations, including those which took place at the first moment of the panic, reached altogether upwards of eighty or ninety, and it would not be above the mark, if, taking into consideration the arrangements made, one hundred were placed as the number of parties who had been broken by the general depreciation in prices. Several individuals, who were men of large fortune, and who were able to consider themselves comparatively millionaires, lost almost the whole of their previous accumulations, and others, who should have received on the balance of their accounts moneys to the extent of thousands, had, instead, to pay into the house a considerable amount to keep themselves clear and fulfil engagements.

It was not before the end of June that the markets finally recovered from their prostration, or that anything approaching real business activity ensued. Everything, nevertheless, was done by the executive of the Stock Exchange to readjust balances, settle partial compromises, and readmit those who were able to pay the recognized six-and-eightpence in the pound. Numbers, however, were finally ruined, and persons who had lived in the most affluent circumstances were suddenly beggared, being compelled to break up their establishments, and retire either into the country to eke out a miserable existence, or leave for the Continent, through assistance received from friends, to fare eventually as best they could. The honourable conduct of the majority of the members deserves the highest commendation; few instances were mentioned of irregular dealing, and though some individuals were found to have engagements open, in a variety of securities, not in due proportion with their capital or connections, no barefaced attempts at fraud or misappropriation were discovered. It was also greatly to the credit of the more fortunate brokers and jobbers, notwithstanding all more or less suffered from the collapse, that they aided, as far as possible, their more needy brethren; and many cases could be cited in which substantial relief was afforded, even at the risk of individual involvement, during the worst phases of the panic. Among the numerous members of the house, of course an almost illimitable chain of connection exists, and hence it is apparent that assistance, under such circumstances, was often rendered when it was scarcely known whether the party who proffered the aid would in the end, if fluctuation continued, be enabled to fulfil his own contracts.*

But with all this disposition to serve one another, the great difficulty experienced was the possession of the necessary money. Securities of every kind, from the first symptoms of panic, so rapidly depreciated that the banks and the other financial institutions almost refused to make advances, and they themselves in turn being sellers of stock, particularly Turkish, Sardinian, and Mexican, upon which loans to a considerable extent had already been made, the depression became more intense and accelerated the general fall. In fact, it is evident from what has since occurred, that there must have been an enormous speculation, at the advent of the crisis, "open for the rise." Every one had hoped that peace would be preserved, and that as the fine weather appeared, affairs would change, and produce profit and prosperity. The advance of Austria against Sardinia was the precursor of alarm, and this, followed by the equivocal announcement of the relations between Russia and France, terminated in that crash which soon became so painfully apparent. The markets, naturally weak from inactivity and exhaustion—first through disappointed expectations, and, secondly, through the want of confidence in the French Emperor—became deranged, and from the general involvement it was impossible, without sacrifices of the character stated, to effect an extrication which could restore or rearrange the current of business at the Stock Exchange.

The extent of the sudden revulsion in Consols and Railway Securities, at the early period of the panic, may in some degree be estimated by the subjoined table of fluctuations in those securities during April. Although the retrogression to the middle of the month was only partial, the markets had previously presented symptoms of heaviness, with an occasional slight recovery; but the most intense depreciation was occasioned by the events which occurred subsequently to Easter

* It was a common thing at this period to meet brokers and ask them respecting their position. The general reply was, "Well, I hope to pull through; but I fear it will be a difficult affair." Another, "I am safe, but regularly skinned; everything I hold is down some ten, twenty, or thirty per cent., and many of my kind friends will not be able to pay their differences." Another, "Very dreadful, everything gone but my credit." Another, "I have fortunately obtained assistance, and shall escape the hammer this time."

Monday, and the general disturbance in quotations which immediately ensued. The variation in Consols was about 7½ per cent., the fall in reality being 6¼ per cent.; and the Three per Cent. Rentes at Paris drooped to nearly the same extent, the collapse caused by the convulsion being universal. Shares, after being very depreciated, slightly recovered; but the average fall was still 5 to 10 per cent. on English, 3 to 6 per cent. on French, and 1 to 3 per cent. on Indian Shares.

FLUCTUATIONS IN THE STOCK AND SHARE MARKETS DURING THE MONTH OF APRIL, 1859.

	Amount per Share.	Amount Paid.	Price on the 1st of April.	Highest Price during the Month.	Lowest Price during the Month.	Present Price.
		£	£	£	£	£
Consols	—	—	95⅝ to ¾	95⅞	88¼	89½
Exchequer Bills	—	—	35s. pm.	36s. pm.	5s. pm.	18s. to 25s. pm.
RAILWAYS.						
Brighton	Stock.	100	112½	113	104½	105
Caledonian	Stock.	100	82½	82½	69¼	72½
Eastern Counties	Stock.	100	60	60¼	49	55½
Great Northern	Stock.	100	102½	103	96	97
Great Western	Stock.	100	59½	59¾	45½	52½
London and North-Western	Stock.	100	94¾	95¾	83	86½
Midland	Stock.	100	101¾	102⅜	91	96
Lancashire and Yorkshire	Stock.	100	94¾	95	86	86
North Staffordshire	£20	17½	13	13⅝	11½	11¾
South-Eastern	Stock.	100	70½	70¾	59	61
South-Western	Stock.	100	92½	93	85	86
York, Newcastle, and Berwick	Stock.	100	92¼	92½	81	85¼
York and North Midland	Stock.	100	76½	76¾	68	70
Northern of France	£20	16	36½	36¾	33	33
East Indian	Stock.	100	102½	103	88	93

The variations in the principal foreign stocks were also proportionately great. Turkish suffered in the principal proportion, the range in the Old and New Six per Cents. being no less than 33 per cent., while the actual fall was about 24 per cent. The fluctuations in Sardinian reached 17 per cent., and the positive decline was about 15 per cent.; in Austrian the range has proved 20 per cent., and the stock for some time stood at nominal quotations. A variation of 5 per cent. occurred in Mexican, with a recovery of 1 per cent. at the latest moment, and a drop in Russian of 12 or 13 per cent. indicated the severity of the fluctuation. Even Dutch Two-and-a-Half per

Cents. were largely depreciated, the sales on foreign account having been heavy. Altogether, the influence shown to have been exercised on prices by the events which ensued, was considered to have been much more serious and extensive than at any previous period :—

FLUCTUATIONS IN FOREIGN STOCKS DURING THE MONTH OF APRIL, 1859.

	Price on the 1st of April.	Highest Price during the Month.	Lowest Price during the Month.	Present Price.
Mexican Three per Cents.	21½ to ⅝	21⅝	16½	17¼
Russian Four-and-a-Half per Cents.	100	100⅛	87	88
Sardinian Five per Cents.	82	82	65	67
Turkish Old Six per Cents.	90¼ to 90 ex. div.	90¼ ex. div.	57 ex. div.	64 to 68 ex. div.
Ditto New................................	79¾	79¾	47	52 to 58
Austrian Five per Cents.	—	65	45	—
Dutch Two-and-a-Half per Cents....	65⅛	65¾	56	56 to 58

The alterations during May were considerable, the panic having extended into the early period of the month.* A much more favourable appearance was, however, afterwards presented, the alarm having subsided, and a reaction having set in. The range in Consols was 3¾ per cent., principally in the upward direction, the actual rise established, compared with the first of the month, being 3 per cent. Immediately after the Bank advanced the rate of discount to 4½ per cent., and it was found commercial credit was not seriously compromised, there was a revival, and the return to ease which was eventually experienced through the full supplies of gold from Australia, America, and St. Petersburg, stimulated greater activity. Proportional firmness was exhibited in foreign securities, but railway shares presented even increased improvement, the rise being from 4 to 7 per cent. This, after the late depreciation, it was said, might in some respects have been anticipated, particularly in connection with Indian, which were forced forward at all risk.

* The failures that occurred comprised both brokers and jobbers. It has been complained that no correct list of suspensions ever was published. This was impossible, because the uncertainty attending the details of names, and the fact that difficulties experienced one day were surmounted the next, would have rendered such announcement in many cases incorrect and impolitic. Besides, at one period false reports abounded; and instances of compromises were mentioned, which, it is difficult now to say, have been completely adjusted.

FLUCTUATIONS IN THE STOCK AND SHARE MARKETS DURING THE MONTH OF MAY, 1859.

	Amount per Share.	Amount Paid.	Price on the 2nd of May.	Highest Price during the Month.	Lowest Price during the Month.	Present Price.
		£				
Consols	—	—	89¾ to 90¼	92¼	89½	93¼
Exchequer bills	—	—	18s. to 25s. pm.	30s. pm.	15s. pm.	18s. to 22s. pm.
RAILWAYS.						
Brighton	Stock.	100	106½	110	104	109½
Caledonian	Stock.	100	72½	78	71	77½
Eastern Counties	Stock.	100	52½	54½	51	54¼
Great Northern	Stock.	100	95½	100½	93¼	99½
Great Western	Stock.	100	50	54	48½	53½
London and North-Western	Stock.	100	86	89½	83½	88½
Midland	Stock.	100	94	97½	90	97¼
Lancashire and Yorkshire	Stock.	100	85	87¾	83	87½
North Staffordshire	£20	17½	12	13	11¾	13
South-Eastern	Stock.	100	62	64¼	59½	63½
South-Western	Stock.	100	85½	89	84	88¾
York, Newcastle, and Berwick	Stock.	100	84½	87½	81½	87
York and North Midland	Stock.	100	69½	72	67	71½
Northern of France	£20	16	—	35	34⅛	—
East Indian	Stock.	100	93½	99½	90	99

The appended figures present the progress of the recovery, which took place, after the depth of depression was in May, and it will be seen that the rise was very rapid, through the purchases effected by parties who considered it a good opportunity to "get in at low prices." Since that date a further rally has occurred, and the markets at the moment these lines are written show altogether a satisfactory appearance :—

FLUCTUATIONS IN FOREIGN STOCKS DURING THE MONTH OF MAY, 1859.

	Price on the 1st of May.	Highest Price during the Month.	Lowest Price during the Month.	Present Price.
Brazilian Five per Cents.	98 to 96	99½	95	98⅞
Mexican Three per Cents.	16⅝ to 17¼	17½	16	17½
Peruvian Four-and-a-Half per Cents.	80	84	80	84
Portuguese Three per Cents.	40	41¼	38½	41¼
Russian Four-and-a-Half per Cents.	90 to 91	94	90	93½
Sardinian Five per Cents.	68	79	66	78
Spanish Three per Cents.	39	40	36	40
Dutch Four per Cents.	92¾	92¾	87½	92½
Venezuela Five per Cents.	33	38½	33	38½
Austrian Five per Cents.	42	42¼	36½	nominal.
Old Turkish Six per Cents.	66 to 64	67¼	59	66½ to 67
New Turkish Loan	55 to 56	57	48	56 to 55½

The following extracts, culled from the papers at the moment, furnish graphic accounts of the course of events at this particular period :—

Tuesday, April 26, 1859.

The panic symptoms in the markets for Public Securities have not abated. The various quotations have touched to-day a lower point than they had before descended to, but a recovery, which for a short time promised to be permanent, occurred, and Consols, which had been done at $92\frac{7}{8}$, went first to $93\frac{5}{8}$ to $\frac{3}{4}$, and then to $94\frac{1}{8}$. Eventually fresh fluctuation, produced by unfavourable statements and discouraging quotations from Paris, occasioned great uneasiness, and not the least confidence was expressed in the stability of prices. The different phases through which things are now passing, and the serious variations created by every story brought into circulation, prove how completely business is in the hands of the speculators, and, although the address of Lord Derby at the Mansion House was considered to show the anxiety of the Government to preserve peace, it was feared that the effort would scarcely be attended with success. While prices were suffering from the effect of sales, steadily continued by the operators for the fall, it was affirmed that a telegram had arrived intimating a disposition on the part of Austria to suspend hostilities for fourteen days; but though this statement could not be traced to any authentic source it caused numerous purchases, and prices became at once very animated. It was also said that the French Government would not require a new loan, and that the communication to the Corps Legislatif by the Emperor would simply be on the question of additional levies. Still nothing certain transpired, and the improvement was well maintained until a very advanced hour, when irregular transactions, as will be seen, were once more concluded at a serious depreciation. A failure on the Paris Bourse exercised an unfavourable influence on Lombardo-Venetian shares, and those securities were at one period of the day extremely depressed; but after hours a partial change for the better was apparent.

Late in the evening, and when business was wholly suspended, it was asserted in special channels, and seemingly on good authority, that Austria had accepted the mediation of England, which it was hoped would lead to a satisfactory adjustment of the difficulty.

As indicating the excited position of business and the sensitive state of quotations, it should be mentioned that long after the termination of the regular hours operations were entered into among the dealers. At three o'clock, the official closing of the Stock Exchange, Consols were quoted $93\frac{5}{8}$ to $\frac{3}{4}$ for money and the account. Between that period and four, a sudden rally to 94 to $\frac{1}{8}$ had followed. Subsequently weakness was apparent and numerous fluctuations were visible, the reaction that set in carrying down the quotation to $93\frac{1}{2}$ to $\frac{5}{8}$, which was the last price at half-past five o'clock. To show the activity of the operators, many thousands of stock passed in the interval between four and half-past five o'clock, a full market being held in the neighbourhood of New Court, one of the approaches to the Stock Exchange, and the majority who had before purchased resold to cover themselves on the fall that ensued. The prices received from Paris showing heaviness, the rumour regarding the suspension of hostilities for fourteen days was discredited, and hence the sudden and smart decline. Other securities, including Turkish, Sardinian, and Lombardo-Venetian shares, were proportionally affected, though quotations could not with accuracy be obtained.

The demand for accommodation at the Bank to-day was more than usually active. The description of paper offering showed that the impression entertained is that an advance in the rate will not be long delayed. Merchants and others are discounting, with the view of providing against any contingency, and the result is that a rise to 3 per cent. is now daily anticipated. The declension in the stock of bullion, which during the last five or six weeks has been nearly £2,000,000, with the decrease in the reserve of notes, marks the action which has been progressing, notwithstanding predictions to the contrary in more than one quarter. The various banks and discount establishments are transacting a favourable business, but for paper of the most unexceptionable character they only charge the Bank *minimum* of 2½ per cent. For second and third class bills, however, they obtain much better terms, and it is through these operations that the greater proportion of profit is secured.

Wednesday, April 27.

Intense excitement has again prevailed on the Stock Exchange, and prices have presented increased fluctuation. Consols, which left off, at the advanced hour of half-past five yesterday evening, 93½ to ⅝, were before regular business called 91½ to 92. This decline was brought about by the alleged treaty entered into between Russia and France, through which it was supposed the position of England was in a measure compromised, and the consequence was heavy sales, both on *bona fide* and speculative account. The announced acceptance of British mediation by Austria exercised little or no influence, and partial variations continued apparent. Meanwhile, difficulties among brokers and jobbers occurred, and the consequence was, that although an endeavour to sustain prices was made, an immediate reaction to 91½ took place. A doubt was then thrown upon the Russo-French alliance; and as it failed, on inquiry among the St. Petersburg houses, to be confirmed, Consols rallied to 92¼, the highest quotation of the day, and for the moment some animation existed. Nevertheless failures were reported, and the markets altogether exhibited a very sensitive and discouraging appearance, through the belief expressed that war would at once be entered upon, despite every endeavour to conciliate or to make honourable terms of arrangement. Under these circumstances fresh gloom was visible, which was greatly increased when it was stated that Austria had signified her desire to accept the mediation of England, but that France, after recent events, declined. This produced a more severe reaction than ever, and Consols went at once to 91. The regular close of business did not guide transactions at all, and though there was temporary steadiness at this fall the drop subsequently was still serious. A quotation of 90½ was made in the depth of the depression, when it was discovered that panic reigned on the Paris Bourse, and that the Rentes continued to go down. The nearest last price that could be ascertained was 90¾ to 91, and a variety of rumours were circulated with respect to the intentions of the Government, and the measures of defence to be adopted to prevent any undue advantage being taken by France of the existing situation. Up to as late an hour as a quarter to six o'clock, transactions were entered into, the Stock Exchange and neighbouring business localities not having then been deserted.

The state of panic and excitement may be in some slight degree imagined from the circumstance of no less than seven failures having occurred during the day. It was supposed that serious embarrassments would be encountered, particularly among the smaller class of operators, and this has unfortunately proved the case. Four of the suspensions will not be for any considerable amount, but in the instance of one large speculator in Consols, his liabilities are taken at £20,000, though it is said the dividend will be satisfactory. Two firms are included in the list, one an old and very respectable house, the partners being unable to realize their securities. The majority have more or less suffered from the default of principals, who cannot at the moment meet their engagements through the sudden and severe blow quotations of every kind have received. These failures having been announced in the preliminary attempts to arrange the foreign stock and railway share accounts, it is apprehended that other difficulties will be experienced before pay-day arrives, and the result is great anxiety for the termination of the present settlements.

The condition of things was most unprecedented. In many instances good securities were altogether unmarketable, and the effect was of a disastrous character. Variations of 8 to 10 per cent. in Turkish, 3 to 4 per cent. in Sardinian and Russian, and 2 to 3 per cent. in other descriptions, showed the current of operations, the range in shares being from 2 to 5 per cent., with scarcely any purchasers, French and Lombardo-Venetian continuing extremely heavy.

A feature in the course of business proceedings to-day was the indisposition of capitalists and others to make advances upon the different classes of securities on any terms, for the purposes of carrying over, and hence the general and increased fall. Accounts, consequently, had to be closed, and the best arrangements for settlement effected to meet the sudden emergency. It was also confidently stated that the Bank directors would raise the rate of discount to-morrow to 3 per cent., and adopt additional measures of precaution, if necessary, to arrest the efflux of bullion, should the drain be stimulated by the war. The intimation respecting the French loan of £20,000,000, although it was regarded as evidence of the determination of Louis Napoleon to push his views to the extreme, exercised no important effect, because it was felt that no subscription would be made in this country. The Russian loan will also have to be deferred, or at least have to be taken to other markets to complete, if hostilities ensue, and the asserted compact proves to have been formed.

The several phases of the day's intelligence were discussed with great avidity at Lloyd's, the Jerusalem, Baltic, and Jamaica Coffee-houses, the interests involved, and the points in dispute furnishing topics for every class. There was great excitement at Lloyd's and the Baltic, the Russian trade especially manifesting a desire to trace the effect of the supposed change in the relations between that country and France. At Lloyd's, war risks were demanded on steamers and vessels to the Mediterranean, and it is not improbable that they will be extended to other ports if affairs continue to present such a menacing aspect. In Mincing Lane and on 'Change considerable activity was noticeable, saltpetre, tallow, and rice being the articles principally affected.

Austrian Five per Cents., which, on the 15th inst., were dealt in at 63, were sold this morning at 45, showing a decline of 18 per cent., and bargains to any extent would have been impracticable at the price.

An average fall of 8 per cent. took place in Turkish Securities by the comparison of prices in the official list. Ultimately they were quoted still lower, and at irregular margins. The Original Six per Cents. were dealt in at 70, 69, 65, and 69, and were at four o'clock called 60 to 63. The New Six per Cents. were then marked 50 to 54, having been operated in at 57, 61, and 60. There was again an absence of business in the Four per Cent. Guaranteed.

The scrip of the new India Debentures was done at one period at 90½, or 4½ discount, but subsequently recovered to 92½, or 2½ discount.

Thursday, April 28, 1859.

Another day of great excitement and alarm has just been passed. The panic which has prevailed since Tuesday, if possible, increased in intensity, and the condition of business in public securities has been more fluctuating and irregular than at any time since the first development of the unsatisfactory symptoms. The influence exercised by the statements published of the course pursued by the French Emperor, the supposition that the Austrian troops had crossed the Ticino, and the knowledge that the complications might involve England in a war, produced much disquietude, and this having been followed by an advance in the Bank rate of discount to 3½ per cent., or, in fact, a rise of ½ per cent. higher than was generally expected, the full gravity of the situation became appreciated. Besides, it was affirmed that the Government, preparing for the worst, notwithstanding it is questioned whether the express conditions of the treaties with Russia are precisely what have been indicated, have issued orders for raising men for the navy on a £10 bounty, so as to meet whatever emergency may arise. Every description of report was circulated, some being of a favourable, and others of an unfavourable, character, but the depression exhibited was not surmounted, though occasionally a partial rally in prices occurred. The alternations, nevertheless, were very violent throughout the hours of regular business, and as yesterday, up to six o'clock, a market was held in Throgmorton Street, in which bargains could be completed.

The Bank directors were not long occupied in their deliberation upon advancing the rate of discount to 3½ per cent., and it was formally announced early in the day that this would be for the present the price for the negotiation of bills having 95 days to run. The rate before was 2½ per cent. which has been in existence since the 9th of December last. This sudden rise of 1 per cent., it seems, is considered to be fully justified by the course of events, and it has been stated that there is a prospect of a further augmentation. The continuous declension in the stock of bullion, and the apprehension that the drain may be accelerated by the position of relations on the Continent, will account for the promptitude of the measure, though in many quarters it was anticipated that the alteration would at first have been limited to ½ per cent. The rate was last advanced to 3½ per cent. on the 2nd of June, 1853, when the stock of bullion stood at £18,253,934 and the reserve of notes at £8,366,970. According to the last return the position of the bullion was £18,051,375, and the reserve of notes £9,880,240, while the statement to be issued to-morrow will, doubtless, present some additional important variations.

At the commencement of business to-day, the scene witnessed at the Stock Exchange was one of the most painful ever remembered. Those whose memories carry them back to the days of the celebrated Spanish panic, scarcely think that the exhibition then was more distressing. The first list announced was five names, followed by a second of eight, and then four others were declared, making a total, including partners, of nineteen. These, with the suspensions previously intimated, constitute an entire number of between twenty-seven and twenty-nine, comprising the one or two declared early in the week. It was remarked that whereas on the occasion of the great revulsion already referred to the scarcity of money and the sense of alarm extended principally to one market or department, in this instance it was apparent throughout the house, the operators in every species of security being affected. At one moment the condition of things seemed destined to paralyze transactions altogether, and such was the apprehension exhibited, that the further announcement of failures was awaited with the most intense anxiety. Although, with one or two exceptions, the names to-day were of second and third class character, still the great number showed the severity of the blow, and the fatal influence exercised by the immense change in the position of our political relations. In the course of the afternoon there was a greater exhibition of confidence, but the slightest rumour served to create agitation, and to lead to every kind of conjecture with respect to the consequences which it is alleged will follow the Russo-French alliance.

So incessant was the variation in the prices of Consols to-day, that it was with great difficulty the course of change could be traced. The first price was 89, or about 2 per cent. below the last quotation of last night; there was then a fresh decline to 88⅜, the foreign intelligence being considered unsatisfactory. The rise in the Bank rate was subsequently made public, but it produced no further effect. A sudden rally ensued to 89½, and 90 was reached on the statement that the Russian alliance was unfounded. Presently, however, weakness was again perceptible, and the price once more receded to 88⅝. Just before the termination of business, another movement upwards was visible, and the latest official price was 89 to ¼. A considerable amount of operations ultimately took place at quotations varying from 89½ to ¾ to about 90, but once again a sudden collapse ensued to 89, which was the nearest final quotation. The warlike intelligence from Tuscany was not viewed with satisfaction, and the position of Austria seems to become more dangerous every hour.

It is mentioned, with reference to some of the suspensions that have taken place, that the accounts opened by the parties were on a scale of considerable magnitude. Although it is true that the depreciation in prices has been very severe, nothing could warrant such extensive contracts as are alleged to have been shown by the books of one firm that was compelled to announce their inability to meet their payments. Of course it is understood that many of the individuals who have been placed in embarrassment have suffered solely through the default of principals, and there is no question that if they were paid the differences due to them they could have met the whole of their liabilities with punctuality; but with such a state of depression, and a fall in quotations, as in the cases of Turkish and other securities, ranging from 10 to 30 per cent., the difficulty of averting the catastrophe must be generally admitted.

The demand for money after the Bank had given notice of the increase of the rate of discount, was not so great; but still the inquiry was considerable, and the public availed themselves of the opportunity afforded of securing the assistance they required. Out of doors it was said that the rate for 60 days' paper was 3½ per cent., and 95 days', 3¾ to 4 per cent.; but for six months' bills very much higher rates were charged. Indeed, in some quarters the discount establishments refused to take in six months' paper, unless on special terms with those who desired to negotiate it. After all, notwithstanding the demand was well supported, there was not what could be properly designated a pressure, and the supply was ample for all purposes; but through the increase of the distrust it was impossible to induce capitalists to make advances beyond rather limited periods.

An extreme range of 7 to 11 per cent. occurred in Turkish Securities, and the average decline was about 8 per cent. This wide margin was occasioned by the refusal of the dealers to continue transactions, and sales were consequently pressed in every direction.

Friday, April 29.

The effect of the panic at the Stock Exchange is still painfully apparent, and an additional list of twenty-three suspensions has been recorded. Two of the firms comprised in the new failures represent liabilities for large amounts, but they have been brought down through the default of other members. Such is the general confusion that it is scarcely possible to tell who is yet solvent, and although the accounts in foreign securities and railway shares terminated this afternoon, other stoppages are anticipated to-morrow. In consequence of the almost universal embarrassment experienced, two or three very large houses were temporarily compromised, and had it not been for assistance rendered by their bankers, they would have been compelled to intimate their inability to meet their engagements. Not the least question could be entertained in these special cases of the value of the securities they possessed, but the obstacle was the immediate supply of funds to meet the sudden demands occasioned by the emergency. To the close of business this evening it appears that about fifty-two "declarations" have, in all, occurred, the majority, however, being dealers and jobbers, whose debts severally have not been important, although, in the aggregate, they will prove enormous. As showing how seriously the influence of a suspension of this kind works, it was stated that one broker, who struggled through with only a moderate amount of stock open, would, if he had been compelled to default, have brought down four other members. In this manner has the whole of the present panic extended its sway, so that solvent and insolvent have alike suffered from the paralysis which has seized quotations. It yet remains to be seen what will be the process of recovery, particularly as there is still much uncertainty respecting the future course of business.

At one period this morning there was a prospect of quotations taking a more favourable position, and Consols, as well as foreign securities and railway shares, presented a better appearance, but increased gloom was again visible through the entanglement of the accounts and the difficulty manifested in adjusting differences. A considerable supply of bank-notes was required to meet payments, the state of credit having been greatly impaired by the events of the last two days. The fluctuations in prices were extensive, and though Consols opened at 90, or 1 per cent. above the six o'clock quotation of last night, and eventually touched 91, the complexion of the foreign intelligence, with the serious disasters among the operators, produced a reaction till the price returned to 89 to ¼, exhibiting a very partial rally. There was a limited attendance late this evening, and the jobbers only made a fractional variation in the value. A special committee was called to-day, when it was agreed to recommend the immediate arrangement of the Consol account, and preparations were at once commenced, and before the end of the afternoon most of the dealers had adjusted their respective transactions. Under ordinary circumstances, this course of proceeding would not have been entered upon before Monday.

It is a noticeable feature that the banking interest and the public are making investments in stock at the current depreciated quotations. Exchequer Bills were rather firmer this afternoon, but the late decline is attributed to the displacement of capital from these securities, and its outlay in Consols, Reduced, and New Three per Cents.

At this stage, it will be perceived, the course of the panic was nearly exhausted, and affairs were on the turn.

Monday, May 2.

The Stock Markets have at length assumed a much more tranquil appearance, and although the rally in prices has not been important, there was, for the first time since the outbreak of the panic, a cessation of failures. It is nevertheless feared that, before the Consol account is completed, other of the members will have to succumb; the severe fall in quotations, together with the fatal effects of the late movement, having compromised the position of many parties whose differences have yet to be adjusted. The public were buyers of securities to-day, especially those of a leading description; but the advance at first attained was not subsequently fully supported, rumours being circulated in connection with the situation of foreign mercantile houses, whose affairs will be disarranged through the commencement of hostilities. The state to which Austrian finance has been reduced by the occurrences of the last few weeks, must entail serious responsibility upon that Government, and a war prosecuted with an exhausted and deeply embarrassed treasury will tell with double effect upon its trade and external relations. It is not probable that much reliable information will be received from the seat of operations, if communication is to be interrupted by the destruction of the telegraphic wires, though there will be no paucity of reports promulgated with the view of affecting quotations or exaggerating the importance of events as they may take place. Consols were first dealt in this morning at 89½ to ¾ for money and the account, and improved to 90½, the scarcity of stock causing a rather higher quotation for money; but a reaction on sales followed, and the price then drooped to 89⅝. The accounts from the Paris Bourse were not favourable, the value of the Rentes showing a decline of nearly a quarter per cent., and the advance in the corn market indicated that apprehensions were entertained of the probable consequences of the collision with Sardinia. Before the final termination of the afternoon there was again a partial recovery, and the last quotation was 90 to ¼ for money and the account, exhibiting a rise of about a half per cent. It is noticed that although not much animation is imparted to prices by the business transacted, the character of dealings gives them a reality which they have not for some short time possessed.

The tendency in the rates of discount is still of an unfavourable character, and the demand for accommodation has been extensive to-day, through the caution exercised by the banking interest and the bill-brokers. The pressure upon the discount department of the Bank was, therefore, proportionate, and, if the present state of things continues, it is believed that an additional advance in the *minimum* will take place. Although the requirements for railway calls are far from large, still some inquiry will arise to meet the necessary payments on the India Four per Cent. Debentures falling due on the 6th, and it yet remains to be seen whether the Council will make any ameliorative arrangements, so as to relieve the holders from the full weight of the amount then to be discharged. It is suggested that a different adjustment for the periods of the instalments would probably meet the emergency, which, if strictly enforced, through the altered condition of affairs, may put many individuals in the position of defaulters. If a forfeiture were carried out, it would place the operation on a very unsatisfactory footing, and, consequently, the only way to cope with the difficulty would be to make the payments as light as possible by spreading them over an extended period. The depressed prices at which the scrip stands, will, it is presumed, induce the authorities to make some such variation in the understood terms of the tenders. No delay should be suffered in giving notice of any alteration, if one is really contemplated, because the parties interested are very anxious on the point.

Part of the inquiry for money arises from the preparations to meet the paper of the 4th of the month, which, it appears, will represent a considerable amount. It is stated that the bill-brokers are charging extreme rates, the quotation for two months' paper being 3¾ per cent.; three months', 4 per cent.; four months', 4½ per cent.; and five months', 5½ per cent.

The arrangements for liquidating several of the estates of the brokers and jobbers who have failed during the recent disruption of credit at the Stock Exchange are making active progress, and it has already been notified that seven dividends will be immediately distributed, varying from 20*s.* to 5*s.* in the pound, which, considering the enormous depreciation in quotations, is not regarded as altogether unsatisfactory. At present there will be a great difficulty in realizing assets, except those of first-rate character, owing to the absence of animation in business and the indisposition to make anything approaching regular prices for any of the neglected securities. Should a recovery occur better prospects will be apparent for the creditors, one estate now so much depending upon the other for the increase or the diminution of their respective dividends. Notwithstanding the impression that the principal of these firms will be wound up in a short period, several months must elapse before the whole can be brought under effective administration.

Tuesday, May 3.

The markets for Public Securities are gradually assuming a more settled condition, although the extent of business is considerably curtailed. This arises from the indisposition of the public or the general class of operators to enter into enlarged engagements after the serious results produced by the late decline, and the apprehension that the full extent of the mischief created by the panic has not been altogether realized. Prices, except in special instances, must still be affected by sales of securities, which will be necessary to complete the arrangement of differences, and hence an immediate recovery is not to be anticipated. Capitalists, however, are purchasing Consols, and the steady absorption, with the effect of the explanations of the Chancellor of the Exchequer, occasioned an advance this morning of about three-quarters per cent., but a reaction took place before the final close of business. Stock throughout the day was extremely scarce for immediate delivery, and hence the quotation for money was slightly in advance of that for the account. More readiness was also observable to conclude bargains in other descriptions, though the impression that the war must entail a disturbance of no ordinary character arrests the rebound which under different circumstances might speedily follow. Conflicting reports were again spread with respect to the intentions of France and Sardinia; and, curiously enough, at the latest hour, it was said there was the prospect of a peace arrangement. The assertion, however, was not credited, and the further slight decline on the Paris Bourse increased the dulness subsequently apparent. Consols were first dealt in at 90 to ¼, advanced to 91, and then fluctuated between 90⅜ and 90⅝; the last price for money being 90⅜, and for the account 90¼ to ⅜.

The investments by the Government broker of £10,000 per diem, on behalf of the savings' banks, have not yet been suspended.

The principal demand for money has fallen upon the Bank, through the inability of the brokers to obtain their ordinary supplies from the joint-stock and private banks, and consequently there has been much activity in the discount department of that establishment. The joint-stock banks are now limiting their transactions to the negotiation of the paper of their customers, and the result is, that the amount of capital in the open discount market has become comparatively restricted. Expectations are entertained that the directors will further advance their *minimum* terms to four per cent., and through this impression many parties are already providing for future necessities. The payments of to-morrow (the 4th) have been in a great measure already arranged; but there are still other engagements, especially the instalment of the Indian Four per Cent. Debentures, which will absorb an increased amount of money. Although the *minimum* terms of the Bank of England are 3½ per cent., the quotations charged out of doors range in advance of that point, and the operations would be more extended than they are if accommodation could in all instances be secured.

The preparations for the Consol account are still in progress, and every means is being adopted to lighten the pressure which may be experienced when differences come to be adjusted. By mutual concessions it is felt that the brokers and jobbers can assist one another, and hence there is a disposition to encourage even the acceptance of payments to the extent of 10*s*. in the pound, allowing parties who may find themselves in difficulties a short period for punctually completing their engagements in preference to resorting to the painful alternative of "declarations," which have previously exerted such a baneful influence. Through this arrangement it is hoped that the position of some of the dealers who are known to be in difficulty may be ameliorated, and confidence will therefore be less severely tested than it was during the arrangement of the foreign stock and railway share account. It has also been agreed, in accordance with the terms of the following resolution, that assets shall be collected as quickly as possible, and dividends declared *pro rata*, with the object of allowing those interested in the several estates to obtain whatever may be coming to them without unnecessary delay:—

"The committee recommend that dividends due on defaulting estates be paid as early as possible, and put into the joint names of two of the creditors, who will thus be enabled to distribute on the same day to each creditor on each estate rateably as may be advisable."

It is mentioned in connection with the recent disturbing influence of the panic at the Stock Exchange, that the devastation of credit would have been much more extensive but for the timely assistance afforded by some of the leading discount establishments, and it is affirmed that Messrs. Overend and Co. were among those most forward to render the necessary aid in cases where it could be shown that the securities were of an appreciable character, and that operations had been carried out on a sound and legitimate basis. The banks also gave considerable facilities, but it was necessary in some instances to curtail loans, with the view of bringing the transactions which had been extended beyond due limits into manageable compass. The intensity of the effect created by the exaggerated intelligence of the Russo-French alliance will long be remembered in City circles, the position of the dealers who, in many instances sacrificed by their principals, could not provide the whole of the balances required at the latest moment, being of the most embarrassing description.

Wednesday, May 4.

Business at the Stock Exchange was not transacted on an extensive scale. The arrangement of outstanding engagements, and the provision for the payments on the Consol settlement, occupied the chief attention, the events of the last week creating caution. The effect of the publication of the official contradiction of the existence of a treaty, offensive and defensive, between France and Russia, by the *St. Petersburgh Journal*, was counteracted by the advance in the rate of discount on the Continent, the Banks of France, Frankfort, and Bremen, having each increased their terms, the former to 4 per cent., and the latter to 4½ and 5. It was also believed that an advance to at least 4 per cent. will be made by our own Bank directors to-morrow, and, consequently, the prices at which Consols stood in the early part of the morning were not maintained. Investments on behalf of the public were constantly effected, and these sustained values for a period, the variation in the market having been between 90¼ to ½ for money, and 90 to ½ for the account. A failure of a mercantile firm in the Australian trade was then announced, and this, with a rumour that a loan operation to a considerable extent is contemplated by the Government, caused a succession of sales, and the quotation receded to 89⅝ to ⅞ for the account, showing a decline of about three-eighths per cent. After ordinary hours the dealers concluded a few transactions at 89⅝, but the quotation was viewed as irregular.

The great pressure for money still continues on the Bank of England, the discount brokers not employing their funds to any great extent, since their resources are in a measure curtailed by the restriction of facilities in obtaining loans from the joint-stock banks. The applications consequently were exceedingly numerous, and the mercantile public found it necessary to resort to this department for the negotiation of paper, the average terms out of doors being in almost every case above the *minimum* of 3½ per cent. Six months' bills cannot now be discounted except at very advanced terms, and it is said that 5¾ to 6¼ per cent. is the range for that class of paper.

Thursday, May 5.

The Bank directors, at their weekly court to-day, raised the rate of discount from 3½ per cent., at which it was placed last Thursday, to 4½ per cent., showing a further advance of 1 per cent., the augmentation being in the same ratio as that made this day week. The movement was fully anticipated, and the effect, it is thought, will be salutary; but there seems to be an impression that the alarm is obtaining an exaggerated tendency, the fall in prices, and the sacrifice of securities, increasing the general heaviness pervading business. The efflux of bullion from the Bank, and the rapid absorption of remittances as they arrive in this country, prove the serious posture of affairs, though it is difficult to trace the cause of the continuous and extended depreciation. No doubt the "Gazette" return, to be published to-morrow, will exhibit changes of an extensive character, and trade will suffer contraction through the condition of continental politics. The first effect of the intimation was an unfavourable reaction in all classes of securities, from which there was scarcely any recovery, English stocks alone indicating firmness in the later hours of the afternoon. Foreign bonds and railway shares were extremely heavy, the late difficulties attending the adjustment of the accounts, and the inability of dealers to support quotations, rendering the markets again unsettled. Another circumstance that tended to create apprehension was the suspension of Messrs. Arnstein and Eskeles, the important banking firm of Vienna, not from the event itself, looking at the position of finance in Austria, but through the disastrous influence it may exercise in other directions. The public are still purchasers of Consols, Reduced and New, though scarcely on the scale of yesterday, but sufficient to support the quotations, and to give them a firm appearance, particularly for money. For the June account the price is not quite so well maintained, the speculative business being altogether inconsiderable. Consols last night, which closed after hours 89⅝, recovered at the commencement of business to 90 for money and 89⅞ for June. Prices then exhibited dulness, and on the advance in the Bank rate returned to 89¾. Fresh purchases followed for investment, and the progress of the settlement being more satisfactory than was anticipated, the tone of quotations was better, and manifested strength. The foreign political intelligence was not of vital interest; but the prices from Paris also indicated a more favourable appearance, though the latest failed to arrive during ordinary hours. The final quotation of Consols, for money and the account, was 90 to ¼, and at an advanced period of the evening the variation was merely fractional.

In expectation of the advance in the rate of discount, the attendance of those who required accommodation this morning was very numerous, but, although many applications were made, it was announced that no business would be transacted until after the rising of the court. The directors met as customary, and broke up after a short deliberation, when the alteration was notified in due form. A good business was then completed, and the pressure continued throughout the ordinary hours. In the open market operations have been comparatively limited, the weight of transactions being in that particular quarter, and it is remarked that the absence of assistance to the brokers and others is producing an interruption to the general progress of the trade of the country; since the refusal to discount for them is preventing the negotiation of four months' and longer dated paper, which otherwise would be dealt with if the discount houses could secure partial facilities when they require them. The quotation for three months' paper has varied from 4½ to 5 per cent., but in some cases for very prime descriptions a fraction under the Bank rate has been accepted.

The several joint-stock banks have intimated to their customers that their allowance for deposits will in future be 1 per cent. below the rate of discount established by the Bank of England. As usual, Messrs. Overend, Gurney, and Co., and the other private discount houses, have advised their friends that their terms for the present will be 3½ per cent. for money at call, 3¾ per cent. for three days' notice, and 4 per cent. for seven days' notice. The National Discount Company and the London Discount Company have not at present made any alteration; but it is fully expected they will follow the same course. In that case the transactions will date from to-day.

Some anxiety is still manifested to ascertain the final result of the adjustment of the Consol account at the Stock Exchange. All descriptions of compromises are being effected among dealers, owing to the fatal effects of the late fluctuations, and arrangements have been made by which payments are undertaken from 10*s*. to 5*s*. in the pound, with the view of ultimately providing the remainder. By this means additional failures are for the moment prevented, and it is thought that the assistance so extended will generally lighten the difficulties experienced. Even some of the principal of the jobbers have been necessitated to ask indulgence under existing circumstances, and the consequence is that the body of the members have more or less suffered from the influence of the late panic.

SUPPLEMENTAL MATTER.

THE PROGRESS OF THE YEAR 1857.

In tracing the progress of events in the course of this remarkable year, it may be stated that it opened with rather discouraging appearances. The money market had for some short period been tight, and there was little amelioration in the position of general financial affairs in January. The relief afforded by the payment of the dividends and the arrival of Australian gold was wholly temporary, and a revived demand was again apparent, which, it was feared, would eventually lead to an increase in the rate of discount by the Bank directors. The existing *minimum* of 6 per cent. was considered to be comparatively high, and, as they had augmented the terms for advances on stock to 6½ per cent., if the pressure had continued, they would have been obliged to adopt additional restrictions with reference to the negotiation of mercantile paper. For some time it was expected that such a movement would be announced, but the gold withdrawn to supply the requirements of the Irish banks during the late partial panic, having returned, it assisted to support the metallic reserve. As, however, the drain for export to the Continent had set in with renewed severity, and the remittances from Australia and America were not, at this period, extensive, an apprehension was entertained that not only might the stock of bullion be further trenched upon, but that the supplies themselves from these sources would be purchased and sent to Paris.

Under such circumstances, the directors could not follow out any other policy, but it was hoped that the stringency would not be maintained at a level which would either distress commerce or interfere with the prosperity which it was supposed would be initiated by peace. The range of fluctuation in English stocks did not exceed 1 per cent., but it was principally on the side of a decline. The symptoms of weakness which followed the announcement of a rupture with China, and the Persian war, gradually increased, while the renewed demand for money induced many of the speculators for the rise to close their accounts. Indeed this was a prudent course to adopt, the Bank directors having again raised their terms for advances on stock to 6½ per cent. In connection with this movement it was remarked, "the Bank of England have notified that they will decline making advances on Government securities at a lower rate than 6½ per cent. During the recent shutting of the transfer-books the rate for advances of this description was the same as that for discounts—namely, 6 per cent. There is now, however, no reason for affording the accommodation. Very large amounts are in course of repayment, and, instead of these being again lent out to enable persons to hold Consols on speculation, they should, in the present state of the discount business, be held available for commercial purposes. In case of the general market soon becoming easier, a return to the previous rate will probably be adopted, but in former times it was not the custom of the Bank, under any circumstances, to make loans on stock; and although this rule may have been needlessly stringent, it is unquestionably the duty of the Bank never to relax it except when they can do so without prejudice to the mercantile public." At the latest moment Consols exhibited depression, and Exchequer bills were likewise heavy. Foreign securities had not exhibited any great variation, but the market was rather weaker, owing to the general absence of business. All Northern Stocks were lower, and there was apparently not the least disposition to extend transactions. The following were the failures in the month of

JANUARY.

Mr. G. B. Rocca, Mediterranean trade, London.
Messrs. Begby, Wiseman, and Co., merchants, Glasgow.
Messrs. Rice, Harris, and Co., glass trade, Birmingham.

The opening of the Parliamentary campaign, and the favourable change in the weather, produced a satisfactory influence upon financial affairs in February. The budget, although it encountered opposition, was carried by a large majority, and the measures of retrenchment adopted, with the reduction of the income tax, increased the confidence of the public. Notwithstanding some parties inclined to the belief that greater economy might have been exercised, the changes were accepted as an earnest of what the future might produce. While the benefit of diminished taxation was thus experienced, there was confident anticipation that the settlement of the Persian question would be shortly announced, and under these circumstances one great cause of anxiety was allayed. The dispute with China did not create much alarm, and although it might partially restrict trade, the result, it was thought, would eventually prove advantageous to the whole of our European mercantile relations. The Bank directors having announced their intention of making advances on stocks, and also on bills not having more than six months to run, until the payment of the April dividends, at the rate of 6 per cent., increased animation was witnessed in the market for public securities. But, although this was the case, there was no immediate probability of a reduction in the terms of discount, the demand, generally, being well maintained. Arrivals of gold from Australia had lately taken place, to the extent of upwards of £1,000,000, in addition to supplies from America, but the entire sum was absorbed by purchases on behalf of the Bank of France, and those engaged in securing silver from the Continent for remittance to the East and China. Fortunately the stock of bullion in the Bank did not undergo diminution, for, notwithstanding some withdrawals took place, the return of coin from Ireland, Scotland, and the provinces, was sufficient to replace the amount so as nearly to balance the total. The failures in the Greek trade, although rather numerous, had not occasioned alarm, and the private letters from Paris described a strong revival of speculative feeling.

The transactions on the Bourse were extremely large and numerous, and there seemed a general disposition to believe in the approach of another period of inflation. Money was obtainable without much difficulty at 5½ to 5¾ per cent. A reduction in the rate of discount in the bank of Frankfort from 5 to 4 per cent. had just been reported, and a downward movement was also expected at Amsterdam. The French Government were alleged to be on the point of making some new railway concessions. All the continental exchanges continued with a favourable appearance. In each of the principal cities, also, the rate of discount was below the minimum in London. At Hamburg, Frankfort, and Brussels, it was 4 per cent.; at Amsterdam it was 5 per cent.; and at Paris 5¾ per cent. The operations in the English funds were not on a scale of magnitude. Subsequently, however, they increased, and prices were comparatively buoyant. The range of fluctuation was about 2 per cent., but an advance of 1½ per cent. continued supported. The prospect of a settlement of the Persian question, the easier state of money, and the success of the ministry with the budget, were circumstances which influenced business, and caused the more favourable feeling apparent. A rise of from 1 to 2 per cent. occurred in the principal foreign securities. The amount of speculation was not great, but the improvement, which had been gradual, was fairly sustained. Turkish, Russian, and Dutch attracted the greatest attention, although some of the other descriptions were more freely dealt in. The railway share market at length exhibited symptoms of renewed vitality, prices advanced, and business generally improved. Greater confidence was manifested by the public as well as the dealers, and the result was a much more favourable state of business. The demand for accommodation to carry out operations, although large, was not pressing, and the rates consequently were rather lower. The following were the failures during

FEBRUARY.

Messrs. Jennings and Hargreaves, stuff merchants, Bradford.
Mr. G. Sichel, German trade, London.

Messrs. C. Franghiadi and Sons, Greek trade, London.
Messrs. Vuros Brothers, Greek trade, London.
Messrs. Sinanides and Co., Greek trade, London.
Mr. J. Basilio, Greek trade, Manchester.

The course of financial and mercantile affairs during March, was not distinguished by much variety. The pressure in the money market did not diminish, but although the rates at the Bank of England and in Lombard Street were supported at their former elevation, the public sustained little inconvenience. Several gold vessels arrived from Australia, and remittances were received from America, but nearly the whole, if not the whole, were purchased to supply the demand on the Continent, and to provide the large amount of silver forwarded to India and China. But while this was the case, and the withdrawals from the Bank slightly augmented, the stock of bullion was kept in a favourable position by the return of coin from Ireland and Scotland, and the influx of specie from Turkey, the Levant, etc., and consequently the weekly returns did not occasion alarm. The adjustment of the accounts at the Stock Exchange were now regarded with some anxiety, as they temporarily influenced the value of money, and led, during the progress of their arrangement, to an increased call in all quarters. There was little expectation of a return to ease, since the heavy advances made by the Bank would have to be repaid when the April dividends were distributed, and no great surplus could be retained from the further remittances immediately anticipated from Australia. Indeed, it seemed to be believed, by those conversant with the probable future of the money market, that any relaxation could not be looked for previously to the turn of the half year, and that a great deal would then depend upon the progress of the crops and the general state of trade. It was stated that it would be requisite to bear in mind that even a favourable alteration, at this advanced period, would also be subject to the disposition evinced to encourage speculative business, and that if attempts were made to launch numerous joint-stock schemes, they would at once retard and perhaps permanently delay the movement. The bankers and discount establishments were well supplied with capital, but some of the latter incurred losses through the recent failures. This was expected from the competition to secure business, especially since the organization of discount companies and the introduction of several new joint-stock banks. The rates for money on the Continent presented little alteration, with the exception of Hamburg, in which city they gradually advanced to 5½ per cent., owing to the continuous operations in silver. The range of prices in Consols was about one per cent., but the actual decline did not exceed three-eighths. The operations were wholly influenced by the events connected with the dissolution of Parliament, which occurred on the 21st of the month; and although at length there was a trifling recovery, sales on the average preponderated. Foreign securities were dealt in to some extent, but Turkish and Mexican took the lead. A rise in both these descriptions occurred, though the former alone maintained the improvement. Mexican, after advancing about 2½ per cent., receded nearly 1⅓; the treaty with America, through which pecuniary assistance was proposed, having been rejected. In railway shares there was fluctuation to the extent of about £4 or £5 per share. After a general improvement, a reaction ensued, and medium quotations were those eventually current. Annexed were the failures during

MARCH.

Messrs. Swayne and Bovill, merchants and patentees of machinery, London.
Messrs. Cheape and Leslie, East India trade, London.
Mr. O. Foa, seed merchant, London.
Mr. E. Train, merchant, Liverpool.
Messrs. F. W. Stein and Co., silk trade, London.
Messrs. Barnes, Copland, and Co., provision merchants, London.
Messrs. R. James Brown and Co., timber merchants and shipbuilders, Sunderland.

In April, financial and mercantile affairs did not present an encouraging appearance. The increased pressure for money, followed by the restrictive measures of the Bank of England, was attended by an unfavourable influence, and checked activity in all departments. Considering, however, the lengthened prevalence of the stringency, and the small prospects of an immediate change, trade was considered to be in a remarkably healthy state. With money at 6½ per cent., and a steady efflux of the precious metals, the position of credit could scarcely be expected to have been so well maintained, and it was consequently, for the moment, considered a cause for general congratulation. The arrivals of gold from Australia were enormous, reaching upwards of £1,250,000, but only about £350,000 was purchased by the Bank. Contracts were again open for France, and since a demand existed not only to supply the necessities of the Bank of France and of the Credit Mobilier, but also to pay for silver in course of shipment to the East, the greater portion of immediate remittances would be absorbed. It was curious to watch the anxiety manifested to ascertain the destination of any quantity of gold received, and its delivery at the Bank, while the fact of its retention or despatch appeared to be recorded with the greatest minuteness. Money, which was rather easier towards the middle of the month, again became in active request, and the foreign stock and share settlements, which were in progress, with the payments of the 4th of the month, assisted to diminish the amount available in the open market. The rates of accommodation abroad were still high. At Hamburg the rate of discount was maintained at 7 per cent.; in Paris the demand was active at 6 per cent.; in Amsterdam the rate remained at 4 per cent., while in Belgium it was 3½. The prices of English stocks experienced some fluctuation. One and a-half per cent. was the range in Consols, a decline having taken place from 93¾ to 92¼. There was subsequently a reaction of about one-half per cent., but quotations continued unsettled, and a fresh decline seemed not improbable. The pressure for money, and the uncertainty of future prospects, appeared to stimulate speculative sales. Exchequer bills were seriously depressed, having touched 10*s*. discount, but they afterwards recovered to par. Foreign securities also exhibited a relapse. Turkish and Mexican were principally affected, the Turkish Six per Cents. particularly showing an adverse movement. The alleged failure of the project of the National Bank caused some disappointment, and holders who were interested realized. Mexican was weak, and Spanish showed heaviness, the dealers in any description being unprepared to support prices. Railway shares were heavy, and the best classes receded £3 to £4 per share. Quotations opened, at the commencement of the month, with firmness, and there was subsequently a general rise, traffic returns being satisfactory, and the public appearing as buyers. The alteration in the rate of discount by the Bank directors, however, soon caused a change, and with the increased tightness of money, quotations gradually gave way. The following were the failures during

APRIL.

Mr. A. Marks, shipbuilder, Sunderland.
General Wood-cutting Company, timber trade, London.
Mr. William Pitcher, shipbuilder, Northfleet.

The financial and mercantile events of May did not present features of interest. Favourable weather and receipts of specie from Australia and America assisted to keep things in a steady position, but the demand for money was yet sufficient to prevent any reduction in the rates of discount. The Bank having discontinued advances on stock, the speculators could not command the usual facilities, and many, consequently, closed outstanding accounts. Silver continued to be remitted to India and China in large quantities, the proportion taken for Hong Kong and Shanghae being very considerable, and hence gold was absorbed to make the necessary purchases. It was also believed that the directors of the Bank of France had placed contracts at the disposal of their brokers, in case supplies might be required. A few failures occurred, but none of them were considered important, or as in any way indicating a general lapse in credit. English stocks fluctuated about 1⅛ per cent.; and Consols, after descending to 92⅛, rallied to nearly the highest point. Occasional symptoms of ease in the money market were apparent, but they did not prove permanent, and hence the range noticeable. Purchases by the Government broker, who was re-investing the proceeds of the Exchequer bonds lately held by the Commissioners of Savings' Banks, tended to keep the value of the unfunded debt in a firm position. Foreign securities were dealt in to only a limited extent. From about 1 to 2 per cent. was the outside margin of fluctuation, and the dealers, having at first sold for a fall, subsequently operated for a rise. Railway shares declined from the highest prices of the month, but, compared with the opening, they closed about the same. The following failures occurred in

MAY.

Messrs. Barker and Co., timber trade, London.
Messrs. Thomas Biggs and Son, drug and wine merchants, London.
Messrs. Edward J. Hambro and Co., general merchants, London and Newcastle.
Messrs. M'Alpin and Nephew, warehousemen, London.
Mr. G. Dogherty, provision and corn trade, Liverpool.

Notwithstanding the public showed greater confidence in the general prospects of monetary and commercial affairs during June, there was much less activity than might have been supposed, from the fact of the Bank directors having reduced the rate of discount from 6½ to 6 per cent. In addition, the favourable advices of the crops, and the announcement that the authorities of the Bank of France had also lowered their terms for the negotiation of commercial paper from 6 to 5½ per cent., should have exercised a favourable influence, but the public did not appear prepared to promote speculation, especially while the drain to the East continued of so alarming a nature. The returns of the Banks of England and France steadily improved, but the former was in the more satisfactory position; the influx of Australian and American gold having, after providing for the requirements of the Continent, left a respectable surplus, which was purchased on account of the National Establishment. The rate of money at the Stock Exchange touched as high as 7 and 7½ per cent. on English securities, but it subsequently receded to about 6 per cent., with a moderate demand. At Paris the rate was 5½ to 6 per cent., and at Hamburg 7 per cent. The imports of the precious metals during the month amounted to about £4,387,000, and the exports to £5,330,000. The fluctuations in English securities were about 1 per cent., and the transactions were not very numerous. Although a little stimulus was afforded to prices, through the reduction in the rate of discount by the Bank of England and the Bank of France, the purchases subsequently diminished, and there was much stagnation in business. A heavy "bull" account continued open, which prevented any elasticity in prices, and the result was that a few sales kept the market in a quiescent state. The operations in foreign stocks were not extensive, and the only buyers were the Greek speculators, who entertained favourable views with respect to the future. Turkish and Russian exhibited the greatest advance. Spanish American securities presented little change, and Dutch continued steady. The railway share market exhibited some fluctuation, but prices on the average improved; the rise was about 30*s.* to £2 per share. The principal rise was in guaranteed securities, and the heavier descriptions, after advancing, experienced a partial relapse. The following failures occurred in

JUNE.

Messrs. Gotch and Sons, bankers, Kettering.
Messrs. Smith, Hilder, Smith, and Scrivens, bankers, Hastings.
Messrs. Evans, Hoare, and Co., Australian trade, London.
Mr. William Macintosh, manufacturer, Manchester.
Messrs. J. Langton and Co., ship-brokers, Liverpool.

The position of affairs in July was satisfactory, and the large influx of gold from Australia and America enabled the Bank directors to reduce the rate of discount to 5½ per cent. At the same time, the authorities of the Bank of France exhibited a disposition to relax their restrictive measures; and while they lessened the rate of interest on treasury bonds, they also reduced the rate for advances on public securities from 6 to 5½ per cent. These features would, under ordinary circumstances, have assisted an upward movement in prices, but the alarm occasioned by the spread of the insurrection in India, not alone checked a rise, but produced a contrary effect, and stimulated sales on the part of speculators, through which a fall of at least 1 per cent. occurred. Great anxiety was evinced to ascertain the probable result of the outbreak upon the financial resources of the East India Company; and until some information transpired, it was felt that the uneasiness lately apparent would continue. Everything in other respects was favourable to an amelioration of the money market—the accounts of the crops in England, throughout Europe, and America being of the most gratifying character, it was hoped that the rate of discount would recede to 5 per cent., or a fraction below that point. The variation in the English funds did not exceed 1¾ per cent., while the actual decline was about 1 per cent. This was to have been anticipated from the character of the Indian news, and the anxiety to ascertain the prospect of future accounts. The reduction in the Bank's rate of discount and in the increased ease in the money market would have, no doubt, caused a further improvement in the value of Consols, but a heavy "bull" account being open, and the advices by the Overland mail being doubtful, there was a succession of sales, which caused this depression. Although in foreign stocks the transactions were limited, the speculators continued to sell stock. Turkish were rather lower, and there was little expectation of a recovery while Greek failures took place, and it remained the medium for extensive gambling. Russian advanced about 2 per cent., and the quotation was firm at the improvement. Mexican and Peruvian were steady, with an average amount of business. In the railway share market there had been a tendency to decline, but prices were not so low as in the middle of the month, and quotations slightly recovered. East Indian shares dropped nearly 9 per cent., but the receipt of favourable intelligence would, it was thought, occasion a strong rebound. French shares had also been seriously affected, the state of business on the Bourse having caused a general depreciation. The annexed failures occurred in

JULY.

Mr. H. Schwabe, broker, Liverpool.
Mr. J. P. Giustiniani, merchant, London.

The state of mercantile and financial affairs in August was not unsatisfactory. Had the intelligence from India been of a more encouraging character, general circumstances in other respects would have produced greater animation in business. Large arrivals of gold from Australia and America, with the very favourable progress of the harvest, assisted to keep the money market in a more steady position; and although no reduction in the rate of discount had taken place, the general demand on occasions was less urgent. The influx of bullion placed the accounts of the Bank of England in a better position, but the purchases of silver for export to the East, with some operations in gold on account of the Bank of France, prevented any great amelioration in the terms of accommodation. A more than usual number of failures took place this month, but they occurred principally from over-speculation in the provinces, where credit proved to be less sound than in the metropolis. High rates of discount prevailed on the Continent, especially at Hamburg, the current quotations having at one moment reached 7 per cent., but it was now quoted 6 per cent. Operations in the English stock market caused a range of about $1\frac{3}{4}$ per cent., but there was a considerable recovery from the lowest price, so that on the average the decline was scarcely more than $\frac{3}{4}$ per cent. The greater part of the fluctuations were produced by the nature of the advices from India, but the increased abundance of money, with the less unfavourable state of the exchanges, caused greater confidence whenever the first effect of the telegraphic accounts subsided. A sudden rise on purchases by the public occurred about the middle of the month, but a reaction soon ensued, the speculators having again come forward and sold. India stock and bonds continued heavy, but the dealers had not speculated to any great extent. Bank stock and Exchequer bills presented no important variation. Foreign securities fluctuated from 1 to 3 per cent., and at the close showed rather increased firmness. The operators did not adventure largely, though late in the month they showed a greater disposition to purchase. In Turkish, the operations had been on the most extensive scale, but less speculative activity was now visible in this department. The Railway share market suffered to a serious extent, but a slight recovery took place in quotations. Some influence was exerted by the arrangement of the account, money having been in less demand than expected. The operators apparently sold largely for the fall, and being compelled to repurchase, there was a corresponding reaction in prices. The results of the half-yearly meetings, as far as they had transpired, were of a mixed character, but in many cases the management appeared to have improved, and higher dividends were in some cases declared. Great Westerns proved an especial exception, and from $64\frac{1}{4}$ declined to 51. Two or three failures occurred on the Stock Exchange, owing to losses sustained through the fluctuation in prices. The subjoined failures were announced in

AUGUST.

Messrs. Rutty, Hall, and Co., warehousemen, London.
Messrs. Carr Brothers and Co., coalowners, Newcastle.
Messrs. Milrose and Hussey, ironfounders, Dudley.
Messrs. T. Ashmore and Sons, drysalters, London.
Mr. T. H. Hayes, corn factor, Liverpool.
Messrs. Buchanan, Brown, and Co., produce brokers, Liverpool.
Messrs. Dumbell, Son, and Howard, bankers, Isle of Man.
Mr. W. Eassie, contractor, Gloucester.
Mr. T. Stewart, manufacturer, Stockport.
Messrs. Bruford, Dyer, and Co., African trade, Bristol.
Messrs. Jones and Moore, soap manufacturers, Bristol.

In September there was great stagnation in public affairs. Depression, unfortunately, was the order of the day, through the political position of India; and the anxiety evinced to ascertain the nature of the intelligence by each mail, showed the importance attached to the progress of events in that quarter of the globe. The result was, that business continued checked, uneasiness was created, and a general heaviness produced, which, without severely affecting monetary or mercantile interests, prevented activity. The value of money was well supported, and there was evidently no expectation of a decline, the efflux of specie to the Continent, with the advance in the rates of discount at Amsterdam, Hamburg, and Berlin to 6½ per cent., causing some parties, on the contrary, to believe that a prospect exists of a movement in the other direction. Notwithstanding large Australian remittances had been received, the drain to France and Hamburg excited some apprehension, particularly since the disturbances in the money circles of the United States, where pressure and panic prevailed, had caused shipments of gold to be suspended. Trade generally appeared to be scarcely so sound as formerly, and the failures in the provinces, although unimportant, indicated that weak firms could not resist the effects of the late maintenance of the rates for mercantile accommodation. The *minimum* of the Bank of England had not varied, while in Lombard Street the rate was fully 5½ per cent. The range in English stocks was about 1¼ per cent., and a decline of 1 per cent. was established. The fall appeared gradual, with an absence of animation in business, the advices from India continuing to exercise an unfavourable influence upon prices. An increased demand for money likewise tended to create an unsettled feeling. Exchequer bills were not supported, and as the East India Company were understood to be selling, the quotation had descended to 10*s*. discount. Foreign stocks were comparatively inactive, and a decline had occurred in several of the leading descriptions. Peruvian, Mexican, and Turkish were lower, the reaction having been mainly produced by speculative sales. Railway shares fluctuated to the extent of 2 or 4 per cent., and there was still a decline of 1 to 3 per cent. upon the quotations at the commencement of the month. The following were the failures in

SEPTEMBER.

Messrs. Warburton and Omershaw, silk trade, Manchester.
Messrs. H. F. Fardon and Co., soap trade, Bromsgrove.
Mr. A. Crosfield, merchant, Bristol.
Messrs. G. Wyld and Sons, distillers, Bristol.
Messrs. Perren and Co., merchants, Bristol.
Mr. W. Summerskill, silk trade, Manchester.
Mr. Lynd, oil trade, Leeds.
Messrs. Harrison, Watson, and Co., bankers, Hull.
Messrs. Taylor and Bright, corn merchants, Hull.

The state of financial and mercantile affairs in October was considerably influenced by the advance in the Bank rate of discount from 5½ to 8 per cent. The American crisis, with its attendant consequences, including the suspension of banks, and the failure of the principal firms, produced great stringency in the money market, while the raising of large sums by the East India Company assisted to increase the pressure. The value of money on the Continent further advanced, and the terms of the Bank of France had risen to 6½ per cent., which showed that the absorption of capital was general. It was satisfactory, however, to notice that, notwithstanding failures of magnitude in the country, London was comparatively free from embarrassment, and great confidence was felt in the ultimate adjustment of mercantile relations. Shipments of gold were made to America to the extent of about £750,000, and it was thought that a further amount, equal if it did not exceed that total, would be forwarded to meet balances and purchase produce and securities. To the East, by the mail of the 4th, silver to the extent of at least £1,000,000 would be despatched. The Continental rates of money varied 6½ per cent. at Paris, to 8½ per cent. at Hamburg, the effects of the crisis in the United States having been largely experienced in that city. With the Bank of England rate at 8 per cent., and the joint-stock banks in a position to allow 7 per cent. for deposits, no great activity in business could be expected. In English securities this month, the range was about 3¾ per cent., but the actual fall had not exceeded 1¼ per cent. It was to have been presumed that a decline would take place in prices, but the extreme fluctuation, after all, was not important, considering the vicissitudes through which the mercantile community had passed. Exchequer bills were heavy, at about 15*s.* to 10*s.* discount, but the great depreciation occurred in Indian bonds, which were quoted at 40 per cent. discount. Foreign stocks did not vary in any great degree, but the decline in the several descriptions was from 1 to 3 per cent., with moderate transactions. Brazilian, Buenos Ayres, and Dutch receded 2 per cent.; Russian, 2; and Mexican, 1 per cent. Railway shares were of course affected by the extraordinary position of business, sales of the various descriptions having been freely made; the fluctuation being from 2 to 5 per cent. The late pressure brought various descriptions to market, but quotations in a measure slightly recovered.

OCTOBER.

Messrs. J. and J. Beard, silk trade, Manchester.
Messrs. J. Monteith and Co., merchants and manufacturers, Glasgow.
Messrs. Macdonald and Co., merchants and manufacturers, London and Glasgow.
Messrs. Wallace and Co., merchants and manufacturers, Glasgow.
Mr. Hugh Fergusson, stuff merchant, Manchester and Bradford.
Messrs. Ross, Mitchell, and Co., Canadian merchants, London.
Mr. S. F. Stephens, bill-broker, London.
Messrs. Scarratt and Partington, Blackwell Hall, factors, London.
Messrs. Affleck and M'Kerrow, Scotch warehousemen, Manchester.
Mr. W. B. West, haberdasher, Manchester.
Mr. John Little, milliner, Manchester.
Mr. Edmund Whitehead, silk manufacturer, Middleton.
Messrs. Whan, M'Lean, and Co., manufacturers, Glasgow.
Messrs. Auld and Buchanan, merchants and shippers, Glasgow.
Mr. Jas. Condie, writer, Perth.
Messrs. Chas. Smith and Co., provision trade, Manchester.
The Liverpool Borough Bank, bankers, Liverpool.
Messrs. Thornton, Huggins, Ward, and Co., American trade, Huddersfield, etc.
Messrs. J. S. De Wolf and Co., shipowners, Liverpool.
Messrs. Gould and Davis, wine merchants, Liverpool.
Messrs. Robert Morrow, Son, and Garbutt, timber merchants, Liverpool.
Messrs. John Haly and Co., New York and Canadian trade, London.
Messrs. J. Jaffray and Co., ship and insurance brokers, London.
Mr. Caporn, lace dealer, Nottingham.

The course of financial and mercantile affairs during November was of the most chequered character. With a further advance in the rate of discount from 8 to 10 per cent., the occurrence of a crisis, and the suspension of the Bank Charter Act, it might readily be supposed the position of business was not satisfactory, while the number of failures which took place created general gloom and despondency. In this condition of things, the issue of the Government letter produced some benefit, but it did not immediately effect a full resuscitation, the panic having progressed too far to render its operation immediate. The consequences of the suspensions, dangerous as they were, did not prove so pernicious as might have been anticipated, though the character and importance of the houses showed the effect of continuous high rates of discount upon the commercial community. A steady restoration of confidence was expected, but it was nevertheless thought it would, under any circumstances, be slow and progressive, the shock given by the revulsion having created excessive apprehension. The speculators were not so active during this panic as they were in that of 1847, although unfounded rumours were from time to time circulated with respect to the position of houses whose credit remained undoubted. The influx of the precious metals and the advice of additional remittances from Australia would, it was felt, assist to ease the condition of the money market, the recovery in America being likewise calculated to facilitate the movement. At one period during the crisis, it was almost impossible to negotiate paper at all, the charge under the most favourable circumstances being 12 and 15 per cent. Within the last few days of the month greater readiness was manifested to discount, and the quotation by the principal brokers receded to a fraction below 10 per cent. The imports of the precious metals continued on a large scale, and sovereigns were rapidly returning from Scotland and Ireland, the run for gold in those localities having terminated. English securities fluctuated from 3 to 4 per cent., but they eventually showed a general recovery. Of course, great depression took place in the midst of the panic, the operations having been on the adverse side. From $90\frac{1}{2}$ they dropped to $87\frac{7}{8}$, but although the speculators exhibited a desire to carry out further transactions for the fall, the public became purchasers, and created a reaction, which terminated in a general advance. Exchequer bills were quoted 4 discount to par, and Bank stock 215 to 216. Foreign stocks varied to the extent of between 2 and 4 per cent. The operations, even at the most unfavourable period, did not produce a great effect, the speculators, through fears of a sudden change, not having the courage to enter into important transactions. Quotations at the close of the month were generally better, the influence of the latest purchases having proved in a degree beneficial. Railway shares, as might have been supposed, experienced a decline of from 4 to 5 per cent. The dealings showed that some speculative operations occurred through the despondency created by the panic, but the decline, after all, was not extensive, compared with the gravity of the events which had taken place. Joint-stock bank shares suffered from sales, and at one period of the month there was considerable depreciation. The disgraceful reports circulated with respect to the situation of these establishments, and the extent to which it was said the deposits had been trenched upon, caused temporary apprehension, but the alarm proved groundless. The following failures occurred in

NOVEMBER.

Messrs. R. Wilson, Hallett, and Co., merchants, Liverpool.
Messrs. Powles Brothers and Co., Spanish American trade, London.
*Messrs. Naylor, Vickers, and Co., iron and steel merchants, Sheffield, etc.
Messrs. W. Orr and Co., merchants, Liverpool.
Messrs. J. and A. Dennistoun, Cross, and Co., American bankers and exchange agents, London, etc.
The Western Bank of Scotland, bankers, Glasgow.
Messrs Bennoch, Twentyman, and Rigg, silk trade, London.
Messrs. Broadwood and Barclay, West India merchants, London.
Messrs. Hoge and Co., merchants, Liverpool.
*Messrs. Henry Dutilh and Co., merchants, Liverpool.
*Messrs. B. F. Babcock and Co., American trade, Liverpool, etc.
Messrs. Foot and Co., silk manufacturers, London.
*The City of Glasgow Bank, bankers, Glasgow.
Messrs. Sanderson, Sandeman, and Co., bill-brokers, London.
Messrs. Bruce, Wilkinson, and Co., bill-brokers, London.
Messrs. Wilson, Morgan, and Co., wholesale stationers, London.
Messrs. Fitch and Skeet, provision merchants, London.
*Messrs. T. B. Coddington and Co., iron merchants, Liverpool.
Messrs. Mackenzie, Ramsay, and Co., merchants, Dundee.
Messrs. Draper, Pietroni, and Co., Mediterranean trade, London.
Messrs. Stegman and Co., manufacturers, Nottingham.
Messrs. Bowman, Grinnell, and Co., American trade, Liverpool.
Messrs. R. Bainbridge and Co., American trade, London.
Messrs. Munro, Grant, and Co., timber merchants, Swansea.
Messrs. Jellicoe and Wix, Turkey merchants, London.
Messrs. José P. De Sá and Co., Brazilian trade, London.
Messrs. Bardgett and Picard, grain trade, London.
Messrs. F. C. Perry, iron trade, Wolverhampton.
Messrs. W. Riley and Sons, iron trade, Wolverhampton.
The Wolverhampton Iron Company, iron manufacturers, Wolverhampton.
Messrs. Solly Brothers, iron trade, Wolverhampton.
Messrs. Rose, Higgins, and Rose, iron trade, Wolverhampton.
*The Wolverhampton and Staffordshire Banking Company, bankers, Wolverhampton.
Messrs. Hoare, Buxton, and Co., North of Europe trade, London.
Messrs. Edwards and Matthie, produce brokers, London.
Messrs. Clayton and M'Keveringan, shipbuilders, Liverpool.
Messrs. E. Sieveking and Co., North of Europe trade, London.
Messrs. Allen, Smith, and Co., North of Europe trade, London.
Messrs. Svendson and Johnson, North of Europe trade, London.
Messrs. Gorrissen, Hüffel, and Co., American bankers and exchange brokers, London.
Messrs. Brocklesby and Wessels, corn importers, London.
*Messrs. J. R. Thomson and Co., Cape and Australian trade, London.
Messrs. Herman Sillem and Co., North of Europe trade, London.
Messrs. Carr, Josling, and Co., North of Europe trade, London.
Messrs. Alexander Hintz and Co., North of Europe trade, London.
Messrs. Peniston and Marshall, American provision merchants, London.
Mr. Jonathan Bottomley, spinner, Bradford.

Messrs. Godfrey, Pattison, and Co., merchants, Glasgow.
The Northumberland and Durham Bank, bankers, Newcastle-upon-Tyne.
Messrs. Rehder and Boldemann, North of Europe trade, London.
Messrs. T. Morris and Sons, iron trade, Wolverhampton.
Mr. A. Cruickshank, corn trade, Glasgow.
Mr. J. J. Wright, cotton broker, Glasgow.
Messrs. H. Hoffman and Co., German trade, London.
Mr. T. Mellidew, commission agent, London.

NOTE.—The firms marked with an asterisk have resumed business.

During December the progress of the crisis, and the numerous failures which followed, engrossed attention, and caused anxiety to be manifested to trace the results of the existing distress. It was feared early in the month that the pressure might be prolonged through the continuance of mercantile disasters and the disinclination of the banks to afford accommodation, especially as the appearance of the Government letter did not immediately produce the desired relief. Subsequently the influx of specie with the more favourable advices from America and Hamburg created greater confidence, and a partial recovery was then apparent. The Bank of France having twice reduced the rate of discount, till, at length, it descended to 6 per cent. for all classes of mercantile paper (with the prospect of being eventually down to 5 per cent.), the public appeared induced to believe that the favourable reaction had set in, and this was subsequently confirmed by the Bank of England reducing the rate from 10 to 8 per cent. The large arrivals of Australian and American gold, with remittances from Turkey, Russia, etc., rapidly augmented the stock of bullion, and the reserve of notes having reached a satisfactory position, the over issue of £2,000,000 was repaid to the proper department. The crisis was, therefore, considered to have terminated, and the opinion was entertained that, with the restrictions which would necessarily be placed upon business, the accumulation of capital could not fail to be more rapid than ever. The reduction of the rate of discount by the Bank of England, which took place upon the 24th, would, it was presumed, be followed by additional alterations on the favourable side, and some were sanguine enough to predict that, in the course of the next six or seven weeks, 5 or 6 per cent. would be the extreme quotation. Although the failures of the month included a number of important mercantile establishments, the extent of overtrading was fully developed by the nature of the balance-sheets exhibited, and notwithstanding dividends were, in many cases, better than was supposed, still the pernicious effects of late operations were made clearly visible. The panic feeling which pervaded the English stock market checked business, and sales produced a further speedy decline. When, however, money became easier, and a report was circulated that the Bank would shortly be enabled to reduce the rate, great firmness was apparent, and a steady rise occurred of $2\frac{1}{8}$ per cent., at which point quotations were supported. Exchequer bills fluctuated between 5 discount and 2 premium; and eventually stood at par. Bank and India stocks were firmer with a greater amount of business, and affairs in this department appeared altogether better. In foreign stocks there was increased activity, and the speculators purchased freely for the rise. The improvement was most apparent in Turkish, Mexican, and Russian, each of which recovered 2 to 3 per cent., although the operations were of rather a mixed character. Railway shares were more extensively operated in, and notwithstanding the items of traffic showed unfavourable results, purchasers of the various descriptions supported prices. The recovery, on the average, was from 4 to 5 per cent., with a satisfactory market, and the dealers seemed inclined to consider that quotations might yet go higher. French and Belgian shares also improved, though the purchases were not extensive. The subjoined were the failures in

DECEMBER.

Messrs. Hermann Cox and Co., cotton trade, London and Liverpool.
Messrs. Bischoff, Beeret, and Co., merchants, London.
Messrs. Mendes Da Costa and Co., West India trade, London.
Messrs. Keiser and Co., German trade, London.
Messrs. Barber, Rosenauer, and Co., general merchants, London.
Messrs. Hirsch, Strother, and Co., German trade, London.
Mr. G. C. Pim, corn trade, Belfast.
Mr. P. Magee, shipowner, Liverpool.
Messrs. Fredericksen, Clunie, and Co., corn trade, West Hartlepool.
Messrs. F. and A. Bovet, China trade, London.
Messrs. C. A. Jonas and Co., merchants, London.
Messrs. Sewells and Neck, North of Europe trade, London.
Messrs. A. Pelly and Co., North of Europe trade, London.
Messrs. Krell and Cohn, commission agents, London.
Messrs. Hadland and Co., warehousemen, London.
Messrs. Lichtenstein and Co., German trade, London.
Messrs. D. Couvella and Co., Greek trade, London.
Mr. W. R. Urain, iron and chain trade, Newcastle.
Messrs. Ogden and Fergusson, merchants, Newcastle.
Messrs. Heine, Semon, and Co., German bankers and exchange brokers, London.
Messrs. Weinholt, Wehner, and Co., East India and Australia trade, London.
Messrs. T. H. Elmenhorst and Co., German trade, London.
Messrs. Montoya, Saenz, and Co., Spanish American trade, London.
Messrs. Farley, Lavender, and Co., bankers, Worcester.
Mr. G. T. Ward, Smithfield banker, London.
Messrs. H. and M. Taldorph and Co., merchants, London.
Messrs. Rew, Prescott, and Co., North of Europe trade, London.
Messrs. R. Willey and Co., silk mercers, London.
Messrs. S. C. Lister and Co., worsted spinners, Halifax.
Messrs. Saalfield Brothers, German trade, London and Leeds.
Mr. E. Smith, woolstapler, London.
Messrs. W. Cheesbrough and Son, wool merchants, Bradford.
Messrs. W. Yewdall and Co., wool merchants, Rawdon, near Leeds.
Messrs. Jonathan Hills and Sons, bankers, Dartford and Gravesend.
Messrs. Powell and Sons, warehousemen, London.
Messrs. W. Dray and Co., agricultural implement makers, London.
Messrs. Klingenden Brothers, American trade, Liverpool.
Messrs. Charles Nicholson and Co., warehousemen, London.
Messrs. Lloyd Brothers, wholesale picture dealers and exporters, London.

Mr. W. Reid, wholesale grocer, Edinburgh.

Messrs. W. C. Haigh, woollen trade, Bradford.

Messrs. Hans Marcher and Co., Dutch trade, Hull.

Messrs. Stevenson, Viemehren, and Scott, merchants, Newcastle.

Messrs. Greenslade and Co., corn trade, Bristol.

FOREIGN AND AMERICAN FAILURES IN 1857.

Jan. Messrs. Tenney and Co., Boston, United States, carpet dealers.
„ Mr. Peirce Butler, Philadelphia, stock speculator.
Feb. Messrs. Kelly, Townsend, and Co., New York, merchants.
„ Mr. C. B. Fessenden, Boston, merchant.
„ Messrs. Wetherell Brothers, Boston, merchants.
Mar. Messrs. Dodge, Bacon, and Co., London and Newark, United States, merchants and patentees, subsequently resumed.
„ Messrs. Green and Co., bankers, Paris.
„ Messrs. Alexander Frear and Co., New York.
„ Messrs. Blashfield and Co., New York.
April. Mr. Bettman, New York.
„ Messrs. Valie and Co., Paris and Havre, bankers.
May. Messrs. C. Thurneyssen, Paris, banker and stock dealer.
„ Messrs. Whitney, Fenno, and Co., Boston, United States, importers.
„ Messrs. Chapman, Lord, and Hale, Boston, United States, importers.
„ Messrs. Shaw, Sampson, and Bramhall, Boston, United States, importers.
July. M. Giustiniani, Constantinople, merchant.
Oct. M. Nessin Soldal, Marseilles, merchants.
„ Mr. Boskowitz, Pesth, commission merchant.
„ Messrs. Malanotti and Co., Vienna, merchants.
„ Messrs. Balabio and Co., Milan, bankers.

The failures announced to have taken place in the United States, during the crisis in the months of September and October, were said to exceed nine hundred, but the most important establishments appear in the subjoined list. It should be added that many of these establishments ultimately resumed payment:—

Oct. Messrs. E. C. Bates and Co., Boston, merchants.
„ Messrs. Whiting and Hinds, Boston, merchants.
„ Messrs. Brewster and Co., Rochester, bankers.
„ Mr. Joseph Fairweather, St. John's, New Brunswick.
„ The Ohio Life and Trust Company.
„ Messrs. De Laney, Iselin, and Clark, New York, foreign bill brokers.
„ Mr. E. S. Munroe, stock-broker.
„ Mr. John Thompson, bank agent and share dealer.
„ Messrs. Tuttle, Cutting, and Co., New York, grain and flour trade.
„ Messrs. Bates, Griffin, and Livermore, New York, dealers in grain.
„ Messrs. L. W. Kirby and Co., Warren (Pen.), dry goods jobbers.
„ Messrs. Stillman, Allen, and Co., Boston, machinist.
„ Messrs. Saroni and Goodheim, Boston, clothiers.
„ Messrs. Beebe and Co., New York, bankers and bullion dealers.
„ Messrs. J. H. Prentice and Co., New York, hat and fur dealers.
„ Messrs. Adams and Buckingham, New York, flour dealers.
„ Messrs. Breese, Kneeland, and Co., Jersey, car and locomotive manufacturers.
„ The Mechanics' Banking Association, New York.
„ The Ontario County Bank.
„ The Bank of Orleans.
„ Messrs. Hatch and Langdon, Cincinnati.
„ The Rhode Island Central Bank.
„ Messrs. Allen and Son, Providence, Rhode Island, cloth printers.
„ Mr. H. Allen, Providence, Rhode Island.
„ Messrs. Reeves, Abbott, and Co., Pennsylvania, iron manufacturers.
„ Messrs. Reeves, Birch, and Co., Pennsylvania.

Oct. Messrs. Hudson, Robertson, and Fulliam, New York, importers.
" Mr. F. G. Swan, New York, warehouseman.
" Mr. Stephen Colwell, Philadelphia, iron merchant.
" Bank of New Jersey.
" Messrs. M. J. Bell and Co., New York, money dealers.
" Messrs. L. and O. Kirby, New York, importers.
" Messrs. Fitzhugh and Littlejohn, Oswego, carriers.
" Messrs. Grant, Sayles, and Co., produce receivers.
" Messrs. C. H. Stone and Co., Boston.
" Messrs. Conant, Dodge, and Co., New York.
" Mr. Philip Allen Torrs, Providence.
" Mr. Z. Allen, Providence.
" The Huguenot Bank, New Paltz.
" Messrs. Pierre and Houlse, Washington.
" Georgetown Bank of Commerce.
" Messrs. Nasmith and Co., New York.
" Messrs. Sword, Watling, and Co., Philadelphia.
" The Citizen Bank, Cincinnati.
" Messrs. Collard and Hughes, Cincinnati.
" Messrs. Davis, Suydam, Dubois, and Co., Rondoubt, Nova Scotia.
" Messrs. Carpenter and Co., New York.
" Messrs. Ward and Nash, Louisville.
" Messrs. Clark, Dodge, and Co., New York, bankers.
" Messrs. Pittits and Co., New York, importers of hardware and whale oil.
" Messrs. P. Chonteau, jun., and Co., New York, merchants.
" Messrs. Hutchinson, Ziffana, and Co., New York, warehousemen.
" Messrs. Bowen, M'Hamee, and Co., New York, silk importers.
" Messrs. Levy and Co., Philadelphia, importers of manufactured goods.
" Messrs. Lawrence, Stone, and Co., Boston, extensive factory owners.
" The Bank of Pennsylvania.
" The Bank of Baltimore.
" The Girard Bank, Philadelphia.
" Messrs. Bangs Brothers and Co., New York, trade sale auctioneers.
" Messrs. Spencer and Porter, New York, grocers.
" Messrs. L. J. Leny and Co., Philadelphia, dry goods trade.
" Messrs. Smith, Murphy, and Co., Philadelphia, dry goods trade.
" Messrs. C. Hallowell and Co., Philadelphia, dry goods trade.
" Messrs. Fassett and Co., Philadelphia, dry goods dealers.
" Messrs. Hieskell, Hoskin, and Co., Philadelphia, dry goods merchants.
" Messrs. Tennent and Derrickson, Philadelphia, dry goods dealers.
" Messrs. Caleb, Cope, and Co., Philadelphia, dry goods trade.
" Messrs. Persse and Brooks, Philadelphia, paper manufacturers.
" Mr. James Carter, Galena, merchant.
" Messrs. Masterman and Co., New York.
" The Union Straw Works, Foxboro'.
" Messrs. Tyler and Wild, Rochester, shoe dealers.
" Messrs. C. H. Mills and Co., Boston, domestic goods trade.
" The Ellenville Glass Company.
" Messrs. A. S. Lippincott, and Co., Philadelphia.
" Messrs. French, Sisson, and Co., Providence, oil merchants, etc.
" Messrs. J. Woolley, Indianopolis, banker.
" The Miami Valley Bank, Dayton.
" Messrs. Cyrus W. Field, and Co., New York, paper dealers.
" Messrs. J. Farnham and Co., Philadelphia, domestic goods trade
" Messrs. Hocker, Lea, and Co., Philadelphia, dry goods trade.
" Messrs. T. P. Remington, Philadelphia, dry goods trade.
" Messrs. Deal, Millington, and Burt, Philadelphia, dry goods jobbers.
" Messrs. Emmans, Danforth, and Scudder, Boston.
" Messrs. Mason and Co., Taunton, Massachusetts, machine and locomotive manufacturers.
" The Montour Iron Company, Danville.
" The Rock River Bank, Wisconsin.
" The Bank of Elgin, Illinois.
" The Whittenton Cotton Manufactory, Taunton, Massachusetts.

Oct. Messrs. E. J. Tinkham and Co., Chicago.
" Messrs. Darby and Barksworth, St. Louis.
" Messrs. John J. Anderson and Co., St. Louis.
" The Western Bank, Springfield, Massachusetts.
" The Ware Bank, Hampton Falls, New Hampshire.
" The Bank of Central New York, Utica.
" The Exchange Bank, Murfreesborough.
" Bank of Clairborne, Tazewell.
" Bank of Lawrenceburg, Lawrenceburg.
" Bank of Jefferson, Dandridge.
" Bank of Memphis, Memphis.
" Northern Bank of Tennessee, Clarksville.
" The River Bank, Memphis.
" Messrs. Kennet, Dix, and Co., New Orleans.
" The Northern Bank of Mississippi, New Orleans.
" Messrs. Vorhill, Greggs, and Co., New Orleans.
" Mr. S. Fothingham, jun., and Co., Boston.
" Messrs. J. W. Clark, and Co., Boston.
" Messrs. Sweetzer, Gerkin, and Co., Boston.
" Messrs. Chickering and Co., Boston.
" Messrs. Lawrence, Stone, and Co., Boston.
" Mr. J. A. Lowell, Boston.
" Mr. Benjamin Howard, Boston.
" Messrs. Richardson, Kendall, and Co., Boston.
" Messrs. P. Jones, Parson, Cutter, and Co., Boston.
" Messrs. Moses Pond and Co., Boston, stave dealers.
" Messrs. R. N. Wardall, and Co., Philadelphia, agents.
" Messrs. White, Stevens, and Co., Philadelphia.
" Messrs. Bowen, Ely, and M'Cornell, New York.
" Messrs. Gage, Sloam, and Slater, New York.
" Messrs. L. Bauer and Co., New York.
" Messrs. Livingston and Ballard, New York.
" Messrs. Cheteau, Harrison, and Valle, St. Louis.
" Messrs. James H. Lucas and Co., St. Louis.
" Mr. R. H. Brett, Toronto.
" The Exchange Bank, Tennessee.
" The Shelbyville Bank, Tennessee.
" The Lawrenceburg Bank, Tennessee.
" The Trenton Bank, Tennessee.
" Messrs. Dunbarry, Drake, and Co., Cincinnati.
" Messrs. Culner, Hutchings, and Co., Louisville, bankers.
" Messrs. Smidt and Co., Louisville, bankers.
" The Cumberland Coal Company.
" Messrs. Powell, Ramsdell, and Co., Newburg.
" The Powell Bank, Newburg.
" The Bank of Belleville, Illinois.
" Messrs. Moore, Hallenbush, and Co., Quinsay, bankers.
" The Charter Oak Mercantile and Exchange Bank, Hartford.
" Messrs. Yelverton and Walker, New York.
" Messrs. Latner and Co., New York.
" Messrs. Sathen and Church, New York.
" The Farmers' and Citizens' Bank, Williamsburgh.
" The Fox River Bank, Wisconsin.
" Messrs. Hutchinson, Tiffany, and Co., Newburgh.
" Messrs. Powell, Ramsdell, and Co., Newburgh.
" Messrs. Swift, Ransom, and Co., brokers and agents of a Chicago bank.
" The Hudson River Bank, Hudson.
" The Farmers' Bank, Hudson.
" The Old Saratoga Bank.
" The Stark Bank of Vermont.
" The Bass River Bank, Massachusetts.
" The Bank of Leonardsville.
" The Bridgport City Bank, Connecticut.
" The People's Bank, Milwaukie.

Oct. The Badger State Bank, Wisconsin.
,, The Exchange Bank, Bangor, Maine.
,, Hallowell Bank, Hallowell, Maine.
,, Hancock Bank, Ellsworth, Maine.
,, Shipbuilders' Bank, Rocksland, Maine.
,, Maritime Bank, Bangor, Maine.
,, Ellsworth Bank, Ellsworth, Maine.
,, China Bank, China, Maine.
,, Central Bank, Hallowell, Maine.
,, Sandford Bank, Sandford, Maine.
,, Cochituate Bank, Boston, Massachusetts.
,, The Grocers' Bank, Boston, Massachusetts.
,, The Manufacturers' Bank, Georgetown, Massachusetts.
,, The Exeter Bank, Exeter, New Hampshire.
,, The Lancaster Bank, Lancaster, New Hampshire.
,, The Stark Bank, Bennington, Vermont.
,, The South Royalton Bank, South Royalton, Vermont.
,, The Danby Bank, Danby, Vermont.
,, The Eastern Bank, West Killingly, Connecticut.
,, Wooster Bank, Danbury, Connecticut.
,, Woodbury Bank, Woodbury, Connecticut.
,, Hartford County Bank, Hartford, Connecticut.
,, Colchester Bank, Colchester, Connecticut.
,, Messrs. Harper Brothers, New York.
,, Messrs. Wood and Grant, New York.
,, Messrs. Hoppock and Greenwood, New York.
,, Messrs. Connelly and Adams, New York.
,, Mr. G. S. Hilman, New York.
,, Messrs. Buckley and Moore, New York.
,, Messrs. Baptist and White, New York.
,, Messrs. Brummell and Roysters, New York.
,, Messrs. Dykers and Alstyne.
,, Messrs. Hall, Dana, and Co., New York.
,, Messrs. Ludlum, Leggett, and Co., New York.
,, Messrs. L. C. Wilson and Co., New York.
,, Messrs. Cahart, Bacon, and Co., New York.
,, Messrs. Kitcham, Montrose, and Co., New York.
,, The New York and Erie Railroad Company (bills protested).
,, The Michigan Central Railway Company (on its floating debt).
,, Mr. J. A. Genin, New York.
,, Mr. H. A. Coit, New York.
,, Messrs. Leymorn and Co., New York.
,, Messrs. Corlies and Co., New York.
,, Messrs. Corning and Co., New York.
,, Messrs. Fenton, Lee, and Co., New York.
,, Messrs. Willetts and Co., New York.
,, Messrs. Francis Skinner and Co., Boston.
,, Messrs. George, T., and W. Plyman, and Co., Boston.
,, Messrs. Fale, Nash, French, and Co., Boston.
,, Messrs. Oakley and Hawkins, New Orleans.
,, The Albion Cotton Manufacturing Company, Baltimore.
,, Messrs. Dean, King, and Co., St. Louis.
,, Messrs. A. J. Maccrary and Co., St. Louis.
,, Messrs. Collins, Kirby, and Co., St. Louis.
,, Messrs. Fowle, Snowdon, and Co., Alexandria.
,, Messrs. E. F. Whittemore and Co., Toronto.
,, The Bank of South Carolina, South Carolina.
,, Messrs. Hadsworth and Co., Chicago.
,, The Cayuga Steam Furnace Company, Cleveland.
,, Messrs. Harkness and Stead, Providence.
Nov. Messrs. John Munro and Co., Paris, merchants.
,, Messrs. Gallerkamp, Brothers, Amsterdam, merchants.
,, Messrs. Guimáraes and Co., Paris, merchants.
,, Messrs. Bourbon, Du Buit, and Co., Paris, merchants.

Nov. Messrs. Ponson, Phillippe, and Vibert, Paris, merchants.
,, Messrs. Ullberg and Cremer, Hamburg, Swedish trade.
,, Messrs. F. Bloss and Scomburgk, Hamburg, merchants.
,, Messrs. Sassenberg and Meyer, Bremen, merchants.
,, Mr. H. Overbeck, Bremen, merchant.
,, Messrs. Topuz and Co., Smyrna, merchants.

The failures announced to have taken place during the month of November, in the United States, are comprised in the following list :—

Nov. The New Bedford Bank.
,, The Hartford Bank.
,, The Fairhaven Bank.
,, The Augusta Bank.
,, The Worcester Bank.
,, The Portland Bank.
,, The Bank of Manchester.
,, The Petersburg Bank.
,, The Lawrence Bank.
,, The Methuen Bank.
,, The Bank of Bangor.
,, The Trenton Bank.
,, The Brooklyn Bank.
,, The Albany Bank.
,, The Pacific Mills Corporation, Lawrence.
,, The Louisville Central Bank, New Orleans.
,, Messrs. J. L. Johnson and Co., New Orleans.
,, The Bridgeport Bank, Connecticut.
,, The Nashville Bank.
,, The Missouri Iron Works, Wheeling, Virginia.
,, Messrs. Pringle, Cook, and Lunhart, Brownsville, Pennsylvania.
,, Messrs. Harper, Brothers, New York, publishers.
,, Messrs. J. H. Colton and Co., New York, publishers.
,, Messrs. H. Cowperthwaite and Co., New York, publishers.
,, Messrs. John P. Jewett and Co., New York, publishers.
,, Mr. J. S. Redfield, New York, publisher.
,, Mr. P. S. Cozzens, New York, publisher.
,, Messrs. Miller, Orton, and Co., New York, publishers.
,, Mr. Richard Marsh, New York, publisher.
,, Messrs. J. M. Emerson and Co., New York, publishers.
,, Messrs. Miller and Curtis, New York, publishers.
,, Messrs. Bangs, Brother, and Co., New York, publishers.
,, Messrs. G. P. Putnam and Co., New York, publishers.
,, Messrs. Sanford and Swords, New York, publishers.
,, Messrs. H. W. Derby and Co., New York, publishers.
,, Messrs. Fowlers and Wells, New York, publishers.
,, Messrs. Robb, Hallett and Co., New York and New Orleans, exchange brokers.
,, Messrs. Lord, Warren, Evans, and Co., New York.
,, Messrs. Ross, Mitchell, and Co., Toronto.
,, The Dunneil Manufacturing Company, Providence.
,, Messrs. Winslow, Lanier and Co., New York.
,, Messrs. Frost and Forest, New York.
,, Messrs. S. and T. Laurence, New York.
,, Messrs. Sampson, Baldwin, and Co., New York.
,, Messrs. W. Greenhow and Co., Boston.
,, Mr. J. M. Holden, Boston.
,, Mr. George Forsaith, Boston.
,, Messrs. Keith and Thornton, Boston.
,, Messrs. J. W. Carter and Co., Boston.
,, Mr. W. P. B. Brooks, Boston.
,, Messrs. Tesson and Dangem, St. Louis.
,, Messrs. J. K. Doherty and Co., New Orleans.
,, Messrs. Raiqueland Co., New Orleans.
,, The Jeffersonville Branch of the State Bank of Indiana, St. Louisville.

Nov. Messrs. Warrell, Coates, and Co., Philadelphia, importers.
„ Messrs. John Hooper, Son, and Co., Philadelphia, dry goods dealers.
„ Messrs. C. and A. Ives, Detroit, bankers.
Dec. Messrs. Thomas and Martin, Philadelphia.
„ The Central Bank of Fredericton.
„ Messrs. Sather and Church, San Francisco.
„ Messrs. Fiske, Sather, and Church, Sacramento.
„ Messrs. Crocker, Sturgess, and Goodall, Boston.

Before the final termination of the American panic, the banks of New York, Pennsylvania, and other States suspended specie payments, but subsequently, and after general arrangement and reorganization, resumed.

THE COMMERCIAL CRISIS IN HAMBURG.

LIST OF FAILURES IN HAMBURG IN BANKRUPTCY (HANDELSGERICHT), FROM NOVEMBER, 1857, TO 3RD MAY, 1858.

Amsberg, Julius.
Ballheimer, Johannes.
Behn, Conrad, and Co.
Berndes and Sandtmann.
Blass, Ferdinand, and Schomburgk.
Bock, C.
Bonne, Anton.
Brandt, Ad., and Co.
Brauer, Heinr Sohn.
Bull, N. R.
Dittmer, F.
Drost, William, and Co.
Christen, Martin Heinr, Theodor.
Christie and Co.
Cordes and Gronemeyer.
Eckstein, F. J., and Co.
Elfeld, Ernst.
Feddersen, Aug., and Co.
Fuchs and Co.
Gaedechens and Boreremann.
Glückstadt, R. M.
Goldschmidt, Samuel Hertz.
Heidsick, H.
Hertz, Levin, and Söhne.
Hesse, H. F.
Heymann, H. J.
Hodges, H. R.
Hoffman, J. M., St.
Horwitz, B. and N.
Kall, J.
Kleissen, George.
Langmack, J. P.
Morrison, William.
Müller, A. (Hempel and Co.)
Müller, M. and J.
Neuber, Carl.
Peters, J. and W.
Pincus, J.
Plinck, J. B.
Plomer, Hunt and Co.
Prahl and Wübbe.
Roscon, Berend, Jr.
Ruden, Gustave.
Rüppel, Ant. Max.
Sanne, J. L.
Schneider, H. J. H. T.
Schwark, B. H. F.
Steen, D. F.
Stockfisch and Co.
Tauer, Carl.
Teves, Th.
Thiel, E. and Co.
Woldum, J. B.
Zahn and Vivie.
Zeiss, Gebrüder.

LIST OF FAILURES IN HAMBURG, UNDER INSPECTION (UNTER ADMINISTRATOREN), FROM NOVEMBER, 1857, TO 3RD MAY, 1858.

(Firms marked ** *paid in full;* those marked * *have compounded.*)

Ed. Ahlers and Sohn.
D. Arnow.
Bach and Raspe.
Heinrich Bachmann.
Bachof and Overweg.
Barbeck and Wall.
Georg Behre and Co.
P. J. H. Berger Naecf.
Bing, Gebrüder, and Co.
Laue, Bödeker Nachf.
Hugo, Bohres, and Co.
Fr. Brandt and Lüttjshann.**
Carl Busch.
Busse and Halske.
M. F. Claren.
John Christiansen and Co.
Sigvardt, Colberg, and Co.
Custer, Brunswig, and Co.
E. O. Deneken and Co.
Dependorf and Co.
Dittmer and Koch.
Droop and Co.

F. Duncker and Co.
Jos. Edelheim Nachf.
J. J. H. Eschels.
A. and G. Eulert.
M. C. Faurschon and Co.
Edu Ferber.
Fesser and Vialhaack.
Gentz and Schlutz.
Rudolph Goedell.
Goerne and Co.
Moritz D. Goldschmidt.
Eduard Grimm.
J. F. Grosz.
Gross and Lantzius.
Ferd Guntrum.*
H. Hald.
Hamer, Carl, jr.
Siegfried Hannover.
P. N. Hansen and Johannsen.
Carl F. L. Harder.
P. E. Hartenfels Sohne.
H. Hartwig and Wilckens.
A. Heilbut.
Holbech and Sommer.
Holterman and Co.
Huber and Haupt.
Hughes, Stobart, and Co.
J. C. Jacobj and Sohn.
F. H. Jacobsen and Co.
C. T. Jacoby and Co.
H. Heinr Jansen.**
Jepsen and Reimer.
John and Seeger.
C. J. Johns Sohne.
Herrm Kellmann.
Klein and Co.
G. Kohrs and Co.
C. F. Kruckenberg.
Lafargue and Hülssen.
Chs. Lavy and Co.
Von Leesen and Co.
Louis E. Levy.
Phillip Joachim Levy.
Jul Lohmer and Uhde.
L. F. Lorent am Ende and Co.
G. H. Lutze and Co.
Brenny Mainzer.
Mankiervicz and Frahm.
H. E. I. C. Marquardt.
Ludwig Friedrich Mathies.
Meyer, Gebrüder, and Co.
Chr. Nic Meyer.
Ferd. F. Meyer.
Hermann Meyer.
P. A. Milberg.
Mohrmann and Herrnbrodt.
Gebrüder Möller.
Otto Möller.
Joseph Munk.
J. Nathan & Co., M. Fortlouis Nachf.*
A. Ohlendorf and Co.
L. Peine.
Pietzcker and Sohn.
E. F. Pinckernelle.
Predohl and Co.
Reichmann and Wilckens.
Reimarus Nachf.
J. D. F. Rieck.
Riecke, Behrens, and Co.
Fr. Reidel and Co.
Rohde and Wolff.
Fr. Robbelen.
C. H. Rover.
Gebrüder Ruben.
Saalfeld, Gebrüder.
H. W. Shellr.
C. Shemmel.
Schenck and Co.**
J. C. Schlüter and Co.**
Theodor Schmidt.
J. A. Schmidt Sohne.
Schnell and Meyer.
T. B. Schnitler.
W. Scholvien.**
L. Schoop and Co.
Anthon Schröder and Co.
Chr. Matth. Schröder and Co.
Joh. Ant. Schröder and Co.
Oct. Rud. Schröeder and Eiffe.
F. O. Schuback.
Emil Schubart.
Schulte and Schemmann.
Seeler, Wolff, and Co.
Wilhelm Seitz.
Sieveking and Co.
J. B. Spengel.
J. A. Spetzler and Co.
H. and P. Spiro.
Rud. Stoffert.**
Strube and Niebuhr.
Suse and Co.
C. Trobitius and Co.
Ullberg and Cramer.
Vogt and Schmidt.
Wagener and Enet.
D. S. Warburg Wittwe and Sohne.
Conrad Warnecke.**
Jens Weile.
Ad. Weiszflog and Cordes.
Theod. Werlich.**
Gustav. Wieler.
H. F. Wilcken.**
Thed. Wille.
Willwater and Co.
H. Theoder Winckler.
Winckler and Nagel.
P. C. Winterhoff and Piper.
Witte and Kümmel.**
W. S. Wolff and Karpeles.
Wolfson and David.
M. L. Wurzburg and Co.
A. Zacharias and Wendt.
Eduard Zodich and Co.
Ziel Balzer and Co.
Th. Zimmermann.

SUSPENSIONS IN 1858.

Although the effects of the crisis may be said to have passed at this date, credit was in an unsettled state and many failures took place.

Jan. Messrs. Richard H. Whitfield and Co., London, West India and general commission merchants.
,, Messrs Thomas Callander and Co., Glasgow, hide and leather factors.
,, Messrs. Mitchell, Miller, and Ogilvie, Glasgow, wholesale warehousemen.
,, Mr. John Ewan, Dundee, manufacturer and export merchant.
,, Messrs. William Clapperton and Co., Glasgow, general merchants.
,, Messrs. James Bannatyne and Son, Limerick, general merchants.
,, Messrs. Arthur and Co., Glasgow, warehousemen.
,, Messrs. Joseph Bainbridge and Son, Rotherhithe, timber merchants.
,, Messrs. Bishop and Gissing, London, wholesale stationers.
,, Messrs. Charles Walton and Son, ship owners, and ship and insurance brokers.
,, Mr. A. Duclos, Marseilles, South American hide trade.
,, Messrs. Matthew Plummer and Co., Newcastle, general merchants.
,, Messrs. Clarke, Plummer, and Co., Newcastle, flax merchants.
,, Messrs. Smith, Russell, and Co., Louisville, United States, soap and candle manufacturers.
,, Messrs. James Shaw and Co., Huddersfield, woollen manufacturers.
Feb. Mr. James Odier, Paris, banker.
,, Mr. Charles Hartmann, Hanover, banker.
,, Messrs. William Gilmour and Co., Glasgow, woollen trade.
,, Messrs. Oak and Snow, Blandford, bankers.
,, Messrs. T. Wilson and Co., Baltimore, United States, produce merchants.
,, Mr. Edward Gwyer, Bristol, African trade.
,, Messrs. Prudel and Co., Marseilles, South American trade.
Mar. Mr. A. R. Lafone, Liverpool, River Plate trade.
,, Mr. H. P. Maples, London, ship and insurance broker.
April. Messrs. Maitland, Ewing, and Co., London, East India and China trade.*
,, Messrs. Newcomen, Noble, and Co., London, East India trade.
,, Sir G. E. Hodgkinson and Co., London, ship owners and brokers.
,, Robert Browne and Co., London, Australian trade.
May. Mr. Lafone, Buenos Ayres, merchant.
,, Messrs. Zimmerman, Frazier, and Co., Buenos Ayres, merchants.
,, Messrs. Pedro, Peyke, and Co., Bahia, German trade.
,, Messrs. Felix Rignon and Co., Turin, silk manufacturers.
,, Messrs. Groothoff and Schultz, Hamburg, merchants.
June. Messrs. Rawson, Sons, and Co., London, East India and China trade.
,, Mr. William Patterson, Bristol, shipbuilder.
,, Messrs. Fenn, Kemm, and Fenn, London, wholesale grocery trade.
,, Messrs. Bristow, Warren, and Harrison, London, wholesale grocery trade.
,, Messrs. Skeen and Freeman, London, timber brokers.
July. M. Domingo Ferreira, Rio Janeiro, merchant.
,, Messrs. Larpent, Saunders, and Co., Calcutta, merchants.
,, Messrs. Astley, Wilson, and Co., Rio Janeiro, merchants.
,, Mr. Charles Snewin, London, timber merchant.
,, Mr. E. A. Skeen, London, timber merchant.
,, Messrs. Astley, Williams, and Co., Liverpool merchants.
,, Messrs. Cox, Brothers, Liverpool, iron merchants.
,, Mr. W. Jeffries, Dudley, iron manufacturer.
,, Messrs. Seaman and Keen, London, silk trade.
Aug. Messrs. Hyde, Hodge, and Co., London, mahogany importers.
,, Messrs. C. Overweg and Co., Hamburg, merchants.
,, Messrs. Allan, Deffell, and Co., Calcutta, merchants.
,, Messrs. Langlois and Co., Calcutta, merchants.
,, Messrs. Portelli, Schembri, and Co., London, Mediterranean trade.

* This firm soon paid in full and resumed active operations.

Aug. Mr. Duncan Gibb, Liverpool, merchant and shipowner.
„ Messrs. Rudolf, Jung, and Co., Paris, silk trade.
Sept. Messrs. J. Carmichael and Co., Liverpool, Honduras and general merchants.
„ Messrs. Archibald Montgomery and Co., London, Honduras and Australian trade.
„ Messrs. J. Plowes and Co., London, Brazilian trade.
„ Messrs. Pardoe, Hoomans, and Pardoe, Kidderminster, carpet manufacturers.
„ Mr. J. G. Moffatt, Birmingham, brassfounder.
„ Messrs. George Chambers and Co., London, pin and needle manufacturers.
„ Messrs. Ascoli, Hartwig, and Co., Manchester, merchants.
„ Mr. Forll, Milan, produce trade.
Oct. M. Duhant, Lille, sugar speculator.
„ Messrs. Wm. Arnold and Sons, London, hemp manufacturers.
Nov. Messrs. Plowes, Son, and Co., Rio Janeiro, merchants.
„ Messrs. W. J. Grey and Son, Newcastle, coalfitters.
„ Messrs. Pickworth and Walker, Sheffield, builders.
„ Messrs. Cowan and Bigg, London and Newcastle, ship and insurance brokers.
„ Messrs. James Hyde and Co., Honduras, merchants.
„ Messrs. James Davies and Son, London, boot and shoe manufacturers.
Dec. Messrs. Hicks and Gadsden, London, American merchants.
„ Messrs. Metcalfe and Co., West Ham, distillers.

THE BANK INDEMNITY BILL.

SPEECH OF SIR G. C. LEWIS, THE CHANCELLOR OF THE EXCHEQUER, ON THE INTRODUCTION OF THE BANK INDEMNITY BILL, IN THE HOUSE OF COMMONS, DECEMBER 4, 1857.

The CHANCELLOR of the EXCHEQUER said,—I rise for the purpose of bringing under the attention of the House the question which, as has been stated in the extract from the gracious Speech from the Throne, has led to the assembling of Parliament before the usual time. In order that the committee may fully understand the nature of the act of her Majesty's Government which has induced the directors of the Bank of England to infringe the provisions of the statute, and thus rendered it obligatory upon me to ask the House to give its consent to a Bill of Indemnity, it will be my duty to call their attention to the existing state of the law. Sir Robert Peel, in the course of his long political career, induced the Legislature to agree to two important Acts relating to the currency, which were in point of time separated from each other by a quarter of a century. The first of these Acts, as is well known to the committee, was passed in 1819. It restored the currency to its proper metallic basis, put an end to the state of things existing during the war—a state of suspension of specie payments, and established the convertibility of a bank-note. After that Act had been passed, the power of the Bank of England to issue notes was unlimited in point of extent. They were only subject to the condition of being payable in gold on demand; and the power of establishing country banks was also unlimited, subject to the privilege which then existed in favour of the Bank of England from the practice in reference to six partners. These banks when established could create any quantity of bank-notes. The same power existed both in Ireland and Scotland. Any person could under that law establish a bank and issue notes payable to bearer on demand, provided only that those notes were paid in specie upon presentation at the bank. That was the state of the law from 1819 to 1844, the date of Sir R. Peel's second important Act on the currency. That second Act, he stated to the House, was to be considered as a complement of the first; he intended it to secure, in fact, that convertibility of the note which was made legal only by the Act of 1819; and the provisions which he introduced were to the following effect:—In the first place, he enabled the Bank of England to issue notes on securities to the extent of £14,000,000, together with any further sum which might accrue from the lapsed circulation of the English country banks, as to which some detailed provisions were introduced into the Act. An Order in Council under those provisions was passed about two years ago, authorizing the Bank to add to that portion of its issue which rests upon securities the sum of £475,000. Therefore at present the issues which the Bank of England can make upon securities—and it is not allowed to make any issues except upon Government securities—is £14,475,000. In addition, the law allows the Bank of England to make an amount of issue equivalent to the bullion which may be in its coffers. The Act of 1844 further introduced a provision which seems to me, although in some manner connected with them, yet to be essentially different in its character from the other provisions—I mean that for dividing the Bank of England into two departments, the issue and the banking department. Gentlemen who are familiar with the mode in which the weekly returns of the bank are published, are aware that the assets and liabilities are always classed under those two divisions. I shall presently have to call the attention of the committee to the practical working of that system with reference to the measure to which I am about to ask their assent. I should add that with regard to English country banks the Act of 1844 contained this important provision—it prohibited any bank which might after the passing of that Act be established in England from issuing bank-notes. It limited, also, the note issues of the existing banks to the amount which was then in circulation. It contained no provision requiring them to issue notes on securities, nor did it allow them to exceed the fixed limit by issuing notes against bullion. Therefore, under the Act of 1844, the condition of the English country banks is this :—No new country banks can issue bank-notes at all, and the country banks existing at the time of the passing of that Act are limited in their issue to the circulation which they then possessed. In the following year was passed an Act with respect to Scotland and Ireland, which was intended as a part of the same measure. Its effect was to limit the existing banks of Ireland and Scotland to their then issues as in the case of the country banks of England. It did not require that those issues should be guarded by securities of any kind, but it authorized them to make issues in excess of their certified circulation against a deposit of bullion. Therefore, in Ireland and Scotland there is a certain amount of paper which is limited under the Act, in addition to which the banks issue a certain portion against sovereigns. The committee ought to bear in mind the practical result of this compound system upon the circulation as it exists at the present moment. On the 11th of April last the notes of the Bank of England in the hands of the public amounted to £19,752,045. At the same date the note circulation of

The English country banks was	£6,920,000
The Scotch banks	3,832,000
The Bank of Ireland	3,557,000
And other banks in Ireland	3,596,000
Total................................	£17,905,000

Adding that amount to the circulation of the Bank of England, we may take the paper circulation of the United Kingdom to be in round numbers about £38,000,000. Of that £38,000,000 the only portion which is strictly covered by bullion is that which is issued by the Bank of England in addition to the £14,475,000 which they issue upon securities. The whole of the circulation of the English country banks and the chief part of the circulation of the Bank of Ireland and the Scotch and Irish country banks is, so far as the law is concerned, not represented by bullion. The security which the law takes for that portion of the circulation is the legal convertibility of the note—the liability of the banker to pay in specie when the note is presented. This is the existing state of the circulation under the different Acts which regulate our currency ; and it is essential to bear in mind, with reference to the solvency of the Bank of England and the importance of keeping a large bullion reserve in that central establishment, the considerable amount of the provincial issue. When Sir R. Peel introduced the measure of 1844, he stated that one of the main objects which he had in view was to provide security against the excessive issue of paper, and thereby to guard against the recurrence from time to time of those commercial panics under which the country had at different intervals suffered previously to 1844. At the same time he stated that he did not propound his measure as a panacea, or as an infallible guard against such panics, because he said that the issue of bank paper was only one of their causes, and he recognized the fact that commercial panics might arise from causes which did not lie within the scope of his legislation. But in 1847, after the commercial crisis of that year, when it became requisite for the Government to authorize the Bank, if the necessity should arise, to exceed the limits fixed by the Act of 1844, Sir R. Peel, in the discussion which arose, delivered his opinions more fully on the subject than he had previously done, and he gave a definition, corrected by experience, of his own views as to the objects of the Act of 1844, which, as it contains an authentic exposition of his opinions, I trust the committee will permit me to read. Sir R. Peel upon the 3rd of December, 1847, said :—

"I say, then, that the bill of 1844 had a triple object. Its first object was that in which I admit it has failed, viz., to prevent, by early and gradual means, severe and sudden contraction, and the panic and confusion inseparable from it; but the bill had two other objects, of at least equal importance—the one to maintain and guarantee the convertibility of the paper currency into gold, the other to prevent the difficulties which arise at all times from undue speculation being aggravated by the abuse of paper credit in the form of promissory notes. In these two objects my belief is that the bill has completely succeeded. My belief is that you have had a guarantee for the maintenance of the principle of convertibility such as you never had before; my belief also is, that whatever difficulties you are now suffering, from a combination of various causes, those difficulties would have been greatly aggravated if you had not wisely taken the precaution of checking the unlimited issue of the notes of the Bank of England, of joint-stock banks, and private banks."

That was Sir Robert Peel's definition of the Act after he had become aware of the events of 1847. I may also call attention to the remarks that my right hon. friend near me (Sir C. Wood), who then held the office which I now hold, made upon the measure then brought before the House, as showing that in his view the Act was never considered as a specific against commercial panics. He said :—

"A complaint has been made that this Act has not preserved us from commercial convulsion. I think that those who expected such an effect from it much miscalculated the motives by which persons engaged in commercial transactions are actuated. Certainly, for one, I never held out any such expectations to the country. I stated most distinctly that I did not contemplate that such would be the effect of the Act. . . . . . The same effects precisely might have occurred under a metallic circulation as have occurred under our mixed currency, and the advocates of the Act of 1844 never professed that its operation would produce any other result than that which might have happened under a metallic circulation."

I think that this reference will satisfy the committee that the authors of the Act of 1844 did not contemplate its proving a panacea or a complete guarantee against the occurrence of commercial crises, which are produced by a variety of causes, upon which no legislation, and certainly no legislation confined to the currency, could operate. I will now briefly advert to the peculiar crisis of 1847. It was described by my right hon. friend, in moving for a committee on the subject, as having been caused by the absorption of available capital for the purchase of corn and the construction of railways acting upon a very unsound state of commercial affairs. It arose from internal causes, from an internal demand for a large and extraordinary amount of capital, and had no connection, as in the case which we are now considering, with our foreign trade. In 1847, as in the present year, her Majesty's Government, considering the urgency of the commercial crisis, came to the conclusion to issue a letter to the directors of the Bank, authorizing them in certain contingencies to exceed the limits fixed by the law. The cause of issuing that letter was stated to be the want of confidence in the commercial classes which then existed, and its object was also defined to be to put an end to the present alarm. Although the cause of the panic in 1847 was different from that of the panic of this year, yet the object aimed at by the issuing of the Government letter in those two years was identically the same. The Bank directors, however, did not find themselves under the necessity of overstepping the law of 1847. Confidence was re-established by the issuing of the letter. Its influence upon the commercial classes was immediate, and the Bank directors were enabled to navigate their ship through their difficulties without exceeding the limits of the law. Nevertheless, Parliament was assembled before the usual time. The attention of Parliament was called to the matter by her Majesty's Speech, but as no Indemnity Bill was required the direct assent of Parliament to a bill was not obtained. Nevertheless, a committee of inquiry into the Act was appointed upon the motion of my right hon. friend. That motion gave rise to a debate, but I believe that it was unanimously assented to by the House. Sir Robert Peel, who was the author of the Act, expressed himself in terms the most unambiguous, in approbation of the sanction which had been given by the Government to the suspension for a time of the limits fixed by the Act. I shall quote some of his words on that occasion, and I think it is of importance that the committee should bear them in mind, inasmuch as the distinctness with which he spoke proves that he, who was certainly better qualified than any other person either in or out of this House, to speak with authority upon the policy of the Act, was satisfied of the propriety of the step then taken. He said:—

"But when there occurs a state of panic—a state which cannot be foreseen or provided against by law—which cannot be reasoned with—the Government must assume a power to prevent the consequences which may occur. There is the necessity for a discretion which I think was properly exercised in the present instance. It was better to authorize a violation of the law than to run the risk of the consequences which might have ensued if no intervention on the part of the Government had taken place. . . . . Sir, I think that the Government were justified in issuing that letter. I think that, having issued it, they acted with the strictest regard to constitutional principle in forthwith summoning Parliament."

That was the judgment, if I may say so, which Sir Robert Peel pronounced upon that question. But the matter did not rest merely with the appointment of a committee, and the approbation by individual members of Parliament of the step taken by the Government. On the 17th of February, 1848, Mr. Herries, who was then a member of this House, moved two resolutions in a committee of the whole House, the first of which confirmed, in a most distinct manner, the policy of the step. It was in these terms:—

"That, looking to the state of distress which has for some time prevailed among the commercial classes, and to the general feeling of distrust and alarm by which the embarrassments of trade have been aggravated, it is the opinion of this House that her Majesty's ministers were justified, during the recess of Parliament, in recommending to the Bank of England (for the purpose of restoring confidence) a course of proceeding at variance with the restrictions imposed by the Act of the 7th and 8th of Victoria, cap. 32.

"That this House will resolve itself into a committee upon the said Act."

That resolution was put and carried without a division, from which it is manifest that the House, upon a consideration of the subject, explicitly approved the act of the Government. But a select committee was appointed to inquire into the Act of Parliament, and several most able and competent witnesses gave evidence before them. In their first report, dated the 8th of June, 1848, they said:—

"The issue of that letter was no doubt an extraordinary exercise of power on the part of the Government, but the House has decided that in the peculiar circumstances of the period they were justified in taking that step."

Therefore the committee—perhaps, it is a solecism to say so—adopted the resolution of the House; at all events, the committee, after great consideration and inquiry, referred to the previous resolution of the House as embodying their own opinion. Having thus, I fear at too great length, but with a minuteness which I thought the importance of the subject required, stated what took place in 1847, I will now, with the permission of the House, call their attention to the events which are now immediately under our consideration. With regard to the commercial crisis into which we have lately entered, and from which I think we can scarcely be said yet to have emerged, it may be observed that no symptom of an alarming or even of a threatening character (if we compare the bullion and reserve notes of 1856 and 1857) manifested itself until about the 18th of October. From that day a perceptible deterioration began. The bullion in the Bank fell between the 10th and the 18th of November from £10,110,000 to £6,484,000; while the reserve notes fell during that period from about £4,500,000 to about £1,400,000. But before that time it cannot be said that anything occurred which could create uneasiness in the course of affairs at the Bank. The immediate cause of the crisis was, as must be within the knowledge of most of the hon. gentlemen who are now present, unconnected either with the management of the Bank or with any issues of paper money in this country. It grew out of the derangement of the American trade, supervening upon some previous inconveniences created by the Indian revolt, the shipment of bullion, the disturbance of the Indian trade, and, to a certain extent, speculation upon the Continent. However, it may be said that the almost exclusive cause of the commercial distress was the derangement of the American trade. For some months previously to November the foreign drain of bullion was accompanied by an adverse state of the exchanges and a high rate of discount at the Bank. On the 2nd of April it was at 6½ per cent., on the 18th of June it was 6 per cent., on the 16th of July 5½ per cent., on the 8th of October it was raised to 6 per cent., on the 12th of October to 7 per cent., on the 19th of October to 8 per cent., on the 5th of November to 9 per cent., and on the 9th of November to 10 per cent. These figures show that high rates of discount had prevailed, the natural consequence of an adverse state of the exchanges; but the operations of the Bank and of the discount houses in this country had been successful in reversing that state of things, and in producing a favourable state of exchanges before the issue of the Government Letter. At that time the foreign drain of gold had been checked, and the exchanges were favourable. Well, although some uneasiness might still exist as to the state of the Bank, nothing at that moment seemed to indicate any immediate probability of the Government being called upon to exercise any extraordinary powers; and as the right hon. gentleman (Mr. Disraeli) referred yesterday to the state of our trade during the month of October as a circumstance which ought to have induced her Majesty's Government to hesitate before they recommended the prorogation of Parliament until the end of December, I would beg leave to remind him that even if the state of trade had been more unfavourable than it actually was, there was no symptom of commercial distress which would have rendered it necessary for the Government to resort to any extraordinary act. It was only in the event of their being called upon by any urgent and unforeseen necessity to authorize a suspension of the law that it was likely to be necessary that, upon merely commercial grounds, they should advise her Majesty immediately to assemble Parliament. In the event of that improbable contingency—as we thought it—arising, a constitutional power existed, which could be exercised at any moment, of summoning Parliament by proclamation to meet in fourteen days. That is an explanation of the charge of want of caution and foresight on the part of the Government brought forward by the right hon. gentleman. Well, the first event which precipitated matters in the late commercial crisis was the failure of the Borough Bank of Liverpool. I wish to call the attention of the committee to the fact that that was not a joint-stock bank, and that it was not a bank of issue. I believe there is not a single bank of issue in that town, but the Borough Bank was not a bank of issue; it was a corporation of a general character, and the amount of its liabilities was very considerable. As far, therefore, as that establishment was concerned, the consequences produced by its failure cannot in any way be traced to an excess in the issue of bank paper. That failure took place about the 27th of October, and the next event of importance was the failure of Messrs. Dennistoun on the 7th of November. That house was extensively engaged in the American trade. I believe that it was a house conducted with prudence, and possessed of large capital, but it was suddenly forced to suspend payment by the misfortunes arising out of the convulsions that had taken place in America, which were attended with much greater distress than had occurred in this country, and accompanied by a general suspension of cash payments. The house of Messrs. Dennistoun failed, as I have said, on the 7th of November for upwards of £2,000,000. About two days afterwards, I believe on the 9th of November, a large joint-stock bank in Scotland failed—the Western Bank, which had numerous branches. Within a day or two afterwards another large joint-stock bank—the City of Glasgow Bank—also failed. The joint circulation of these banks was about £800,000, of which about one-half was the certified circulation they were authorized to make upon securities, and the remaining half upon bullion. Their deposits were about £9,000,000, showing that it was not their note circulation, but the magnitude of their other transactions and liabilities which led to their unfortunate failure. (Hear, hear.) I may mention with regard to the Western Bank of Scotland that there had been no run at all upon it previous to its failure, and that when its doors were closed not a single note had been presented for payment out of the usual course, thus showing that it was not the discredit of the notes of the bank, but a failure to meet its large engagements of other descriptions which led to its unfortunate fall. The first sign of pressure more immediately affecting the mercantile classes in London, and approaching nearer to the Bank of England, was the failure of the house of Messrs. Sanderson and Co., bill-brokers, which took place on the 10th of November. Such an event could not occur without exercising a serious influence upon the credit of other similar establishments in this city. The publication of the Bank returns for the week ending Wednesday the 4th of November, which showed a considerable reduction of the Bank reserve, was a circumstance which, combined with the other facts I have mentioned, could not but create uneasiness and alarm, and call attention to the diminishing resources of the Bank. The alarm increased in London, and I think I cannot adduce a clearer or more compendious proof of the existence and extent of that alarm than by referring to the increase of the bankers' deposits at the Bank of England in the week from the 4th to the 12th of November. It may at first sight seem somewhat paradoxical to the committee to refer to the increase of bankers' deposits in the Bank of England as a proof of alarm (Hear, hear); but it must be borne in mind that the Bank of England is the bank in which the London and other bankers keep their deposits, and therefore, when they increase those deposits suddenly and extensively, it is a proof that they are guarding themselves against a run upon their funds, that they expect demands from their customers, and are, in fact, if I may so say, hoarding their resources. The increase of bankers' deposits at the Bank of England is, therefore, a recognized test of the existence of alarm. In the week from November 4 to November 12 these deposits increased by no less than £1,970,000. But there was another decisive proof of the existence of alarm in this city—namely, the general cessation of all discounting except by the Bank of England. At that time it was scarcely possible to obtain an advance, even upon bills which were undoubtedly good, either from bill-brokers or bankers; their operations were for the moment universally suspended, and the Bank of England was the only establishment in London at which discounts could take place. That circumstance may be referred to as a tolerably conclusive proof of the general prevalence of alarm. I may mention that at that time a single discounting house paid in one day £800,000, £700,000 to depositors, and £100,000 to discounters—a sum, I apprehend, almost unparalleled in operations of that kind. I will now call the attention of the committee to the state of the Bank on the day before the recent Government Letter was issued, and to its state on the day previous to the issue of a similar letter in 1847, and I think the committee will see that with regard to every point indicative of the strength of its resources, the position of the Bank in 1857 was less favourable than its position in 1847. The inference to which I wish to lead the committee is, that if the Government were justified by the opinion of Parliament, of the public, and of competent judges, in issuing the letter in 1847, then *a fortiori* they were justified this year in issuing a similar letter. The following is a statement of the position of the Bank of England in October, 1847, and November, 1857:—

	October 23, 1847.	November 11, 1857.
	£	£
Bullion	8,313,000	7,171,000
Reserve of notes in Banking Department .	1,547,000	957,000
Reserve of coin	447,000	504,000
Deposits (Private)	8,580,000	12,935,000
Private Securities	19,467,000	26,113,000

Therefore I think the committee will see that whatever were the grounds for the interference of the Government in 1847, the grounds were still stronger in the present year. There is another most material point of difference between the two cases, which perhaps even still more justifies the interference of the Government in the present year. In 1847 the turning point had been reached before the issue of the letter—the state of things had begun to improve. It is true that the issue of the letter, in the opinion of competent judges, accelerated that process; but the improvement had already manifested itself, and the worst had been seen before it was issued. Now, in the present year the direct contrary is the fact. We had not reached the worst when the letter was issued; and, whatever opinion may be formed as to the beneficial effects of the letter, there is no doubt that if it had not been written, if this authority had not been given, we should have seen a worse state of things after the 12th of November than before it. (Hear, hear.) Now, the fact which seems to me material as constituting the difference between the two years is the progress of the London discounts at the Bank in 1847 and 1857. On the 23rd of October, 1847, the London discounts at the Bank were £7,762,000. The amount remained stationary till about the 4th of November, when it fell to £7,500,000; and a few days later it sank even lower. In the present year, on the other hand, those discounts amounted on the 9th of November to £7,256,000. They went on regularly increasing, till on the 27th of November they reached £12,022,000, and on the 1st of December they still stood at £11,961,000. This shows that the sum under discount at the Bank of England was greater at the end of that period than at its commencement, whereas in the corresponding period of 1847 that amount had considerably diminished. Therefore, I infer from these facts that, whereas the worst of the crisis had been overcome before the letter of 1847 was issued, the worst had not been overcome when the letter of this year was issued. (Hear.) I trust, Sir, that I have established to the satisfaction of the committee that the course taken by the Government in 1847 was maturely considered and deliberately approved by Parliament—that the authors and promoters of the measure approved this proceeding, and that it constituted a precedent by which subsequent Governments could with propriety regulate their conduct. In the present case the causes of the panic were, no doubt, different, but the result was the same; the circumstances with which the Government had to deal were similar, the urgency was greater, and whatever reason justified the resort to extraordinary measures in the year 1847, will be found to apply, with greater cogency and greater conclusiveness, to the case of the present year. But the existence of a precedent has not grown out of mere accident from the occurrence of circumstances which had not been foreseen when the original measure was prepared. I have the evidence of Mr. Cotton, the Governor of the Bank at the time when the bill of 1844 was under consideration, and that gentleman stated to the committees of both Houses of Parliament, that the precise contingency which occurred in 1847, as also in the present year, was foreseen by Sir Robert Peel, who appeared to have contemplated a mode of meeting it exactly identical with that which has been adopted in both years. That seems to me an important part of the case; and I will therefore read an extract on this point from Mr. Cotton's evidence before the committee of this House, which sat in 1848. Similar evidence was given by him before the committee of the House of Lords. Mr. Cotton was thus examined:—

"In the case of a panic, do you think that a contraction of the circulation is any cure for it?—I think it is, inasmuch as it increases the value of money and induces people who have hoarded money to part with it.

"Do you think, even in the case of an internal panic, the proper mode of cure is a contraction of the circulation?—The proper mode of cure is a contraction of the circulation; whether it is worth while to submit to the remedy is another question. I can easily conceive that there may be monetary crises when it is necessary for the Government to interfere, and to do as was done in October.

"You contemplate, as I understand you, that under the operation of the Act circumstances may arise which will call for the interference of Government?—I did contemplate it, and I believe that I shall not be guilty of a breach of confidence if I state that the subject was discussed very fully with the First Lord of the Treasury and the Chancellor of the Exchequer during the consideration of the provisions of the Bill; and when the subject was again pressed on him, Sir Robert Peel expressed his opinion to me in these terms:—'My confidence is unshaken, that we are taking all the precautions which legislation can prudently take against the recurrence of a monetary crisis. It may occur in spite of our precautions, and if it does, and if it be necessary to assume a grave responsibility for the purpose of meeting it, I dare say men will be found willing to assume such a responsibility.'

"It was contemplated that circumstances might occur which would render it necessary to suspend the limit of the Act of 1844. Can you state what were the sorts of events that you had then in contemplation?—I should say events similar in effect to the events of 1825, when there was entire discredit of all the country bank circulation, and a panic which brought down, not merely an immense number of bankers, but others who were men of undoubted property."

I think that passage ought to convince this committee that, in referring to the proceedings of 1847, as constituting a Parliamentary precedent of the utmost authority, we have every security that the subject was fully considered both in its character and its consequences, and that legislation was left in its present state upon mature deliberation. The right hon. gentleman (Mr. Disraeli) put a question last night which he had a perfect right to ask, and to which he is entitled to have an answer. It was:—"Whether the issue of this letter was called for by the Bank directors, and whether pressure was applied to the Government in order to obtain it?" Now, Sir, I state distinctly that the issue of that letter was the spontaneous act of her Majesty's Government (Hear, hear); that they proceeded upon their own deliberate and conscientious view of the circumstances of the case (Hear, hear); that they were not urged by deputations, by the representations of mercantile bodies, nor were they implored to interfere by the directors of the Bank of England. No men invested with a public and official responsibility could have acted in a more honourable and conscientious manner than the directors of that corporation. They showed themselves throughout ready to carry into effect all the obligations which the law cast upon them, and they merely said that they were willing to act upon any advice which the Government might tender to them, but that they were prepared to obey the provisions of the law. Sir, the Government were guided by their view of the facts which I have disclosed to the committee—facts the corresponding circumstances of which were deliberately reviewed by the Parliament of 1847, and were considered by them as constituting a legitimate ground of interference. These facts existed in a far more intense degree in November, 1857, than in October, 1847. Our justification therefore is, that we acted in consequence, not of the pressure of the mercantile body, but in consequence of the pressure of facts, and that we should have been guilty of a grave dereliction of duty if, in so grave and urgent a state of circumstances, we had hesitated to take upon ourselves the responsibility which was assumed by the Government. (Cheers.) I do not at all seek to screen myself under the existence of mercantile pressure. (Hear, hear.) If the act which we have done is right, let the House indemnify the Bank directors, let them indemnify those who advised them. If it is wrong, let them refuse that indemnity. We do not seek to shelter ourselves under any pretence of having been coerced into the act. (Hear, hear.) The committee is in possession of a numerical return from the Bank, fully exhibiting the state of things on each day from the 11th of November, and it is unnecessary that I should trouble them with a minute analysis of these figures. I will only say that, unlike what occurred in the year 1847, the Bank directors immediately found themselves compelled to act upon the authority given by the Government, and to make issues in excess of the limit fixed by the Act of Parliament. They first issued £1,000,000 of notes from the issue to the banking department, placing against that sum an equal amount of securities; and on a subsequent day they made a similar transfer. The first issue took place on the 12th of November, and the second on the subsequent day. The effect of that, technically, is that the Bank of England has infringed the provisions of the Act to the extent of £2,000,000 sterling; but the committee should bear in mind that, although owing to the peculiar manner in which the accounts of the Bank are kept the law is technically infringed to that amount, at no time has the Bank issued so much as £2,000,000 to the public. It has merely passed £2,000,000 from one department of the establishment to another. I have before me a statement of the amount in excess which was from day to day issued to the public. On the 13th of November, for instance, the excess of notes in the hands of the public was £186,000, on the following day £622,000, and so on until on the 20th of November it reached the maximum of £928,000. If the committee will deduct the amount of the reserve of notes in the banking department from the £2,000,000 transferred from the issue department they will see that the difference is the figures which I have given. That is the real amount which has reached the public. On the 21st of November that amount was £617,000; on the 23rd, £397,000; on the 24th, £317,000; on the 25th it had fallen to £81,000; on the three following days there was an increase, but on the 30th of November it had fallen to £15,000; and on the 1st of December the note reserve was more than £2,000,000; therefore, strictly speaking, there was then no issue in excess of the limit. I need not, however, remind the committee that the operations of the Bank being now divided into two departments it was impossible to work the banking department without an issue of notes from the issue department, and that that transfer was technically a violation of the law; but the Committee should bear in mind that the real and essential violation of the Act has been confined to the amounts which I have just read. It may be asked what would have been the result if the letter of the Government had not been issued. That is a matter about which we can only form conjectures. The actual result of issuing that letter must, I think, be admitted to have been favourable. It diminished alarm and restored confidence (Hear, hear); and it did not in the smallest degree endanger the convertibility of the note (Cheers), because a favourable turn of the exchanges had taken place, and there was at that time no fear of a foreign drain of gold. Therefore I cannot think that at that moment the convertibility of the note was endangered by the step which was authorized by the Government. If that authority had not been given, it is certain that the only measure to which the Bank could have resorted for its own protection would have been the immediate and total cessation of discounts. (Hear, hear, from the Opposition benches.) Indeed, could they have foreseen the impossibility of the Government taking such a step, they might, perhaps, have diminished their discounts a day earlier. At all events, they must at the time the letter was issued, have entirely stopped their discounts. Whether so decided a step as that on the part of the Bank of England might not have led to the extension of discredit among establishments of a similar nature engaged in similar operations of banking, I leave it to the committee to determine. (Opposition cheers.) It is not for me to give any positive opinions upon such a subject, but if such discredit had arisen the committee must see what would have been the effect upon the banking department of the Bank of England by leading to the withdrawal of deposits. (Hear, hear.) Now, as by law the two departments of that Bank are segregated the one from the other, and the assets of the issue cannot be applied to the relief of the banking department, the Bank of England might have found itself in the painful position of postponing payment to depositors (Hear, hear), or of applying a portion of the assets of the issue department—because its assets would undoubtedly have been sufficient

(Hear, hear)—to the discharge of the liabilities of the banking department. I think it probable that such a contingency might not have arisen; nevertheless, it was within the range of possibility. Events, as I have learned from painful experience—events march very rapidly during a commercial crisis. (Hear, hear.) Persons who are loaded with the responsibility of action in moments of that sort must be prepared for them, and I leave it to the committee to judge whether, with such a contingency before them, her Majesty's Government would have been justified in withholding the letter which they sent to the Bank. (Hear, hear.) I am fully conscious of the gravity of the step which the Government took. The deliberate infraction of a fundamental law governing the currency is, as this committee will readily believe, a step which no persons responsible for their acts would lightly take. (Hear, hear.) I do not in the smallest degree seek to diminish the importance of the proceeding resorted to by her Majesty's Ministers; at the same time, let it be remembered that the actual excess hitherto committed by the Bank has been limited to the amount which I have stated, and that the effect of the authority which was given was simply to enlarge the amount of the issues upon securities. Now, if the entire circulation of this country had been metallic, I could have understood any objection which might have been entertained to an authority being given to a corporation like the Bank of England to issue £2,000,000 of bank-notes; but, inasmuch as they were authorized by the Act of 1844 to issue £14,000,000 upon securities, and as that Act contained a provision for further increasing that amount, in accordance with which it has by an Order in Council been increased to £14,500,000, the only effect of the authority given by the Government was in a guarded manner, under the close observation of the Government and of Parliament, and to meet a momentary emergency, to increase the issues of that description. No new principle was introduced by that authority. It has been said that the Government have authorized the Bank to commit an act equivalent to a debasement of the currency, that what we have done is almost equal to permitting a repudiation of contracts, and that the sanctity of property was invaded by this additional issue of bank-notes. I entirely dispute the correctness of that view. (Hear, hear.) If any doubt had existed as to the instantaneous convertibility of the additional notes issued under our authority, there would have been some ground for the charge which has been made against the Government; but, inasmuch as the additional issue by the Bank was not sufficient to affect the value of the notes, and each note was convertible into gold upon demand, it cannot be said with justice that the Government have depreciated the currency, authorized the repudiation of contracts, or struck at the security and sanctity of property. (Hear, hear.) I wish I had it in my power to say that this commercial crisis was definitively terminated, and that we could look forward with entire confidence to the Bank not having occasion to avail itself of the authority which we have given to it. I think it highly probable that the directors may not have occasion again to exceed the statutory limit; but, at the same time, I cannot assure the House that no such contingency can be expected to occur. Perhaps, in some respects, it would have been more convenient to Parliament if it could have looked back upon a completed commercial crisis before it was called upon to indemnify the Bank directors. The Government thought, however, that it was their duty, having authorized an infraction of the law, at once to lay the matter before Parliament, and hence the celerity with which the Legislature has been assembled. The pressure still continuing, and the wounds caused by the crisis being yet unhealed, it will be necessary for me to propose in the Bill a prospective power to the Bank to exceed the limits of the charter for a period of 28 days after the meeting of Parliament which will occur immediately subsequent to the recess. (Hear.) I have taken that as a convenient period with a view to the revision of Parliament, if revision should be necessary, and I trust that if the House should be willing to pass the Indemnity Bill it will not refuse to allow me to insert a clause to the effect I have stated. (Hear, hear.) I shall also ask the House in the course of the evening to agree to the reappointment of the committee on the Bank Acts which sat last session. That committee examined many important witnesses, but its inquiries were not completed, and at the end of the session it presented the following report to the House:—

"Your committee have agreed to report to the House the evidence which they have received; and, as their investigation of the subject referred to them is still incomplete, they beg leave to recommend that a select committee be appointed to resume the inquiry in the ensuing session of Parliament."

We should, as a matter of course, have acted upon that recommendation and have moved the reappointment of the committee, but it has been said that what has occurred during the last few months has weakened the reasons for that course. It seems to me, on the contrary, that the commercial crisis and the act of the Government have strengthened rather than weakened the reasons for the reappointment of the committee. (Hear, hear.) It must be obvious that if the House is desirous of reconsidering any of the details of the Act of 1844, it cannot do so with greater advantage than after a full inquiry before a select committee, and after the examination of competent witnesses upon all the different branches of the subject. The general portion of the subject was fully investigated last session, but so much of it as relates to the country banks of England and the banks of Scotland and Ireland was not entered upon, and I think it must be obvious, after the events of the last few months, that there is a great deal in connection both with the country banks of England, and the banks of Scotland and Ireland, which well deserves careful investigation. (Hear, hear.) Moreover, I confess I doubt whether it would be prudent, when men are smarting under the painful embarrassments of a commercial crisis, when that crisis is still incomplete, when its consequences have not yet been fully developed, to select that particular moment for revising your currency laws. Whether the proposal was made by the Government or by independent members, I cannot think that we should deliberate upon a most difficult subject, a subject which has perplexed the minds of some of the ablest men whom this country has produced, under circumstances likely to lead to a safe and beneficial conclusion. Therefore, as I should have moved the reappointment of the committee of last year under ordinary circumstances, it seems to me that the extraordinary circumstances under which we are now assembled render such a step still more expedient. (Hear, hear.) By what I have said the House I think must be led to the inference that the commercial crisis has not arisen from any abuses of the currency—from any excessive issues of bank paper in this country. (Hear, hear.) Whatever its causes may be, they stand, it seems, wholly aloof from the proper and direct policy of the Act of 1844. (Hear, hear.) Let us first take the Bank of England, the most important part of this subject. Can it be said with truth that the directors of the Bank of England have shown any want of foresight, any want of prudence, any want of sagacity, or any want of firmness during the past year ? I confess it appears to me, having had necessary occasion to watch their proceedings with vigilance, that they have not failed in any of the duties incumbent upon them. (Hear, hear.) Let us compare, as a test of excessive issues on their part, their notes in the hands of the public with their private securities during the year 1856-7. The following is a statement of their notes with the public, including Bank post-bills, at various periods of 1856 :—

October	4	£21,885,000	October	25	£21,412,000
,,	11	£21,501,000	November	1	£21,483,000
,,	18	£22,141,000	,,	8	£21,149,000

For the present year, when as we are told commercial discredit has been caused by excessive issues of Bank paper, the returns are as follows :—

October	3	£20,825,000	October	31	£21,184,000
,,	10	£20,863,000	November	4	£21,080,000
,,	17	£21,052,000	,,	11	£21,036,000
,,	24	£20,586,000			

So that, in point of fact, the amount of notes in the hands of the public has been actually less this year than in 1856. Let us now compare with the issues of notes and Bank post-bills the private securities of the Bank at corresponding periods of 1856-7. The following is a return of the private securities at the latter end of each of these years :—

1856.			1857.		
October	4	£21,582,000	October	3	£21,835,000
November	8	£18,626,000	November	11	£26,113,000

Therefore that shows that there was not any excessive issue of paper on the part of the Bank of England, and that the amount of Bank of England notes has not been excessively or unduly increased. With respect to the country banks, can it be said that any portion of this discredit is traceable to their over issue ? In the first place, take the Borough Bank of Liverpool. That is not even a bank of issue. Then the large joint-stock bank which has lately unfortunately fallen—the Northumberland and Durham District Bank ; that, though a bank of very extensive operations, and though it has failed for a large sum of money, was likewise not a bank of issue, and, therefore, whatever its imprudences or misfortunes, they were in no way connected with an abuse of the power of issuing paper. (Hear, hear.) With respect to the Scotch banks which have failed, both of those undoubtedly were banks of issue, but their notes bore a very small proportion to their entire liabilities, and about half were covered by gold; and no person aware of the facts, or who will make himself acquainted with them, can fail to come to this conclusion, that the failure of these banks is wholly independent of all questions relating to paper currency. Therefore, as far as the direct operation of the Act of 1844 is concerned, as far as its object was to prevent the excessive issue of paper either by the Bank of England or by the country banks of England, or by the banks of Scotland and Ireland, it must be admitted that the policy of that Act, and of the Act of 1845, was successful. Those who impugn the policy of those Acts must say that at all events they were inoperative, or that in spite of them the currency was in a wise and sound state; and those who are friendly to the policy of those Acts will attribute the sound state of the currency to their direct operation. But, whatever these parties may think, the main fact cannot be disputed, that our paper currency was and has remained up to the present moment in a perfectly sound state. (Hear, hear.) It is quite conceivable, as my right hon. friend remarked in 1847, that a commercial panic might arise if the circulation were purely metallic. There might be undue credit and unsound speculation, and these causes might produce the same effects as are observable at the present moment in the town of Hamburg, where there is no paper circulation, where the only circulation is metallic, and where there exist all the acute symptoms of a commercial crisis. (Hear, hear.) Now, with respect to the operation of the Act of 1844, I am desirous of drawing the attention of the committee to the distinction between the ordinary operations of the Act in those circumstances which were contemplated by its authors, and its operations at a moment of panic or crisis. If we take its operations under ordinary circumstances, the limit imposed on the Bank of England, and on the country banks, and on Irish and Scotch banks, operates in the way of prevention. It does not check the ordinary habitual operations of commerce, but it is like a punishment in the penal law which by its preventive consequences determines men's conduct. But when we come to a state of panic all these circumstances are reversed. Men's minds are then in an extraordinary state. Mutual distrust prevails through the community, and a law which may be perfectly suited to the ordinary conditions of trade, in which it may operate beneficially when men are left under the influence of ordinary motives, may at a time of panic be found to produce effects the reverse of beneficial. The question then arises whether some extraordinary power should not be required to suspend the operation of such a law. The present law does two things. It imposes a limit on the issue of the Bank of England, and exhibits the accounts of the Bank of England in a peculiar form, and likewise requires that those accounts should be published once a week. Now, wherever you impose a limit, there is no question that the existence of that limit, provided it makes itself felt at a moment of crisis, must increase the alarm. People feel at the moment that a peril presses on them, they begin to calculate how much remains of that fund to which they look for assistance in times of commercial difficulty, and in whatever way you fix the limit, whether by Act of Parliament, or, as Mr. Tooke proposed, by a sort of usage, or, as in France, by the discretion of the Government acting on the Bank of France, there is no doubt that in moments of crisis the limit must aggravate the alarm. (Hear, hear.) I can bear witness, from my own observation, that such was its operation in 1847. There is no doubt, in my opinion, that when a crisis once sets in, when you have an extraordinary state of distrust quite different from the state in which men ordinarily live (because it is by mutual trust and confidence that commerce is carried on), when you get into a state of apprehension, and every man suspects his neighbour, then undoubtedly any limit which is imposed on the paper circulation must aggravate the alarm. In that way I distinguish between the ordinary and the extraordinary operation of this Act. (Hear, hear.) Well, then, the manner in which the assets of the Bank are distributed, by which the whole bullion is set against the paper issue of the Bank, and by which only a small portion of reserve notes is placed in the banking department in a very conspicuous manner against the deposits, concentrates the attention of the commercial public upon the amount of gold in reserve. They watch its diminution from day to day, and when it has reached a certain point each person desires to get a share of it as long as it lasts. (Hear, hear.) The banking account tends, then, by that accident, to aggravate in a commercial crisis the public alarm. (Hear, hear.) These circumstances appear to me to show that in its extraordinary operations this Act may tend to aggravate panic, though in ordinary times it tends to promote a sound state of circulation, and, so far as the state of the currency is concerned, to avert panics; but when a panic arises from other causes, at that particular crisis, though the panic may only last a few days, it aggravates the alarm. It appears to me that at moments of that sort we can only resort to the discretion of the Government for the time being; but I shall be happy to hear if any gentleman has any other remedy to offer for those rare, I am happy to say, but, nevertheless, recurring occasions. It seems to me you can only resort to the discretion of the Government in some form or other. Now, there is a question on which the Government have carefully deliberated, and on which after deliberation they have not thought it their duty to make any proposition to the House. The question is one which will be considered by the committee to be appointed, I hope, by this House. The question is whether it is preferable to leave the power of dealing with these emergencies vested in the executive Government for the time being, to be exercised by them on their responsibility to Parliament, and subject to the duty of assembling Parliament at an early period after the exercise of the power, of submitting to it the whole facts of the case, and asking for an indemnity,—the question, I repeat, is whether we should, in conformity with the opinion of the Commons' committee of 1847, leave the matter in that state which undoubtedly involves an infraction of the law, though the wound which may be done to the constitution is healed by constitutional and Parliamentary means, or whether we should adopt the conclusion of the committee of the House of Lords in the same year, who thought it irregular to grant such a power to the executive Government, and that Parliament ought to engraft into the Act itself a power of suspension. The Bill the introduction of which I hope the House will sanction, and introduced at a time when the events which have called for its introduction are still recent, while the crisis in its consequences is not fully developed, will not contain any distinct provision upon that subject. It appears to me, indeed, that the question is one of importance in a constitutional point of view rather than as involving any great difference in its practical results. If, after the proceedings which took place in 1847, and if Parliament should come to a similar conclusion now, the Government, acting under the influence of two such precedents, might find their course facilitated and guided, while, upon the other hand, if a clause were inserted in the present Bill authorizing the Government in cases of emergency and in times of panic to relax the limit, but containing a clause similar to that which is contained in the Act which enables Government to embody the Militia, but which provides that they shall call Parliament together within fourteen days, I think the practical result of such a state of things would be the same as those of the present system, because I think that a Government would be apt to look upon the distinction as a mere formal one, and would feel the same reluctance to act upon the power afforded them, except in cases of extreme emergency. While, therefore, I think that the subject is more important as a constitutional question than from any difference in the practical result of the two systems, there are other reasons why I think that a clause conferring such a power upon the Government ought not to be introduced into the present Bill. Two select committees have sat to inquire into the subject, one a committee of this House, and the other a committee of the House of Lords, and, after fully hearing and considering evidence on the subject, those committees arrived at diametrically opposite conclusions. (Hear, hear.) It is a subject, therefore, with regard to which we may anticipate a great difference of opinion, and the Government would be glad to hear the views entertained by leading members of this and of the other House of Parliament before they can consider it their duty to submit any measure with regard to it. Now, sir, there are questions relating to the operation of the law in Ireland and Scotland, which are worthy of notice, and I think that it may become the House to consider whether some amendments in the existing law may not be made with great propriety—(loud cries of "Hear, hear")—to consider, for example, whether there could be any objection to providing that the Bank of England note should be a legal tender over the whole of the United Kingdom. (Cheers.) Also, it might become us to consider whether, in order to prevent the inconvenience arising from drains of gold from this country under the present system, it might not be advisable to allow the Irish and Scotch banks to issue notes against Bank of England notes. (Hear, hear.) There is another subject to which I adverted last year in moving the appointment of the committee to inquire into the operations of the Act of 1844 which has been the subject of much popular discussion at the present period—I mean the policy of permanently increasing for the future the limit of the issue of the Bank of England. (Hear, hear.) It is said by some persons that if the Bank of England had been authorized by law in the beginning of November to issue £16,500,000 instead of £14,500,000, and to give them such a power does not involve any difference of principle from the Act of 1844, for it was quite within the views of the promoters of that Act, the crisis might have been got over without any extraordinary step being taken by the Government. I think, then, as far as I understand the case, that is a fair representation of the views of many gentlemen. (Hear, hear). Now, sir, I wish in answer to that view to call the attention of gentlemen who entertain it to this fact:—If the bullion reserve of the Bank of England had been what it was at the commencement of the crisis, and if the Bank of England had been empowered by law to issue £16,500,000 instead of £14,500,000, undoubtedly no extraordinary interference might have been necessary. (Hear, hear.) But at the same time I can quite understand that, inasmuch as I believe that in a monetary panic the existence of a limit tends to increase alarm, such interference might have been necessary. (Hear, hear.) Now, if the Bank of England had not kept the same bullion in reserve which the law now requires, if it had merely issued £2,000,000 more upon securities, and diminished the reserve by £2,000,000, which, under such a state of the law, it would be enabled to do, the Bank of England would have had no greater facility than it has now had, but the reserve of bullion would be less by £2,000,000. (Hear, hear.) I think that this consideration must tend to convince gentlemen who hold that view that the resort of increasing the credit of the issue of the Bank from £14,500,000 to £16,500,000 might not be quite so effectual a remedy in times of panic as they imagined. (Hear, hear.) I regret that I have been compelled to trouble the House at such great length upon this subject, but inasmuch as these questions were the real occasion which induced the Government to recommend her Majesty to call Parliament together at this early period, and inasmuch as they have occupied a great amount of public attention, inasmuch as they have grown out of an important crisis which must lead to important results, inasmuch as the step which has been taken by the Government was an extraordinary and extra-legal step—for we stand before Parliament, I will not say in the position of culprits, but in the position of having advised the Corporation of the Bank of England to make a departure from the existing law—inasmuch also as we are now called upon to ask Parliament to indemnify the Bank of England and ourselves, I trust that I may be forgiven if I have felt it my duty to make a statement of such length to the House. (Cheers.)

SPEECH OF MR. B. DISRAELI, ON THE SAME OCCASION, IN REPLY.

Mr. Disraeli said,—I trust that in discussing this question we shall abstain from that vague declamation which has occasionally been indulged in upon both sides of the House, and among others by the noble lord who has just spoken, although I admit that many of his observations, like much that falls from his lips, were worthy of our most serious attention. What are we asked to do to-night? It is no ordinary demand that is made upon the House of Commons. We are asked to consent to a Bill of Indemnity for an infraction of the law, anthorized by the Government in consequence of the commercial distress and distrust that have prevailed, and that unfortunately still prevail. We have heard a great deal of committees upon commercial distress being advisable. Permit me, though very briefly, to lay before you some suggestions as to what may be the causes of the present commercial distress and distrust. The distress is general. It is not confined to this country, where it is extensive. All know that it prevails in the United States. It is not peculiar to the United States, as some gentlemen who have addressed you seem to infer. It is European. It prevails in Germany, in Austria, in Prussia, in Denmark, in Sweden, and what is occurring in Hamburg at this moment must be fresh in the recollection and the fears of many members of this House. Again, it is to be recollected that the commercial transactions of Europe are carried on by an amount of capital that is not to be counted by millions, but by hundreds of millions. It is, therefore, a very great error—and here I entirely agree with the noble lord the member for the City—to suppose that the management of a limited portion of the currency of England could have occasioned this immense disaster, and these wide consequences, which really have originated, not from the mismanagement of the currency of this country, but from the mismanagement of the capital of Europe. (Hear.) Now, there is one important point which we must bear in mind in considering the Act of 1844. That Act is an Act to regulate the currency of this country. The purport of it is in my mind irreproachable. Its object is one which every man ought to wish to secure, and the means by which that object is sought to be obtained are in many respects worthy of public admiration. But is it or is it not the fact that this Act, which really is intended only to regulate the currency of this country, to establish and maintain the convertibility of the notes of the Bank of England, has nevertheless the effect—as I believe, the unintended effect—of greatly aggravating commercial distress and distrust when they are occasioned by the misapplication and mismanagement, not of the currency of England, but of the capital of Europe? (Hear, hear.) Now, I apprehend that that point has never been met throughout all these observations, which I have listened to, and I apprehend that it is a position which can be established. I say it can be shown that it is in consequence of the machinery of the Act of 1844, though intended merely as a currency Act, and to have a limited application to objects, no doubt, of paramount importance, that the managers of the currency of this country have been forced to treat in an identical manner and by the same means two circumstances totally opposite in their character, and both exercising a most powerful and injurious influence on the commerce and capital of this country—that is, a foreign drain and a domestic drain. I say that the Act of 1844, from the manner in which it is framed, forces those who regulate the currency of this country to apply to a domestic drain exactly the same treatment which is applied to a foreign drain; while, if there is anything which can be established by argument and an appeal to facts, it is that the two sets of circumstances should be treated in an exactly contrary manner, and be encountered by means exactly opposite. (Hear, hear.) If that point be established, and if that be the sound position, then surely we may fairly consider a question of that kind without being accused by the noble lord of wishing to tamper with the standard of value, to inundate the country with paper money, or of desiring to recur to obsolete modes of practice, and to what took place in this country with respect to the circulation, when, in fact, the same circumstances did not exist, and when there were not the same means of regulating the currency. The noble lord has made a very great appeal to-night to the disastrous year of 1825, when he informed us that 700 banks were broken in the same morning, I believe. ("Hear hear," and a laugh.) The noble lord was not then the Prime Minister of this country, and though no man is better acquainted with its history than the noble lord, yet I think that on this occasion he has trusted too carelessly to his memory; for at that disastrous period, not 700 country banks, but 73 banks altogether, in London and in the country, broke, and by no means were the majority of these banks of issue. (Hear, hear.) Therefore I do not think that that illustrative warning bears very much on the question we ought to consider this evening. ("Hear," and laughter.) But this year of 1825—this phantom of ancient distress constantly raised in our debates—has really very little to do with the modern principles of banking, and the means by which the currency of this country is now carried on. It is very easy to abuse the Bank directors of that day for what occurred, but we must remember that in 1825 it was quite impossible for the directors of the Bank of England to avail themselves of those resources which you now are always ready to remind them are at their disposal to guard their accounts by raising the rate of discounts. In 1825 you allowed the usury laws to prevail, and it was impossible for the Bank to have a higher rate of discount than 5 per cent. Therefore the year 1825 refers to circumstances which no longer prevail in our laws and manners, and I protest against the quotation of the year 1825 for the future in any currency debate. (Hear.) Now let us refer to the point we have before us. We are in a state of great commercial distress and distrust. I entirely agree with the noble lord on this point, that to suppose that the great commercial distress which now prevails could have been occasioned by the mismanagement of the currency of this or any other country is like supposing that a great nobleman could be ruined by the mismanagement of his pocket-money. (Hear, hear.) The fact is we are mistaking, and have for a long time too easily mistaken, in all these discussions, capital for currency. But though the commercial distress and distrust are entirely occasioned by the management and application in this country, in America, and in the European nations, of the great mass of floating capital which prevails, still, at the same time, we cannot conceal from ourselves that we have a law in this country intended only to act on our currency, and which, so far as the principle it asserts is concerned, I hope the vast majority of this House will maintain, but which, whenever commercial distress and distrust occur, is found entirely unendurable, oppressive, and impracticable, and on every occasion of emergency we are obliged to suspend it. (Hear, hear.) The noble lord who has just addressed the House, and who has addressed it with peculiar authority on this subject, since he was the very first minister who recommended the violation of the statute, has referred to the two occasions in a brief space of time when this Act was suspended; but, if he would refresh his recollection from the evidence taken before the committee of last year, to which constant reference has been made during this debate, he will find that practically, or virtually at least, the suspension of this Act is not to be confined to those two instances, memorable as they are; because in the autumn of 1856 there was a very great pressure, and the precious metals were leaving this country—one of the causes, by-the-by, of our present embarrassment, and consequent on the expenditure for the Russian war. The pressure on the money-market was extreme; the rate of discount was raised—and wisely raised—by the directors of the Bank of England, and by the prudent and courageous manner in which in that instance, as subsequently, they managed their affairs, the danger for the moment was averted. But we have the evidence of Mr. Chapman, and that is no inconsiderable testimony, as all acquainted with the subject will admit. Well, he by whom the force of a money pressure is always felt, who represents the greatest discount house in Lombard Street, says that had it not been for some private information which reached him, to the effect that in case of extreme pressure there would be an interference on the part of the Government, he should at that moment have given up the idea of struggling any further, and it was only on that tacit understanding that he went on with his business. (Hear, hear.) Remember who gave that evidence. He is the very individual whose representations, if I am rightly informed, in 1857 as well as in 1847, induced the Minister of this country to recommend the suspension of this Act. (Hear, hear.) The question naturally arises what is the effect of allowing the currency of this country to be regulated by an Act which we are in a continual state of being prepared to suspend? So entirely do I admire the object of that Act, so anxious am I myself to hope that a large majority of this House is prepared to uphold the practicable and virtual convertibility of the note, that I should be most anxious to interpret the machinery of the Act in a manner the most favourable to its continuance, and to the non-necessity of alteration. I would pass over one crisis, in which its provisions had been suspended, with what I should consider a wise indulgence, but it becomes the committee deeply to consider whether it will sanction a chronic state of suspension—whether we are to have an Act of such a character as always either to be suspended, or to be in that state in which those who are acting under it knew that when an emergency arises it will not be put in practice. Is that wise? (Hear.) "But," says the noble lord, "on the whole that is the course I recommend." The noble lord loves a precedent. He set a precedent in this instance himself, and I can easily understand the tender regret with which the noble lord would at the same time witness that we had turned from that precedent, and as it were passed some censure, mild, I am sure, on the sagacity of his conduct in 1847. But the noble lord says that this Act is a considerable check on what he calls over-speculation and an inflated currency, which in times of prosperity abound. I am sorry to hear from the noble lord that our periods of prosperity are always necessarily to be accompanied by such distress, but when the noble lord so fairly tells us what are the consequences of increased circulation—when he tells us that, unless we have these considerable checks, the currency will be greatly increased—there will be no check on the issues of the Bank—a crisis will arise, the circulation will be inflated, and there will, in fact, be no standard of value, I may be permitted to say to the noble lord, with all respect, that though these were opinions once held by persons of considerable reputation, and for aught I know may be still adhered to by persons of considerable repute, yet they are conclusions the authenticity of which is entirely impugned by authorities of a more novel nature than perhaps the noble lord may be aware of, and of no less high a character than any to which the noble lord could appeal. Many high authorities have held that the depreciation of a bank-note convertible at par is a simple impossibility, and they have also held that it is quite out of the power of any bank to issue beyond the requirements of the country, and what is still more important, is that it is not in the power of any one to prove that any issue at any time has ever affected the prices of articles. Nay, more, we have conclusive evidence in our possession that the reverse is the fact. Now, really, Sir, under these circumstances I must protest against the noble lord so freely, and in that style of declamation which he can command even upon a currency question, lending his authority to a principle which I believe to be quite fallacious, and to conclusions which I believe to be untenable, and to details which I think he will find himself totally unable to substantiate. "But," says the noble lord, "this Act is a check upon the over-speculation of the day, and although there may be some objection to the mode in which that check is exercised, still the benefit which results from it is indubitable." But when the noble lord talks of this Act operating as a beneficial check, while the Government may interfere with the period of its being carried out, let me ask the House to consider one point which I think has not been fairly taken into consideration during the course of this debate. If the House be prepared to sanction and countenance this dispensing power to be exerted at the arbitrary will of the Minister of the day, how does it propose to recognize the difference of position between those firms who fall victims before the dispensing power is exercised? (Hear, hear.) Are those who hold on a favoured twenty-four hours struck upon the clock of the Royal Exchange of London to dominate over those with whom they ought to be fellow-victims? (Hear, hear.) Let me remind the House that when we discuss this question, and perhaps discuss it too easily as regards those who have fallen, it might perhaps be as well to go a little into detail, for it is only through detail that the House can realize some idea of the consequences of this fear and distrust in London and the other great marts of commercial enterprise. Now, Sir, here is a document with which I have been furnished, and I will answer for its being correct and accurate, although I cannot answer for its being complete, because there are the names of many victims with regard to whom the amounts for which they have fallen have not yet been ascertained; but here is a document of desolation in my hand, the contents of which I think will surprise even those gentlemen connected with the commerce and monetary transactions of the country who have addressed us to-night. It is a return of the number of firms which have fallen between the 7th of September and the 12th of November, a period of little more than two months, and among those firms there are some of no inconsiderable importance, and some which ranked, as far as capital can be a test of position, in the highest class in mercantile repute. In that brief period I find no less than eighty-five firms fell which represented a capital of £42,000,000 sterling, and, with the exception of firms representing £8,000,000 of money, those mercantile establishments fell before the letter of November, some of them a few days, even hours, before that letter. (Hear, hear.) Now if we are to leave the law in this state, I really think that it becomes us calmly, but earnestly and seriously, to consider what is the operation of the policy, what the state of the administration in this country which thus conducts and controls its commerce. I have heard statesmen to-night, and among them the noble lord the member for London, who do not shrink from the responsibility of exercising this dispensing power. I must say that the right hon. the Chancellor of the Exchequer showed more modesty, and displayed his reluctance to exercise such a responsibility, by intimating the possibility upon his part of making some suggestion which would avoid a responsibility so terrible. A man must have great confidence in himself, in his position, and in his character who, under circumstances such as I have described, when eighty-five firms representing a capital of £40,000,000 had fallen within so short a period, would, as a a Minister of the Crown, have no reluctance in deciding when to interfere. (Hear, hear.) I am deeply conscious that every English statesman is sincerely anxious to do his duty to his country, and I can readily conceive the terrible anxiety with which a Minister under such circumstances might consult some eminent capitalist, perhaps one of his principal supporters, and one whose mercantile knowledge and intelligence had been of great assistance, I won't say in preparing a budget, but in furnishing details for the assistance of a Minister in framing his financial scheme. I am not making any personal allusions, but I am only supposing the general position in which a Minister may be placed. Well, then, suppose the Minister to consult such an individual. It may be of the utmost importance to a man engaged in commercial pursuits that the suspending power should be delayed for twenty-four hours, or precipitated by that period, in order to destroy a rival or to save himself. The consequences of such a course I need not dwell upon. I know that such a case may not happen, but would it be wise in us to sanction a law which might, under any circumstances, lead to consequences of such a character? I am sure that the House will reflect deeply upon the difficulty which now surrounds us, that they will recollect the contents of the startling document to which I have alluded, and that they will feel that it is no child's play to bear responsibility upon such a subject—a subject which I say is the last with regard to which the responsibility of a Minister of the Crown ought to be exercised. It is a responsibility almost beyond the endurance of any individual, and one which is totally alien to the character of the constitution of this country. (Hear, hear.) You have thought a dispensing power with regard to civil and political rights unbearable, and you have changed a dynasty rather than submit to it, and are you prepared now to extend to a Minister such a power with regard to a subject which in this country touches you, perhaps, more nearly than either civil or political rights? Where commercial and monetary interests are concerned, are you prepared to submit quietly to a dispensing power exercised arbitrarily at the discretion of the Minister, the exercise of which, even when most virtuously employed, may be unjust and ruinous in its results? (Cheers.) I think, therefore, that it is most desirable that we should take some course to prevent a repetition of the transactions of 1847 and 1857. I think the time has arrived in which we ought to exert ourselves to solve a difficulty which may be great, but one which we ought not to evade, but which we ought to endeavour to surmount; and I regret that I see no disposition on the part of the Government to undertake the task. That such appears the case I regret the more, because I think that all that has been said upon the subject during the present session might have tended to prove to Her Majesty's Ministers that there was no desire in any quarter to throw any obstacle in the way of the full exercise of their discretion, but that, on the contrary, there was every wish to assist them to a solution of the difficulty which would redound to their honour. (Hear, hear.) We are to-night asked to agree to a Bill of Indemnity; now, I said yesterday, perhaps rather inadvertently, that my consent to that Bill would depend upon the line taken by the Government with respect to the Act itself. I forgot that the Bill was not a Bill of Indemnity for Ministers alone, but also an indemnity for the Bank directors. Now, I don't think that any one can impugn the conduct of the Bank directors. There is no evidence before us that the Bank directors required this interference; and in 1847, not only did they not solicit the Government for a suspension of the Act; on the contrary, they did not avail themselves of its suspension, but disapproved the course which had been adopted, and maintained the convertibility of the note. I have reason to believe that in the present instance the directors of the Bank have acted in the same magnanimous spirit; and if, at the suggestion of the Government, they have consented to take an illegal step, I think, whatever may be the differences of opinion on the general subject, it would be cowardly and vindictive to throw any obstacle in the way of their indemnity. So far, therefore, as the Bill of Indemnity is asked, I am prepared to accede to the request of the Government, but further than that I cannot afford my consent. I think the Government have taken an erroneous step, notwithstanding the high authority just expressed in their favour by the noble lord, in recommending that the House should again appoint the committee to consider this question. I must repeat my belief that there is no subject which can solicit the attention and consideration of public men on which there exists even such a surplusage of knowledge and information as upon the principles on which the currency of this country should be established, and on which it has of late years been carried on, and that any attempt to postpone the solution of the grave difficulties which surround it by further inquiry will lead to great public inconvenience, possibly to great public disaster. I think we ought to show no hesitation in meeting those difficulties, that we ought to give our best consideration not only to the passing, but the immediate passing, of some measure which will prevent the future necessity of appealing to this House for Bills of Indemnity to her Majesty's Ministers for such acts as that in respect of which they now deem it necessary to ask us to support them. I have endeavoured to call the attention of the House to what I think the vital point in this Bill which requires our consideration. I am for supporting the spirit of the Bill so far as it maintains the convertibility of the notes which the Bank issues. I would do that completely, sincerely, and in no spirit of equivocation, but believing as I do that the law by its enactments forcing the Bank to treat a foreign and a domestic drain by the same means, not only occasions but aggravates distress and distrust among the commercial classes of this country, I think we ought to meet that difficulty by remedying the law in that respect. Let the House realize to itself one great difference between a domestic and a foreign drain. With a foreign drain we have learnt by experience efficiently to cope. We have never had a greater strain on our resources in that respect than we have had of late years, but by that powerful arm which the usury laws now place on the Bank of England we have been enabled to cope with that efficiently; and I therefore think there is no man in this country who has now any fear of ultimately disastrous consequences of a foreign drain. A foreign drain is a very unimpassioned, calculating transaction. It is founded on a knowledge of our resources, and upon the necessities of those who in foreign countries require our aid, and, however inexorable the demand, it is one which is acted on almost in an instant by the results of reason and calculation. But the moment you come to a domestic drain you have to encounter elements of a very different character. The moral quality, as well as the material element, enters into the calculation of the domestic drain. The passions of fear and hope govern mankind; the mart and the exchange are not free from the influence of those passions, and when you have circumstances in which a whole nation is panic struck you cannot hope by the aid of economic science successfully to encounter them. It is therefore the duty of Parliament to remedy the deficiencies of the law in that respect. But I cannot believe we have any chance of obtaining that remedy if we consent to the course suggested by her Majesty's Ministers. In 1840 the whole question of a bank of issue was gone into by a committee which, whether we look to the reputation of the members who formed it, or the high character of the witnesses called before it, produced a volume inferior in interest to none in this House. In 1847, after the panic, you had a committee to inquire into the same subject and into the influence of this very Act of 1844. At the beginning of last year a third committee was appointed, and its transactions are before you. What more can you want? All the more distinguished writers of the day on public and political economy have favoured the country with their speculations on this subject. There is no new writer before us, and I think therefore the time has come when we ought to settle this question. What is the use of a Government if it cannot settle this question? (A laugh.) What is the use of a House of Commons if it cannot animate and inspire a Government? (Laughter, and an ironical cheer.) There are some men who think that members of Parliament are only made to form Parliamentary committees; but it is only a preparatory, though a very useful, training to accumulate intelligence in that way, to sift evidence and to draw conclusions. There is something higher and more necessary; that is, to recommend a policy; and I think the time has come when we ought to recommend a policy. (Cheers.) Now, the right hon. gentleman the Chancellor of the Exchequer has no wish that there should be any great delay in passing this Bill of Indemnity. I wish to meet him in that respect frankly and in the most friendly spirit; but I hope the right hon. gentleman will not really and practically hold if we give him this Bill of Indemnity that Parliament was only called together for that object. The right hon. gentleman has a motion on the paper for a committee. If he would only settle with me the night on which to take that motion for a committee, I would move to that motion an amendment the purport of which would be, without, of course, pledging myself at this moment to its exact phraseology, to express the opinion of the House that it is expedient to legislate on the subject, and not to refer it to a select committee. (Cheers.) If the right hon. gentleman will say this day week for his committee, I will assist him to the best of my power in carrying his Bill; and on this day week, if he will move the committee, I will endeavour in a manner befitting the importance of the subject to ask the House to consider whether, under all the circumstances of the case, it is expedient that this question should be referred to a select committee, and whether it is not wiser to proceed at once to legislate on the subject. (Hear, hear.)

FLUCTUATIONS IN PRODUCE.*

A STATEMENT of the Highest and Lowest Prices of the Principal Articles of Produce during the past Three Years, with the Stock at each Point of the Respective Years. Also the Highest and Lowest Prices for the first Five Months of this Year, with Comparative Stocks, 5th June:—

SUGAR.

West India, per cwt., Duty 1d., *Low Brown to Fine Yellow.*

		s. d.		s. d.	Stock. Tons.
1855.	Nov.	... 57 0	to	67 0	...13,130
,,	April	... 30 0	,,	41 0	...10,250
1856.	Dec.	... 47 0	,,	54 6	... 6,100
,,	Feb.	... 34 0	,,	45 0	...16,830
1857.	June	... 53 0	,,	63 0	...11,743
,,	Dec.	... 33 6	,,	47 0	...17,290
1858.	Mar.	... 35 0	,,	47 6	... 7,350
,,	June	... 33 0	,,	44 6	...17,070

Stock, 5th inst. 17,070
,, 5th June, 1857 11,743

Mauritius, per cwt., Duty paid, Low Brown to Fine Yellow.

		s. d.		s. d.	Stock. Tons.
1855.	Nov.	... 48 0	to	65 0	... 6,780
,,	April	... 27 0	,,	39 6	...10,590
1856.	Dec.	... 39 0	,,	52 6	... 7,480
,,	Feb.	... 32 0	,,	43 0	...10,850
1857.	June	... 46 6	,,	62 0	... 8,600
,,	Dec.	... 28 0	,,	45 0	... 6,950
1858.	Mar.	... 29 0	,,	47 0	... 7,940
,,	June	... 27 6	,,	44 0	... 9,980

Stock, 5th inst. 9,980
,, 5th June, 1857 8,600

Bengal, Duty paid, per cwt., Ordinary Khaur to Fine White Benares.

		s. d.		s. d.	Stock. Tons.
1855.	Nov.	... 48 0	to	66 0	... 1,870
,,	April	... 26 6	,,	39 0	... 7,300
1856.	Dec.	... 39 0	,,	53 6	... 3,840
,,	Feb.	... 32 0	,,	45 0	... 2,190
1857.	June	... 46 0	,,	62 0	... 3,060
,,	Dec.	... 27 0	,,	50 0	... 6,760
1858.	Mar.	... 28 6	,,	49 6	... 6,400
,,	June	... 26 0	,,	48 0	... 4,400

Stock, 5th inst. 4,400
,, 5th June, 1857 3,060

Stock of all kinds of Raw Sugar:—
th June, 1858 .... 54,880 tons.
5th June, 1857 .... 37,450 ,,

Refined Sugar, per cwt., Brown and Middling Grocery Lumps.

			s. d.		s. d.
1855.	Nov.	....	73 0	to	76 0
,,	April	....	43 6	,,	46 0
1856.	Dec.	....	61 0	,,	63 0
,,	Feb.	....	47 0	,,	50 0
1857.	June	....	70 6	,,	74 0
,,	Dec.	....	55 6	,,	59 0
1858.	Mar.	....	56 0	,,	62 0
,,	June	....	54 0	,,	57 0

MOLASSES.

West India, per cwt.

		s. d.		s. d.	Stock. Tons.
1855.	April	... 17 0	to	21 6	... 770
,,	Nov.	... 17 0	,,	21 6	... 1,490
1856.	Dec.	... 23 6	,,	26 0	... 1,410
,,	Mar.	... 16 6	,,	18 0	... 1,686
1857.	June	... 28 0	,,	29 6	... 1,440
,,	Dec.	... 15 0	,,	16 0	... 7,820
1858.	April	... 15 6	,,	19 0	...10,170
,,	June	... 15 0	,,	17 6	...10,110

Stock, 5th June, 1858 10,110
,, ,, ,, 1857 1,440

COFFEE.

Native Ceylon, per cwt., Good to Fine Ordinary.

		s. d.		s. d.	Stock. Tons.
1855.	Nov.	... 54 0	to	56 0	... 5,100
,,	Jan.	... 44 6	,,	46 6	... 6,700
1856.	Sept.	... 51 0	,,	53 6	... 4,450
,,	June	... 48 0	,,	52 6	... 5,360
1857.	June	... 64 0	,,	66 0	... 3,300
,,	Dec.	... 52 0	,,	53 0	... 4,480
1858.	Feb.	... 53 6	,,	55 6	... 3,750
,,	May	... 45 0	,,	52 0	... 4,120

Stock Ceylon Coffee, 5th inst. ...4,120
,, ,, ,, 1857 2,600

COCOA.

West India, per cwt., Old Gray to Fine Red.

		s. d.		s. d.	Stock. Packages.
1855.	Nov.	... 54 0	to	64 0	... 7,870
,,	April	.. 33 0	,,	42 0	... 3,810
1856.	Dec.	... 66 0	,,	75 0	... 1,420
,,	April	... 34 0	,,	48 0	... 5,110
1857.	Sept.	... 91 0	,,	105 0	... 3,980
,,	Jan.	... 66 0	,,	75 0	... 620
1858.	Feb.	... 58 0	,,	92 0	... 2,670
,,	May	... 48 0	,,	84 0	... 7,500

Stock, 5th June, 1858 7,500
,, ,, ,, 1857 6,650

TEA.

Congou, per lb. on Bond, Low to Fine.

			s. d.		s. d.
1855.	Aug.	....	0 8	to	2 4
,,	Jan.	....	0 9½	,,	1 10
1856.	Dec.	....	0 10¼	,,	2 4
,,	May	.. ..	0 8½	,,	2 4
1857.	Oct.	....	1 2	,,	2 4
,,	Jan.	....	0 11	,,	2 4
1858.	Jan.	....	1 1	,,	2 4
,,	May	....	0 8	,,	2 2

* From these prices may be traced the depression in trade, and the serious sacrifices entailed by the panic.

RUM.

Leeward, per gallon, proof.

	s. d.		s. d.	Stock. Puns.	Stock. Hhds.
1855. Jan.	3 3	to	3 4	...20,120	5,270
,, Apr.	2 0	,,	2 1	...18,486	5,660
1856. Dec.	2 7	,,	2 8	...21,830	3,750
,, Mar.	2 0	,,	0 0	...23,312	7,683
1857. Apr.	2 9	,,	0 0	...19,650	4,260
,, Dec.	1 11	,,	2 0	...20,365	4,170
1858. Jan.	1 11	,,	2 0	...19,580	4,240
,, June	1 10	,,	0 0	...21,780	4,240
Stock, 5th June, 1858				...21,780	4,240
,, ,, ,, 1857				...20,970	5,130

RICE.

Bengal, per cwt., Ordinary to Fine White.

	s. d.		s. d.	Stock. Tons.
1855. Nov.	... 16 0	to	18 6	... 7,980
,, Jan.	... 13 0	,,	15 0	... 5,430
1856. Jan.	... 14 0	,,	17 6	...11,710
,, May	... 9 0	,,	12 6	...27,720
1857. Sept.	... 11 0	,,	15 0	...48,110
,, Dec.	... 8 0	,,	11 0	...62,720
1858. Jan.	... 8 0	,,	11 6	...71,270
,, June	... 7 0	,,	11 0	...92,560
Stock, 5th June, 1858				92,560
,, ,, ,, 1857				36,810

SALTPETRE.

Bengal, per cwt., Ordinary to Fine.

	s. d.		s. d.	Stock. Tons.
1855. Sept.	... 33 0	to	42 6	... 6,880
,, Feb.	... 21 6	,,	27 6	...11,755
1856. Dec.	... 45 0	,,	56 0	... 2,140
,, June	... 28 0	,,	32 0	... 5,230
1857. Sept.	... 55 0	,,	65 0	... 6,760
,, June	... 35 0	,,	41 0	... 5,260
1858. June	... 35 0	,,	46 6	... 5,048
,, April	... 28 0	,,	37 0	... 5,430
Stock, 5th June, 1858				 5,048
,, ,, ,, 1857				 4,245

COCHINEAL.

Honduras, per lb., Low Silvers to Fine Blacks.

	s. d.		s. d.		Stock. Serons.
1855. Feb.	... 2 10	to	4 10	...	6,400
,, Jan.	... 3 5	,,	4 0	...	5,800
1856. Aug.	... 3 7	,,	5 11	abt.	9,000
,, Dec.	... 3 6	,,	5 2	,,	6,780
1857. Oct.	... 3 0	,,	5 10	,,	3,980
,, May	... 2 7	,,	5 6	,,	6,600
1858. Jan.	... 3 0	,,	5 10	av.	8,680
,, May	... 2 5	,,	5 4	,,	6,080
Stock, 5th June, 1858				....	6,080
,, ,, 1857				....	6,500

COTTON

Surat, per lb., Ordinary to Fine.

	d.		d.	Stock. Bales.
1855. June	... $3\frac{1}{2}$	to	$5\frac{1}{2}$	... 70,577
,, April	... $2\frac{1}{2}$	,,	$4\frac{1}{4}$	... 58,920
1856. Oct.	... $4\frac{5}{8}$	to	$5\frac{3}{4}$	... 24,835
,, Jan.	... 3	,,	$4\frac{3}{4}$	... 54,680
1857. Oct.	... $5\frac{3}{4}$	,,	$6\frac{3}{4}$	... 33,538
,, Dec.	... $2\frac{3}{4}$	,,	4	... 33,795
1858. May	... $4\frac{1}{4}$	,,	6	... 45,026
,, Jan.	... $3\frac{3}{4}$	,,	$5\frac{1}{4}$	... 42,630
Stock, 5th June, 1858				 45,026
,, ,, 1857				 37,990

INDIGO.

Bengal, per lb., Ordinary to Fine.

	s. d.		s. d.	Stock. Chests.
1855. Jan.	... 3 6	to	7 6	... 23,060
,, July	... 1 1	,,	7 0	... 22,500
1856. Jan.	... 1 6	,,	7 6	... 12,148
,, Oct.	... 1 0	,,	7 5	... 26,140
1857. Oct.	... 2 6	,,	1 10	... 19,400
,, Jan.	... 1 0	,,	7 9	... 18,300
1858. Jan.	... 2 6	,,	10 0	... 19,065
,, May	... 1 0	,,	9 0	... 22,016
Stock, 5th June, 1858				 22,016
,, ,, 1857				 20,634

TALLOW.

St. Petersburg Y. C., per cwt.

	s. d.		s. d.	Stock. Casks.
1855. Nov.	... 72 0	to	72 6	... 17,507
,, Mar.	... 47 0	,,	47 6	... 35,532
1856. Jan.	... 68 0	,,	68 6	... 19,339
,, May	... 45 6	,,	45 9	... 16,900
1857. Mar.	... 80 0	,,	0 0	... 16,746
,, Nov.	... 48 0	,,	48 3	... 38,622
1858. April	... 55 6	,,	55 9	... 11,691
,, Feb.	... 52 6	,,	52 9	... 20,825
Stock, end of May, 1858				... 10,500
,, ,, 1857				... 13,009

IRON.

Scotch Pig, per ton.

	s. d.		s. d.
1855. Sept.	85 0	to	85 6
,, April	61 0	,,	61 6
1856. Sept.	79 0	,,	82 0
,, Nov.	70 0	,,	72 0
1857. June	80 0	,,	0 0
,, Nov.	52 6	,,	55 0
1858. Mar.	68 0	,,	72 0
,, May	60 0	,,	0 0

CORN.

Wheat—Gazette Average per quarter.

	s. d.
1855. Dec.	83 1
,, Mar.	66 6
1856. Aug.	77 10
,, Dec.	59 8
1857. July	63 10
,, Dec.	47 5
1858. Jan.	48 9
,, April	43 1

APPENDIX

TO THE

HISTORY OF THE COMMERCIAL CRISIS, 1857—58.

PART I.

SUSPENSIONS IN 1849.

Jan. Messrs. R. Eglington and Co., London, East India trade.
„ Mr. M. Kenrick, Wrexham, banker.
„ Mr. R. M. Lloyd, Wrexham, banker.
„ Messrs. Curtis and Buddendorf, New Orleans, Prussian trade.
July. Messrs. Butterfield, Petersfield, bankers.
Oct. Messrs. C. H. and G. Enderby, Greenwich, merchants.
Dec. Messrs. Ewing and Co., Glasgow, manufacturers.
„ Messrs. Schwartz, Brothers, Hamburg, bankers.

THE ESTATE OF MESSRS. SPEIR AND CO.

The inspectors under the estate of Messrs. Speir and Co., of the East India trade, who were compelled to suspend in the crisis of 1848, in January announced the completion of the payment of 20*s.* in the pound, with interest, to the whole of their creditors.

THE ESTATE OF J. AND G. CAMPBELL AND CO.

Messrs. J. and G. Campbell and Co., of Liverpool, West India and Mexican merchants, who suspended, February, 1848, declared in January a final dividend of 2*s.* 6*d.* in the pound, making, with the previous instalments, a full payment of 20*s.* in the pound.

THE ESTATE OF G. T. BRAINE.

On the 22nd of August, 1849, the annexed circular was published, announcing the final dividend under this estate, making a full payment of 20*s.* in the pound to the creditors. It will be recollected that prior to Mr. Braine's suspension on the 4th of July last year, with engagements to the amount of £350,000, application was made to the Bank of England, and rejected, for assistance to carry him through, and that it was then stated there were surplus assets of no less than £65,000 or £70,000. It is now understood that not only will the entire liquidation of Mr. Braine's engagements have been effected from the *bonâ fide* proceeds of his property, but that little doubt exists of the surplus being nearly double the amount of the estimate originally formed :—

"London, August 22.

"Sir,—I have the pleasure to advise my intention of paying the remainder of your claim upon me, with interest, on the 28th inst.

"At the same time I beg to express my thanks for the indulgence and co-operation I have experienced from yourself and others, through which I have been enabled to accomplish a speedy liquidation without undue sacrifice of property.

"I am, Sir, yours obediently,
"George T. Braine."

THE ESTATE OF COTESWORTH, POWELL, AND PRYOR.

The appended circular was issued by this firm on the 6th December, announcing a remaining dividend of 7*s.* 6*d.*, to make up 20*s.* in the pound, to the creditors :—

"St. Helen's Place, London, December 6, 1849.

"We have the satisfaction to inform you, that we are now prepared to pay you a further and final dividend of 7*s.* 6*d.*, making, with the previous dividends, 20*s.* in the pound. At the same time, we shall pay you the interest which has accrued during the period of our suspension.

"The dividend will be payable at our counting-house on the 12th inst., up to which day interest at 5 per cent. will be calculated; and in the mean time we will thank you to furnish us with a note of the balance of principal and interest due to you, that we may agree the amount. Upon your applying for payment, it will be necessary for you to give up all bills and collateral securities which you hold.

"We remain, your most obedient servants,
"Cotesworth, Powell, and Pryor."

THE ESTATE OF BENSUSAN AND CO.

Court of Bankruptcy, *January* 17.

(*Before Mr. Commissioner Fonblanque.*)

The bankrupts carried on business as extensive general merchants, in Magdalen Row, Great Prescott Street. They failed at the close of 1847, and have been before the Court since November of that year, in which month the fiat was issued. This was the certificate meeting. Mr. Hutton, of Bucklersbury, was trade assignee and accountant to the estate; Mr. Rixon, of King William Street, was solicitor to the assignees ; and Mr. Lawrance appeared for the bankrupts.

The course of trading and present position of the firm will be seen from the following lucid report of Mr. Pennell, the official assignee :—"These bankrupts carried on an extensive business as general merchants. They commenced their balance-sheet on the 1st of January, 1841, with an assumed capital of £1,345. In arriving at this sum they take credit for various debts due to them, including Bensusan and Brandon, £5,986 ; J. Hassan, £4,381 ; and Judah Pariente, £8,158. Upon turning to that portion of the balance-sheet which contains a list of the bad debts still due to the bankrupts, I find these same parties entered for the following sums :—Bensusan and Brandon, £4,469 ; J. Hassan, £11,763 ; Judah Pariente, £5,003, from which it may be reasonably inferred that these bankrupts, instead of possessing any capital, were insolvent so far back as 1841. The course of trading pursued since 1841 has the following results :—

	£	s.	d.
Profits	£58,889	2	5
Trade expenses	26,081	9	10
Losses	60,693	16	2
	£86,775	6	0

Showing an excess of £27,886 3*s.* 7*d.* over the profits. The bankrupts have, during the seven years, in which they pursued a ruinous trade, drawn out of the joint estate, for their own private expenditure, no less a sum than £25,534 7*s.* 10*d.* The upshot of such a course of trading may be anticipated. The total liabilities are £57,962 3*s.* 11*d.*, while the assets do not exceed £1,245 16*s.* 11*d.*, or about 5*d.* in the pound."

Mr. Pennell thus concludes his report:—"I regret to observe, in addition, that the books have been very carelessly kept, and from their imperfect state the assignees have been obliged to require the bankrupts to raise a fresh set of books for the whole period over which their balance-sheet extends. The bankrupts have evinced every disposition to assist the assignees in the elucidation of their accounts; but, looking at their previous neglect, it is not a matter of surprise that they still remain in an unsatisfactory state."

The assignees offered no opposition.

Mr. LAWRANCE addressed the Court on behalf of the bankrupts. He presented the case to his Honour as one of great misfortune. The firm of which the present applicants were members had been established in the city of London for nearly a century, and was for a long period regarded as one of the most respectable and quite as solvent a house as any other in the trade. It was true their losses had been great, but they had not been incurred by excessive speculations, or engaging in wild projects, but arose in the ordinary way of their business. The balance-sheet commenced in 1841, and the Court would perceive that they had a capital at that time of £1,300.

The COMMISSIONER remarked, that he was not satisfied that the capital set down in the balance-sheet was a *bonâ fide* one, and wished to know of what it consisted.

Mr. LAWRANCE replied, that it was composed of debts due to the house at the time, which were then considered good, but which had since become bad. For the unfortunate events which had since happened in the mercantile world, and which destroyed this capital, the bankrupts could not with justice be held accountable. They put down their capital in 1841 at what they then considered a fair estimate, and had a perfect right to do so. It was true the capital, as compared with their present liabilities, seemed small; but the Court would recollect that the chief business of this house was commission, which did not require a large capital. That the trade was not quite a losing one was shown by the fact that the profits were £59,000, whilst the trade expenses did not exceed £26,000. There was another creditable feature in this case, that so far from his clients having gone on recklessly increasing the amount of their liabilities, his Honour would perceive, from the proceedings, that as their means of payment diminished, so did their contraction of fresh debts, and that there was a striking difference between the amount of debts incurred in the months of February and September of the year in which they failed. This arose from no inability on the part of the bankrupts to obtain goods (their character was such that they could, at almost any period before their failure, get credit to a vast extent), but from a conscientious feeling that they ought not to imperil the property of others. The official assignee had remarked in his report that the assets were small. Had the bankrupts been knavish, this would not have been so. They would have got in £20,000 or £30,000 worth of goods a month or two before they stopped, and although their liabilities would have been increased to that amount, the amount of dividend would have been very different. There was no allegation, nor even a suspicion, that property had been clandestinely removed, or that there had been the slightest tinge of fraud, either after the fiat, or during the whole course of their mercantile career. They had not attempted to prop up their credit by accommodation bills, nor to save themselves from ruin by selling goods under cost price, or by false representations to creditors.

The COMMISSIONER remarked that their losses were large.

Mr. LAWRANCE—I urge that as an extenuating circumstance—the cause of my clients' failure. They lost £20,000 by a single house in St. Thomas's, and there are eight bad debts incurred in the ordinary way of trade, which amount in the aggregate to no less than £53,000. They were ruined by that disastrous monetary pressure and commercial embarrassment, which brought down more extensive although not perhaps more ancient or respectable houses. In a valuable work, entitled the *Commercial Crisis*,* which contains an admirable history of that calamitous period, your Honour will see the causes which led to one of the most terrible commercial crises that ever occurred in the annals of this country. The question then was, not which house was likely to fall, but which of them could possibly stand. My clients were obliged to succumb. But, when they saw that stoppage was inevitable, they must be commended for the course they took. They at once submitted their affairs to the administration of this Court. Where the tree fell, there it lay. Since the bankruptcy they have done all they could to assist their creditors; and I beg your Honour to note that no individual creditor opposes, and that the assignees are silent. You will also observe

* Published by Letts, Son, and Steer, 8, Royal Exchange.

that there are no law costs, and no interest. I regret to have to inform you that pecuniary difficulties are not the only sufferings my clients have had to contend with. The hand of misfortune has pressed heavily upon the members of this house. One of them has, since the issue of the fiat, sunk into the grave, and another is physically and mentally paralyzed. The surviving partners now present themselves, and entreat your Honour not to superadd to their past sufferings the stigma so much dreaded by mercantile men—a suspended certificate. They ask you to take into account the whole course of their mercantile career—the causes of their calamity, the assistance which they have given to their creditors since the fiat, the length of time during which they have been before the Court, the absence of any opposition on the part of creditors—and, taking all these things into your consideration, they feel assured that you will not for one day withhold the certificate.

Judgment deferred.

BENSUSAN AND CO.

Court of Bankruptcy, *April* 18.

The bankrupts in this case failed in the panic period of 1847, and to-day was named by the Commissioner for giving his judgment on the question of certificate. Messrs. Rixon and Son appeared for the assignees ; Mr. Lawrance for the bankrupts.

His Honour, in giving judgment, said—That this fiat was issued in 1847 against the four bankrupts, Moses L. Bensusan, Jacob L. Bensusan, Samuel L. Bensusan, and Joshua L. Bensusan. With respect to the first-named bankrupt, who had died since the fiat was issued, there was little to say, but it might be stated that, from what could be ascertained, there appeared every reason to believe that the management of business, considering his age and infirmities, was past his control. It ought, also, to be noticed, that his expenses had been exceedingly moderate ; and, under such circumstances, it might be some consolation to his family and friends to know that the Court, had he lived, would in all probability have granted him his certificate immediately. He was sorry, however, to say, that with regard to the position of the firm generally, the Court could not come to so satisfactory a conclusion. He (the Commissioner) had looked carefully over the voluminous proceedings which had taken place under this estate, and he found that, in January, 1841, the business of this firm with four partners was based upon a nominal capital of £1,345 only—he said a nominal capital, because, in reality, they then possessed no capital at all, but were, in fact, insolvent, since, included in this estimate of capital were three debts, one of £6,000, due from the old firm of Bensusan and Brandon ; a second of £4,000, from the house of Hassan ; and a third of £8,000, from Pariente, making a total of what shortly proved to be £18,000 bad debts. Viewing, then, their situation in this respect, and giving them credit for their assumed amount of £1,345, they were in truth upwards of £16,000 worse than nothing. From 1841 up to the time of their bankruptcy, the amount of profits was stated at £58,000 ; but then the item of trade expenses and losses was put down at no less a sum than £86,000, so that there was a deficiency on that account of £27,000 ; and yet, in spite of that deficiency, the partners had managed to absorb, in the shape of private expenses, the immoderate sum of £25,000. The result of such a course of trading might be easily conceived. This estate, with debts and liabilities to the extent of £58,000, possessed assets, from which not more than £1,200 would be realized ; and therefore, allowing for the costs of investigation, it was impossible that the joint estate could be estimated to pay 4*d*. in the pound. And now to enter more minutely into their conduct as traders. Their books were carelessly kept, so much so, that it had been found utterly useless to depend upon them, and hence, since the bankruptcy, it had been deemed necessary to raise new accounts for the purpose of elucidating their affairs. The bankrupts, associated as they were in family, could not, therefore, be exonerated on that ground, and must take the consequences of that dereliction of duty which attached to rendering insufficient accounts. But there was another objection in relation to accounts which weighed in the mind of the Court, and that was a certain secrecy and mystery which had been observed in keeping the books. One of the brothers complained of not having been allowed to see the accounts, notwithstanding his urgent representations, and he (the Commissioner) must express his opinion that whenever such secrecy was practised, it must be regarded as a badge of misconduct. The reputable manner of keeping books was to keep them in such a manner that every body interested might be well acquainted with their contents, in order that, if error or misstatement arose, it might at once receive correction, and that they might (if ever it were required) reveal to creditors, in a clear and straightforward course, the whole of the trader's transactions. Having considered the way in which the books were kept, attention might properly be directed to the question of capital when the bankrupts commenced their business. If at that date they really considered themselves in a solvent condition, that delusion, if they ever entertained it, was soon dissipated, for in that year Bensusan and Brandon, who owed them £6,000, failed, and their affairs were wound up. This large loss, compared with their estimated capital of £1,300, should have proved a warning to them, and had they acted prudently then, they ought to have made a strict inquiry into their position, and called in their creditors, or, on the other hand, reduced considerably their dealings. But this first warning was not enough—there was a second, for Hassan, with whom they appeared to be connected, failed in 1843, £4,000 in their debt—another warning, which, it might have been thought, would have induced discretion, but no heed was taken of it. A third warning followed, and Pariente, owing the firm £8,000, failed in the year 1843 or 1844. Large as the gross amount of these three losses was, it had apparently created little effect upon the bankrupts. Their duty at such a juncture, especially after the third warning, was to have suspended payment and distributed their estate, whatever might have remained, and, blameable as their conduct might have been considered after the second warning, they would have been less liable to severe condemnation than they were now. On this point alone, two years' suspension would scarce be adequate punishment. But, having looked at their conduct with respect to the joint estate, to carry the inquiry throughout, it was requisite to look at their conduct independently in connection with their separate estates. He (the Commissioner) had already stated that £25,000 had been drawn out of the firm in the shape of private expenditure. Referring to figures, it seemed that Samuel had taken out £8,400; Jacob, £8,600; and Joshua, £6,800. According to these amounts, Samuel had lived at the rate of £1,100 a-year, and Jacob at £1,600 a-year, sums quite unjustifiable. Joshua's expenditure was more moderate, and much beneath his brothers.

Mr. Lawrance—And he has in addition given up £1,500 to his separate estate.

Mr. Commissioner Fonblanque—The dividends are as follows:—Jacob pays 6*d.* on his separate estate; Samuel pays 1*s.* 5*d.*; and Joshua 10*s.* in the pound, with probably in the latter case a further distribution. The father's expenditure was moderate in the extreme, not being at the rate of more than £250 a-year, and his separate estate would pay 15*s.* in the pound. He (the Commissioner), looking at these facts, could only come to the conclusion that the private expenses of every one of the surviving bankrupts were wholly unjustifiable; Joshua's least so, Jacob's most so. The application of trust funds was a painful topic to advert to, and Samuel stood in that respect condemned, for he had misappropriated between £4,000 and £5,000 which belonged to his wife and children. In another breach of trust connected with the family of Abraham Levi, Samuel and Jacob were each identified, and their conduct deserved strong reprehension. Whenever cases of breaches of trust came under his (the Commissioner's) notice, he was inclined to exercise the administration of the law with severity. It was not only the serious evil inflicted upon the parties properly entitled to the funds misappropriated that was to be considered, but also the means such acts afforded for raising fictitious credit; and he would never suffer breaches of trust to go unpunished, because persons when reduced to extreme exigencies of the kind, must be fully aware of the desperate nature of their position. Reviewing, therefore, the conduct of the bankrupts, so far as concerned their joint estate, and also their separate estates, and taking into consideration that Joshua, although as culpable as the rest with regard to the joint estate, stood the lowest in the scale of private expenses, and was not implicated in the breaches of trust, and regulating his decision in that direction, he should award (recollecting the length of time which had elapsed since the issue of the fiat) a suspension of his certificate for ten months; Jacob, whose expenditure was the most excessive, and who had been a party to the breaches of trust, his certificate must be suspended eighteen months; and Samuel, who, among other misconduct, stood most implicated in breaches of trust, must be adjourned for two years. His Honour concluded by saying that as, with the exception of the breaches of trust, he did not find frauds to have been committed on creditors, he should give the bankrupts protection; but he added that when his mind was first brought to bear upon the question, he had doubted the propriety of this course, or whether he ought not to have refused granting certificates, particularly in the case of Samuel, altogether.

THE ESTATE OF CRUIKSHANK, MELVILLE, AND CO.

COURT OF BANKRUPTCY, *June* 6.

The bankrupts are thus described :—Patrick Cruikshank, John Melville, and William Fauntleroy Street, of Austin Friars, carrying on business under the title of Cruikshank, Melville, and Co. The case includes two failures—that of Cruikshank, Melville, and Co., who stopped in 1845; and that of Melville and Co., who continued to trade up to the end of 1847. The fiat bears date December 19, 1848, the interval having been engaged in an ineffectual attempt to wind-up the affairs of the bankrupts under a deed of inspection. The Governor and Company of the Bank of England, who proved on a bill discounted by them, were the petitioning creditors, and it was stated during the proceedings that this was the first fiat ever issued on their petition. The bankrupts passed their examination on the 8th of May. It appears from the balance-sheet that the depreciation in the bankrupts' West Indian produce was enormous; and the loss by the sale of "plant" in the West India islands was also excessive. The following item is in the "proceedings":—"Stevens and Crosby's patent machinery for hydraulic sugar-press, which cost £4,000: worth nothing more than the price of old iron." Messrs. Trueman and Cook, of Mincing Lane, are large creditors. There is the following entry with respect to their claim:—"The ruinous state of the West Indian property, and the depreciation of produce in the calamitous state of trade between 1845 and the present period, have caused the ruinous loss which will result on their accounts, whereby, instead of Messrs Trueman and Cook being more than covered by their securities, it is expected that, after realizing the whole, they will be creditors under this estate for upwards of £100,000." The whole amount of the claim of Messrs. Trueman and Cook is £125,460, and the principal security held by them was various estates and plantations in the West Indies.

Amongst the property set down in the balance-sheet are "50 shares, of £20 each, in mines in Cornwall, £1,094; Wylam's Patent Fuel, £1,592; shares in other mines, £198." Under the head of losses are the following items:—"Adventures to Calcutta, £3,162; adventures on wool, £1,083; adventures on cotton, £195." Amongst the property given up to the assignees were:—"Furniture and effects of 28, Eastbourne Terrace, £161 5*s*.; 20 Dendre Valley Railway shares, of no value; 50 Newport and Abergavenny ditto, of no value; 30 Southampton Dock shares, of no value; life interest in Richmond estate, St. Vincent's, mortgaged to Government for £8,150; to Mrs. Cruikshank's trustees for marriage settlement, a life policy, premium £115 per annum, with £115 arrears; and to Lady Anne Cruikshank for her annuity of £500, and arrears of same, £750; plate deposited, value £145."

The debit to the Bank of England arose in this way:—The bankrupt (Street) drew upon Messrs. Trueman and Cook for £8,200; the latter accepted the bill, which was discounted by the Bank to the bankrupt. The proceeds, however, were at once handed over to Messrs. Trueman and Cook, in part payment of their debt. The estate of Messrs. Trueman and Cook having paid 12*s*. 6*d*. in the pound, the Bank only proved for the remainder, or for £4,140, including interest.

This was the certificate meeting.

Mr. Lawrance appeared for the bankrupts; and Mr. Denton, of Messrs. Freshfield's, for the assignees; Mr. Coleman attended as accountant on behalf of the Bank of England; Mr. Wryghte prepared the balance-sheet.

Mr. LAWRANCE said he believed there would be no opposition to the certificate on the part of the assignees.

The COMMISSIONER—Are the assignees present?

Mr. DENTON—On the part of the Bank I am not instructed to offer any opposition. The Bank is satisfied to leave the case in your Honour's hands. The official assignee has drawn up a report, and by that report the Bank is willing to abide.

The COMMISSIONER—But I would like to know what the feelings and opinions of the assignees are respecting the conduct of the bankrupts. I am desirous of ascertaining whether or not the assignees are satisfied with the conduct of the bankrupts.

Mr. DENTON—I will send for Mr. Elsey.

Mr. John Green Elsey, of the discount department of the Bank, soon after arrived, when

The COMMISSIONER directed the report of the official assignees to be read aloud.

The following is the report of the official assignees :—

"The above bankrupts commenced business as merchants under the firm of Cruikshank, Melville, and Co., on the 1st of January, 1840, with a capital amounting to £31,204 2*s.* 9*d.*, which was increased the following year to £55,364 15*s.* 9*d.* They continued to carry on their business as East and West India merchants, trading also largely with New South Wales, until the latter end of the year 1845, when they commenced winding-up; and with the exception of a bill of exchange for £7,800, drawn by Cruikshank, Melville, and Co., upon Messrs. Trueman and Co., bearing date September 3, 1847 (which bill was discounted by the Bank of England, who are now the holders, and the petitioning creditors under this fiat), it does not appear that any new mercantile transaction was entered into by this firm. On or about the 7th of July, 1845, the bankrupts, by deed, assigned to Messrs. Trueman and Cook (or rather to James Fairlie, Thomas Depnall, and Robert James Rouse, two of whom were clerks in Messrs. Trueman and Cook's employ) property and debts valued at that time at £131,510 5*s.* 2*d.* (which may be considered the greater portion of their estate), in trust, to secure the payment of the debt due by the bankrupts to Messrs. Trueman and Cook, and which was taken at £89,037 4*s.* 1*d.*, the amount due on the 1st of January, 1845. Under this deed the trustees were authorized to make advances for the cultivation of the estate, and to realize at their discretion; and after payment of the debt due to Messrs. Trueman and Cook, the residue to be held in trust for the bankrupts or their creditors; but notwithstanding this assignment, the bankrupts, in their books, treated the properties and debts so assigned, as their own, the accounts being continued as though no such assignment had been made.

"The trustees have realized a portion of the securities, and the proceeds have been paid to Messrs. Trueman and Cook, who have (since the fiat) tendered an account, which has been adopted by the bankrupts, showing an excess of payments beyond receipts, thereby increasing their debt, which now amounts to £117,660 10*s.* 4*d.*

"The joint balance-sheet of Cruikshank, Melville, and Co. commences on the 1st of January, 1845, with a capital of £36,313 15*s.* 1*d.*, as appears by the original account raised in the books at that period, and which is analyzed as follows, viz.:—

CAPITAL ACCOUNT, *January*, 1845.

DEBTOR.

	£	s.	d.
To sundry creditors	£396,239	7	6
Balance carried down	36,313	15	1
	£432,553	2	7
Capital of P. Cruikshank	£12,400	16	11
Ditto of J. Melville	7,625	17	1
Ditto of R. Ramsay (who died in 1846)	4,887	9	7
Ditto of W. H. Street	8,213	17	11
Reserve fund, or profits not divided	3,185	13	7
	£36,313	15	1

CREDITOR.

	£	s.	d.
By cash in hand	£3,650	13	9
Bills receivable	27,758	16	0
Good debts	246,789	14	6
Consignments abroad	10,300	0	2
Produce in hand	12,140	5	7
Ships and shipments	39,852	5	0
Patents	5,410	16	9
Estates	76,652	17	3
Miscellaneous	8,881	4	7
Premises at Austin Friars	1,126	9	0
	£432,553	2	7
Balance brought down, being capital	£36,315	15	1

"It is difficult to ascertain at this period whether the bankrupts were justified in assuming that the debts and properties for which they take credit in their capital account, were good to the full extent at that date. The realization shows the following result, viz.:—

	£	s.	d.
1. Debtors, taken at	£246,789	14	6
2. Consignments	10,300	0	2
3. Produce	12,140	5	7
4. Ships and shipments	39,832	5	0
5. Patents	5,410	16	9
6. Estates, etc.	76,652	17	3
	£391,125	19	3

The first item (debts) has since realized about £150,000; the second (consignments), £5,700; the third (produce), £9,106; the fourth (ships, etc.), £20,300; the fifth (patents), worthless. The supposed value of the estates is £12,000, making in all £197,106, which, being deducted from the above £391,125 19*s*. 3*d*., shows a deficiency in value of £194,019 19*s*. 3*d*.—subject to a reduction by receipts from debtors still outstanding. In explanation of the great deficiency arising in the realization of the above debts and properties, the bankrupts state, that in consequence of the depreciation in the value of West India sugars, and the admission of slave-grown sugar in 1846 and 1847, many of the debtors, to whom they had made advances on account of their crops grown on the West India estates, became totally bad by reason of the produce not paying the expenses of cultivation, and the estate becoming of mere nominal value; and this statement is to a great extent borne out by the balance-sheet, which shows the following result:—

"The subjoined balance-sheet is dated December 19, 1848:—

DEBTOR.

To creditors unsecured				£40,741	12	3
Ditto secured	£142,243	0	5			
Less estimated value of securities	20,000	0	0			
				122,243	0	5
Total creditors				£162,984	12	8
Capital, January 1, 1845	£36,313	15	1			
Since added by partners	20,843	8	5			
				57,157	3	6
Melville and Co., being payment by that firm on account of Cruikshank, Melville, and Co.				9,071	6	4
				£229,213	2	6
Profits				nil		
Check on bankers gone				67	0	1
Difference in books				2	3	3
				£229,282	5	10

CREDITOR.

By good debts	£271	10	10
Doubtful debts, estimated to realize	365	2	3
Property	100	0	0
Total estimated assets	£736	13	1
Banker's balance	nil		
Claimed by bankers	13	1	10
Losses	196,785	6	10
	£197,535	1	9
Charges of trade	25,967	4	11
Drawings of partners	5,779	19	
	£229,282	5	10

"The losses which form the principal features in the balance-sheet, as accounting for the total absorption of the capital and deficiency of assets are thus analyzed:—

	£	s.	d.
By West India estates, being a difference between their cost and present value	£90,535	2	1
Ships	20,573	2	8
Bad debts	68,359	15	3
Adventures to Sydney and elsewhere	10,560	16	7
Merchandise, etc.	6,756	10	3
	£196,785	6	10

The charges of trade consist of clerks' salaries and rent of business premises, and also include a sum of £18,689 8*s.* 3*d.* for interest on loans and discount of bills.

"The firm of Melville and Co. commenced business on the 1st of July, 1846 (no dissolution of the firm of Cruikshank, Melville, and Co. having taken place), and consisted of John Melville and William Fauntleroy Street. This firm, as well as Cruikshank, Melville, and Co., carried on business at 13, Austin Friars, but abandoning the West Indian business, confined their operations, with some exceptions, to the East India and Sydney trade. It will be seen by the balance-sheet the capital of the bankrupts was nominal; but it appears that they were in negotiation with Mr. H. H. Oddie, to take his son, Mr. John Oddie, into partnership, on his attaining his majority; and by this arrangement (which was not finally completed), a sum of £14,635 8*s.* 9*d.* was paid in by Mr. H. H. Oddie, as part of the intended capital of £20,000, but is now claimed against the estate. The balance-sheet, which is supported by the books, shows the following results:—

December 19, 1848.

				£	s.	d.
To creditors unsecured				£27,884	19	7
Ditto secured	£23,544	7	3			
Less estimated value of securities	£23,544	7	3	0	0	0
H. H. Oddie, paid in by him as part of intended capital				14,635	8	9
				£42,520	8	4
Liabilities, being acceptances of the bankrupts against consignments shipped on drawer's account				38,224	9	4
Capital of J. Melville	£247	14	0			
Ditto of W. F. Street	242	14	11			
Ditto of J. Oddie	250	0	0			
				740	8	11
Profits				0	0	0
				£81,485	6	7

CREDITOR.

	£	s.	d.
By good debtors	£246	7	1
Doubtful debts expected to realize	500	0	0
Property	81	0	0
	£827	7	1
Losses	£29,099	4	1
Cruikshank, Melville, and Co.'s balance due by them	11,287	9	5
Liabilities	38,224	9	4
Charges of Trade	1,898	2	4
Partners' drawings	148	14	4
	£81,485	6	7

"The losses, which in this case form a considerable item, are thus analyzed, viz.:—

	£	s.	d.
By adventures to Calcutta	£4,247	7	9
West India Indigo, etc.	15,004	8	7
Shares in mines, etc.	2,881	2	7
Bad debts	6,966	5	2
	£29,099	4	1

"The debtor balance against the firm of Cruikshank, Melville, and Co. arises from accounts which the firm of Melville and Co. adopted, and which were transferred from the books of Cruikshank, Melville, and Co., to Melville and Co. The charges of trade consist of salaries to clerks, law costs, and rent of business premises. The drawings of the partners, it will be seen, are small, amounting to £148 14*s.* 4*d.* The books of the firm are well kept and balanced.

"E. Watkin Edwards."

The Commissioner—Were all the creditors aware of this assignment to Messrs. Trueman and Cook?

Mr. Brown (one of the assignees)—I was.

The Commissioner—Were the creditors generally?

Mr. Brown—I cannot say.

The Commissioner—Was due notice given of it?

Mr. Lawrance—I should say, generally speaking, the creditors were aware of it.

The Commissioner—There seems to have been some strange conduct on the part of the bankrupts. It appears that after the assignment was executed, the bankrupts drew a bill upon Messrs. Trueman and Cook for £8,000, which bill having been duly accepted by the latter, was discounted by the Bank of England. That does not seem in accordance with the usual custom of trade.

Mr. Lawrance—But I beg to remind your Honour that when my clients drew upon Messrs. Trueman and Cook there was produce of theirs in the hands of Messrs. Trueman and Cook worth upwards of £10,000. The bill was, in fact, drawn by Mr. Street without the cognizance of the other partners, so that whatever blame may attach to the transaction, Messrs. Cruikshank and Melville are not at least morally answerable.

The Commissioner said the act must be considered as the act of the firm.

Mr. Lawrance—It is clear, at all events, that the bankrupts had an honest intention in drawing the bill, and could have had no other, for the moment the Bank of England discounted it, the proceeds were handed over to Messrs. Trueman and Cook, to whom at that period, as well as at the present, the bankrupts were largely indebted. Messrs. Trueman and Cook, owing to the adversity of the time, were in great want of money themselves; they naturally pressed the bankrupts for the liquidation of at least a portion of their debt, and as Messrs. Trueman and Cook had been their best friends, and had always been ready to advance money on their plantations and growing crops, the bankrupts could not easily have refused to take this course.

Mr. Brown, in reply to the Commissioner, again stated that he had no opposition to offer.

His Honour insisted to know something more respecting the assignment to Messrs Trueman and Cook.

Mr. Lawrance said the assignment was not for Messrs. Trueman and Cook's exclusive benefit. They were the managers of the bankrupts' West India property, and made from time to time the requisite advances. The result was so far from being advantageous to Messrs. Trueman and Cook, that whereas they were only creditors for £89,000 when the deed was executed (July, 1845), their debt was by reason of that assignment increased to £117,000. In short, those gentlemen were losers to the extent of nearly £30,000 by this transaction.

The Official Assignee—That sum includes interest.

Mr. Lawrance—As regards the bill drawn by Mr. Street on Messrs. Trueman and Cook, and discounted by the Bank of England, I may add that the Bank has been paid 12*s.* in the pound upon it out of the estate of Messrs. Trueman and Cook, and a further dividend is expected. As regarded the general aspect of the case, he would make a few observations. The bankrupts, finding the West Indian trade a most unprofitable one, determined to give it up. At the time they did so they anticipated losses from their West Indian property, but to nothing like the extent which subsequently occurred. Cruikshank withdrew from the firm, but Melville and Co. determined upon carrying on those branches of the business which they thought least liable to fluctuation. What chiefly induced them to take this course was, that a gentleman named Oddie, with whom they were intimate, offered to place his son in the firm, and to advance £20,000 as his capital. He held in his hand a deed, dated July, 1846, made between Messrs. Melville and Street and John Oddie, which was a deed of co-partnership between the parties just mentioned, but which was not executed, because John Oddie had not then attained his majority. Oddie's father died, and the aspect of affairs of the house of Melville being altered for the worse, he never became a partner, but advanced to the firm about £14,000, for which sum he was now a creditor against the estate. In 1847 the bankrupts stopped payment, finding it impossible to hold out longer. From that period their affairs were virtually in the hands of their creditors until the fiat was issued at the instance of the Bank of England.

Mr. Denton—The affairs of Melville, Cruikshank, and Co. were not in the hands of their creditors, but in the hands of Trueman and Cook. It was for the purpose of investigation that the fiat was issued.

The Commissioner—Am I now to understand that the Bank is satisfied with the conduct of the bankrupts as regards Trueman and Cook? Are they satisfied with their conduct as merchants and traders?

Mr. Denton said the Bank did not wish to pursue a course different from that pursued by the body of the creditors. Their main object was to obtain a full and searching investigation, and for that purpose the fiat was issued. The information of the assignment of property to Messrs. Trueman and Cook came upon the Bank by surprise. They procured inquiry, the result of which was now before the Court, and would be satisfied with his Honour's decision upon the facts before him.

Mr. Lawrance said it was plain there was no collusion, or the slightest tinge of anything like dishonest or dishonourable conduct on the part of his clients, for if it were otherwise, the Bank would have undoubtedly opposed them to the last.

The Commissioner—Do you think that the Bank of England would have discounted that bill, of which they were the drawers, if they knew that they had assigned all their property over to a creditor, and that they were utterly insolvent?

Mr. Lawrance—Probably not. The bill was in all likelihood discounted in consequence of the acceptors being Messrs. Trueman and Cook, who had then unlimited credit.

The Commissioner—The Bank did not know it was an accommodation bill; but the Bank did not merely look to the names of Messrs. Trueman and Cook, but to those of the bankrupts also. He believed it was their custom to regard every name on a bill.

Mr. Lawrance said he believed the names of his clients stood high in the Bank, but Messrs. Trueman and Cook stood higher still.

Mr. Elsey was understood to say that when the bill was discounted, the Bank was not aware of the position of the bankrupts' affairs; and he was also understood to say (but he spoke in so low a tone that he was not distinctly heard) that the Bank considered the bankruptcy rather the result of adverse circumstances, over which they had no control, than of actual misconduct.

Mr. Lawrance remarked that his clients had opened an account at the Bank for a considerable time. It would be, no doubt, better if the transaction to which the Court had referred had not taken place; but, after all, it was an isolated occurrence. Messrs. Trueman and Cook had value in their hands; it was to satisfy their just claim the money was obtained; only one of his clients was a party to it; and although he admitted there might be irregularity and impropriety in the transaction, there was nothing worse. But he had now to call the attention of the Court to the main features of this truly lamentable case. Here was a house of first-rate respectability, which had carried on a most extensive trade with different parts of the world —which started with a large capital—which for more than twenty years carried on a most prosperous business—which could not be accused of wild or excessive speculation—which kept its books in the most regular manner—and which had a name and a credit scarcely second to those of any British merchant—here was this eminent firm completely crushed and ruined. And ruined by what? By an act of the Legislature. Did he state anything novel—was the case of this house singular? Not at all; but the wonder would have been if, amongst the general wreck, his clients' credit and fortunes could have survived. He would refer his Honour to an admirable work which he had had often occasion to quote—the *Commercial Crisis*, by Mr. Evans, and in the second edition of that work his Honour would perceive the awful extent of the calamity and the enormous sacrifice of property that had taken place in consequence of the Sugar Bill. He would read to his Honour the following authentic tabular return, showing the depreciation which had occurred in West India property:—

	Slavery value.	After abolition.	After abolition of apprenticeship.	Since passing Sugar Bill of 1845.
	£	£	£	£
Windsor Forest Estate	120,000	60,000	45,000	5,000
La Grange Estate	65,000	32,000	26,000	5,000
Belle Plaine Estate	55,000	27,500	23,000	3,500
Rabacca Estate	80,000	30,000	20,000	6,000
Prospect Estate	70,000	25,000	17,000	3,000
Richmond Hill Estate.........	45,000	20,000	15,000	5,000
	435,000	194,500	146,000	27,500

Slavery value..	£435,000
Estimated present value	27,500
	£407,500

Or equal to 93½ per cent. on original value.

Mr. Lawrance concluded by reminding the Court of the monetary pressure at the latter period of the bankrupts' trading, when even Mr. Gurney, the banker, according to his evidence before the Parliamentary Committee, quoted in the *Commercial Crisis,* declared he was obliged to pay 9 per cent. for money. Taking all these things into account—recollecting that the creditors were satisfied with the conduct of his clients—the long interval that had elapsed since their stoppage—the regularity of their book-keeping—the high and honourable reputation they had long borne—the disgrace that would attach to even the shortest period of suspension—bearing in mind that they were the victims of legislation, and that if protection had been continued to our colonies, this house, so far from being insolvent, would be extremely wealthy; and also remembering that since the fiat they had given every assistance to their creditors, he felt satisfied the Court would at once grant the certificate.

The Commissioner said that everybody must lament the severe losses those gentlemen had sustained in the course of their trading. If he were to consult his own feelings, nothing would afford him greater pleasure than to grant the certificates at once, and allow them forthwith to resume their position as merchants; at the same time, he must not forget what was due to the public, and to the principles upon which trade ought to be regulated. The bill transaction with Messrs. Trueman and Cook, at a time when the bankrupts were winding-up their affairs, seemed a most unmercantile one, and required consideration. The high position the bankrupts occupied made it the more necessary to scrutinize anything doubtful in their conduct, and if the charge were established, to award a proper measure of punishment. He did not mean to say that he would not take into account the effect of a suspension of the certificate upon such men, nor the causes which had been mainly productive of their losses. As to what had been said of the "ruinous policy" of Parliament in removing protection from our colonies, he had nothing to do, but might remark that it frequently happened that legislative enactments which inflicted individual suffering were productive of great public advantage.

Judgment deferred

Court of Bankruptcy, *August* 8.

(*Before Mr. Commissioner Holroyd.*)

His Honour gave judgment in this case this morning as follows:—The judgment in this case has been somewhat deferred, in consequence of the bankrupts wishing to lay before the Court a memorial or further statement explanatory of the assignment to Messrs. Trueman and Cook, and of the discount transactions with the Bank of England. The statement has been communicated to the solicitors of the Bank of England and of Messrs. Trueman and Cook. The latter assent to its reception, and the Bank does not oppose it. Now, of those cases in which the Court of Bankruptcy is from time to time called upon, either by creditors or from what appears on the balance-sheet, to refuse or suspend a bankrupt's trader's certificate, there are many which require careful examination and anxious thought, in order rightly to discriminate between accident and fraud, misfortune and misconduct; but I have rarely had any question to dispose of which has given me so much painful reflection as the case upon which I am now to pronounce judgment. I have felt and been impressed with the acknowledged respectability and the previous high commercial position of the parties before the Court, as well as with the heavy and severe losses which they undoubtedly have sustained. On these grounds I more deeply lament the obligation upon me to add that there are circumstances in this case exhibiting conduct inconsistent with that openness and frankness of dealing which are the great and peculiar characteristics of a British merchant, and which a banker more especially expects at the hands of a merchant by whom he is solicited for pecuniary accommodation. Indeed, the merchants of this country, as a body, are a state and degree of persons not only to be respected, but to be regarded at all times with the most lively interest. It has been truly said of our merchants, "they are *vena porta*, and if they flourish not, a kingdom may have good limbs, but will have empty veins and nourish little." Hence the shock given to credit, and the consequent alarm and anxiety which for a time prevailed after the late extensive failures when (to use the words of a great man applied on another occasion) "it was in truth a crisis to try men's souls; for a while all was uncertainty and consternation; all were seen fluttering about like birds in an eclipse or a thunder-storm; no man could tell whom he might trust; nay, worse still, no man could tell of whom he might ask anything." At this critical juncture her Majesty's Government fortunately interposed its hand, and by a temporary measure of a remedial nature, restored the confidence which is necessary for carrying on the ordinary dealings of trade. With reference to this period, valuable information is contained in a recent publication, the *Commercial Crisis*, 1847-48. Mr. D. Morier Evans's work furnishes an interesting though painful memorial of the combination of causes which led to the downfall of so many of the large commercial establishments in this kingdom. There is matter, moreover, in the same work which may be of use, and operate as an instructive warning to the mercantile and manufacturing community for time to come. It will not admit of doubt that when a want of confidence is once created, some may find themselves unable to meet their engagements, although they may be perfectly solvent; but comparing the liabilities of many of the failed firms with their assets, I question whether it would not have been much better for themselves and their creditors if a greater part of them had been obliged to stop much sooner than they actually did. The danger of a departure from sound principles in the conduct of commercial affairs—the system of over-trading—the locking up of capital—the disregard of the necessity of a due proportion between capital and credit—the making no sufficient distinction between borrowed and real capital in the nature and amount of risk incurred—the abuse of the facility afforded for procuring aid through the discount of bills, by incurring liabilities in this way beyond all prudent bounds—the illegitimate mode of obtaining advances upon forthcoming crops, and upon goods in many cases before they were shipped, and in others before they were manufactured—these several circumstances, not to mention the princely establishments in some instances kept up, are disclosures apparent from the balance-sheets and their results of the majority of the houses which fell in the city of London during that memorable period. Lord Bacon says, "Young men in the conduct and manage of actions embrace more than they can hold, stir more than they can quiet, and fly to the end without consideration of the means." May not a like moral and intellectual blindness be aptly attributed to some of our too sanguine and speculative merchants in their late commercial career? The assertion I am now going to make may possibly, to some persons, seem paradoxical at first sight. I deplore the sufferings occasioned by our late difficulties, and sympathize with those who have been unfortunately exposed to privations, and to which I fear many are still subject; but, at the same time, I believe that our past misfortunes afford a circumstance favourable to our future hopes. Undoubtedly the most likely ways and means are not always effectual for the attaining of their end; there are secret workings in human affairs which overrule all human contrivance, and counterplot the wisest of our counsels, "The race is not to the swift, nor the battle to the strong, neither yet bread to the wise, nor yet riches to men of understanding, nor yet power to men of skill; time and chance happeneth to them all." But, however inclined we may be to think that the distresses of the commercial interests have been in a great measure occasioned by circumstances over which the merchant had no control, a sober review of past occurrences must convince us of the imprudent abandonment by many of our men of business of the ordinary rules of commercial reasoning. They have at least gained experience by what has happened, and "experience is the mother of wisdom," and we are taught sagacity by exposure to misfortune. I look, then, for greater prudence in commercial projects, and a sounder system of trade; above all, I trust that merchants and traders will be more prone to remember, and act upon, the admonition, that paper credit must ever bear a due proportion to the quantum of their existing capital, or, in other words, to their real intrinsic ability or wealth. By thus keeping the accommodation system within its proper legitimate limits, and preventing an undue stimulus to trade, with its attendant reaction—by thus using the legitimate means we may hope for good success, and be expectant of a more permanent increase of that commerce, which, "whilst it is the main source of strength and power to this country, contributes in no less degree to the happiness and civilization of mankind."

> "Instructed ships shall sail to quick commerce,
> By which remotest regions are allied,
> Which makes one city of the universe,
> Where some may gain and all may be supplied."

But to proceed with the case before the Court. The firm of Cruikshank, Melville, and Street was one in the list of houses whose failures were announced during the eventful time of which I have been speaking. They carried on business as East and West India merchants. Messrs. Trueman and Cook, colonial brokers, in the years 1843 and 1844 were largely under advances to the bankrupts, and on the 31st of December, 1844, the debt due to Trueman and Cook amounted in round numbers to £89,000. As security for this there were from time to time deposited title deeds and other documents relating to several plantations, ships, properties, and undertakings by letter in the nature of equitable charges. The total amount of the creditors of Cruikshank and Co. on the 1st of January, 1845, was about £396,000; but taking their debts and properties at that time (consisting of good debts, consignments abroad, produce in hand, ships and shipments, patents and estates) at the value assumed in the capital account, there appears a balance in favour of the firm amounting to about £36,000, and although in general the value of the property is over estimated by the owner, I cannot say that the bankrupts were not justified in estimating the debts due to them and their properties as good to the full extent for which they take credit in their capital account on the 1st of January, 1845; but the realization since, actual and expected, shows the sad and ruinous loss of above £190,000, or nearly 50 per cent. The large amount of the liabilities of the house of Cruikshank, Melville, and Co., in January, 1845, is, I think, thus accounted for:—The bankrupts say, "The productiveness of West India estates had been reduced, by the emancipation of the negroes and by the abolition of the apprenticeship, to an unprecedentedly low point, and their recovery from this great depression, although in progress, had been slower than had been anticipated either by the Government or by those interested in these properties, thus rendering a much larger advance of capital necessary than had been expected." Now, it is matter of history that "from 1842 discounts had been easy and money plentiful; the funds maintained a high rate, low interest could only be obtained. In 1844 it was remarked that there had been a larger continuance of a plentiful supply of money than had occurred in the memory of the oldest capitalists." This state of the money-market, and a hope of the revival of colonial affairs, might naturally induce the bankrupts to go on longer and to a greater extent with borrowed means than they would have done could they have foreseen or even conjectured the calamitous course of events which the veil of futurity at that time concealed from them. The official assignee in his report says:—"In explanation of the great deficiency arising in the realization of the above debts and properties of the bankrupts, they state that in consequence of the depression in the value of West India sugars, and the admission of slave-grown sugar in 1846-47, many of the debtors to whom they had made advances on account of their crops grown on the West India sugar estates, became totally bad by reason of the produce not paying the expenses of cultivation, and the estates becoming of mere nominal value, and this statement is, to a great extent, borne out by the balance-sheet, which shows the following result"—then, after giving the debtor and creditor account on the balance-sheet, he adds, "the losses which form the principal features in the balance-sheet, as accounting for the total absorption of the capital and deficiency of assets, are thus analyzed :—

By West India estates, being a difference between their cost and present value	£90,535	2	1
Ships	20,573	2	8
Bad debts	68,359	15	3
Adventures to Sydney and elsewhere	10,560	16	7
Merchandise	6,756	10	3
	£196,785	6	10

I find, then, at the time of this bankruptcy, the firm of Cruikshank, Melville, and Co., having been in a course of liquidation for above three years, owe about £192,000, and their loss, as I have before stated, is above £196,000. The capital of Mr. Cruikshank in January, 1845, and added since, was about £25,006, part being borrowed on security of property; of Mr. Melville, between £17,000 and £18,000 all his own; and of Mr. Street (principally borrowed), between £9,000 and £10,000. The trade expenses (including £18,689 for interest on Trueman and Cook's debt), amount to £25,967 up to July, 1846, when the firm of Melville and Co. commenced. The drawings of the partners during the same period are as follows:—P. Cruikshank, £2,234; J. Melville, £1,133; and W. F. Street, £1,014. The assets under this estate for creditors will be nothing. Under the partnership of Melville and Co., formed the 1st of July, 1846, after about fifteen months' trading, I find the unsecured creditors amount to £27,884, in addition to £14,635 claimed against the estate, being the sum advanced by Mr. Oddie as part of the intended capital of £20,000 of his son, then a minor, who, when of age, was to have joined the firm. The greater portion of this sum was applied in discharge of debts of the old firm of Cruikshank, Melville, and Co., which were adopted by Melville and Co. to secure to their house certain mercantile accounts. The liabilities are £38,224; the losses about £29,000. The official assignee says, "The lease of the premises in Austin Friars, which was given up by Messrs. Trueman and Cook, has produced £300, and that there is no chance of any other assets." Under Mr. Cruikshank's separate estate, I find :—

The creditors unsecured	£32,291	0	0
Creditors holding security	27,704	0	0
Liabilities	20,871	0	0
His capital, January 1, 1845	27,380	0	0
His losses, about	44,000	0	0

About £160 has been realized under this estate, and nothing more is expected. Under Mr. Melville's separate estate I find :—

The creditors (secured and unsecured), £33,510; securities valued at £14,262, leaving unsecured	£19,248	0	0
Liabilities	9,000	0	0
Capital, January 1, 1845	2,790	0	0
Losses	3,858	0	0
Property given up, estimated at £1,429, but realized about	1,100	0	0

Under Mr. Street's separate estate, I find :—

The creditors unsecured	£31,535	0	0
Creditors holding security	1,280	0	0
Liabilities	7,009	0	0
His capital, January 1, 1845	482	0	0
Losses, about	10,848	0	0
Bad debts	12,083	0	0
His property, about	1,500	0	0
Good debts	123	0	0

Of this about £930 has been realized, and nothing more is expected.

Now, looking at the figures of the different items in the several balance-sheets to which I have just adverted, I think we have but an unfavourable comment on the latter part of the bankrupts' mercantile course; nevertheless, under the peculiar circumstances of the times, and considering the period which has been suffered to elapse since their failure before the issuing of a fiat, I would gladly have confined my attention to the losses sustained by the bankrupts, and in such view I would fain have treated this case simply as an ineffectual attempt by a merchant to wind up his affairs *dehors* the Court of Bankruptcy; but, though inclined to make great allowance for gentlemen struggling against a heavy weight of adversity in their affairs, the course which the bankrupts pursued, upon finding themselves placed in such difficulty as not to be able to meet their ordinary engagements in the usual course of business, compels me to go further. On or about the 7th of July, 1845, the bankrupts, by deed, assigned to Messrs. Trueman and Cook, or, rather, to James Fairlie, Thomas Depnall, and Robert James Rouse, two of whom were clerks in Messrs. Trueman and Cook's employ, property and debts valued at that time at £131,510 5*s*. 2*d*. (which may be considered the greater portion of their estate), in trust to secure the payment of the debt due by the bankrupts to Messrs. Trueman and Cook, and which was taken at £89,037 4*s*. 1*d*., the amount due on the 1st of January, 1845. Under this deed the trustees were authorized to make advances for the cultivation of the estates, and to realize, at their discretion, and after payment of the debt due to Messrs. Trueman and Cook, the residue to be held in trust for the bankrupts or their creditors; but, notwithstanding this assignment, the bankrupts, in their books, treated the properties and debts so assigned as their own, the accounts being continued as though no such assignment had been made. The trustees have realized a portion of the securities, and the proceeds have been paid to Messrs. Trueman and Cook, who have, since the fiat, tendered an account, which has been adopted by the bankrupts, showing an excess of payments beyond receipts, thereby increasing their debt, which now amounts to £117,600 10*s*. 4*d*. In the memorial which the bankrupts have laid before me, they are desirous of drawing my attention to the chief motives and reasons of this assignment, and that it should not be judged of by the result; to the importance of carrying on effectively and continuously the cultivation of the West India properties—that by the assignment in trust this object was considered to be secured for the benefit of all concerned, as well as of Messrs. Trueman and Cook, and that it was fully expected, at the time it was executed, that the properties would not only eventually supply the means of paying all claims, but would leave a surplus for the partners. In this view, they say it was considered most desirable to obtain the hearty co-operation of Messrs. Trueman and Cook; and that this was really effected, the further heavy advances made by them subsequently afford ample evidence. They further say:—"This assignment, although not formally announced by circular, was known to our bankers, to the proprietors of those West India estates upon which we had been obliged to make such heavy advances, to nearly the whole of our unsecured creditors, and very generally in the money-market; indeed, we may say that the support we were receiving from Messrs. Trueman and Cook, pending the expected recovery of the West India estates, was all but universally known." The bankrupts state that "the assignment to Trueman and Cook was not wholly a new transaction, but was, to a great extent, the merely putting into a formal legal shape several equitable assignments, which had been made to them at various dates previously, for advances which they had made to us; neither was it an assignment of our entire estate and effects, but, on the contrary, we had other assets in hand, and forthcoming, from which we did actually make payment to other creditors." Notwithstanding the further explanation thus given by the bankrupts, it appears to me, looking at the schedule of properties assigned by that deed, and regarding it in connection with the assignment of property to Martin, Stone, and Co. (their bankers), that the statement of the official assignee is fully borne out. I think there can be no doubt but that the bulk of the bankrupts' property passed from them under the assignment of Trueman and Cook, and it is also clear that the assignment conveyed to Trueman and Cook much property over which they had no previous equitable charge, and herein the bankrupts were not only following a mistaken judgment, but they proceed to carry out their views in a culpable manner, and, although success would have admitted of no examination, the contrary allows of no justification of the means. The moment the bankrupts signed that deed their independence as merchants was gone. Amongst the propety conveyed, I observe the premises of the bankrupts in Austin Friars, in which they carried on their business. Considering the state of the bankrupts' affairs at the time the assignment was made to Trueman and Cook, and regarding the assignment itself with reference to the circumstances of the transaction at the time, I think the bankrupts ought not to have taken that step without first calling their creditors together, and obtaining their consent to such a course; but, having taken it, they are chargeable with another breach of commercial duty, in treating the properties and debts so assigned as their own, and continuing the accounts as if no such assignment had been made. The Bank of England, too, were ignorant of this assignment; the circumstance came upon them by surprise. Now, the intention of Messrs. Cruikshank and Co., in making this assignment, may have been good, and a good intention may in some measure extenuate; but it will not be an excuse to justify a deviation from the plain and obvious rules of duty. To clamber over fences of duty, to break through hedges of right, to trespass upon hallowed enclosures, may seem the most short and compendious way of getting "where one would be, but doth not a man venture breaking his neck, or scratching his face, incurring mischief and trouble thereby?" The bankrupts, having thus divested themselves of the bulk of their property, afterwards carry on their business in their own names, apparently as free agents, and on their own account, and although, as between themselves, it might be understood they were to enter into no new engagement, still, in all contracts concerning negotiable paper, as well as in any other usual mercantile transaction, any one partner had full power to bind the firm, the partners, by their course of proceeding, gave reciprocally to each other the power of acting, the one for the other, in the partnership business, in the same manner as if no assignment had been made to Trueman and Cook; thus each partner was the cause or occasion of confidence or credit being subsequently reposed in any one of the partners in any matter within the scope of the partnership concerns. It was during this period, namely, in August, 1847, that Mr. Street, without consulting his partners, took the bill for £7,800, which was drawn by him in the name of his firm (Cruikshank, Melville, and Street), and accepted by Trueman and Cook, to the Bank of England for discount. The bill was discounted, and the amount handed over to Trueman and Cook for the purposes of the assignment. In the memorial to which I have referred, the bankrupts make a further statement, more fully detailing the circumstances under which the bill for £7,800 was drawn. The explanation there given, however, does not appear to me to absolve the bankrupts from culpability. Mr. Street wishes that the responsibility of that act should rest with him, as the bill was drawn and sent into the Bank of England for discount without his consulting Mr. Cruikshank or Mr. Melville respecting it; but, for the reason which I have before given, although the main weight of blame may lie upon Mr. Street, Mr. Cruikshank and Mr. Melville cannot altogether exculpate themselves at the expense of their partner. With respect to the discount of that bill, the bankrupts say that "Mr. Street being aware how very generally it was known in the City that Cruikshank, Melville, and Co. had been for nearly two years in liquidation, and, also, that during that liquidation, they were being supported by Messrs. Trueman and Cook, he considered that the bill was discounted by the Bank entirely in reliance upon the names of Trueman and Cook, whose credit stood very high, as is evidenced by the fact that, at the time of their suspension, the Bank held their acceptances to the extent of more than £60,000. Judging, however, of the discount of that bill by the ordinary rules applicable to such transactions, and from which I feel in no degree released by what Mr. Street has said upon the matter, and in whose reasoning I think there is more art than solidity, I cannot come to any other conclusion than that the Directors of the Bank of England were in reality deceived; and the deception was, I think, as real, though it might not have been so palpable, as if the bankrupts had made a direct false representation of the state of their affairs. When Mr. Street applied to the Bank of England to discount that bill, I think he must have felt that the basis of the transaction was a belief by the Bank of the responsibility of his firm (the drawers of the bill), as well as of Trueman and Cook (the acceptors). "To all moral purposes, and, therefore, as to veracity, speech and action are the same, speech being only a mode of action." Then, again, what public establishment, or even private bankers (unless otherwise involved with the parties), would think of discounting a bill on the credit, wholly or in part, of any house, if they knew it to be in course of liquidation, and to have made an assignment of the bulk of its property primarily for the benefit of one creditor? Indeed, to a question put by the Court, it was stated that the Bank of England would not have discounted that bill if they had been informed that Cruikshank and Co. were winding up, and had assigned their property to Trueman and Cook. Here the *suppressio veri* was calculated to produce the same mischief, and is consequently open to the same censure, as the *suggestio falsi*. In viewing this transaction with the Bank of England, it should also be remembered that that establishment is generally resorted to in periods of distress, or, when any circumstances occur to occasion a pressure in the money-market, she then becomes, as it were, a bank of support. Her advances are looked for in seasons of distress and difficulty, and are required to be prompt. Surely, then, the most perfect good faith should be observed by persons applying to her for aid. Her freedom of action and power to render assistance should be in no way impeded by suspicions, on her part, of the *bonâ fides* of the parties who seek her help. No falsehood or deceit is to be endured in any contract, least of all with a body ready to promote the interests of the mercantile classes on such trying occasions. There is another circumstance affecting Messrs. Melville and Street. I allude to the arrangement, in July, 1846, for the new partnership with the son of Mr. Oddie, with the obligations of the old firm still upon them, and the more questionable effect of the deed to Trueman and Cook hanging over them. I should observe, however, that, at a meeting of the creditors of this firm, held in April, 1848, it was resolved that the estate should be wound up by the partners under inspection; and it would not have come before this Court but for the fiat against the original firm. Such, then, are the circumstances of this distressing case. Mr. Lawrance, in his observations to the Court on behalf of the bankrupts, remarked strongly on the effect which a suspension of the certificate must have upon the minds of gentlemen in their position in life. I am asked, too, by the bankrupts to consider the long period of time during which they have been prevented from employing themselves usefully for themselves and their families, and I am reminded that no creditor opposes the application for their certificate, and they submit whether either transaction (which I have observed upon), if viewed in the most unfavourable light, can become a subject for the censure of the Court, where not called for by the creditors themselves. Now, although I am sensible that the same degree of punishment will operate with greater severity on one class than on another, and though I am willing to consider the length of time the bankrupts have been deprived of employment, and to give some weight to the absence of opposition by creditors, I must consider the influence which the example of our great merchants has upon the industrious part of the nation—that they are necessarily the leaders and conductors of the whole industry of every nation. If offences be committed by them, can it be contended that they deserve more sympathy than ignorant or inferior men? I am bound also to bear in mind that this fiat was issued by the Bank of England for the purpose of investigation, and although the Bank of England does not oppose the allowance of the certificate, but thinks it more consistent with her duty, as a public body, to hold, as it were a neutral position, taking no part, directly or indirectly, either for or against the bankrupts, it is the duty of the Court to give judgment upon the result of that examination. Before I conclude, let me observe that, in deciding upon the rectitude or obliquity of a man's commercial conduct, it is of the greatest importance that the Court should be governed by known and general rules, as the standard of right and wrong. An action which would be adjudged an offence in the small trader cannot be deemed a venial transgression in the merchant. The same sort of actions must be generally permitted or generally forbidden. How else can we expect to produce a proper influence or effect upon the conduct of others? So, in visiting any commercial delinquency, we must look to general consequences rather than particular consequences, and the general consequence of any action must be estimated by asking what would be the consequence or mischief from the general impunity or toleration of actions of the same sort. Applying this question to the assignment to Trueman and Cook, and to the course of proceeding under it, and to the discount transaction with the Bank of England, to mercantile men the answer must be obvious—*res ipsa loquiter*. Moved by these several considerations, after maturely weighing all the circumstances of this case, the Court feels bound to adjudge that the certificates of the bankrupts be suspended—that of Mr. Street for twelve months, and of Mr. Cruikshank and Mr. Melville respectively for six months, from the day of hearing the application.

The Court, on application, granted protection.

THE ESTATE OF BARCLAY BROTHERS AND CO.

A circular was issued on the 13th March to the creditors of Messrs. Barclay Brothers and Co., whose suspension, with liabilities for £400,000, took place on the 13th October, 1847, announcing a second and final dividend of 2*s.* 6*d.* in the pound. The circumstances under which this further payment is to be effected reflect honour upon the family connections of the house, and are described in the following minute accompanying the notice :—

"At a meeting of the inspectors of the estate of Barclay Brothers and Co., the usual statement of realizations and liabilities was laid before the meeting, from which it appeared that there were no funds in hand from which any dividend could be paid, nor any reasonable hope of further realization to any extent; and that the Mauritius property having, in the first instance, failed to meet the debts of the Mauritius creditors, there was but little prospect of any surplus to the general creditors from that property.

"An offer, on the part of Messrs. Hedworth and Alexander Barclay, was brought under consideration.

"This offer was to give up all claim to dividends on their debts, and to put at the disposal of the estate a sum required to pay a further and final dividend of 5*s*. 6*d*. to the other creditors.

"The offer was found to involve a total sacrifice, on the part of Messrs. Hedworth and Alexander Barclay, of about £40,000.

"The inspectors were of opinion that it would be greatly to the advantage of the creditors to accept this offer, and that, on payment of it, an immediate release should be given to the partners in the late firm.

"It was, therefore, the opinion of the inspectors that Messrs. Barclay Brothers and Co. should announce the offer, and their preparation to pay the 2*s*. 6*d*. in the pound, on the necessary release being executed by the creditors.

(Signed) "J. G. Hoare.
"Kirkman D. Hodgson.
"J. H. Pelly."

THE ESTATE OF MESSRS. TRUEMAN AND COOK.

A circular was issued on the 13th July, by the inspectors of the estate of Messrs. Trueman and Cook, whose suspension took place in November, 1847, with liabilities for £577,547, stating that a further dividend of 1*s*. in the pound will be paid on the 31st inst., whereby the liabilities of the firm will be reduced to £60,366, and recommending for acceptance a proposal by Mr. Cook to take the remaining assets in charge, rendering himself responsible, whatever amount they may realize, for a further dividend of 1*s*., to be paid on the 30th of June, 1851. The Bank of England, it appears, have agreed to the recommendation, and the concurrence of the other creditors is fully expected. When the final dividend of 1*s*. shall have been paid, the total received by the creditors will then have amounted to 14*s*. in the pound, and the liabilities will have been reduced to about £40,000, while a further diminution will still be possible, owing to the intention stated by Mr. Cook, that, if the assets transferred to him should yield anything beyond the amount necessary for the proposed dividend, he will apply it for the equal benefit of the creditors. By collateral realizations, many of the claimants on this estate have now received, in all, considerably more than 14*s*. in the pound, and, looking at the circumstances which have occurred since the suspension, the small dividends from some of the chief debtors, by whose failure the firm was originally brought down, and the depreciation in the value of colonial produce consequent upon the continental disturbances, there is reason, perhaps, to consider that the result has been less unfortunate than might at one time have been feared. The application of Mr. Cook to the administration of the assets is stated by the inspectors to have been unceasing.

Annexed are the official documents:—

"At a meeting of the inspectors of the affairs of Messrs. Trueman and Cook, held on the 19th June, 1850, the report already sent by Mr. J. E. Coleman, with the sanction of the inspectors, to the Bank of England, came under consideration.

"It was deemed expedient to declare a further dividend, as suggested by the report, of 1*s*. in the pound, whereby the liabilities, originally £577,547 10*s*. 1*d*., will be reduced to £60,366 8*s*. 10*d*., Mr. Cook offering, as on the last occasion, to make the bills in hand available for the purpose.

"A dividend was therefore directed to be announced to the creditors, to be paid on the 31st day of July next.

"The inspectors then took into their consideration the remaining assets of the concern, and the means of bringing the affairs to an early close.

"They felt this to be desirable to the creditors, and due to Mr. Cook, whose application to the administration has been unceasing.

"It appeared, by reference to the inspectorship-deed, that, on the inspectors certifying that the surviving members of the firm in liquidation ought to be discharged from the further administration of the estate, the remaining assets are to be transferred to trustees for the creditors, to be named by the inspectors, or by the members of the late firm, if the inspectors decline to name a trustee.

"The inspectors felt themselves called upon to certify accordingly, and such certificate was therefore indorsed upon the inspectorship-deeds.

"The inspectors were prepared to have named a trustee, but they preferred, in the first instance, to give the creditors the opportunity of suggesting one or more trustees, who might be satisfactory to themselves.

"But, on further review of the assets, and the probable realization, the inspectors concurred with the report, in thinking that it would not be prudent to assume such realization at more than 1*s*. in the pound, in addition to the dividend now declared, and they came to a conclusion that a very considerable delay would take place, and much uncertainty must exist as to the result; and it appeared to the inspectors that it would probably be more satisfactory to the creditors, if Mr. Cook would take upon himself to make a proposal for the final close of the affairs, so as to save the necessity of any trust. And they recommended that he should take upon himself the risk of the assets, and engage to pay, at an early period in the ensuing year, a final dividend of 1*s*.

"Mr. Cook having assented to this recommendation of the inspectors, the solicitors of the estate were directed to communicate with the creditors the result of this meeting, and if the inspectors' recommendation were adopted by the creditors, Mr. Cook engaged to pay such final dividend on the 30th of June, 1851.

"Mr. Cook declared his determination, if the above recommendation were adopted, to keep an account of the assets, and, if there should be any surplus, to apply it for the equal benefit of the creditors.

"The inspectors did not consider that Mr. Cook ought to come under any legal obligation to that effect; but that it was quite right that his intention should be made known to the creditors.

"If the creditors preferred the assignment to a trustee, the solicitors were to communicate with them as to the parties to take the trust; or, if the creditors had no choice, the inspectors were ready to name a trustee.

"The solicitors were directed by the inspectors to allow any creditor to inspect Mr. Coleman's report.

(Signed) "JOSHUA BATES,
"HENRY BLYTH,
"ERIC ERICHSEN, } *Inspectors*."

"We have been directed by the inspectors of the affairs of Messrs. Trueman and Cook to submit to your consideration a copy of the report which is enclosed, and which embodies the recommendation which we are authorized to bring under the notice of the creditors.

"This report was, shortly after its date, submitted to the Bank of England, as the largest creditors, and we have to inform you that we have received a letter from the Bank, dated this day, which conveys to us the acquiescence of the Bank in the arrangement proposed by the inspectors.

"We shall be glad to be informed whether you concur with the Bank, so that the arrangement may at once be carried into effect.

"We feel it due to Mr. Cook to add our testimony to the services and attention which he has rendered to the estate, although, by the inspectorship-deed, he has been virtually discharged from all the liabilities of the late firm since January, 1848.

"We beg further to state, that we have a copy of Mr. Coleman's report to the Bank, which is open to your perusal whenever it may suit you to call here. And we understand that the original is also at the Bank, and will be shown to any creditor who may prefer to apply there.

"We are, your obedient, humble servants,

(Signed) "OLIVERSON, DENBY, AND LAVIE.

"Frederick's Place, July 11th, 1850."

PART II.

SUSPENSIONS IN 1850.

Feb. Messrs. Erichsen and Co., London, corn trade.
June. Messrs. Nash and Neale, Reigate, bankers.
„ Messrs. D. Braggiotti and Co., London, merchants.
Aug. Messrs. E. C. Meyer and Co., Hamburg, merchants.
„ Messrs. Herbert and Co., Hamburg, merchants.
„ Mr. J. F. Hinck, Hamburg, merchant.
Sept. Messrs. Luydam, Sage, and Co., New York, merchants.
„ Messrs. Hunter and Co., New York, merchants.
Oct. Messrs. Garnett, Balch, and Co., Boston, merchants.
„ Messrs. M'Williams and Gregory, New York, merchants.
„ Messrs. Henley, M'Knight, and Co., Sacramento, merchants.
„ Messrs. Mebass and Co., Sacramento, merchants.
Nov. Mr. H. M. Naglee, San Francisco, merchant.
„ Messrs. Simmons, Hutchinson, and Co., San Francisco, merchants.
„ Messrs. H. Howison and Farley, San Francisco, merchants.
„ Messrs. Johnson, and Co., San Francisco, merchants.

THE ESTATE OF D. BRAGGIOTTI AND CO.

A meeting of the creditors of Messrs. D. Braggiotti and Co., whose failure was mentioned on the 26th June, took place on July 11, 1850, when, after the examination of a balance-sheet, prepared by Mr. J. E. Coleman, it was unanimously agreed to liquidate the estate under a deed of inspectorship. The statement presented showed that the gross liabilities amounted to £113,900, which will be reduced, by securities held and bills expected to run off, to about £30,000. The greater part of the assets being produce, it is hoped that, if good market prices are realized, a dividend of 4*s.* 6*d.* or 5*s.* may be eventually distributed.

SUSPENSIONS IN 1851.

Mar. Messrs. James Deacon and Sons, London, linen factors.
„ Messrs. Oliver, Worthington, and Co., London, warehousemen.
„ Messrs. Austens and Spicer, New York, produce dealers.
April. Messrs. Bulcher and Carstenjen, London, Dutch merchants.
May. Mr. N. Poutz, Liverpool, general merchant.
Aug. Messrs. Castelli, Giustiniani, and Co., London, Greek merchants.
„ Messrs. Fraser and Lightfoot, London, East India merchants.
„ Messrs. S. Rucker and Sons, London, colonial brokers.
Sept. Messrs. Spencer Ashlin and Co., London, corn factors and merchants.
„ Messrs. Campbell, Arnott, and Co., Liverpool, American merchants.
„ Messrs. W. M. Neil, and Co., Liverpool, corn merchants.
„ Messrs. Clark and Voigt, London, Ionian merchants.
„ Messrs. Peter Clark and Co., Zante, etc., Ionian merchants.
„ Messrs. Maitland, Fawkes, and Co., London, colonial brokers.
„ Messrs. Eggers and Taylor, Liverpool, cotton merchants.
Oct. Messrs. Ezponda, Corredor, and Co., London, Spanish merchants.
„ Messrs. Slater and Robertson, Glasgow, general merchants.
„ Messrs. Chesebrough, Stearns, and Co., New York, silk and general merchants.
„ Monmouthshire and Glamorganshire Banking Company, Monmouth bankers.
„ Messrs. F. S. and D. Lathrop, New York, dry goods jobbers.
„ Messrs. Thompson and Co., New York, carpet manufacturers.

Oct. Commercial Bank of Perth, Amboy, United States.
„ Peoples' Bank of Paterson, New Jersey, United States.
„ Bank of Salisbury, Salisbury, Maryland, United States.
„ Monmouthshire Newport Old Bank, Monmouth.
„ Messrs. Paul and Dastis, London, wine merchants.
„ Messrs. Kirkman and Brown, London, colonial brokers.
„ Messrs. Ewbank and Gray, London, merchants.
„ Mr. W. H. Imlay, Brooklyn, New York, president of the Atlantic Dry Dock Company.
„ Messrs. Perkins, Brooks, and Co., New York, dry goods dealers.
„ Mr. William G. Brown, Buffalo, United States, merchant.
Nov. Messrs. Forman and Hadow, London, colonial brokers.
„ Messrs. F. F. Braggiotti and Co., Manchester, Greek merchants.
„ Mr. Francis Chambers, London, East and West India agent.
„ Messrs. Quarles Harris and Sons, London, wine trade.
„ Sunderland Joint-stock Bank, Sunderland, bankers.
„ Messrs. Jewett and Prescott, Boston, dry goods dealers.
„ Messrs. James Briggs and Brothers, New Hampshire, United States, woollen manufacturers.
„ Messrs. W. Burger and Co., New York, druggists.
„ Messrs. Suter, Symington, and Robinson, New York, general merchants.
„ Messrs. Seaman and Muir, New York, dry goods dealers.
„ Messrs. Dexter, Harrington, and Co., New York, leather dealers.
„ Messrs. P. R. Southwick and Co., New York, leather dealers.
„ Messrs. J. S. Gould and Co., Salem, United States, iron merchants.
„ Mr. David Pingee, Salem, United States, railway and manufacturing establishment.
„ The Maryland Coal Company, Maryland.
„ Messrs. E. J. Barton and Co., New York, paper manufacturers.
„ Messrs. J. M. Rutherford and Co., Louisville, United States, general merchants.
„ Columbus Insurance Company, Cincinnati, United States.
„ Messrs. Crosby and Carkey, Norwich, Connecticut, United States, wool merchants.
„ Messrs. B. Upton and Son, Salem, Massachusetts, United States, merchants.
Dec. Messrs. Luckie Brothers and Co., London, West India merchants.
„ Messrs. Platt, Sons, and Casson, Liverpool, general merchants.
„ Messrs. Claypole and Son, Liverpool, West India merchants.
„ Messrs. John Cabbell and Co., Glasgow, West India merchants.
„ Messrs. E. Fyffe and Son, London, East and West India merchants.
„ Messrs. Hicks and Co., New York, shipping and general merchants.
„ Messrs. Hill, M'Lean and Co., New Orleans, cotton brokers.
„ Messrs. Mixer and Pitman, Boston, United States, oil and tallow merchants.
„ Messrs. Beach, Case, and Co., New York, merchants.
„ Messrs. John Bacon and Son, New York, iron merchants.
„ Messrs. J. B. Smith and Co., New York, silk merchants.

SUSPENSIONS IN 1852.

Jan. Messrs. James Bult, Son, and Co., London, bullion dealers.
„ Mr. D. A. Scanavi, London, Greek merchant.
„ Messrs. Seaward, Capel, and Co., London, engineers.
„ Messrs. Lantz and Co., Petersburg, general merchants.
„ Messrs. Thatcher, Tucker, and Co., New York, domestic wares merchants.
„ Messrs. J. B. Adams and Co., Cork, provision merchants.
„ Messrs. M'Ewan and Co., Glasgow, tea merchants and sugar refiners.
„ Messrs. Donald Mackay, Hadow, and Co., London, East India merchants.
„ Messrs. Daw, Wilson, and Herrimam, New York, wholesale grocers.
„ Mr. Glendie Broake, New Orleans, cotton merchant.
Feb. Messrs. T. S. and W. Hardy, Cork, corn and provision trade.
„ Mr. Denny Lane, Cork, corn and provision trade.
„ Mr. J. Dunbar, Cork, corn and provision trade.
„ Messrs. M. J. Wilson and Co., Liverpool, general merchants.
„ Messrs. Foxall and Co., Dublin, East India merchants.
„ Messrs. Maunsal, White, and Co., New Orleans, cotton merchants.

Feb. Messrs. Mills, M'Dowall, and Co., New York, general merchants.
„ Messrs. W. B. Hutton and Co., London, African merchants.
„ Messrs. C. and B. Hooper, London, leather factors.
„ Messrs. D. C. Mackey and Co., Calcutta, East India merchants.
„ Messrs. Rosetto, Carati, and Co., London, Greek merchants.
„ Messrs. Keith, Shoobridge, and Co., London, shawl printers.
Mar. Messrs. Carleton and Co., New York, dry goods importers.
„ Messrs. Moulton, Barker, and Helfier, New York, dry goods importers.
„ Mr. A. A. Lackersteen, London, East India merchant.
„ Messrs. Ritchie Brothers, London, West India merchants.
„ Mr. J. G. Lacy, of the firm of Lacy and Reynolds, London, gum manufacturer.
„ Messrs. M. Retemayer and Co., Liverpool, merchants.
„ Messrs. Maeri and Co., Corfu, merchants.
„ Messrs. Godfrey, Ouseley, and Co., Savannah, cotton merchants.
„ Messrs. Thomas Trierson and Co., Savannah.
„ Messrs. Hamilton and Hardman, Savannah, cotton merchants.
„ Messrs. R. C. Whetmore and Co., New York, hardware merchants.
May. Mr. John Birse, Dundee, flax merchant.
„ Messrs. Sinclair and Boyd, Belfast, East and West India merchants.
June. Messrs. Dixon, Walne, and Co., Liverpool, Baltic merchants.
„ Messrs. Thomas Apt and Co., London, linseed crushers.
„ Mr. Knowles, London, Norwegian merchant.
Oct. Messrs. Walter Logan and Co., London, Spanish-American merchants.
„ Messrs. Emanuel and Son, Hamburg, merchants.
„ Messrs. Dunscomb, Cook, and Co., New York, Spanish-American and West Indian merchants.
„ Messrs. Johnson and Travers, Baltimore, Spanish-American and West Indian merchants.
Nov. John Ruck and Co., London, provision trade.
„ Jos. Stuart and Son, London, provision trade.
„ C. H. Harber, London, provision trade.

THE ESTATE OF MESSRS. CASTELLI AND CO.

A meeting of the creditors of Messrs. Castelli, Giustiniani, and Co., whose suspension took place on the 16th of August, was held on the 24th of September, 1851, and was very numerously attended. Mr. Le Breton presided, and among the several solicitors present, representing different mercantile firms, were Mr. Tilson, Mr. Ellis, and Mr. Hobler, while Mr. Lavie attended on behalf of the estate. After the usual *pro formâ* proceedings, the following statement was read, prepared by Mr. J. E. Coleman, the accountant:—

STATEMENT OF THE AFFAIRS OF CASTELLI, GIUSTINIANI, AND CO., BURY COURT, ST. MARY AXE, MERCHANTS, AUGUST 16, 1851.

		Liabilities as per the Creditors' own Accounts.	Liabilities as per Messrs. Castelli's Estimates of Securities.
To bills payable, uncovered		£102,076 13 10	£102,076 13 10
To bills payable, against which goods are hypothecated to the full amount of claim	£51,173 12 11		
Estimated value of goods	55,246 12 11		
Surplus to *contra*	£4,073 0 0		
To bills payable, against which goods are hypothecated	£11,430 1 1		
Estimated value of goods	10,295 6 11		
		1,134 14 2	1,134 14 2
To creditors on open accounts		19,854 2 11	19,854 2 11
Carried forward		£123,065 10 11	£123,065 10 11

Brought forward		£123,065 10 11	£123,065 10 11
To creditors on accounts current holding security, after deducting estimated value of securities		30,520 14 7	18,710 0 8
To liabilities on bills receivable on account of Rucker and Son		53,100 0 0	53,100 0 0
To ditto on general account, the whole of which, it is expected, will be duly honoured at maturity	£112,285 3 8		
Total liabilities		£206,686 5 6	
Total liabilities as per Castelli and Co.'s estimates			£194,875 11 7

		Assets.	Assets as per Messrs. Castelli's Estimates of Assets.
By debtors, good		£1,960 15 1	£1,960 15 1
By debtors, doubtful	£460 14 8		
By surplus from goods hypothecated, as *per contra*		4,073 0 0	4,073 0 0
By surplus from goods held by creditors as security		1,339 3 11	11,049 16 0
By sundry property		7,628 0 0	7,628 0 0
Total assets		£15,000 19 0	
Total assets as per Castelli and Co.'s estimates			£24,711 11 1

In alluding to the position of the accounts, Mr. Coleman explained that the difference between the estimates of assets, one showing 1*s*. 6*d*., the other 2*s*. 6*d*. in the pound, arose from the value put upon produce by the parties who had made advances, and the value subsequently revised by Mr. Castelli. In the case of cotton, for instance, in which the estate was interested to the extent of about 24,000 bales, some small sales had lately been effected at an improvement of nearly 10 per cent. compared with the estimate previously formed. Parties who had made advances to the firm, and who were originally creditors, found themselves now under these circumstances to be debtors. The assumed value of the whole of the produce connected with Messrs. Castelli and Co. was nearly £500,000, the cotton being estimated at about £200,000. Mr. Hobler complained of alleged irregular commercial conduct on the part of Castelli and Co., with reference to the cargoes of wheat obtained from Alexandria, and also with regard to some bill transactions; but these allegations, it was generally contended, ought not to be entered into at a meeting called to settle preliminaries, especially as they might, after all, be capable of full explanation. Mr. Chapman said it was evident the great point for consideration on this occasion was as to the best means of managing for the benefit of the creditors. According to what had been stated, there was nearly half a million of property involved, and it was the course to be pursued for the realization of the surplus that ought now properly to occupy attention. Other incidental matters could subsequently receive notice. His opinion was, that the safest plan would be to select persons well acquainted with mercantile affairs for superintending a liquidation, and to allow the estate to be wound up under inspection, administration in bankruptcy not being applicable to its position. It was also his impression, that when the affairs of Messrs. Castelli and Co. should come to be thoroughly investigated, it would be found that they had been conducted in an honourable and straightforward manner. This proposal did not meet unanimous approval, Mr. Kettlewell, representing the Ionian Bank, and another creditor, objecting to so speedy a decision. Mr. Lavie intimated that if some of the creditors should be hereafter found to disagree with the steps that might now be taken, another meeting could be convened. Mr. Kettlewell was in favour of a week being given to consider the statement laid before the creditors. On the other hand, Mr. Chapman and Mr. Cunliffe thought that the interest of every one concerned would dictate the necessity of coming to a conclusion as soon as possible. Mr. Tilson thought it would be prudent, perhaps, to ask Mr. Coleman to give them some information of the way in which these very heavy losses were sustained, since it might obviate apparent difficulties in arriving at a result. Mr. Coleman stated that he had not yet been able to draw out a perfect balance-sheet, but from the examination he had already prosecuted, it was quite certain the firm of Castelli and Co. possessed a capital of £35,000 in 1849. Their losses on produce, however, in 1850 and 1851 amounted to £149,800, the chief articles being cotton, corn, coffee, sugar, silk, and cochineal. When his investigation should be terminated he would be enabled to show the course of the whole trading, but in the meantime it would only be fair to announce that the drawings from the firm had been very small. The question of private property having been alluded to, it was inquired how many partners were in the house. The answer, that Mr. Frank Castelli alone represented it, seemed to excite strong remark upon the part of the creditors. The appearance of the name of Mr. Giustiniani in the firm was canvassed, and it was then elicited that he had retired, but that the dissolution had not been inserted in the *Gazette*. Mr. Lavie said this point was open to discussion, but the proposed resolutions would not interfere with the rights against other parties. Partnership *en commandite* had existed, the facts connected with which could be inquired into. Mr. Chapman regarded this matter as one of great importance, and said he trusted it would be closely probed. The question, however, in the hands of Mr. Lavie would doubtless receive due consideration, and, believing this, he should press for the adoption of the plan of inspectorship. Mr. Hobler contended that bankruptcy was the only satisfactory process through which the creditors could hope to see their rights tested, whether with regard to partnership law or a due investigation of property. Under inspectorship no authority whatever could be exercised, either to compel parties to disclose the condition of their affairs, or to render strict justice to their creditors. The dividend in this case would more likely be 1*s.* than 2*s.* 6*d.*, and bankruptcy administration was therefore the right course. Mr. Cunliffe advocated the system of inspectorship where economy was required. Mr. Ellis and Mr. Tilson considered that as a preliminary step a choice of inspectors might be made, and bankruptcy could eventually be enforced if it were thought necessary. With respect to Mr. Giustiniani, it was stated that even the bankruptcy laws would not reach him, should the creditors wish to carry them out, as he was a resident at Leghorn. As to proceedings against other parties, if liable, that was a subject deserving serious consideration. No complaint could, however, be made concerning the disposition of the *en commandite* capital, not the least portion of it having been withdrawn, it having been fairly absorbed in the business of the house. After some further discussion, it was agreed by the majority to pass resolutions adopting the plan of inspection, Mr. Kettlewell permitting his name to be inserted on the understanding that the directors of the Ionian Bank assent to the proceedings. "If they should not, the *Gazette*," he said, "would certainly follow." Annexed are the resolutions carried in official order:—

"At a meeting of the creditors of Messrs. Castelli, Giustiniani, and Co., held at No. 8, Frederick's Place, on Wednesday, the 24th day of September, 1851, Mr. Le Breton in the chair—

"A statement of the assets and liabilities was read to the meeting by Mr. Coleman, the accountant, and the meeting having come to the conclusion that it was expedient that the affairs should be wound up under a deed of arrangement, with the supervision of inspectors, it was, therefore, resolved—

"1. That Mr. Whately, of the firm of Le Breton, Whately, and Co., and Mr. William Kettlewell, of the Ionian Bank, be requested to be inspectors, with power to appoint a third inspector.

"2. That a deed of arrangement be prepared, with proper provisions for the winding-up the affairs under the above-mentioned inspectors, and for securing the due liquidation and the early distribution of the assets *pro ratâ* among the creditors, and according to the rules of administration in bankruptcy, and containing a covenant by Mr. Castelli to carry on such liquidation under the inspectors, and covenants on the part of the creditors not to sue Mr. Castelli for twelve months; and that the inspectors shall have power to enlarge that time if they shall deem it necessary; and that at the expiration of such period, or of such enlarged period, and upon Mr. Castelli assigning the outstanding assets to trustees for distribution among the creditors, he shall be released from all claims. The deed shall also contain such other provisions as are usual in deeds of inspectorship, and as shall be approved by the inspectors on behalf of the creditors.

"3. The creditors acceding to this arrangement shall not be prejudiced as to any securities or liens they may be entitled to, or as to their rights against third parties.

"4. That the signature of this memorandum and of the deed of arrangement shall not prejudice the rights, remedies, or position of any creditor, against or in respect of the parties at any time members of the firm of Castelli, Giustiniani, and Co.

"5. That the inspectors may be at liberty, if they shall think fit, to release to Mr. Castelli his furniture, and such other private property at his residence as they may think fit.

"6. That in the meantime, until the deed of arrangement shall be executed, a copy of these resolutions may be submitted to the creditors, and shall form a memorandum of arrangement for the administration of the affairs.

"7. The thanks of the meeting to the chairman were then moved by Mr. Cunliffe, seconded by Mr. Chapman, and voted unanimously."

The estate was ultimately wound up in bankruptcy.

THE ESTATE OF RUCKER AND SONS,

COLONIAL BROKERS.

A meeting of the creditors of Messrs. Rucker and Sons, whose failure took place on the 25th of August, was held on the 5th of September, 1851, Mr. D. B. Chapman, of the firm of Overend, Gurney, and Co., presiding. After the usual preliminaries the following statement of the affairs of the estate was presented by Mr. J. E. Coleman, the accountant :—

DEBTOR.

To creditors on acceptances				£158,606	1	2
To creditors on open accounts				116,898	4	4
To creditors holding securities (but which securities will not cover advances to the full extent)				9,655	13	8
				£285,159	19	2

CREDITOR.

By cash at banker's				£6,189	4	10
By bills receivable				525	5	6
By sundry debtors, good (the greater portion of these are secured by produce, etc.)				127,207	0	0
By debts secured by liens on West Indian estates, estimated to yield				7,600	0	0
By cash to receive on account of sundry debtors, viz., estimated surplus from produce beyond advances received thereon				33,500	0	0
Prompts	£87,957	0	0			
Less, to be paid to those entitled thereto	70,427	0	0			
				17,530	0	0
By private property of the late Mr. Rucker and of Mr. S. Rucker, jun.				22,400	0	0
				£214,951	10	4
Less, sundry small bills (to be paid in full) drawn against produce to arrive, etc.	£634	5	10			
Salaries and charges to be paid in full	1,000	0	0			
				1,634	5	10
				£213,317	4	6

In explaining the circumstances of the suspension, Mr. Freshfield referred to the reports first circulated of its having been caused by the failure of Messrs. Castelli and Co. Although the two firms were connected in business, Messrs. Rucker were not compelled to suspend by the stoppage of Castelli and Co., but in consequence of the discredit brought upon them from the rumours propagated with respect to it. The engagements of Rucker and Sons, it was well known, were principally in connection with money taken in deposit, money borrowed upon produce, and running acceptances. Since the period of suspension those transactions had become considerably narrowed, owing to the redemption of securities and the consequent cancelment of acceptances; and there was also the probability of a further large reduction from the same causes. Under the general estimate made, there was a good prospect of the creditors receiving 15*s*. in the pound. Mr. Coleman, in speaking of the character of the assets, alluded to the favourable and compact position in which they stood. Little doubt was entertained of their ultimate realization, and, with reference to the produce, it was understood that the parties who represented the interest identified with that portion of the estate would not dispose of it except at fair market prices. It was therefore hoped that in the course of four months the whole of the available property might be distributed. Mr. Freshfield said it must be clear to all concerned, that the principal feature in the estate, as shown by the accounts, was the tangible nature of the assets. First, there was the cash at the banker's; then the debts represented by produce, which might almost be regarded as money. The debts secured by liens on West Indian estates had been put down at a sum which it was fairly calculated they would yield; and, although the value of West India property was liable to great depreciation, it was nevertheless thought that in this case the estimate would be found on the safe side. With regard to the item of "cash to receive on account of sundry debtors, namely, estimated surplus from produce beyond advances received thereon, £33,500," every reason existed for presuming that such a sum would be realized in full, even if no excess accrued, because the produce was virtually pledged for less than its real value, and hence, as it was not likely to be forced upon the market, there was scarcely room for apprehending an unfavourable result. The remaining asset of consequence was that of the private property of Messrs. Rucker and Sons, stated at £22,400, making in gross £213,317 to pay £285,159, or, as already stated, 15*s*. in the pound, with every expectation of a speedy realization. With this object it was proposed to pay the first dividend of 2*s*. 6*d*. in about ten days, the next on the 1st of November, and the others in regular order, as nearly as could be calculated, on the 1st of each succeeding month; sometimes 2*s*. 6*d*. and sometimes more or less, as the property might be rendered available. Mr. Chapman, as chairman, in reviewing the statements made by Mr. Freshfield, thought it must be satisfactory to the creditors to find that the assets, so substantial in their appearance, were considered capable of producing 15*s*. in the pound. He bore testimony to the assumed tangible nature of the leading items, but particularly to the private property at Wandsworth, and then submitted the course that ought to be adopted for the final liquidation of the liabilities. Considering the position of the firm, and also the circumstances which had transpired, especially with reference to the produce to be realized, he thought that the most advisable proceeding would be to allow the partners to manage the business of liquidation under the superintendence of inspectors, the latter representing the leading creditors. This proposal at once met the unanimous consent of the parties present, but before the resolutions were put, a question was asked respecting the capital of the house. Mr. Freshfield was understood to remark that the capital had been merged in losses. It had not recently been as high as £100,000; the sum was nearer £50,000. The resolutions were then unanimously passed for winding up the affairs of the house under inspection, and after a discussion regarding the desirableness of the whole of the produce held as security passing through the hands of Messrs. Rucker and Sons for sale, so as to keep the markets in a proper state, as well as to promote the business of the firm in its subsequent career, a vote of thanks was carried to the chairman, and the meeting separated. Annexed are the resolutions agreed to in official order:—

"At a meeting of the creditors of Messrs. Rucker and Co, held at the counting-house in Tower Street, on Friday, the 5th of September, Mr. David Barclay Chapman in the chair, Mr. Coleman, the accountant, having produced and read to the meeting a statement of the debts and assets, it was thereupon resolved—

"1. That the house should be liquidated by the partners, under the inspection of the following gentlemen, viz., Mr. David Barclay Chapman, Mr. Edward Harnage, and Mr. Christian Turck.

"2. That the salaries of clerks and servants be paid in full, and that the house be authorized to pay small debts, not exceeding in the aggregate £1,000.

"3. That a deed be forthwith prepared, by which the partners shall covenant to liquidate the affairs of the house, and divide the proceeds among the creditors rateably and in proportion to their several debts, under inspection, observing in such liquidation the rules of administration adopted in bankruptcy; that such deed shall contain covenants by the creditors not to sue the partners for twelve months, and that the inspectors shall have power to enlarge that time, if they shall deem it necessary; and that, at the expiration of such period, or of such enlarged period, and upon the partners assigning the outstanding assets to trustees for distribution among the creditors, the partners shall be released from all claims. The deed shall also contain such other provisions as are usual in deeds of inspection, to be settled by the inspectors on behalf of the creditors.

"4. That creditors acceding to this arrangement shall not be prejudiced as to any securities or liens they may be entitled to, or as to their rights against third parties.

"5. That the private property of the partners, after payment of their separate liabilities, be applied in the payment of the debts of the firm, except furniture and such other private property as the inspectors shall think proper to release to the partners.

"6. The partners having undertaken to conduct the liquidation without any personal remuneration, it is resolved that they shall be at liberty to transact business as brokers on their covenanting not to use, either directly or indirectly, any of the existing assets of the firm, and to incur no new engagements which could by any possibility be thrown on the existing assets, but such business shall be discontinued if the inspectors shall certify that it ought not to be continued.

"D. B. Chapman.

"Resolved—That the thanks of the meeting be given to Mr. David Barclay Chapman for taking the chair, and for his conduct therein."

THE ESTATE OF SPENCER ASHLIN AND CO.,

CORN TRADE.

At a meeting held on October 13, 1851, of the creditors of Messrs. Spencer Ashlin and Co., who failed in the corn trade on the 13th of September, Mr. De Bruyn in the chair, the following statement was read by Mr. Quilter, the accountant:—

Statement of Affairs, September 11, 1851.

DEBTOR.

To sundry creditors, unsecured		£69,380 5 9
To sundry creditors secured by consignments, etc., on hand	£87,916 18 2	
Deduct value of consignments on hand	73,554 14 10	14,362 3 4
To sundry liabilities, viz.:—		
On bills receivable	213,873 2 3	
Less estimated to be bad, and included in statement of creditors unsecured	49,122 9 10	
Considered good	£164,750 12 5	
		£83,742 9 1

CREDITOR.

By cash and bills on hand, viz.:—		
At Glyn and Co.'s	£4,071 10 5	
Bills receivable, considered good	7,192 17 8	£11,264 8 1
Ditto, considered bad, £1633 8*s.* 6*d.*		
By railway shares, being the sum since realized on sale thereof		1,192 10 0
By sundry debtors, considered good		1,426 12 0
Ditto, considered doubtful and bad, £32,484 1*s.* 1*d.*		
By surplus security in the hands of creditors		60 0 0
		£13,943 10 1
By deficiency		69,798 19 0
		£83,742 9 1

In elucidation of the accounts, Mr. Quilter read a lengthy report, showing the position of the firm from its commencement, with the capital introduced and the course of trading. From this it appeared that the deficiency, as presented in the above statement, arose from £41,000 losses on grain speculations, and the remainder through bad debts. Mr. M'Leod intimated that the estate showed a dividend of 3*s.* 4*d.* in the pound; but it was proposed on behalf of Mr. Spencer Ashlin, who alone constituted the firm, to pay 3*s.* 6*d.*, his friends having promised to assist him to some extent. Questions had been mooted of preference payments made by the house on the eve of its failure to an amount of about £4,600, of which £3,000 had already been recovered, and it was thought that the balance could be arranged without litigation. Mr. Ashlin, it was stated, was ready to follow any course proposed by his creditors for liquidating his affairs, and had already placed himself in a position to become amenable to bankruptcy administration, if that alternative were deemed desirable. The assets already realized were in the custody of Messrs. Glyn and Co., to whom also would be transferred what other sums might be obtained for the benefit of the creditors. A discussion was then entered into respecting some transactions between Spencer Ashlin and Co.; Alsop and Co., of Dublin; and Mr. Neil, of Liverpool, the merits of which will have hereafter to be determined, the National Bank of Ireland being interested in the issue. When the resolutions usually adopted at meetings of this description came to be definitively proposed, Mr. Murray, as representing Messrs. Glyn and Co. and Messrs. Overend, Gurney, and Co., considered further investigation necessary. He therefore suggested an adjournment, which was finally adopted, Mr. J. G. Elsey, of the discount department of the Bank, being appointed, in conjunction with a second party, to be named by Mr. Murray, to investigate the various points, and to ascertain, if possible, their actual position.

The adjourned meeting of the creditors of Messrs. Spencer Ashlin and Co., who failed in the corn trade on the 13th of September, was held on the 28th of October, to receive the report of Mr. Murray and Mr. J. G. Elsey, the parties named to investigate the affairs of the firm. Previously to this document being presented, it was explained by Mr. M'Leod, Mr. Ashlin's solicitor, that from some misunderstanding of what had transpired on the former occasion, it appeared to have been considered that 3*s.* 6*d.* in the pound was the amount to which the creditors would have been limited, had they adopted the proposition then made. Such, however, was not precisely the case, the intention having been to secure 3*s.* 6*d.*, and rateably divide any further assets which might have been realized. The report read by Mr. Murray detailed the course of investigation, which had been principally directed to some transactions in grain regarded as constituting preference arrangements. The debts and liabilities were taken at £85,193, and the assets were on the former occasion placed as representing £15,469. Increasing this latter sum by the amount of the preference claims and the value of household furniture, it would reach £17,762. The gross assets, under these circumstances, showed a dividend of 4*s.* in the pound with the small surplus of about £395. It was therefore the opinion of Mr. Murray—on behalf of the largest creditors, Messrs. Glyn and Co. and Messrs. Overend, Gurney, and Co.—that a dividend of 4*s.* in the pound, 3*s.* being at once paid and the remaining 1*s.* secured at an early date, should be accepted. A long discussion followed relative to the mercantile career of Mr. Spencer Ashlin, especially in connection with Mr. Alsop of Dublin and Mr. Neil of Liverpool. Mr. M'Leod said the question of the trading of Mr. Ashlin could not in all respects be defended, but the real point at issue was the best course to be adopted for the benefit of the creditors. The accounts prepared by Messrs. Quilter and Co., it was intimated, showed in an accurate and detailed shape the whole of the transactions, Mr. Murray himself bearing testimony to their completeness in this and every other particular. Mr. M'Leod was distinctly of opinion that Mr. Ashlin, assisted by his friends, in concurring with the proposal to pay 4*s.* in the pound, was doing the utmost he could for his creditors, since it was, after all, probable that in the realization of the estate a deficiency of from £2,000 to £3,000 might occur. The representatives of the National Bank of Ireland announced that they were unable to give their assent immediately to the terms proposed; but finding that the annexed resolution was unanimously carried, they promised to consult the board of directors on the subject, and to communicate their final decision without delay:—

"At a meeting of the creditors of Messrs. Spencer Ashlin and Co., held at the office of Messrs. M'Leod and Stenning, of 13, London Street, Fenchurch Street, on the 28th of October, 1851, it was resolved—

"That the report made by Mr. Murray be received and confirmed, and that 4*s.* in the pound be paid and received in full of all claims by the respective creditors of

Mr. Ashlin, as follows, *i.e.*, 3*s*. in the pound to be paid down on the signing of this resolution by all the creditors, or their assent thereto being obtained, and 1*s*. in the pound to be secured to the satisfaction of Mr. Murray's clients, payable at three months from this date."

The estate was ultimately wound up in bankruptcy, and, on unfavourable disclosures, the bankrupt quitted the country.

THE ESTATE OF QUARLES HARRIS AND CO.,

OPORTO (WINE) TRADE.

A meeting of the creditors of Messrs. Quarles Harris and Co., engaged in the wine trade, who failed on the 28th of November, 1851, was held on the 28th of January, 1852, when the following statement, prepared by Messrs. Quilter, Ball, and Co., was presented :—

QUARLES HARRIS AND SONS' STATEMENT OF AFFAIRS, NOVEMBER 28, 1851.

DEBTOR.

To creditors unsecured, viz. :—		
Trade accounts	£5,215 1 7	
Acceptances drawn from Oporto	13,709 15 1	
Ditto in London	9,400 0 0	
Cash creditors, as per statement	14,850 8 2	
		£43,175 4 10
To creditors partially secured, viz. :—		
Amount of claims	5,997 9 2	
Estimated value of securities	5,514 0 0	
		483 9 2
To creditors fully secured, viz. :—		
Estimated value of securities	41,489 8 4	
Amount of claims	35,695 7 11	
Estimated surplus carried to credit as an asset *per contra*	5,794 0 5	
To liabilities on bills receivable—		
Considered good	42,680 6 7	
Considered bad	4,606 16 1	
Less securities held	43 13 7	
		4,563 2 6
To acceptances in hands of Oporto house, not expected to be claimed against this estate	6,000 0 0	
		£48,221 16 6

CREDITOR.

By bills receivable—on hand :—			
Considered good		£935 11 0	
Considered bad	0 0 0		
By unaccepted bills	£1259 10 2		
Estimated to produce		1,000 0 0	
			1,935 11 0

By sundry debtors, viz. :—

	Good.	Doubtful.	Bad.
London	£3,466 10 11	£167 9 7	£2,180 5 4
Country	3,050 6 1	2,177 3 4	4,185 14 4
Irish	820 17 3	553 18 0	1,154 5 9
Swedish	1,105 13 8	3 15 0	Nil.
Foreign	846 8 5	543 5 10	556 11 10
	£9,289 16 4	£3,445 11 9	£8,076 17 3

Carried forward £1,935 11 0

Brought forward		£1,935 11 0
Debtors considered good		9,289 16 1
„ considered doubtful £3,445 11 9		
Estimated to produce, at 5*s.* in the pound		861 7 11
Estimated bad £8076 17 3		
By stock on hand—		
On docks and with country agents		4,496 0 0
By surplus securities—		
In the hands of creditors, as *per contra*		5,794 0 0
By wines on consignment—		
At Stockholm, £809 10*s.*, estimated to produce		500 0 0
At New York, estimated produce of Terragona wine (doubtful)		2,400 0 0
		£25,276 15 0
DEDUCT.		
Rent, taxes, salaries, and sundry small creditors to be paid in full	£600 0 0	
Reserved, to retire a bill of £2000	255 0 0	
Expenses of realizing stock and of liquidating the estate	1000 0 0	
		1,855 0 0
		£23,421 15 0

In reviewing the statement, it was explained that it showed a dividend of about 10*s.* in the pound, but that Messrs. Quarles Harris and Co. were unable to offer any definite terms for a secured composition. Under these circumstances, they were willing to pursue any course considered desirable by their creditors, and were only anxious to realize the property to the best advantage. The stock of wine, it was suggested, should be disposed of gradually, instead of at auction. In answer to questions, it was intimated that the books of the firm had not been balanced subsequently to 1844, and that it was insolvent in 1848 to the extent of £20,000, taking the stock at selling prices. The securities placed with creditors, and the whole of the transactions connected with them, had been thoroughly investigated, and not the least suspicion existed but that they had been appropriated in a perfectly legal manner. With regard to the firm in Oporto, it appeared that a desire had been expressed among the local friends of Mr. James Harris to make arrangements, if possible, for him to carry on business; while the creditors in this country, out of consideration for Mr. Quarles Harris, as well as for their own ultimate advantage, announced themselves favourable to a liquidation of the estate by inspection. Were it proposed to adopt a different proceeding, legal difficulties would doubtless arise, which might, in the end, decrease the available assets, and interfere with the general prospects of their realization. Some further explanations having been given respecting bills running between the two houses, it was moved and carried that a liquidation should be effected in the manner proposed. Subjoined is an abstract of the resolution passed:—

"At a meeting of the creditors of Quarles Harris and Sons, held at the chambers of Messrs. Quilter, Ball, and Co., on the 24th day of January, 1852, Mr. Noble in the chair, a statement of the affairs of the house having been laid before the creditors, it was resolved—

"That a letter of license be granted to Mr. Quarles Harris to liquidate the estate, under the inspection of Mr. Tanqueray, Mr. Complin, Mr. Soares, and Mr. Gardener, on behalf of the creditors, and that the usual deed for carrying such arrangement into effect be prepared for the signature of the creditors, such deed to contain a provision for a release to the Messrs. Harris, on payment of a dividend equal to 7*s.* 6*d.* in the pound, or as soon as the inspectors shall certify that the conduct of those gentlemen, in the liquidation of the estate, has been satisfactory to them.

"JOHN HATT NOBLE, *Chairman.*

JAS. M. BURGHES.	FREDERICK GIESLER AND CO.
E. T. COMPLIN.	BIBBENS, BLOGDEN, AND STOVINS.
H. M. WYATT.	PYE, FIELD, AND TANQUERAY.
THOMAS BARRETT.	JOHN DAY, SON, AND WATSON.
M. S. SOARES.	MARTIN, STONE, AND CO."

THE ESTATE OF MESSRS. EWBANK AND GRAY,

MERCHANTS.

This house failed, but the meeting of creditors was kept strictly private.

STATEMENT OF AFFAIRS, SEPTEMBER 30, 1851.

DEBTOR.

To creditors unsecured		£5,373 16 11
To creditors partially secured		2,111 7 9
To amount of claims	£17,810 7 9	
To securities valued at	15,699 0 0	
	£2,111 7 9	
To creditors fully secured		0 0 0
To claims	£19,692 0 6	
To value of securities	23,881 5 6	
Asset *contra*	£4,189 5 0	
To bills payable		9,001 3 9
To liabilities on endorsements		20,207 16 3
		£36,694 4 8

CREDITOR.

By book debts, good	£1,418 19 4	
By cash	1 11 8	
By surplus value of securities, *contra*	4,189 5 0	
Value of stock	6,747 6 0	
	£12,357 2 0	
Less small debts to be paid in full	617 8 11	
		£11,739 13 1
By profits on adventures, estimated		2,200 0 0
By estimated profits on cleaning and dressing stock and purchases made		3,680 0 0
		£17,619 13 1

THE ESTATE OF EZPONDA, CORREDOR, AND CO.,

SPANISH TRADE.

The accompanying statement and report, respecting the affairs of Messrs. Ezponda, Corredor, and Co., engaged in the Spanish trade, who suspended payment on the 1st October, 1851, were subsequently transmitted to their principal creditors. It will be perceived that expectations are entertained, not only of a liquidation being effected in full, but also of a surplus remaining for the benefit of the partners.

STATEMENT OF THE AFFAIRS OF MESSRS. EZPONDA, CORREDOR, AND CO., OF AUSTIN FRIARS, LONDON, OCTOBER 29, 1851.

DEBTOR.

To creditors holding security	£32,910 17 11
Acceptances to draughts from Spain and Carthagena, for which negotiations have been effected to take the same up as they become due	32,692 0 0
Carried forward	£65,502 17 11

Brought forward		£65,502 17 11
Ditto, for which the drawers will provide		10,579 0 5
Creditors on open accounts in Spain		25,028 11 8
Bills payable, being acceptances for account of drawers, who will provide for the same as they arrive at maturity; and on account of others who are debtors to a larger amount than the draughts now returned on them		14,361 5 0
		£115,471 15 0
To bills payable, not provided for	£15,060 12 1	
Sundry creditors on open accounts	3,621 0 4	
		18,681 12 5
		£134,153 7 5

CREDITOR.

		Debit balances as per books of the house.	Debit balances after giving credit for the unpaid draughts.
By debtors, good	£30,807 11 2	£30,807 11 2	
Less acceptances not paid	14,261 5 0		
	£16,546 6 2		£16,546 6 2
By debtors (doubtful)	423 13 5		
By ditto (bad)	210 17 7		
By property		3,279 10 0	3,279 10 0
By securities held by creditors	38,040 0 0	38,040 0 0	
Creditors as *per contra*	32,910 17 11		
	£5,129 2 1		5,129 2 1
By balance due from Corredor and Co., of Carthagena	£66,695 17 1	66,695 17 1	
Less bills ... £32,692 0 0			
Ditto ... 10,579 0 5			
	43,271 0 5		
	£23,424 16 8		
This balance will be discharged by the payment of	25,028 11 8		
Which the Carthagena house will make in general settlement with the Spanish creditors, the result on which will be a balance in its favour of	£1,603 15 0		
By capital in the house at Havannah, and goods consigned for sale there, in charge of your special correspondent, lately sent out	31,702 6 0	31,702 6 0	31,702 6 0
			£56,657 4 3
Deduct the amount which will be due to the house at Carthagena after payment of all claims in Spain, as above	1,603 15 0		1,603 15 0
Carried forward		£170,525 4 3	£55,053 9 3

Brought forward		£170,525 4 3	£55,053 9 3
Bills payable, and not provided for, as *per contra*...	£18,851 12 5		18,851 12 5
Deduct amounts owing		134,323 7 5	
Surplus		£36,201 16 10	£36,201 16 10

"To Messrs. Ezponda, Corredor, and Co., Austin Friars.

"36, *Coleman Street*, *London*, *Oct.* 31, 1851.

"Gentlemen,—Having perfected the examination of your books, and investigated the correspondence and documents necessary to make a statement of your liabilities and assets, as well as to advise upon a mode of liquidation, I now have the pleasure of placing the results before you, together with some few observations that presented themselves to me during such inquiry.

By the statement annexed hereto, your liabilities in gross are ...		£134,323 7 5
Which will be reduced as follows:—		
By the sale of produce now hypothecated against advances, and which advances are included in the above amount	£32,910 17 11	
By your acceptances to draughts from Spain and Carthagena, and for which negotiations have been effected to take up the same as they become due	32,692 0 0	
By your acceptances to draughts from Spain and Carthagena, for which the 8 drawers will provide	10,579 0 5	
By open creditors in Spain, the settlement of which will also be made there	25,028 11 8	
By acceptances on account of drawers who will provide for the same as they arrive at maturity, and on account of others who are debtors to you to a larger amount than the draughts now returned on them	14,261 5 0	
		115,471 15 0
Leaving liabilities to be provided for in England		£18,851 12 5
By the same statement the assets in your books here amount to		£170,525 4 3
Reduced by your unpaid acceptances, which will have to be taken up by the drawers, as above	£32,692 0 0	
Ditto	14,261 5 0	
Ditto	10,579 0 5	
	£57,532 5 5	
By property held for advances on security ...	32,910 17 11	
And by open accounts in Spain, as above ...	25,028 11 8	
	£115,471 15 0	115,471 15 0
		£55,053 9 3
Consisting of debtors (principally) in Spain and Carthagena		£16,546 6 2
Goods purchased and ready for shipment ...	£3,279 10 0	
Surplus from hypothecated goods when sold...	5,129 2 1	
Carried forward	£8,408 12 1	£16,546 6 2

Brought forward	£8,408 12 1	£16,546 6 2
Capital at the house at Havannah, and goods consigned for sale there, in charge of your special correspondent, lately sent out	£31,702 6 0	
		40,110 18 1
		£56,657 4 3
The balance of £23,424 16*s.* 8*d.*, due from the house of Carthagena, will be discharged by the payments that they will make in general settlement with the Spanish creditors, and a balance will then be due to them of		1,603 15 0
		£55,053 9 3
To pay liabilities as above		18,851 12 5
Leaving surplus		£36,201 16 10

"In treating with the items forming the preceding statement, I have satisfied myself, by communication with the various parties who have made the advances, that such advances will be fully covered by the goods that they hold, and, further, that the surplus arising therefrom, as set forth in this statement, is not over-estimated; that the fact of a large quantity of the bills which have been returned to Spain having been met by third parties, such parties having further agreed to take up all the remaining bills that are drawn on account of the Carthagena house; that the open creditors, to the extent of four-fifths, are parties who have a direct interest in carrying out a prompt liquidation; that the draughts accepted for account of others have been to this time met by the drawers for whose account they were so accepted, and but little doubt can be entertained of the remainder being, in like manner, honoured at maturity. In respect of the assets, although a considerable portion of the debts that are due from parties in Spain will in all probability be remitted in the usual course, still I do not think it would be prudent to calculate on the receipt thereof until the parties who have agreed to retire your acceptances in that country have reimbursed themselves, and, therefore, the only items which I calculate upon as really available for the purposes of your liquidation here are the surplus proceeds of goods hypothecated, the goods that are now ready for shipment, and which will be shortly forthcoming, and the balance due from Havannah, amounting in all to £40,110 18*s.* 1*d.*, showing a sufficiency to pay the full liabilities here of £18,851 12*s.* 5*d.*, and to leave a surplus of £21,259 5*s.* 8*d.*, exclusive of the above-mentioned debits in Spain, etc. In regard to the proceeds to be received from Havannah, I may remark, that I consider your plan, in sending out a person with powers to arrange the sales of property, and remit the proceeds here without delay, has been very judicious, and will no doubt greatly facilitate the completion of your contemplated arrangement. I have also perused with much attention the grants and leases of the several Escorials fields of lead ore in Linaris, which are apparently of great extent and value, and of the foundries and furnaces in Carthagena, together with the various extracts of the letters relating to their several capabilities and productions. It would be presumptuous on my part to approximate any value of these properties, but I do not think much doubt can exist of their workings being ample to pay off all liabilities in a period of six or eight months, particularly when it is borne in mind that there is sufficient ore and fuel now ready at these foundries for the next three months' smelting. In regard to the black lead mines of Marbella, in Benahavis (in which you possess an interest to the extent of two-thirds), looking at the beneficial results of the workings of these mines, and the amount of proceeds which you have received from the same, my impression is, that the property is of a valuable character. I may here observe, that I have relied upon the information that you have given me, that the Escorials fields of ore are free from charge or incumbrance, except to those parties for the payments which they have undertaken to make on the bills returned to Spain; but with regard to the mines of Marbella, the production by you of your shares therein proves to me that the same is free from lien, and, consequently, unencumbered. It further appears to me, that the arrangements which you made with the several parties in Spain, for the provision of funds to meet the outlay and expenses attending the working of the ores, before you commenced operations in Carthagena, were ample to have carried the same through, had such engagements been duly met, and that it is the non-fulfilment of these engagements alone that has caused your suspension. Viewing all the circumstances connected herewith, for the purpose of advising as to the most speedy course for the liquidation of all your liabilities, I think it most desirable, and, indeed, essential, that sufficient shipping should at all times be ready for the transit of fuel to Carthagena, and of the lead from thence to such ports as may be deemed most prudent to ship to. I therefore consider that it would be most desirable for you to obtain the sanction of your creditors here to a letter of license for the same period as I calculate will be occupied in the liquidation in Spain; such license to continue from this date to the 1st of July, 1852, with liberty for you to pay the outstanding freights, and to contract further charters for the purpose of fully carrying on the realization of the productions, and thus both provide and hasten the means for liquidation. In conclusion, I have to press upon your attention the fact that I have not included, in any way, in my statement of your assets, the value of your interests either in the fields of Escorials or in the blacklead mines of Marbella, consequently, whatever may be the estimated or intrinsic worth of these evidently valuable properties must be added to the surplus already shown by me.

"I am, etc.,

"J. E. Coleman."

THE ESTATE OF WALTER LOGAN AND CO.,

SPANISH-AMERICAN TRADE.

The following statement was issued on the 6th October, 1852, by Messrs. Walter Logan and Co., explaining the circumstances of their suspension, which was announced on the 5th instant:—

"46, *Lime Street, London, October* 6, 1852.

"In consequence of the absence of remittances from either of our houses in Lima, Bogota, or Panama, we are under the painful necessity of suspending our payments, and we feel it our duty to explain as briefly as possible the circumstances which have led to this unfortunate result.

"In February, 1851, Mr. Portes arrived here from Lima, bringing with him very strong letters of recommendation, together with specie of the value of upwards of £13,000, of which he represented that about two-thirds belonged to himself, and the remainder to friends, who had advanced the same to him, in order to create a capital of such amount as would induce us to join him in a partnership to be established at Lima and Panama. These circumstances, combined with the favourable accounts which he gave of the trade in those countries, induced us to confirm the proposed arrangements, which resulted in the establishment of the firm of Portes, Logan, and Co., of Lima.

"Upon the completion of the partnership, Mr. Portes, together with our Mr. Bonitto, visited Germany and other parts of the Continent to select goods suitable to the South American markets; and having effected purchases in those places, as well as in England, shipments, to the extent of upwards of £33,000 were made to Lima, and Mr. Portes, attended by Mr. Hulsenbeck (a clerk in our establishment, to act in case of accident to Mr. Portes), went out in charge of them, having first engaged to remit regularly to meet engagements incurred by us on account of these shipments.

"In July last, in consequence of the non-receipt of any remittances, and the uncertainty as to the course of action of Mr. Portes, we considered it right to send our tried and trusted partner, Mr. J. N. Boñitto, to South America, to obtain intelligence of Mr. Portes, to ascertain the true position of affairs, and to facilitate the realization of all property there.

"In the meantime, in order to meet our current engagements, we obtained advances on the security of the return, proceeds, etc., coming to us from our outward shipments.

"The mail which arrived here on the 30th ultimo has brought us advices from Mr. Bonitto, at Panama, which unhappily leave no doubt that Mr. Portes has absconded with a large amount of specie and other valuables, although, until we get more detailed particulars from Mr. Bonitto (after he has arrived at Lima), we are unable to state the actual amount of deficiency.

"Under these distressing circumstances, we deem it a duty to place our affairs at once in the hands of our creditors, so that the available assets may be secured and held for the benefit of all parties concerned; and, in order to carry out such views, we have consulted Messrs. Oliverson, Lavie, and Peachey, the solicitors, under whose advice we are now acting; and the accompanying statement, prepared since Monday by Mr. Coleman, the accountant, will convey to you the general position of our affairs.

"More detailed information is expected from Lima by the next mail, when a full statement of the accounts of the house will be laid before you.

"We are your most obedient servants,

"WALTER LOGAN AND CO."

STATEMENT OF THE AFFAIRS OF WALTER LOGAN AND CO., OF NO. 46, LIME STREET, LONDON, OCTOBER 4, 1852.

DEBTOR.

To creditors on open accounts				£19,965	0	0
To creditors on acceptances				7,536	0	0
To creditors on securities of shipments	£35,540	0	0			
To creditors secured by sundry property	3,729	10	0			
To creditors not covered				795	0	0
To estimated liabilities on bills receivable				3,816	0	0
				£32,112	0	0

CREDITOR.

By cash, bills, etc., in hand							£250	0	0
By sundry debtors, good							574	0	0
By sundry debtors, doubtful							4,310	0	0
By cost of property shipped, on which advances have been obtained	£47,901	0	0						
By debt from D. Logan and Co.	726	0	0						
By debt from Portes, Logan, and Co.	5,676	0	0						
By debt due from Amador, Sanudo, and Co., of Bogota	*3,376	0	0						
By debt due from Ezponda, Corredor, and Co.	†3,154	0	0						
Total amount of securities	£60,833	0	0						
Liable to payment of advances *per contra*, amounting to	35,540	0	0						
	£25,293	0	0	£25,293	0	0			
Less the two debts as above	£6,530	0	0						
Balance between cost of shipments and advance received on account of same	‡18,763	0	0						
Carried forward				£25,293	0	0	£5,134	0	0

* Amador, Sanudo, and Co. have required time to pay their engagements in full.
† The estate of Ezponda, Corredor, and Co. is now under liquidation.
‡ This amount of £18,763 will be affected by the advances obtained, and sales made by Mr. Portes, with which he has absconded, as well as by the fluctuations of the markets and the contingencies attendant thereon, in consequence of the conduct of Mr. Portes.

Brought forward				£25,293	0	0	£5,134	0	0
By securities with creditors *contra*	£3,829	10	0						
Less claims	3,729	10	0						
							100	0	0
By balance of claim in respect of steamer that was lost (now under arbitration)							1,524	0	0
By D. Logan, old account							5,267	0	0
By balance at bankers, £436 1*s.* 5*d.*, which will be retained against bills under discount.				£25,293	0	0	£12,025	0	0

THE ESTATE OF MR. FRANCIS CHAMBERS,

EAST INDIA TRADE.

A meeting of the creditors of Mr. Francis Chambers, connected with the East India trade, who failed on the 7th of November, 1851, was held on the 8th of December, Mr. F. Mellersh in the chair, when the annexed statement was presented by Mr. J. E. Coleman, the accountant:—

STATEMENT OF THE AFFAIRS OF FRANCIS CHAMBERS, OF ST. DUNSTAN'S HILL, TOWER STREET, EAST INDIA MERCHANT, ETC., NOVEMBER 6, 1851.

DEBTOR.

To creditors				£48,024	3	4
Ditto holding security for deficiency, after deducting the estimated value of securities held				7,438	7	5
Bills payable	£10,933	19	4			
Against which there are debit balances	9,137	5	5			
				1,796	13	11
Liabilities on bills receivable				2,085	5	6
Underwriting accounts				1,857	16	1
Total liabilities				£61,202	6	3

CREDITOR.

By debtors (good)				£5764	5	8
Ditto (doubtful), £6931 0*s.* 3*d.*, estimated to produce three-fourths in the pound				1,155	3	9
Stock of wines and other property				3,317	3	0
Surplus from securities held by creditors				1,141	19	6
Furniture	£500					
				£11,378	11	11
Less debts under £20, to be paid in full	£180	14	2			
Rents and salaries, etc.	130	0	0			
				310	14	2
Total assets				£11,067	17	9
By securities held for N. de St. Croix, estimated at				720	0	0
By ship "Fortescue," ninth voyage				1,104	11	1

With regard to the statement of account, Mr. Coleman explained that the last two items on the credit side were placed in the manner they had been, because in the case of the securities held for Mr. St. Croix those securities were intended to have passed for advances made; but as they had not in reality been given, they were retained as being "in the order and disposition of" Mr. Chambers, while, with respect to the freight of the ship "Fortescue," it was extremely doubtful what would eventually be recovered. The great deficiency between debts and assets, as presented by the general position of the estate, could be clearly traced, and it appeared that it chiefly arose from bad and doubtful debts, £20,000; losses on general adventures, £25,398; and losses on ships, £2,904. In 1847, Mr. Chambers had, according to his books, £6,364 to the credit of his capital account; his profits had been £9,543; his trading expenditure, £3,090; and his personal expenditure, including payments of premium on life policies, £6,090. The whole of his transactions had been thoroughly investigated, and in all cases where advances had been made on securities it was found, with the exception already alluded to, that the securities, as stated, had passed at the time the advances were made, so that no preferences had been given to creditors. On the subject of the ultimate realization of assets, Mr. Coleman said it was necessary to inform the meeting that they were of a precarious nature, inasmuch as the principal item of good debts, £5,760, consisted of balances on shipping accounts and vessels with which Mr. Chambers was connected. In some cases he possessed the protection afforded by mortgages, but in others he would have to depend upon the captains and parties representing his interest. After some conversation relative to the business transactions between Mr. Chambers and Messrs. Forman and Hadow, the question of the manner in which the estate should be liquidated was brought forward. The wish that a definite offer should be made having been expressed by the majority, Mr. Coleman stated he had already consulted Mr. Chambers on that point, who considered that he might, with some assistance from friends, be enabled to pay his creditors 3*s.* 6*d.* in the pound. To complete this arrangement, however, he would require time, and hence, if concurred in, it was proposed to offer 1*s.* in a month, 1*s.* 3*d.* in four months, and 1*s.* 3*d.* in eight months, furnishing security for the last payment. A resolution to this effect having been prepared and carried, the chief creditors signed it before the meeting broke up.

THE ESTATE OF JAMES BULT, SON, AND CO.,

GOLDSMITHS AND BULLION DEALERS.

A meeting of the creditors of Messrs. James Bult, Son, and Co., engaged in business as goldsmiths and bullion dealers, whose suspension was announced on the 1st January, 1852, took place on the 18th February, when the following statement, prepared by Mr. Quilter, the accountant, was presented :—

JAMES BULT, SON, AND CO.—STATEMENT OF AFFAIRS, JANUARY 1, 1852.

DEBTOR.

To sundry creditors		£45,671 18 10
To sundry creditors—		
In respect of R. Williamson's draughts on us		41,725 0 0
To liabilities on acceptances of Bourke's draughts		1,672 18 2
To liabilities on bills receivable as per statement—		
Considered good	£28,938 8 5	
Considered bad		4,260 0 0
		£93,329 17 0

CREDITOR.

By cash, viz. :—		
In house	£709 18 10	
At Robarts and Co.'s	19,345 5 5	
Carried forward		£20,055 4 3

Brought forward		£20,055 4 3
By bills receivable in hand—		
Considered good	£1,966 0 2	
Considered doubtful, £2,500, estimated at...	600 0 0	
		2,566 0 2
By bullion in house		1,211 1 0
By sundry debtors—		
Considered good	143 7 5	
Considered doubtful, £804 16*s*. 6*d*., estimated at..	400 0 0	
		543 7 5
Considered bad	£1,039 15 0	
By R. Williamson..	£166,805 15 3	
Estimated to produce		22,515 13 8
		£46,891 6 6
Deduct (to be paid in full), viz. :—		
Rent, taxes, salaries, expenses, etc.	£500 0 0	
Sundry creditors under £20 each...............	150 14 1	
		650 14 1
		£46,240 12 5

In explaining the above accounts, which show about 10*s*. in the pound, Mr. Maynard, the legal adviser of Messrs. Bult, detailed the causes which led to their suspension. These consisted solely in the large advances made, through a series of years, on behalf of collieries and iron works in North Staffordshire (the property of Messrs. Williamson), which, in addition to having absorbed all their available resources, had left them with a considerable amount of liability. The ultimate realization of the estate depended in a measure upon the proceeds to be gained from an arrangement effected in connection with those properties, and, therefore, mutual co-operation had been necessary to bring about a settlement. The assets of Messrs. Bult and Co. already in hand would pay a dividend of 5*s*. in the pound; and, if the terms with their chief debtors could be arranged, a further sum of 5*s*. would be secured. The creditors, under these circumstances, would be asked to accept a composition of 10*s*. in the pound, it being understood that one portion of the Staffordshire works would pay 2*s*. 3*d*. and the other 6*s*. 8*d*. in the pound. To effect this, Messrs. Williamson had gained assistance from their friends, and the funds requisite would, it was understood, be provided within a fortnight. The affairs of the collieries and iron works had been thoroughly investigated, and, encumbered as they were with mortgages and other liens, the arrangement proposed was regarded as the best that could be obtained. The feeling of the creditors was evidently favourable to the acceptance of the composition named, and after a short discussion, in the course of which Mr. L. Moseley and others expressed sympathy for the position of the firm, the annexed resolution was at once adopted :—

"At a meeting of the creditors of Messrs. James Bult, Son, and Co., held the 18th day of February, 1852, at the offices of Messrs. Crowder and Maynard, 57, Coleman Street, a proposition having been submitted to the meeting that, upon payment by Messrs. James Bult, Son, and Co., of a composition of 10*s*. in the pound upon the amount of all their debts above £20, they should be released from further liability; and, further, that the holders of bills, bearing their names and drawn or endorsed by Mr. Robert Williamson, should receive, in addition, a composition of 2*s*. 3*d*. in the pound upon the amount thereof; and that the holders of bills drawn, accepted, or endorsed by the Goldendale Company and the Stonetrough Colliery should receive a compensation of 6*s* 8*d*. in the pound upon the amount, and that thereupon the bills should be cancelled and all parties released : it was resolved—That the above proposition be accepted, provided that the compositions be paid within one month from this date."

THE ESTATE OF MACKEY, HADOW, AND CO.,

EAST INDIA TRADE.

A meeting of the creditors of Messrs. Donald Mackey, Hadow, and Co., engaged in the East India trade, whose suspension was announced on the 22nd January, 1852, took place February 3rd, when the subjoined statement, prepared by Mr. J. E. Coleman, the accountant, was presented :—

STATEMENT OF AFFAIRS OF MESSRS. DONALD MACKEY, HADOW, AND CO., OF NO. 6, BROAD STREET BUILDINGS, IN THE CITY OF LONDON, MERCHANTS, FROM MARCH 25, 1848, TO JANUARY 22, 1852:—

DEBTOR.

	£	s.	d.	£	s.	d.
To creditors on open balances				£3,842	14	1
To creditors on bills payable	£10,374	4	4			
Less to be secured in Ceylon by produce	3,500	0	0			
				6,874	4	4
To creditors holding security to the full amount of their claims	£5,364	19	9			
Estimated value of securities held	5,624	1	6			
Surplus to *contra*	£259	1	9			
To creditors partly secured	£420	0	0			
Estimated value of securities held	360	15	9			
				59	4	3
To bills accepted on account of D. C. Mackey's produce account				7,300	0	0
To creditors on bills payable	£1,944	15	11			
Less amount of balance due from the drawers of such bills	937	15	2			
				1,007	0	9
Amount of liabilities expected to rank on this estate				£19,083	3	5
To profits				10,660	1	1
To liability on bills receivable, outstanding, £35,365 13*s.* 4*d.*, the whole of which it is expected will be duly met at maturity				£29,743	4	6

CREDITOR.

	£	s.	d.	£	s.	d.
By cash at bankers	£47	19	6			
By cash in hand	10	19	3			
				£58	18	9
By debtors (good)				1,986	8	3
By debtors (doubtful)	£2,027	11	0			
By profit to be received on part charter of ship				500	0	0
By estimated surplus from securities *contra*				259	1	9
By advances on produce to come from coffee estates in Ceylon				1,251	14	6
By debts due from consigners for advances made upon goods consigned to D. C. Mackey and Co., of Calcutta				3,283	10	2
By balance due from D. C. Mackey and Co., of Calcutta				2,133	3	10
By ditto, D. C. Mackey's produce account				7,300	0	0
By losses				3,816	12	2
By doubtful debts (as above)				2,027	11	0
Carried forward				£22,617	0	5

	Brought forward		£22,617 0 5
By expenses and charges			2,716 2 2
By amount drawn out by D. C. Mackey		£2,943 8 7	
By ditto J. S. Hadow		1,466 13 4	
			4,410 1 11
			£29,743 4 6

Before the accounts were discussed, Mr. Fisher, as the legal adviser of the firm, intimated that Mr. Donald Mackey, who is at present in India, was the sole surviving partner, Mr. Hadow having died in 1849. The business in London since Mr. Mackey's departure had been carried on by Mr. Drysdale, who was deputed to act by a general power of attorney. Under these circumstances it was suggested that, after the accounts had been explained, it would be better to make arrangements for collecting the assets until advices shall have been received from Mr. Mackey respecting the course he proposes to pursue. With regard to the statement, it was mentioned by Mr. Coleman that it had been prepared from the books, which were accurately kept. It was not possible to form an estimate of the ultimate realization, but several items of the assets, there was reason to expect, might produce the amount set against them. Some of the creditors complained of the firm having been carried on by Mr. Donald Mackey, under the title of Mackey, Hadow, and Co., after the death of Mr. Hadow, especially as it now transpired that Mr. Drysdale was not a partner; others, however, exonerated the latter, who had never misled any one regarding the nature of his connection with the business. Mr. Creaton, a partner in the Calcutta house, was asked relative to the prospects of that establishment, and replied that when he left India it was in a sound condition. After further inquiries as to the probable early return of Mr. Mackey to this country, and the period when it was likely accounts could be received from the Calcutta and Moulmein houses, the following resolution was passed, and the meeting adjourned :—

"At a meeting of the creditors of Messrs. Donald Mackey, Hadow, and Co., merchants, held this 3rd of February, 1852, at No. 6, Broad Street Buildings, London, it being represented to this meeting that Mr. Donald Campbell Mackey is the only surviving partner, and that he is now in India, attending to the affairs of Donald Campbell Mackey and Co., of Calcutta, in which house he is also a partner, and it appearing that no definite arrangement as to the liquidation of this estate can be made until Mr. Mackey has been communicated with, it was, after a statement of the affairs had been laid before the meeting by Mr. Coleman, the accountant, resolved that Mr. George Drysdale, who has, since February last, when Mr. Mackey left England, conducted the affairs of the house under power of attorney from Mr. Mackey, be requested to continue his services, under the inspection of Mr. Duncan Dunbar and Mr. T. R. Brown. That all assets, as they shall be got in, shall be paid to an account at Messrs. Glyn's, in the names of the said inspectors and of Mr. Drysdale, who shall be at liberty to allow and pay all necessary outgoings and expenses. That this meeting be adjourned, to give Mr. Mackey an opportunity of making a proposition to the creditors for the ultimate liquidation of the estate. That a copy of these resolutions be sent out immediately to Mr. Mackey."

The house in Calcutta having failed, and Mr. Mackey having sought the benefit of the Insolvent Court in the Presidency, all further proceedings here are regarded as concluded.

THE ESTATE OF KEITH, SHOOBRIDGE, AND CO.,

SHAWL PRINTERS.

A meeting of the creditors of Messrs. Keith, Shoobridge, and Co., shawl printers, who suspended on the 21st February, 1852, was held on the 8th of March, Mr Piot in the chair, when the following statement, prepared by Mr. H. Chatteris, the accountant, was presented :—

STATEMENT OF AFFAIRS OF DANIEL KEITH AND THOMAS SHOOBRIDGE, TRADING AS KEITH, SHOOBRIDGE, AND CO., NO. 124, WOOD STREET, FEBRUARY 23, 1852.

DEBTOR.

To sundry creditors unsecured		£18,256 8 0
To ditto partially secured	£11,248 6 2	
Who being printers claim a lien upon goods in their hands, amounting to	3,904 19 11	
		7,343 6 3
To sundry creditors secured	£12,364 6 9	
Bills and goods deposited in their hands	12,959 12 6	
Surplus carried to *contra* side	£525 5 9	
To Edmund Upton, liability on his acceptances in favour of Keith, Shoobridge, and Co.	3,071 14 0	
Less estimated value of property in his hands applicable to the payment of these bills	1,348 15 0	
		1,722 19 0
To liability on accommodation bill accounts	£15,979 2 5	
Out of which there will be proved against this estate		10,654 16 11
To liability on Keith, Shoobridge, and Co.'s acceptances which should be provided for by the drawers	2,852 6 3	
Out of which it is estimated there will be proved against this estate		1,881 19 0
To liability on bills receivable	18,851 18 6	
Out of which it is estimated there will be proved against this estate		500 0 0
		£40,359 9 2

CREDITOR.

By stock in trade at cost price as per stock-books		£4,306 9 11
By sundry debtors considered good		1,098 15 6
By ditto, bad	£950 7 1	
By surplus property deposited with creditors *per contra*		595 5 9
By cash at banker's		26 11 8
By warehouse fixtures, say		70 0 0
By Daniel Keith's property, being the amount of capital standing to his credit with Keith and Co.		1,651 3 9
		7,748 6 7
Less rent, taxes, and salaries to be paid in full		99 6 8
		7,648 19 11
By deficiency		32,710 9 3
		£40,359 9 2

With reference to the accounts, it was explained that the assets showed 3*s*. 9½*d*. in the pound, and that Messrs. Keith, Shoobridge, and Co. proposed to offer a composition of 2*s*. 6*d*. The cause of suspension was said to have been losses in business spreading over several years. In June, 1848, the firm was insolvent to the extent of £3,735, and in June, 1849, the deficiency had increased to £7,000. The trading was subsequently carried on, though no periodical balance was taken until the 20th ultimo, when the house suspended with a deficiency of £32,700. To account for this the losses on sales were stated at £8,044 ; trade expenses, £6,830 ; less on accommodation bills, £7,540 ; drawings of Mr. Keith, £1,252 ; drawings of Mr. Shoobridge, £1,517 ; and bad debts, £950. When this state of affairs was presented, and when, moreover, it was intimated that there were specific drawings by Mr. Keith for which he could not give a satisfactory explanation, the creditors generally refused to listen to the offer of a composition, and contended it was a proper case for the *Gazette.* Another point which created a strong impression was the allegation of preference payments having been made to certain creditors for £3,400 (a sum equal to one-half of the existing assets) a day after the house had suspended. Under these circumstances it was deemed desirable to pass the annexed resolution, which having been carried, Messrs. Keith and Shoobridge signed the required declaration of insolvency:—

"At a meeting of the creditors of Messrs. Keith, Shoobridge, and Co., of 124, Wood Street, Cheapside, held this 8th day of March, 1852, at the offices of Mr. H. Chatteris, accountant, Gresham Street—Resolved, that in the opinion of this meeting the estate of Keith, Shoobridge, and Co. can only be satisfactorily administered in bankruptcy, and that to facilitate such administration those gentlemen be requested to sign and deposit with the chairman a declaration of insolvency."

The estate was, consequently, wound up in bankruptcy.

THE ESTATE OF MESSRS. RITCHIE BROTHERS,

WEST INDIA TRADE.

A meeting of the creditors of Messrs. Ritchie Brothers, engaged in the West India trade, who failed on the 11th of March, 1852, was held on the 2nd of June, Mr. Kirkman Hodgson in the chair, when the following statement, prepared by Messrs. Quilter, Ball, and Co., the accountants, was presented:—

STATEMENT OF THE AFFAIRS OF RITCHIE BROTHERS, MARCH 11, 1852.

DEBTOR.

To trade creditors— Unsecured		£4,642 14 5
Partially secured, viz.:— Amount of claims	£10,464 2 10	
Less estimated to be covered to the extent of	5,672 11 11	
		4,791 10 11
To cash creditors	53,127 15 1	
Less estimated value of securities	9,369 10 6	
		43,758 4 7
To creditors fully secured, viz.:— Estimated value of property held	6,312 16 6	
Less amount of claims	4,646 13 0	
Taken as an asset *per contra*	1,666 3 6	
To liabilities on acceptances to be provided for by the drawers	68,449 8 0	
To liabilities on bills receivable, considered good	28,408 8 2	
		£53,192 9 11

CREDITOR.

By cash in hand		£167 0 1
By sundry debtors— Considered good		3,381 7 4
Considered doubtful	11,373 13 8	
Estimated to produce		1,000 0 0
Considered bad	9,149 0 2	
By consignments to New York		632 8 3
By surplus securities in hands of creditors *per contra*		1,666 3 6
Carried forward		£6,846 19 2

Brought forward		£6,846 19 2
Deduct, creditors to be paid in full		158 18 1
		£6,688 1 1
By balance deficiency		£53,192 9 11

The following statement accounts for the deficiency:—

STATEMENT ACCOUNTING FOR THE DEFICIENCY FROM JANUARY 1, 1845, TO MARCH 11, 1852.

DEBTOR.

To deficiency, 11th March, 1852, as per Statement of Affairs		£46,504 8 10
To capital, 1st January, 1845, viz.:—		
Thomas Ritchie	£10,000 0 0	
Walter W. Ritchie	10,000 0 0	
		20,000 0 0
To sundry profits, viz.:—		
Commission	£16,764 13 10	
Discount	845 12 4	
Foreign exchanges	27 3 2	
Fire insurance	18 15 5	
		17,656 4 9
To profit on adventures, viz.:—		
Manilla cheroots	£22 3 2	
Lima wood	18 1 0	
Shirting to Shanghai	53 18 7	
China raw silk	298 13 7	
Tobacco	172 5 4	
		565 1 8
To suspense account		67 3 10
		£84,792 19 1

CREDITOR.

By expenses, viz.:—		
Counting-house expenses	£6,483 12 4	
Postage account	238 2 2	
Insurance do.	28 4 2	
Freight do.	163 16 3	
Charges do.	593 3 0	
		£7,506 17 11
By losses on adventures, viz.:—		
Cotton	£4,699 12 3	
Madder	1,186 19 8	
Coffee	1,968 17 3	
Sugar and molasses	1,683 12 5	
Ginghams, etc.	338 11 1	
Cocoa	917 15 1	
Tea	533 3 4	
Pepper	531 13 5	
Gambia	103 7 7	
Swedish steel	135 8 7	
Cinnamon	200 1 5	
Camphor	800 4 11	
Scotch pig-iron	1,300 0 10	
Flour	415 2 8	
Sugar candy	151 6 8	
Carried forward	£14,965 17 2	£7,506 17 11

Brought forward	£14,965 17 2	£7,506 17 11
Merchandise	563 5 11	
Sundries	386 19 11	
		15,916 3 0
By Messrs. Dauncey and Latham—		
Liquidation account	£2,581 7 10	
Guarantee do.	3,811 15 4	
		6,393 3 2
By sundry losses		194 17 2
By bad debts		30,930 1 3
By partners' drawings, viz.:—		
Thomas Ritchie / Walter W. Ritchie		22,884 10 7
By open balances		967 6 0
		£84,792 19 1

With regard to the estate, it was explained by Mr. Ball that the assets showed a dividend of between 2*s*. 6*d*. and 3*s*. in the pound. Messrs. Ritchie Brothers succeeded the firm of Moens, Dauncey, and Latham, and introduced, when they commenced, in January, 1845, a capital of £20,000. They had, however, been unfortunate in the majority of their adventures, and had, through the losses occasioned by these and bad debts, been compelled to suspend. Their chief operations had been in cotton, madder, coffee, sugar, pig-iron, camphor, and cinnamon. It was also intimated that the family creditors represent upwards of £47,000, so that the liabilities incurred in trade are proportionately small. Not the least discussion took place after this explanation, and it was unanimously agreed to wind-up the affairs under inspection. Mr. Kirkman Hodgson is not a creditor, but attended on behalf of the family connection of Messrs. Ritchie Brothers. Annexed is the resolution adopted :—

"At a meeting of creditors of Ritchie Brothers, held on the 2nd of June, 1852, Mr. Kirkman D. Hodgson in the chair, a statement of the affairs of the house having been laid before the creditors, it was resolved—

"That a letter of license be granted to Messrs. Ritchie Brothers to liquidate the estate under the inspection of Mr. C. H. Gray and Mr. Kirkman Hodgson on behalf of the creditors, and that the usual deed for carrying such arrangements into effect be prepared for the signature of the creditors, such deed to contain a provision for a release to Messrs. Ritchie Brothers as soon as the inspectors shall certify that the conduct of those gentlemen in the liquidation of the estate has been satisfactory to them."

THE ESTATE OF C. AND B. HOOPER,

LEATHER FACTORS.

A meeting of the creditors of Messrs. C. and B. Hooper, leather factors, who suspended on the 7th of February, 1852, was held on the 2nd of March, when the following statement was presented by Mr. Broom, the accountant :—

STATEMENT OF THE AFFAIRS OF CLEEVE AND BENJAMIN HOOPER, FEBRUARY 9, 1852.

DEBTOR.

To sundry creditors		£17,390 18 11
To sundry liabilities	£47,509 16 5	
		£17,390 18 11

CREDITOR.

By cash at the bankers		£828 12 6
N.B. The balance at the bankers is subject to any unpaid bills which they may hold, also to two claims of Mr. Nelson for £506 4*s*. 8*d*., and £200 to Mr. Alderman Challis.		
By stock of leather, hides, etc., estimated at		904 13 4
By debts		4,580 2 11
Considered good	£3,613 11 11	
Considered doubtful (£1,933 2*s*. 1*d*., at 10*s*. in the pound)	966 11 0	
	£4,580 2 11	
By horses, carts, fixtures, etc., estimated at		500 0 0
By property, life policies, and reversion		845 0 0
N.B. This property is chiefly held by a client of Messrs. Slee and Robinson, to secure an advance of £4,000 made to Henry Hacker. It is presumed that Mr. Hacker's separate property on mortgage will be nearly sufficient to satisfy this claim.		
By freehold house and tanyard at Aldborough, Norfolk, value beyond existing mortgage		50 0 0
By lease of premises in Seething Lane (six years unexpired), estimated at		500 0 0
N.B. Deposited with the bankers to secure them against any unpaid bills. Will be forfeited in case of bankruptcy.		
By C. Hooper's furniture (exclusive of that portion which is included in the marriage settlement and private debts—about £50)	£200 0 0	
By B. Hooper's furniture	400 0 0	
		600 0 0
		£8,808 8 9

Before details were entered upon, Mr. Rutherford, the legal adviser of Messrs. Hooper, explained the circumstances attending their suspension. Although the firm were in favourable credit, and not the least suspicion had been entertained of their insolvency, the accounts on being examined showed their stoppage to be unavoidable. Mr. Broom, in furnishing information respecting the position of the estate, mentioned that the partners, owing to an inability to estimate the proportion of liabilities likely to come against them, were not prepared to make an offer for a composition. Already a number of bills had run off, and others would doubtless be provided for, but as they were connected with a variety of parties it was impossible to estimate the ultimate result. The whole of the paper arose, however, from *bonâ fide* transactions. Messrs. Hooper commenced business in 1845, succeeding their mother, to whom they agreed to pay £250 for twelve years, or, in other terms, to allow for the purchase of the good-will £3,000. Half of the amount had been paid, and the remainder constituted a legal debt against the firm; although it ought to be relieved, if not from all, from a portion of the demand, since the old estate had been fed at the cost of the new. To account for the deficiency of £8,000, and the profits returned at £19,000, making together £27,000, it was stated that the bad debts had absorbed £14,476, while the drawings of Mr. C. Hooper were £5,700, and those of Mr. B. Hooper £6,700, including payments for premiums of assurance, etc. Their discount account showed no extravagant or usurious charges, their bills having passed through their bankers or the leading discount houses. A good deal of discussion took place on the prospects of the liquidation, and the limited amount of assets shown. It was alleged that Messrs. Hooper had not, at the period of their suspension, a knowledge that they were in so embarrassed a position, a statement they had drawn out themselves inducing them to believe they possessed nearly 20*s*. in the pound. On the proposal that the estate be wound up under inspection, some conversation ensued relative to the possibility of getting it satisfactorily settled, and an objection was raised to bankruptcy on account of the expense. Eventually the appended resolutions were carried, and a number of creditors signed them before leaving. In the course of the proceedings a full statement of the circumstances attending the non-payment of certain checks at the bankers the day the firm stopped was made, with the view of clearing the character of Messrs. Hooper from an unfavourable impression created with respect to those transactions :—

"At a meeting of the creditors of Messrs. C. and B. Hooper, held at the George and Vulture Tavern, the 2nd of March, 1852 (Mr. Nesbitt in the chair), a statement of the assets and liabilities was read to the meeting by Mr. Broom, the accountant, of the firm of Broom and Bagshaw, and the meeting having come to the conclusion that it was expedient that the affairs should be wound up under a deed of arrangement, with the supervision of inspectors, it was therefore resolved—

"1. That Mr. Thomas Laurence (of the firm of Streatfeild, Laurence, and Mortimore), Mr. Richard Hill Fisher (of the firm of Fisher and King), and Mr. Robert Nesbitt (of the firm of A., R., and A. Nesbitt) having consented, are hereby appointed inspectors.

"2. That a deed of arrangement be prepared, with proper provisions for winding up the affairs of the firm, under the control of the above-named inspectors, and for securing the collection and early distribution of the assets, and that when there shall be a clear net amount in hand sufficient to pay to all the creditors 2*s.* 6*d.* in the pound upon the amount of their respective debts, the same shall be immediately paid to them without any preference or priority, according to the rules of administration in bankruptcy, and containing covenants by Messrs. C. and B. Hooper that they shall carry on such liquidation under the inspectors, and covenants on the part of the creditors not to sue Messrs. C. and B. Hooper for three months, and that the inspectors shall have power to enlarge that time, if they shall deem it advisable, to a period not exceeding an additional six months, and that at the expiration of such period, or of such enlarged period, and upon Messrs. Hooper assigning the outstanding assets to trustees for distribution among the creditors, they shall be released from all claims. The deed of arrangement shall also contain such other provisions as are usual in deeds of inspectorship, and as shall be approved of by the inspectors on behalf of the creditors.

"3. The creditors, according to this arrangement, shall not be prejudiced as to any securities or liens they may hold, or as to their rights against third parties.

"4. The inspectors to have power to allow for maintenance and expenses such reasonable amounts as they in their discretion shall see fit.

"5. That in the meantime, until the deed of arrangement shall be executed, these resolutions shall form a memorandum of arrangement for the administration of the affairs."

PART III.

SUSPENSIONS IN 1853.

Jan. Messrs. Collmann and Stolterfoht, London, merchants.
" Mr. E. Werthemann, Amsterdam, merchant.
Feb. Messrs. C. Riva and Co., St. Petersburg, merchants.
" Mr. T. Rawack, Hamburg, merchant.
Mar. Mr. Lemuel Goddard, iron and sperm-oil merchant. (Subsequently resumed.)
April. Mr. John Attwood, London, metal merchant.
" Messrs. Wentzell and Co., Rotterdam, merchants.
May. Messrs. Denny, Clark, and Co., London (liquidated from differences between the partners), East India merchants.
" Messrs. Chiriaco and Co., London, Greek merchants.
June. Mr. B. Mirasyedi, Manchester, Greek merchant.
" Mr. A. R. Homersham, London, wool merchant.
" Messrs. A. Wrampe and Co., London, Baltic merchants.
Aug. Mr. W. G. Maltass, Smyrna, merchant.
Oct. Messrs. Schultz and Wahnschaffe, Hamburg, merchants.
Nov. Isle of Man Bank (Messrs. Holmes and Co.)

SUSPENSIONS IN 1854.

Jan. Messrs. G. and G. Buono, Naples, merchants.
" Messrs. Thomson Brothers and Sons, Clitheroe, calico printers.
Mar. Messrs. Dickson and Co., Glasgow, Australian merchants.
" Messrs. Gladstone, Bond, and Co., Manchester, general brokers.
" Mr. Thomas M'Gregor, London, woollen warehouseman.
" Messrs. Warwick, Harrison, and Co., London, shawl and silk manufacturers.
" Messrs. Benjamin Elkin and Sons, London, West India and Australian merchants.
" Messrs. P. Monteaux and Co., Paris and London, foreign exchange dealers.
" Messrs. Lampronti and Co., Florence, merchants.
" Messrs. Möller and Burroughs, London, foreign exchange dealers.
" Messrs. Leroy, De Chabrol, and Co., Paris, bankers.
April. Mr. C. R. Moate, London, metal broker.
" Mr. Ilja Stephanoff, St. Petersburg, cotton dealer.
" Mr. M. Jensen, Riga, broker.
" Mr. S. Alexeyeff, Moscow, merchant.
" Mr. T. Mathias, Moscow, merchant.
" Mr. C. Kyber, Moscow, merchant.
May. Sir Evan Mackenzie and Co., London, merchants.
" Messrs. Brown, Coultate, and Co., Manchester, cotton brokers.
" Mr. P. C. Salvago, Manchester, Greek merchant.
" Messrs. Scaravaglio and Peloso, Genoa, merchants.
June. Mr. Julius Steding, Moscow, merchant.
" Mr. M. Goddard, Birmingham, American shipper.
" Messrs. Davidson and Gordon, London, produce brokers and metal agents.
" Messrs. Cole Brothers, London, East India merchants and metal brokers.
July. Messrs. H. W. Lord and Co., London, East Indian and colonial brokers.
" Mr. Spiridone Gopcevitch, Trieste, merchant. (Resumed in a few days, assisted by Bank of Vienna.)
" Mr. Mark Gopcevitch, London, merchant. (Resumed in a few days, assisted by Bank of Vienna.)
" Messrs. R. and G. L. Schuyler, New York, railway agents.
" Mr. J. Tucker, Philadelphia, president of Reading Railroad.
" Mr. J. Lawton, Havannah, merchant.
" Messrs. Crawford and Echarte, Havannah, merchants.

Aug. Messrs. Pegler Brothers, London, Spanish merchants. (A. Peglar resumed January 6, 1855.)
,, Messrs. Morpurgo and Co., London, Italian merchants.
,, Messrs. Werthemstein Sons, Vienna, merchants.
,, Mr. R. C. Sercombe, London, corn merchant.
,, Messrs. T. Freen and Co., London, manufacturers of Roman cement.
Sept. Messrs. David Scott, Richmond, and Co., Manchester, merchants.
,, Messrs. Currie, Dale, and Co., London.
,, Messrs. Deane, Youle, and Co., Liverpool, South American merchants.
,, Messrs. Lukin and Skuratoff, Moscow, cotton manufacturers.
,, Mr. J. Osteried, Moscow, calico printer.
,, Mr. H. J. Bolotoff, Shuya, calico printer.
,, Mr. G. A. Ullich, Trieste, merchant.
,, Mr. James Mitchell, Bradford, woollen manufacturer.
Oct. Mr. Edward Oliver, Liverpool, shipowner.
,, Mr. James M'Henry and Co., Liverpool, merchants.
Nov. Messrs. Allen and Anderson, London, American grain and provision merchants.
,, Messrs. Lemon and Co., Brentwood, bankers.
,, Messrs. Brown and Son, Arbroath, tanners.
,, Messrs. Clay and Gillman, London, merchants.
,, Messrs. Ellis and Sturgis, Cincinnati, merchants.
,, Messrs. T. S. Goodman and Co., Cincinnati, merchants.
,, Messrs. J. R. Morton and Co., Cincinnati, merchants.
,, Messrs. Smead, Collard, and Co., Cincinnati, merchants.
,, Citizens' Bank, Cincinnati.
,, Canal Bank of Cleveland, Ohio.
,, City Bank of Columbus, Ohio.
,, Bank of Circleville, Ohio.
,, Exchange Bank of Buffalo, New York.
,, Mr. H. D King's bank, Pittsburgh, Pennsylvania.
,, Nearly all the Banks of Indiana.
,, Mr. Henry Meiggs, San Francisco, merchant.
,, Messrs. William Price and Co., Quebec, merchants.
,, Messrs. Reeves, Buck, and Co., Philadelphia, iron manufacturers.
Dec. Messrs. John Benson and Co., New York, sugar refiners.
,, Mr. Dennis Harris, New York, sugar refiners.
,, Messrs. Patterson, Adams, and Co., New York, tobacco merchants.
,, Messrs. Selden, Withers, and Co., Washington, merchants.
,, Messrs. Carter and Co., London, shipowners.

THE ESTATE OF MESSRS. COLLMANN AND STOLTERFOHT.

A meeting of the creditors of Messrs. Collmann and Stolterfoht, of London and Liverpool, engaged in the American trade, who suspended on the 9th of January, 1853, was held on the 4th of February, Mr. H. Göschen presiding, when the following statement, prepared by Mr. J. E. Coleman, the accountant, was presented :—

STATEMENT OF THE AFFAIRS OF MESSRS. COLLMANN AND STOLTERFOHT, JANUARY 10, 1853.

LIABILITIES.

Creditors on open balances		£37,469 1 7
Acceptances out at date of suspension	£401,431 14 0	
Of which there has been, and will be, withdrawn by other parties	278,673 3 3	
	£122,758 10 9	
Deduct assets specially applicable to part of these bills	6,416 11 11	
		116,341 18 10
Total (carried forward)		£153,811 0 5

Brought forward				£153,811 0 5
Liabilities on Bills receivable discounted, all of which are expected to be honoured at maturity			£195,669 19 10	
ASSETS.				
Cash in hand and produce realized			£36,887 1 3	
Deduct charges, etc.			397 0 0	
			£36,490 1 3	
Good debts, and produce to be realized in a short time			37,864 16 2	
Other assets not immediately available			24,800 17 10	
				99,155 15 3
				£54,655 5 2
Capital, 1st January, 1852		£30,044 17 11		
Profit for the year 1852...	£12,961 9 10			
Less expenditure of the three partners	4,529 5 4			
		8,432 4 6		
			£38,477 2 5	
Deduct bad and doubtful debts			11,390 3 2	
				27,086 19 3
Deficiency				£81,742 4 5
Loss by R. F. Pries			£81,742 4 5	

The above assets are stated irrespective of all recoveries that may be made from the estate of R. F. Pries, or any other doubtful debt.

Mr. James Freshfield, in explaining the accounts, first alluded to the circumstances attending the suspension, and particularly to the delinquencies of Mr. Pries (now under sentence of transportation for life), through which that event had been brought about. There being nothing on the side of Mr. Turck, the acting partner, to conceal with reference to the connection, that gentleman immediately, on discovering that Pries's cheque had been dishonoured, commenced an investigation, and, having ascertained the nature of the forgeries committed upon them, he resolved that the house should at once stop payment, in order to protect the interests of the general creditors. The firm had enjoyed the highest credit up to the moment of suspension, and, had it not been for the heavy loss sustained through the frauds of Pries, it would have possessed a considerable surplus beyond all engagements. From the state of the accounts, it would be perceived that the assets amounted to £99,155, against liabilities to the extent of £153,811, and that they thus showed a dividend of about 12*s.* 6*d.* in the pound. Taking into consideration the expense of liquidation, and other contingencies, that full sum might not be realized; but, at all events, there was every prospect of 10*s.* being paid. The nature of the assets was generally favourable, especially one item of £36,490, consisting of money immediately available, which would of itself be nearly sufficient to pay a dividend of 5*s.* in the pound, although it could not be distributed until some further progress had been made in the withdrawal of liabilities. The other items of £37,864, good debts, produce, etc., and £24,800, assets, not immediately available, although they might take time for realization, were believed to be in every respect good. The difference between £99,000, estimated as the assets, and £153,000, as liabilities, left £54,000 in the shape of deficiency. In accounting for this, it was necessary to state, that the loss by the transactions with Pries, subject to whatever might be secured under his bankruptcy, was £81,742, and, deducting from that amount the capital of the firm of Collmann and Stolterfoht, originally £38,000, but reduced by bad debts to £27,086, the result would be as stated. The suspension, disastrous as it was, exhibited less discouraging features than had been expected, and it now remained to decide as to the plan of liquidation. In answer to inquiries, Mr. Coleman, the accountant, mentioned that the London and Liverpool houses had been included in the statement as one establishment, and that all transactions out of which might arise any liabilities with the firms of Faber and Bierwith, and H. Stuker, of New York, E. Weithermann, of Amsterdam, and J. Hirtch and Co., of Rotterdam, had been properly taken into account. Of the acceptances out at the date of suspension, representing £278,000, which would be withdrawn by other parties, £155,000 had already been arranged, and there was every hope that the remainder would shortly be provided for. The opinion of the meeting having been taken on the desirableness of the estate being liquidated under inspection, the annexed resolutions were passed, it being intimated, however, that the gentlemen accepting the appointment of inspectors were not creditors, but acting as the representatives of other houses. In the course of the proceedings, Mr. Freshfield read to the creditors a narrative of the origin of the connection of Messrs. Collmann and Stolterfoht with R. F. Pries, and the progress of their transactions to the date of the discovery of the forgeries, with the view of placing before the creditors and the public every information on the subject.

"At a meeting of the creditors of Messrs. Collmann and Stolterfoht, held at the counting-house in Broad Street Buildings, on Friday, the 4th February, 1853, Henry Göschen, Esq., in the chair, Mr. Coleman, the accountant, having produced and read to the meeting a statement of the debts and assets, it was thereupon resolved unanimously:—

"1. That it is the opinion of the meeting that the house should be liquidated by the partners, under the inspection of the following gentlemen, viz., Henry Göschen, Esq., and John Cunliffe Pickersgill, Esq.

"2. That the salaries of clerks and servants be paid in full.

"3. That a deed be forthwith prepared, by which the partners shall covenant to liquidate the affairs of the house, and divide the proceeds among the creditors rateably, and in proportion to their several debts, under inspection, observing in such liquidation the rules of administration adopted in bankruptcy. That such deed shall contain covenants by the creditors not to sue the partners for twelve months, and that the inspectors shall have power to enlarge that time if they shall deem it necessary, and that, at the expiration of such period, or of such enlarged period, or sooner, if the inspectors shall certify that the liquidation has proceeded sufficiently, and upon the partners executing such assignment as the inspectors may require of the outstanding assets to trustees for distribution among the creditors, the partners shall be released from all claims. The deed shall also contain such other provisions as are usual in deeds of inspection, to be settled by the inspectors on behalf of the creditors.

"4. That creditors acceding to this arrangement shall not be prejudiced as to any securities or lien they may be entitled to, or as to their rights against third parties.

"5. That the private property of the partners, after payment of their separate liabilities, be applied in the payment of the debts of the firm, according to the rules of distribution in bankruptcy, and the inspectors shall have power to make the partners such allowances as they shall think just for their services.

"6. The inspectors shall have power to allow all or any of the partners to transact business on their own account, on their covenanting not to use, either directly or indirectly, any of the existing assets of the firm, and to incur no new engagements which could by any possibility be thrown on the existing assets."

THE BANKRUPTCY OF MR. W. BERNARD ANDERSON.*

APPROXIMATE STATEMENT OF DEBTS AND ASSETS LAID BEFORE THE COMMISSIONER, JANUARY 12, 1854.

DEBTOR.

To sundry creditors (unsecured) about		£8,500 0 0
To ditto (disputed)	£900 0 0	
Carried forward	£900 0 0	£8,500 0 0

* This bankrupt was tried and convicted of fraud in his business as a commission agent.

Brought forward	£900	0	0	£8,500	0	0
To liabilities	4,401	18	3			
Not expected to be proved	£5,301	18	3			
To creditors (secured) about				22,000	0	0
To ditto (partly secured) about	£5,000	0	0			
Less securities held	600	0	0			
				4,400	0	0
				£34,900	0	0

CREDITOR.

By property seized by the official assignee	£2,500	0	0
By debtors, good and doubtful	3,850	0	0
By warrants, and other property held by creditors	24,800	0	0
	£31,150	0	0

In addition to the above, there are consignments and open contracts, from which it is expected a considerable sum will be realized for the benefit of the creditors, of which a correct estimate cannot at present be made.

THE ESTATE OF MESSRS. WARWICK, HARRISON, AND WARWICK.

A meeting of the creditors of Messrs. Warwick, Harrison, and Warwick, who failed on the 15th of March, 1854, was held at the Guildhall Coffee-house, on the 23rd of March, when the annexed balance-sheet, prepared by Messrs. Parrington and Ladbury, was presented. It will be noticed that the estate shows debts and liabilities amounting to £37,888, while the assets at cost price reach £37,372. An offer of 16*s.* in the pound was made and accepted, the instalments to be liquidated at three, six, or nine months, with security for the final payment, and the partners, it is expected, will be able to carry through the arrangement. The house, it will be noticed, was engaged in bill transactions with Mr. Thomas M'Gregor.

DEBTOR.

To creditors, as per list	£34,237	5	1			
Less debts of creditors holding securities	2,257	16	11			
				£31,979	8	2
To liability on bills drawn by Messrs. T. M'Gregor and Co., upon and accepted by the firm				2,431	4	0
The firm has received value on these bills to the extent of £1,525 1*s.* 6*d.*						
Liability on bills drawn or accepted by the firm, amounting to £5,646 1*s.*, for which the firm has received value				2,542	16	1
Liability on a bill accepted by the firm for a separate debt of Mr. Charles Warwick				898	10	7
Liability on bills receivable discounted, £6,337 3*s.* 4*d.*, of which it is supposed will be proved on this estate				36	17	5
				£37,888	16	3

CREDITOR.

By stock in trade, at cost	£25,001	5	10			
By ditto, at printers, etc.	568	3	9			
				£25,569	9	7
By book debts, good				10,227	11	0
By ditto, bad, £171 8*s.*, estimated at				53	12	11
By cash in hand and at bankers'				147	3	0
Carried forward				£35,997	16	6

Brought forward		£35,997 16 6
By bills on hand		80 19 1
By furniture		278 4 3
By leases and fixtures of premises, Nos. 131, 132, 133, Cheapside, held for unexpired terms of 1¼, 11½, and 19½ years, respectively, at a total rental of £530 per annum, cost	£3,505 3 1	
Mortgaged to the Provident Life Office for	2,016 13 4	
		1,488 9 9
		£37,845 9 7
Less rent, salaries, etc.		473 2 11
		£37,372 6 8

THE ESTATE OF MR. THOMAS M'GREGOR.

A meeting of the creditors of Mr. Thomas M'Gregor, woollen warehouseman, who failed on the 14th of March, 1854, was held on the 28th of March, when the subjoined statement, prepared by Messrs. Parrington and Ladbury, accountants, was presented. A proposal of 6s. 8*d.* in the pound, by instalments, spreading over twelve months, was unanimously accepted, and a committee was appointed to inspect the estate, with the view of determining whether security should be required for the final payment. The losses of Mr. M'Gregor by adverse speculations in English stock, railway, and other securities, it is alleged, have amounted to a large sum, while his accommodation bill transactions, which doubtless include some of these, also represent a considerable amount. It will be noticed in his accounts, that operations in "mining shares" are distinctly alluded to.

DEBTOR.

To trade creditors		£31,442 3 9
To cash creditors	£4,153 3 0	
Less secured	3.925 0 3	
		228 2 9
To liability on bills receivable, amounting to £26.253 1*s.* 6*d.*, of which is supposed will be returned	£6,187 19 7	
Less cash held by a creditor	167 4 7	
		6,020 15 0
To liability on bills accepted for mutual accommodation of T. M'Gregor and of other parties, £11,288 10*s.* 5*d.*, of which will be claimed*		10,633 16 5
To liability in respect of bills accepted for mining shares, etc., £11,603 18*s.* 11*d.*, of which will be claimed		2,506 19 8
To liability on bills accepted wholly for T. M'Gregor's accommodation		2,778 12 4
To bills accepted on account of losses		3,522 1 9
		£57,132 11 8

CREDITOR.

By stock in trade, at cost	£15,744 7 4
By book debts, good	6,566 18 7
By ditto, bad, £3,568 17*s.* 5*d.*, estimated at	413 2 4
By furniture	350 0 0
By fixtures	498 6 0
By bills in hand, £3,712 3*s.* 9*d.*, estimated at	2,145 5 10
Carried forward	£25,718 0 1

* Value £5,926 18*s.* 2*d.* has been given in respect of these bills.

		£	s.	d.
Brought forward..........		£25,718	0	1
By cash in hand		604	4	10
By leases of premises, 149 and 150, Cheapside, and sundry shares and life policy held by a creditor, and supposed to cover his claim, as *per contra*, £3,925 0*s*. 3*d*.				
By lease of premises, Holly Lodge, held for 49 years, at £23, cost £60.				
By shares in bank, etc.		170	0	0
		£26,492	4	11
Less rent, taxes, and salaries		676	7	10
		£25,815	17	1

THE ESTATE OF MESSRS. B. ELKIN AND SONS.

A numerous meeting of the creditors of Messrs. B. Elkin and Sons, lately engaged in the West Indian and Australian trade, who failed on the 20th of March, 1854, was held on the 17th, and, by adjournment, on the 25th of April, Mr. Byass in the chair, when the following statement, prepared by Messrs. Quilter, Ball, and Co., was presented :—

DEBTOR.

To sundry creditors (unsecured)				£89,789	16	5
To creditors (partially secured)—						
Amount of claim	£33,738	17	6			
Estimated value of securities	24,815	0	0			
Also claims, lien on bills of lading received since the suspension, to the extent of £3,280, *per contra*.				8,923	17	6
To creditor (fully secured)	£1,426	5	0			
Liabilities, viz. :—						
On acceptances against consignments, £19,449 13*s*. 3*d*., estimated to be covered, except to the extent of............	£296	10	9			
On acceptances to be provided for by the drawers, viz. :—Considered good, £49,518 19*s*. 5*d*.; amount which it is considered will be claimed on this estate	21,736	15	4			
On guarantee to accept certain drafts from Vienna	3,300	0	0			
On bills receivable—out-standing, £65,860 11*s*. 2*d*.; amount which it is considered will be claimed on this estate	9,386	10	5			
				34,719	16	6
				£133,433	10	5

CREDITOR.

By bills receivable on hand				£173	0	0
By sundry debtors—						
Considered good	£26,894	17	1			
Ditto, doubtful, £13,465 2*s*. 8*d*., estimated to realize about	4,000	0	0			
				30,894	17	1
Bad, £21,010 16*s*. 5*d*.						
A portion of remittances on account of these assets to the extent of £3,289 is claimed by creditor partially secured, *per contra*.						
Carried forward				£31,067	17	1

	£	s.	d.
Brought forward	£31,067	17	1
By consignments outstanding, estimated to produce	7,206	6	11
By store in Trinidad, estimated value	1,200	0	0
By advances on property in Demerara	57,439	15	11
	£96,913	19	11
Deduct rent, taxes, creditors under £20, and sundries to be paid in full	385	7	8
	£96,528	12	3
By deficiency	36,904	18	2
	£133,433	10	5

Previously to the examination of the accounts, the chairman stated that Messrs. Elkin and Sons, under the sanction of some of the principal creditors, had, immediately after their suspension, granted a power of attorney to a responsible party, resident in Demerara, and that Mr. Isaac Elkin had proceeded thither by the last packet to protect the interests of the estate in that colony, the dependencies there being nearly £60,000. The nominal figures show a dividend of about 14*s.* 6*d.*, and, irrespective of the Demerara advances, the assets will pay 5*s.* 10*d.* in the pound. The sufficiency of the security for the Demerara claim is doubtful, but it was mentioned that remittances have been received since the failure. The whole amount already realized on general account is £12,000, of which about £3,000 is claimed by an individual creditor. The exertions of Mr. Byass and Mr. Davies, in temporarily advising Messrs. Elkin and Sons, were specially noticed. The majority of the creditors having expressed an opinion in favour of winding-up under inspection, Mr. Gurney Hoare proposed a resolution, to the effect that a letter of license be granted to the firm, with the view of allowing time for the development of the West India assets. At the adjourned meeting, the accounts having been fully investigated, it was agreed to adopt this proposition. The only parties objecting were the representatives of the London Joint-Stock Bank, who considered that the irregular conduct of Messrs. Elkin, in their bill transactions, rendered them deserving of passing through the ordeal of bankruptcy. It was intimated that there are no separate estates by which the joint one is likely to be augmented.

BANKRUPTCY OF SIR EVAN MACKENZIE AND CO.

In the matter of Sir Evan Mackenzie, Bart., Robert Cameron, and James Holmes Boyle, of Levant House, St. Helen's Place, Bishopsgate Street, merchants.

Joint Balance-sheet, from March 7, 1849, to May 18, 1854, the Date of the Adjudication.

DEBTOR.

	£	s.	d.	£	s.	d.
Creditors unsecured	£15,106	13	6			
Ditto	77	12	0			
				£15,184	5	6
Creditors holding security				5,714	1	2
Creditors holding security on the joint estate, and also on the separate estate of Sir E. Mackenzie...				9,916	12	1
Creditors to be paid in full	£39	0	0			
Liabilities	4,506	15	9			
Ditto, partly secured	2,777	6	4			
Ditto, partly secured by H. S. Boyle and Co.	3,522	2	0			
				10,845	4	1
Capital brought in by Sir Evan Mackenzie				15,000	0	0
Profits				26,939	8	5
				£83,599	11	3

CREDITOR.

	£	s.	d.	£	s.	d.
By debtors, good				£1,713	7	5
By ditto, doubtful				340	17	8
By ditto, bad, £5,406 11*s.* 1*d.* carried to losses below.						
By property	£60	0	0			
Less creditors to be paid in full	39	5	0			
				20	15	0
By property held by liability creditors				2,801	13	10
By property held by creditors				5,440	0	0
By consignments to Calcutta and Sydney, cost				10,653	18	7
By liabilities as *per contra*				10,805	4	1
By trade expenses				4,455	8	0½
Amount drawn out by partners—						
Sir E. Mackenzie	4,084	5	9			
Robert Cameron	4,043	6	4			
J. H. Boyle	2,637	0	9			
				10,764	12	10
By losses	£31,196	7	8½			
By bad debts	5,406	11	1			
				36,602	18	9½
By difference in balancing				0	15	0
				£83,599	11	3

BANKRUPTCY OF MR. HENRY PEARCE.

THE BALANCE-SHEET OF HENRY PEARCE, OF DIGSWELL HOUSE, WELWYN, HERTS, MERCHANT, AND OF 8, FINSBURY PLACE, FROM APRIL 30, 1853, TO DATE OF FIAT, JUNE 17, 1854.

DEBTOR.

	£	s.	d.	£	s.	d.
To creditors unsecured				£53,642	0	8
To creditors partially secured—						
Amount of claim	£5,106	11	6			
Security held	1,588	2	6			
				3,518	9	0
To creditors fully secured—						
Security held	6,540	0	0			
Claim	4,573	9	2			
Surplus to cash account	1,966	10	10			
				£57,160	9	8
To receipts from trustees of my wife's marriage settlement				908	14	10
To fees as director, and dividends on stock and shares, in various companies				1,016	0	0
To profit credited in Pearce and Child's account				1,186	18	11
				£60,272	3	5

CREDITOR.

	£	s.	d.
By debtors	£410	0	0
By property	6,018	15	0
By surplus from creditors holding security, *per contra*	1,966	10	10
	£8,395	5	10
Less sundry amounts to be paid in full	792	6	6
	£7,602	19	4
By personal expenses, etc.	3,579	0	10
Carried forward	£11,182	0	2

	£	s.	d.
Brought forward	£11,182	0	2
By interest and premiums on life policies	1,807	6	2
By annuity	430	0	0
By losses	3,382	17	2
By depreciation on property	3,506	5	10
By capital to the credit of my account in the firm of Pearce and Child, June, 1854	32,367	13	1
(The value of this balance depends entirely upon the valuation of estates and debts in the West Indies.)			
By deficiency, April, 1853	1,596	0	10
By liability on reversionary interest now included in credit, as *per contra*	6,000	0	0
	£60,272	3	3

THE ESTATE OF MESSRS. DAVID SCOTT, RICHMOND, AND CO.

The suspension of Messrs. David Scott, Richmond, and Co., of Manchester, took place on the 14th of September, 1854, and a meeting of their creditors was held on the 22nd, when the following statement was presented :—

LIABILITIES.

	£	s.	d.
To book debts	£42,924	7	1
To liabilities—bills in the hands of third parties	150,370	16	11
	£193,295	4	0

ASSETS.

	£	s.	d.
By book debts owing to the concern	£47,884	6	1
By debts owing to the concern, for which acceptances have been given	27,472	16	3
By goods in the hands of consignees	8,688	0	10
By stock of goods, fixtures, and miscellaneous stock in warehouse	5,411	17	1
By miscellaneous property	480	0	0
By bills and cash on hand	3,412	16	1
By available security deposited	2,000	0	0
	£95,349	16	4

With regard to this balance-sheet, it is observed:—

"In addition to the liabilities above enumerated, there are bills to the amount of £245,000 running, in which the firm is interested, but these are all good and will be duly retired at maturity; so that, although the failure involves a total amount of £438,295, the actual sum with which the creditors will have to deal is that set down in the balance-sheet above given, viz., £193,295 4*s*. The liabilities on bills in the hands of third parties is the item in which the creditors are the most interested, inasmuch as in that item is involved the paper bearing the names of the three other firms which have since stopped payment, and on the realization of their assets depends materially the position of this estate. Besides the assets enumerated, there is Mr. Scott's private estate which will probably produce from £10,000 to £15,000 for the general estate, but not more than that."

THE ESTATE OF MR. EDWARD OLIVER.

A meeting of the creditors of Mr. E. Oliver, merchant and shipowner, of Liverpool, who suspended on the 4th of October, 1854, was held on the 27th, when the following statement of his affairs was submitted, and unanimously agreed to:—

LIABILITIES.

Acceptances out at the date of suspension		£710,724 0 0	
Of which are expected to be withdrawn and retired by the drawers, or other parties		65,823 11 3	
			£644,900 8 9
Creditors on open balances			62,267 0 4
Bills receivable, estimated bad....................		221,063 10 1	
Deduct securities in hands of bankers............		70,916 15 5	
			150,146 14 8
			£857,314 3 9

ASSETS.

Bills on hand considered good....................		£44,366 11 7	
Debts estimated good		23,125 15 8	
Timber cargoes, deducting advances		22,293 6 10	
Insurances in New York ...	£43,900 0 0		
Less lien of Arthur Leary ...	15,021 15 9		
		28,878 4 3	
Palm-oil sold to arrive		32,600 0 0	
Ships..............................	413,131 15 0		
Freights	165,611 0 2		
		578,742 15 2	
House at Wavertree		1,000 0 0	
12 shares in Peel River Company		48 0 0	
1 share in Australian Agricultural Company...		40 0 0	
Collieries at Upholl, and estimated at...........		3,000 0 0	
Estimated surplus in the hands of Messrs. Goodhall, Chilton, and Co.		10,000 0 0	
			744,094 13 6
Deficiency			£113,219 10 3

The following gentlemen, who had acted as temporary trustees, were unanimously confirmed in their appointment:—Messrs. Robert Crosbie, Robert Rankin, Daniel Campbell, John S. De Wolf, and W. J. Fernie. The solicitor for Mr. Oliver intimated that he and his client were confident that, with judicious management, the estate would pay 20*s.* in the pound, and leave a handsome balance. The difference between the surplus originally shown by Mr. Oliver and the present deficit of £113,219 10*s.* 3*d.* was stated to have partially arisen from the amount of bills likely to be dishonoured, and from the large depreciation in the value of ships and timber. It was distinctly mentioned, in answer to a question, that the statement presented to the meeting had been made up in the manner most certain to be realized, so that the public might not be misled.

The extent and importance of the failure of Mr. Oliver may be seen from the following list of creditors, to each of which a number is affixed instead of a name, each number representing a firm :—

LIST OF MR. OLIVER'S CREDITORS, SECURED AND UNSECURED.

No.		No.	
1	£120,000	15	£10,000
2	60,000	16	10,000
3	60,000	17	10,000
4	45,000	18	9,000
5	32,000	19	8,000
6	25,000	20	7,000
7	25,000	21	6,000
8	25,000	22	5,000
9	25,000	23	4,000
10	20,000	24	1,000
11	20,000	The public	132,000
12	15,000		
13	15,000	Total	£700,000
14	11,000		

THE ESTATE OF MR. J. M'HENRY.

A meeting of the creditors of Mr. James M'Henry, who failed on the 4th of October, 1854, was held at Liverpool on the 22nd of November, when an adjournment took place, to afford time for a final consideration of the course ultimately to be adopted. According to the subjoined statement, presented by Mr. Frye, the accountant, the assets showed a dividend of 1*s*. 6*d*. in the pound; but this, it was expected, might be increased to 2*s*. or 3*s*., as the liabilities were likely to be largely diminished by the bill-holders electing to prove against Messrs. Allen and Anderson as partners. The enormous deficiency of £307,857 was represented as having principally occurred through bad debts in the United States. It was finally determined that the estate should be wound up under a deed of inspection.

LIABILITIES.

On acceptances		£14,940 0 9
On acceptances to be retired by other parties	£35,873 0*s*. 2*d*.	
On book debts		16,754 9 6
On foreign debts		1,510 16 11
On bills receivable		268,208 10 1
On 12 drafts of Edward Oliver, accepted by J. M'Henry	£30,964 10 6	
On Wann and M'Birney's three drafts, J. M'Henry, in favour of E. Oliver	5,000 0 0	35,964 10 6
		337,378 7 9

ASSETS.

Claims on book debts considered good		£22,316 13 5
Valuation of doubtful debts		393 8 2
Ditto, stock on hand		3,155 0 0
Ditto, working implements		540 6 4
Mortgage on the "Ava"		600 0 0
100 shares in the Lancashire Insurance Company, valued at		350 0 0
J. Curtis's promissory note, due October 27		157 16 11
Valuation of office fixtures		250 0 0
Cash on hand		300 0 0
5 policies of insurance on the lives of G. Longmore, James Longmore, and James Taylor, valued at		150 0 0
Estimated proceeds to be received from William M'Kee and Co., Philadelphia, of linens shipped on account of S. Turney	£1000 0 0	
Less part to be applied in retiring S. Turney's acceptance for this amount	600 0 0	400 0 0
S. Turney's acceptance, due November 16, in hands of J. M'Henry	600 0 0	
Less surplus receipts expected from W. M'Kee and Co.	400 0 0	
	£200 0 0	
Less amount due to him on book debt	106 8 6	93 11 6
Valuation of residue of book debts due to J. M'Henry in America		2,000 0 0
		£30,706 16 4
Less payments to be made in full		1,185 10 8
		£29,521 5 8
Deficiency		307,857 2 1
		£337,378 7 9

THE ESTATE OF MESSRS. ALLEN AND ANDERSON.

A meeting of the creditors of Messrs. Allen and Anderson, engaged in the American trade, who failed on the 6th of October, 1854, was held on the 15th of November, Mr. D. B. Chapman presiding, when the following statement, prepared by Messrs. Quilter, Ball, and Co., was submitted :—

STATEMENT OF AFFAIRS, OCTOBER 6, 1854.

DEBTOR.

To sundry trade and cash creditors			£100,514 9 3
To liabilities, viz. :—			
On account of M'Henry ...	£102,035 8 3		
On account of others	16,817 6 7		
		£118,852 14 10	
Further on account of M'Henry, £111,883 10*s*., of which claims will probably arise upon		54,182 11 0	
			173,035 5 10
To liabilities on indorsements, etc., which it is estimated will run off and not be claimed on the estate		£265,715 14 5	
			£273,549 15 1

CREDITOR.

By bills receivable on hand		£2,989 5 2
By sundry debtors, viz. :—		
On consignment accounts	£19,166 17 0	
On current accounts	32,538 12 7	
		51,705 9 7
By sundry policies and loans on security		9,291 18 8
By stock		9,284 0 0
By estimated surplus value of cotton		7,500 0 0
		£80,770 13 5
Deduct preferential claims		1,106 12 7
		£79,664 0 10

With regard to the general result of the accounts, it was explained that the assets showed a dividend of about 5*s*. 6*d*. in the pound, and that there was every expectation of 5*s*. at least being realized. The liabilities might, it was stated, be increased by the sum of £26,000, and diminished to the extent of £40,000. The connection of the firm of Allen and Anderson, of London, with that of M'Henry, of Liverpool, arose from the extension of general credit facilities on the part of the former, under the impression that the affairs of the Liverpool house were in a much stronger position. It was mentioned that there were certain liabilities in the United States, which, it was anticipated, would be fully covered by arrangements already made. Mr. Murray, the legal representative of Messrs. Allen and Anderson, after calling the attention of the creditors to the nature of the assets and to the prospect of an early distribution, referred to the position of the holders of bills having the names of M'Henry and Allen and Anderson. In consequence of the connection of the latter firm with Mr. M'Henry, the holders of the bills (£173,035), in the event of bankruptcy, would have to elect against which estate they would make their proof; and the intention of the resolutions, and of the deed of inspection, would be, to reserve the rights of the bill-holders to such an election, and, in fact, to give them all the advantages to which they were entitled. After some conversation, it was generally agreed to wind up the estate under inspection, and the annexed resolutions were carried. It was stated in the course of the proceedings that, so great was the confidence reposed in the stability of Messrs. Allen and Anderson, their own managing clerk, who possessed every opportunity of ascertaining their course of business, was a creditor for £23,000, the principal and interest of capital deposited in their hands.

At a meeting of the creditors of Messrs. Allen and Anderson, held on Wednesday, the 15th day of November, 1854, Mr. D. B. Chapman in the chair, a statement of the affairs having been submitted, the following resolutions were moved, seconded, and adopted :—

"That it is expedient the affairs should be wound up under inspection, and that the following gentlemen be requested to act as inspectors, viz., Mr. Richard Sanderson, Mr. John Green Elsey, and Mr. Raymond Pelley; and that a proper deed of inspectorship, providing for the winding-up, administration, and distribution of the joint estate of the said firm, and the separate estate of the partners thereof, according to the rules in bankruptcy, and as if bankruptcy had taken place on the 6th ult., and with all usual clauses, be prepared and approved by the inspectors; and that a dividend be paid as early as possible upon such deed being signed, and further dividends from time to time when £5 per cent. is in hand. That no creditor signing or assenting to these resolutions shall be in any way prejudiced with respect to his rights or remedies against third persons, or with respect to any security or lien he may have for his particular debt; and such signature or assent shall be subject to the consent of such third persons, where any consent is necessary. That in the mean time, until such deed of inspectorship shall be completed, these resolutions shall be, and be deemed to be, a memorandum of arrangement within the meaning of the Bankrupt Law Consolidation Act, 1849."

THE ESTATE OF MESSRS. CARTER AND CO.

A meeting of the creditors of Messrs. Carter and Co., shipowners and brokers, who failed on the 16th of December, 1854, took place on the 14th of February, 1855, Mr. Cumming presiding, when the following statement was submitted by Mr. Quilter, the accountant:—

Statement of Affairs, December 16, 1854.

DEBTOR.

	£	s.	d.	£	s.	d.
To sundry creditors, unsecured—						
On current accounts	£39,698	14	8			
On underwriter's	38,470	10	1	£78,169	4	9
To sundry creditors having security—						
Amount of security	27,556	15	5			
Amount of claim	19,320	7	3			
Surplus *per contra*	8,236	8	2			
Amount of claim	1,163	8	5			
Amount of security	698	2	2	465	6	3
To liabilities—						
On acceptances running, some of which have since run off, and the securities held against them surrendered	83,336	0	2			
In respect of which claims may arise on this estate to the extent of				10,000	0	0
On indorsements	130,315	4	0			
Most of which have since run off, but claims may arise to the extent of				5,000	0	0
				£93,634	11	0

CREDITOR.

	£	s.	d.	£	s.	d.
By sundry debtors—						
Considered good	£66,220	12	1			
Considered doubtful	2,063	0	5			
Considered bad	287	19	1			
	68,571	11	7			
Deduct estimated losses	7,071	11	7			
				61,500	0	0
By sundry property— Consisting of shares in ships, consignments outstanding, etc., £29,070 9s. 3d.—estimated at				26,500	0	0
By surplus security in hands of creditors—						
Amount of security	27,556	15	5			
Less amount of claim	19,320	7	3			
				8,236	8	2
				£96,236	8	2
By liabilities *per contra*, which, although forming present claims on the estate, may ultimately be made good in full ...				£15,000	0	0

It was explained that the delay in calling the creditors together had arisen from a desire to complete certain arrangements with parties in Liverpool, to diminish the general liabilities; and that, these having been effected, the total had been reduced at least £50,000. In the account presented, the assets showed a surplus of £3,000, after paying 20*s.* in the pound, with interest; and this amount, it was stated, might, with management, be realized. If a debt due to the estate in Montreal was paid—the firm not having suspended, but being unable to make remittances through the refusal of local banks to discount—the assets would be considerably increased. With regard to the vessels possessed by Carter and Co., or those they had an interest in, it was stated that several were in the employment of the Government, at advantageous rates of freight. A final settlement of the accounts of the old house of Robert Carter and Co. had not yet been effected, but it was expected some amount would have to pass to the new firm. In answer to questions, it was mentioned that the business was lucrative, although, under recent management, it had been too rapidly extended. The conduct of all parties, including the bankers, who were secured for a large overdrawn account, was stated to have been satisfactory—slight concessions having been necessary to bring affairs within their present compass. The proposal of Carter and Co. to pay 20*s.* in the pound by instalments, but without security, was then brought forward. It was intimated that an attempt had been made to induce a relative to become responsible to the extent of £10,000, on the presumption that an amount almost equal would be involved in the liquidation of the old firm, but it had not proved successful. After a short conversation, it was agreed that the estate should be wound up under inspection, with the view of carrying out in the best possible shape the proposal made.

Annexed is a formal minute of the proceedings:—

"At a meeting of the creditors of Messrs. Carter, and Co., held at the offices of Messrs. Quilter, Ball, and Co., 57, Coleman Street, this 14th day of February, 1855, Mr. Cumming in the chair, Mr. Quilter read a statement of their affairs, and gave explanations respecting the same, from which it appeared that these gentlemen's liabilities amounted to the sum of £93,634 11*s.* (including liabilities to the amount of £15,000, which ought to be paid by other parties), and that their assets amounted to £96,236 8*s.* 2*d.*; and he informed the meeting that they considered themselves able to pay all their creditors in full, with interest, if time were granted them, and that they proposed to pay by instalments, in the following manner, viz.:—3*s.* 4*d.* in one month, 3*s.* 4*d.* in three months, 3*s.* 4*d.* in five months, 3*s.* 4*d.* in seven months, 3*s.* 4*d.* in nine months, and 3*s.* 4*d.* in twelve months, all from the 1st of March.

"Resolved unanimously, that the estate be wound up under the inspection of Mr. Pitcairn, for the purpose of carrying out the foregoing proposition."

SUSPENSIONS IN 1855.

Jan. Messrs. Morewood and Rogers, London, iron merchants.
" Messrs. Abbott, Nottingham, and Co., London, shawl trade.
" Messrs. Kesteven Brothers, London, woollen trade.
" Messrs. M. Hetherington and Co., London, American trade.
" Messrs. Aubertin Brothers, London, general merchants.
" Messrs. Krohn and Co., London, general merchants.
" Messrs. Lonergan and Co., London, Spanish and West Indian trade.
" Messrs. Rogers, Lowrey, and Co., London, warehousemen.
" Messrs. Home, Eager, and Co., London, Cape and Australian trade.
" Messrs. Turiff and Sharp, Glasgow, iron trade.
" Messrs. Millers and Thompson, Liverpool, shipowners and brokers.
" Messrs. Spence and Co., Liverpool, iron trade.
" Messrs. Boyd, Lawson, and Co., Glasgow, iron trade.
" Messrs. Wadsworth and Sheldon, New York, bankers.
" Mr. Matthew, New Orleans, provision trade.
" Messrs. Brander and M'Kenna, New Orleans, cotton trade.
" Messrs. Seagrave and Steeve, Providence, merchants.
" Messrs. Belcher Brothers, St. Louis, sugar refiners.
" Mr. W. G. Ray, New York, produce broker.
" Messrs. Foster and Stephenson, New York, produce brokers.
" Messrs. Farwell and Co., Boston, merchants.
" Messrs. Horn and Sergeant, Detroit, bankers.
" Messrs. Hill and Co., Detroit, bankers.
" Messrs. G. Lorimer, and Co., Detroit, bankers.
" Messrs. Howard, Smith, and Co., Detroit, bankers.
" Messrs. Capron and Lathrop, Connecticut, bankers.
" Messrs. Walton, Viney, and Co., London, Cape and Australian trade.
" Mr. F. Bell, East Jarrow, Tyne, chemical manufacturer.
" Messrs. Swanwick and Johnson, Manchester, calico printers.
" Messrs. Kirk and Furniss, Liverpool, general merchants.
Feb. Messrs. Keen, Rippon, and Co., London, warehousemen.
" Messrs. Gibson, Ord, and Co., Manchester, commission merchants.
" Messrs. Page and Bacon, St. Louis, bankers, and general financial agents.
" Messrs. Pickett, M'Murdo, and Co., New Orleans, bankers.
" Messrs. A. J. Wright and Co., New Orleans, produce dealers.
" Messrs. Sweeny, Green, and Co., New Orleans, cotton factors.
" Messrs. Fellows and Co., New Orleans, dealers in cotton and tobacco.
" Messrs. G. B. Morewood and Co., New York, agents.
" Messrs. Le Mesurier and Co., Quebec, timber merchants.
" Mr. G. T. Braine, London, East India trade.
" Mr. A. Hamilton, New Orleans, iron trade.
" Messrs. Wells, Fargo, and Co., San Francisco, bankers.
" Messrs. Page, Bacon, and Co., San Francisco, bankers.
" Messrs. Adams and Co., San Francisco, bankers.
" The Miners' Exchange Bank, San Francisco, bankers.
" The Robinson's Savings Bank, San Francisco, bankers.
Mar. Messrs. T. R. and W. Browning and Co., London, timber trade.
" Messrs. Anthony Nichol and Son, London, general merchants.
" Messrs. Fletcher, Rose, and Co., Birmingham, iron trade.
" Mr. Thomas Spencer, Birmingham, iron trade.
" Mr. Selby, Birmingham, iron trade.
" Messrs. S. O. Nelson and Co., New Orleans, cotton factors.
" Messrs. Oldfield, Allen, and Co., Huddersfield, woollen manufacturers.
May. Messrs. Mellors and Russell, Liverpool, South American trade.
" Messrs. Coutts and Parkinson, Newcastle-upon-Tyne, engineers.
" Messrs. Davies and Co., Birmingham, iron trade.
" Messrs. Toy and Co., Birmingham, iron trade.
June. Messrs. Strahan, Paul, and Bates, London, bankers.
" Messrs. J. Heywood and Co., Nottingham and Derby, iron works.
" Messrs. Whitehouse and Jefferies, West Bromwich, iron works.
" Messrs. Hickman and Co., Bilston, iron trade.
" Mr. J. Spencer, Birmingham, iron works.

July. Messrs. Hinde and Co., Birmingham, merchants.
„ Messrs. Latham Brothers, Liverpool, South American trade.
„ Messrs. Adamson, Watts, and M'Kechnie, Melbourne, merchants.
Aug. Messrs. J. Walker and Co., Leeds, manufacturers and merchants.
Sept. Messrs. C. J. Mare and Co., Blackwall, shipbuilders.
Oct. Messrs. De Lisle, Janvrin, and Co., London, bankers and merchants.
„ Messrs. Lewis, Reis, and Co., London, Spanish and Sicilian trade.
Nov. The Toll End Furnaces, Birmingham, iron trade.
„ Messrs. E. V. Blythe and Co., Birmingham, nail trade.
„ Messrs. Goddard Brothers, Birmingham, iron trade.
Dec. Mr. J. Chatham, Manchester, spinner.
„ Messrs. Bond, Gregson, and Co., Manchester, spinners and manufacturers.
„ Mr. J. Currie, Glasgow, grain mills.
„ The Grocers' Bank, Boston, United States.

THE ESTATE OF MESSRS. MOREWOOD AND ROGERS.

The annexed circular was issued by the trustees appointed to wind-up the estate of Messrs. Morewood and Rogers, who suspended on the 3rd of January, 1855, with liabilities for £180,000. The premises and stock having been sold for cash to new parties, the business is carried on as usual:—

"5, *Martin's Lane, Cannon Street, Jan.* 27.

"DEAR SIR,—We beg to inform you, that by virtue of the authority placed in our hands by the creditors of Messrs. Morewood and Rogers, to wind-up, as trustees, the estate of that partnership, we have proceeded in our duty as promptly as the intricate character of their extensive transactions permitted, and by our directions the lease of Dowgate Dock Wharf, standing at the annual rent of £750, has been agreed to be sold for £1,000. The entire stock there (entered on their books at about £12,000) has been sold for £10,000, net, and the several patents (the principal one having but two years to run) have been sold for the sum of £5,000; the whole to highly respectable persons, for cash, to be paid on the 1st proximo; thus both realizing what we considered the full value of those properties, and freeing the estate from the liabilities and expenses of the heavy establishment in Upper Thames Street from that date.

"Messrs. Morewood and Rogers some time since entered into an agreement for the purchase of an estate in Glamorganshire for £180,000, and paid a deposit of £25,000. The vendor has, in consideration of the sum of £2,500, allowed him for legal expenses and losses, consented to forego the forfeiture of £10,000, agreed to be paid in the event of non-completion of the purchase, and to return £22,500 on or before the 20th of March next. The vendor has also to pay for some furniture, estimated to produce from £1,000 to £1,500, and to permit a sale of some farming stock on the premises; and Messrs. Morewood and Rogers, under our direction, have signed an agreement with the vendor to carry out this arrangement.

"We therefore have the pleasure to state, that the first dividend of 2*s*. 6*d*. in the pound will be paid immediately after the receipt of the £22,000 on the 20th of March, and we expect, with the large balance remaining, and other current receipts, that a second 2*s*. 6*d*. in the pound will almost immediately follow, if, indeed, we shall not be able to make it at the same time. We are also taking steps to realize as expeditiously as possible the other assets in this country which stand in the books of Messrs. Morewood and Rogers at £41,000, but being to some extent consignments, and in other respects debts abroad, we will not yet estimate at more than £25,000 certain.

"With respect to the stock in Australia, taken at cost price, it amounts, above all bills drawn upon it, to over £80,000, but it only can be realized by gradual means, and in the regular course of business. We have accounts by the Great Britain and Overland Mail up to the 23rd of November quite as satisfactory as we could expect, and no necessity had arisen to force sales, without considerable profit, of any portion of the stock up to that time.

"The assignment must, of course, now be completed, and we shall be obliged if you will immediately send the full and precise particulars of your account against the estate to Mr. A. J. Pollock, at 16, Crooked Lane, London.

"We are, dear sir, your obedient servants,

"J. WALKER.
"W. BIRD."

THE BANKRUPTCY OF SAMUEL MORRITZ KROHN,

Of Bread Street, Cheapside, in the City of London, Merchant, Dealer, and Chapman.

JOINT BALANCE-SHEET OF KROHN BROTHERS, FROM JANUARY 1, 1853, TO JANUARY 11, 1855.

DEBTOR.

To creditors unsecured				£47,672	14	5
Ditto holding security				3,022	7	9
Ditto to be paid in full				85	16	4
Liabilities as *per contra*				15,286	5	7
Profit				10,454	2	9
Difference in balancing				21	16	8
				£76,543	3	6

CREDITOR.

By debtors—						
Good	£1,571	14	1			
Doubtful	3,505	19	9			
				£5,077	13	10
Ditto (Krohn and Co., of Melbourne)				32,000	0	0
Property				4,086	14	7
Ditto held by creditor				2,730	14	0
Bills receivable held by creditors				3,162	15	0
Partners' drawings—						
S. M. Krohn	£1,455	11	3			
M. A. Krohn	792	3	5			
	2,247	14	8			
Trade charges	4,520	1	3			
Interest	1,055	0	3			
Commission	407	16	4			
				8,230	12	6
Law costs				97	3	5
Losses	1,782	16	3			
Bad debts	514	19	3			
				2,297	15	6
Deficiency, 1st January, 1853				3,573	9	1
Liabilities as *per contra*				15,286	5	7
				£76,543	3	6

THE ESTATE OF MESSRS. ROGERS, LOWREY, HOLYLAND, AND CO.

A meeting of the creditors of Messrs. Rogers, Lowrey, Holyland, and Co., warehousemen, who failed on the 18th of January, 1855, was held on the 31st of the same month, Mr. Bennoch in the chair, when it was agreed, after lengthened explanations, to accept a dividend of 15*s.* in the pound, in the following instalments:—5*s.* at three months, 4*s.* at five months, 3*s.* at seven months, and 3*s.* at nine months; a committee of creditors, consisting of the Chairman, and Mr. D. Evans, and Mr. Tucker, being appointed to determine the nature of the security for the last payment. The following balance-sheet, prepared by Messrs. Quilter, Ball, and Co., shows that the assets of the estate were estimated at nearly 20*s.* in the pound, and that the losses had been principally incurred through the failure of Messrs. Home, Eagar, and Co., and shipments to Australia:—

STATEMENT OF AFFAIRS, JANUARY 17, 1855.

DEBTOR.

To sundry creditors—											
On trade account				£86,290	7	5					
On cash	£35,650	2	1								
Deduct estimated value of lease, etc., held by Messrs. Barclay and Co.	2,500	0	0								
				33,150	2	1					
								£119,440	9	6	
To sundry creditors—											
In respect of salaries and debts under £10 each				6,644	13	2					
Less estimated proportion payable in full, deducted *per contra*				5,000	0	0					
								1,644	13	2	
To liabilities on bills receivable—											
Considered good				67,909	12	4					
Doubtful				6,870	1	7					
Bad								3,752	2	4	
To liabilities on bills receivable—											
In respect of sales effected to Cape and Australian merchants				20,077	3	2					
In respect of goods consigned to Cape and Australia				17,720	18	8					
								37,798	1	10	
								£162,635	6	10	

CREDITOR.

By cash and bills on hand							£728	4	0
Sundry debtors—									
Considered good				£29,543	14	4			
Considered doubtful	£3,014	7	6						
Considered bad	4,373	12	8						
Estimated value				3,000	0	0			
On shipment account, considered good				4,415	6	2			
							36,959	0	6
By stock—									
Consisting of silks, English and foreign ribands, cloths, stuffs, bandannas, shawls, haberdashery, cambrics, muslins, etc., amounting, at cost value, as per stock-books, to							89,021	6	2
By lease, fixtures, etc. :—									
Held by Messrs. Barclay and Co., deducted *per contra*				2,500	0	0			
By furniture and trade utensils—									
Estimated value thereof							1,000	0	0
							£127,708	10	8
Deduct salaries and debts under £10, estimated as payable in full							5,000	0	0
Carried forward							£122,708	10	8

	Brought forward..............	£122,708 10 8
By contingent assets—		
For the ultimate excess of dividend which the estates of both acceptors and drawers may yield above 20*s*. in the pound, in respect of the Cape and Australian sales, amounting to	£20,077 3 2	
Ditto, consignments, ditto	17,720 18 8	
		37,798 1 10
Of which no reliable estimate can be made.		
		£160,506 12 6

Annexed is a formal minute of the resolutions:—

"At a meeting of creditors of Messrs. Rogers, Lowrey, Holyland, and Co., of Watling Street, warehousemen, holden at the Guildhall Coffee-house, King Street, London, on the 31st of January, Mr. Bennoch in the chair, Mr. Quilter read a statement of Messrs. Rogers, Lowrey, Holyland, and Co.'s affairs, gave explanation respecting the same, and informed the meeting that Messrs. Rogers and Co. proposed to pay a composition of 15*s*. in the pound, by instalments of 5*s*. in the pound at three months, 4*s*. in the pound at five months, 3*s*. in the pound at seven months, and 3*s*. in the pound at nine months, and all from the 15th of February, 1855. Resolved—That such offer of compromise be accepted, but that Messrs. Rogers and Co. be required to find security for the last instalment, or a portion thereof, either in shape of a direct guarantee, or by creditors postponing the receipt of dividends until the composition be paid, or any other description of security, and, in either case, to the satisfaction of Messrs. Francis Bennoch, David Evans, and Henry Tucker, who are hereby appointed a committee for that purpose. Creditors holding security of other parties to accept the composition without discharging such parties, and subject to their assent being obtained when necessary; and creditors holding private security on Messrs. Rogers and Co.'s own property not to prejudice such security, and to receive composition on the amount not covered by it. Promissory notes for the composition to be delivered on or before the 15th of February, or within such further time as the committee may, in writing, name, and, on delivery, the creditors to execute a deed of composition and release, to be approved of by Messrs. Reed, Langford, and Marsden, on their behalf; the expense of such deed, and all other expenses attending the investigation and arrangement of Messrs. Rogers, Lowrey, and Co.'s affairs, to be paid by them. In case the committee should be of opinion that the composition cannot be carried through, the estate to be administered in bankruptcy; and the members of Messrs. Rogers and Co.'s firm are requested to sign a declaration of insolvency, and deposit the same with the Chairman, in order that it may be filed, and proceedings for obtaining adjudication in bankruptcy taken by the committee, should they think it desirable."

THE ESTATE OF MESSRS. WALTON, VINEY, AND CO.

A meeting of the creditors of Messrs. Walton, Viney, and Co., whose failure in the Australian and Cape trade, with liabilities for about £40,000, took place on the 29th of January, 1855, was held on the 19th of February, when, after the consideration of a statement presented by Messrs. Quilter, Ball, and Co., it was agreed the estate should be wound-up under inspection. It appeared that the suspension of the firm had been necessitated through the non-remittance of proceeds of consignments by shippers to whom advances had been made. The only partner in this country was Mr. Viney, who introduced £7,100 into the business about twelve months previous. Annexed is the statement presented, showing the debts and liabilities:—

STATEMENT OF AFFAIRS, JANUARY 29, 1855.

DEBTOR.

To sundry creditors unsecured		£3,036 8 3
To creditors partially secured—		
Amount of claim	£2,602 14 0	
Estimated value of security	1,716 11 7	
		886 2 5
Carried forward		£3,922 10 8

	£	s.	d.	£	s.	d.
Brought forward				£3,922	10	8
To creditors fully secured	£1,404	5	1			
Deducted from advances to shippers, *per contra*.						
To liabilities on acceptances against consignments—						
Considered good	16,583	9	2			
Considered doubtful	7,247	8	4			
Less amount estimated to be recoverable from shipments	3,000	0	0			
				4,247	8	4
To liabilities on acceptances held by creditors, in excess of balance due—						
Considered good	1,606	7	11			
To liabilities on bills receivable—						
Considered good	6,666	8	2			
Considered doubtful	597	9	8			
But covered by insurances,						
				£8,169	19	0

CREDITOR.

	£	s.	d.	£	s.	d.
By cash in hand				£321	4	10
By sundry debtors—						
Considered good				291	13	7
Considered doubtful	£58	7	2			
By shares in English, Scottish, and Australian Bank, valued at				210	0	0
By advances to shippers on consignments to Australia and the Cape	£12,089	16	10			
Estimated to produce	6,500	0	0			
Less creditors fully assured, *per contra*	1,404	5	1			
				5,095	14	11
				£5,918	13	4
Deduct—Creditors payable in full				129	19	1
				£5,788	14	3
By Walton, Bushell, and Co., Cape of Good Hope, balance at their debit	£2,867	17	9			
By coals at the Cape, estimated at	5,000	0	0			
				7,867	17	9

This asset depends on the state of the Cape firm's affairs.

The following statement accounts for the present position of the firm :—

DEBTOR.

	£	s.	d.	£	s.	d.
To liabilities, as per statement of affairs				£8,169	19	
To partners' capital accounts, viz. :—						
At credit of R. Walton	£5,512	10	0			
,, J. Viney, Jun.	7,106	16	11			
,, S. Bushell	5,512	10	0			
	£18,131	16	11			
Add amount at credit of S. Bushell's drawing account	1,135	9	7			
				19,267	6	6
Carried forward				£27,437	5	6

			£	s.	d.
Brought forward			£27,437	5	6
Balance at credit of account with Constable, Bushell, and Co., Sydney			2,392	18	1
			£29,830	3	7

CREDITOR.

By assets, as per statement of affairs				£5,788	14	
By Walton, Bushell, and Co., Cape of Good Hope, balance at their debit, and estimated value of coals at the Cape				7,867	17	9
By partners' drawing accounts—						
Balance at debit of R. Walton..................	£2,348	4	10			
" " J. Viney, Jun.	480	1	4			
				2,828	6	2
Losses by bad liabilities and advances, and difference between estimated value and cost of sundry assets, etc.....................				13,345	5	5
				£29,830	3	7

THE BANKRUPTCY OF CURRIE, DALE, AND CO.

THE BALANCE-SHEET OF WILLIAM CLOSE CURRIE, TRADING UNDER THE FIRM OF CURRIE, DALE, AND CO., AS MERCHANT AND AGENT, FROM JUNE 1, 1852, TO FEBRUARY 3, 1855, THE DATE OF THE ADJUDICATION.

DEBTOR.

To sundry creditors..............				£44,361	10	10
To Norman Brothers and Co., Calcutta, for balance owing to them	£11,859	10	10			
Deduct—Secured by shipments consigned to them, account sales of which have not yet been rendered—cost	£7,662	1	9			
				4,197	9	1
To C. and J. Norman, London, for balance owing them, and current account..............	£4,991	12	1			
Deduct—Remainder of John Mackie and Palatine, on joint account with C. and J. Norman, consigned to Norman Brothers, the proceeds of which will be remitted by them to C. and J. Norman, their correspondents in London, to protect them for advances made to Currie, Dale, and Co....................................	1,286	3	9			
				3,705	8	4
To creditors holding security				418	8	6
To creditors on liabilities, £18,602 16*s.* 8*d.*, of which this sum will be claimed on the estate				3,800	0	0
To the late firm of Currie and Co..........................				538	14	7
To commission account				3,338	18	9
To suspense account				1,435	1	2
				£61,795	11	3

CREDITOR.

By cash in hand	£1,518	13	2
By debts considered good	4,665	13	2
By items in suspense*..........................	27,999	2	8
Carried forward	£34,183	9	0

* There is great uncertainty as to the real value of the items appearing as credit in the books of this estate. The result can only be known on the realization.

Brought forward..............	£34,183 9 0
Cash in the hands of the Commercial Bank, claimed by them to be held against bad bills discounted by them; but their right to do so is disputed....................	775 13 3
By charges incidental to the business	7,115 12 0
By C. W. Currie's drawing out	3,617 12 0
By losses	12,303 5 0
By liabilities *per contra*	3,800 0 0
	£61,795 11 3

THE ESTATE OF MR. G. T. BRAINE.

A meeting of the creditors of Mr. G. T. Braine, whose stoppage in the East India trade was announced on the 28th of February, 1855, was held on the 20th of March, at which a statement, prepared by Mr. Coleman, the accountant, was presented, showing debts to the amount of about £306,000, of which £230,000 were to secured, and £76,000 to unsecured creditors. Mr. Braine attended, and it was stated by Mr. Lavie, the solicitor, on his behalf, that the assets to meet the open liabilities of £76,000 would scarcely yield 2*s.* in the pound; but that Mr. Braine proposed, with assistance, to pay 2*s.* 6*d.* at once, and a further sum of 2*s.* 6*d.* in twelve months, upon a release being granted him, such release not to be valid in case of any default in the payment of the second 2*s.* 6*d.* Much discussion took place on the way in which the mass of the property had been disposed of, to secure specific claims and to the injury of the general creditors, while it also appeared that the chief losses incurred had been in connection with mining and other speculations. The capital with which Mr. Braine recommenced business, after the crisis of 1847, was stated to have been £53,000. After hearing the respective statements, the meeting adjourned, to consider whether the offered composition should be accepted, or the estate be carried into the Court of Bankruptcy. It was finally arranged that Mr. Braine should be allowed to carry out, if possible, his proposition.

THE ESTATE OF MESSRS. T. R. AND W. BROWNING AND CO.

A meeting of the creditors of Messrs. T. R. and W. Browning and Co., engaged in the timber trade, who failed on the 3rd March, 1855, was held on the 6th March, when the following statement was presented, showing the debts and liabilities to be £95,820, and the assets £54,450. The estimated amount of dividend was 11*s.* 4*d.* in the pound; but this, it was intimated, might be slightly reduced by losses incurred since the preparation of the accounts. Before resolving in what manner the estate should be wound-up, it was agreed to appoint an accountant (Mr. Turquand) to investigate, and also a committee to report, at a future date, on the general position of the affairs of the house. The absence of capital, and late extensive speculations, appear to have created its embarrassments. This report was subsequently presented, and the firm liquidated under inspection.

BALANCE-SHEET, FEBRUARY 13, 1855.

DEBTOR.

Creditors unsecured		£37,061 8 0
Creditors holding security		53,759 10 9
Liabilities on indorsed bills	£49,520 5 4	
On which claims may arise to the extent of		5,000 0 0
		£95,820 18 9

CREDITOR.

Amount of cash in hand				£684	0	3
" stock				30,698	2	11
" loans due				336	10	0
" bills not yet due	£4,399	8	6			
" bills over-due	987	15	5			
" book debts	19,587	1	1			
	£24,974	5	0			
Deduct for loss in collecting 10 per cent.	2,497	8	6			
				22,476	16	6
Surplus due from Hanbury and Co.				1,247	4	10
Surplus due from W. H. Surman				252	2	7
				£55,694	17	1
Preferential claims for rent stock at docks—London and Liverpool	£1,000	0	0			
Rent at Waterloo Bridge Wharf	245	0	0			
				1,245	0	0
				£54,449	17	1
Deficiency				41,371	1	8
				£95,820	18	9

THE ESTATE OF MESSRS. ANTHONY NICHOL AND SON.

A meeting of the creditors of Mr. J. L. Nichol, trading under the firm of Anthony Nichol and Son, who failed on the 7th of March, 1855, was held on the 22nd of March, when the following statement was presented by Mr. J. E. Coleman, the accountant :—

DEBTOR.

To creditors on open accounts				£5,045	3	1
To ditto on bills payable	£53,577	7	2			
Less amount to be retired by the drawers	39,411	11	2			
				14,165	16	0
To creditors partly secured	£21,967	5	3			
Less security held	17,540	0	0			
				4,427	5	3
To creditors fully secured—Security held	£5,150	0	0			
To claims	4,899	2	2			
To *contra*	£250	17	10			
To liability on bills receivable	28,418	7	1			
Of which it is expected claims on the estate amount to				327	12	0
To liability on claims upon unfulfilled contracts				1,597	0	0
				£25,562	16	4

CREDITOR.

By cash	£214	19	11
By debtors (good)	1,992	12	5
By debtors (doubtful), £347 11*s.* 7*d.*, estimated at	100	0	0
By assets, stock, etc.	1,323	16	11
By surplus from security with creditors, as *per contra*	250	17	10
By fixtures, household furniture, etc.	550	0	0
By yellow metal at Oporto consigned	70	0	0
	£4,502	7	1
Deduct sundries to be paid in full	559	18	7
	£3,942	8	6

It was explained that, owing to some outstanding liabilities, the amount of which, until certain bills became due, could not be ascertained, Mr. Nichol was not in a position to make a positive offer to his creditors. He was, however, prepared to realize the estate, and guarantee assets, through the assistance of friends, to the amount of £3,500 ; any surplus to be likewise for the benefit of his creditors. The losses, it appeared, had been principally incurred through transactions in produce; but when Mr. Nichol assumed the control of the business, shortly before the decease of his father, there was a deficiency of about £2,700. After a lengthened discussion, it was agreed to appoint inspectors to wind-up the estate, the creditors being secured the proposed amount for distribution in the early part of July, previously to which the whole of the liabilities would have run out.

THE ESTATE OF MESSRS. C. J. MARE AND CO.

The suspension of this firm took place early in September, and it was at first presumed that a private arrangement would have been effected. An absolute petition in bankruptcy was, however, subsequently opened, and the firm appeared in the *London Gazette*. A very numerous meeting of the creditors took place on the 27th September, at the Guildhall Coffee-house, Mr. Lee, the official assignee under the bankruptcy, presiding, to concert measures for carrying on the works until further definite proceedings could be adopted. It was intimated that the meeting had been called with the sanction of Mr. Commissioner Holroyd, and that Mr. Mare himself was desirous of using every endeavour to obtain an equitable adjustment for the benefit of all who are interested in his estate. Mr. Lawrance, the solicitor to the petition, explained at length the preliminary steps taken for placing the affairs of Mr. Mare under the operation of the Private Arrangement Act, which, it was at first thought, would have proved available, but difficulties subsequently intervened, and it was found necessary to resort to an absolute bankruptcy. The chief impediment was a creditor for £9,000 or £10,000, who, having placed an execution in the works at Blackwall, refused to withdraw, unless upon immediate payment; and as this creditor was proceeding to sell, through the sheriff, it was considered advisable not to allow a sacrifice of probably £20,000 or £30,000 to meet the claim, but to take the course that had been adopted. There was also an *extent*, at the suit of the Crown, in Mr. Mare's private residence for £3,000, being the amount of a bill for timber from the Royal Forest; but this, under any circumstances, would have to be paid in full, to save the property upon which it was levied, the Crown taking precedence in such cases. The execution upon the works at Blackwall, had it been proceeded with, would not only have sacrificed the proportion of machinery, etc., sold to realize it, but the works would have been stopped, and the whole of the numerous hands thrown out of employment. The adjudication of bankruptcy, with Mr. Mare's consent, having taken place, the first necessary arrangement was the payment of wages, and this was accomplished through the personal exertions of the official assignee, the directors of the Commercial Bank consenting to supply the requisite funds. The last week's wages having been paid, the question of the continuance of the works under the bankruptcy was submitted to Mr. Commissioner Holroyd, who, with a mercantile appreciation of the emergency, consented to the attempt being made, on condition that the creditors should meanwhile be consulted on the question. This result having been arrived at, the directors of the Commercial Bank consented to place to the credit of the official assignee £10,000, to meet the early expenditure, and to be repaid out of the first assets received. The position of the matter stood thus :—The effect of the adjudication was to displace the execution of the creditor, whose debt, though *bonâ fide*, it was contended was not better entitled to a preferential settlement than others, and the works remained uninterrupted. The official assignee possessed sufficient to discharge £5,000 or £6,000 due the next Saturday, and there was enough to liquidate the *extent* from the Crown, so as to avert the loss from a forced sale of the property in Hyde Park Gardens. Assurances having been given of a supply of coal, wood, and rope, every prospect existed of assets being received to keep the general establishment in operation; but, if they did not come in, it was thought other resources were open for assistance. Under these circumstances, it was proposed to appoint a committee of creditors, to advise with the official assignee, in his endeavours to carry on the works until the choice of assignees; thus affording an opportunity for looking into the nature of the various contracts and expenditures, and ascertaining the actual position of this extensive undertaking.

An approximate statement of affairs was then brought forward by Mr. Ball, of the firm of Quilter, Ball, and Co., the accountants; but it was mentioned that, having been prepared only on a short notice, it could not be considered to represent accurately the position of the estate. It showed the liabilities to be £156,400, and the assets £186,000, leaving an estimated surplus of £29,600. Mr. Mare, however, considered his establishment at Blackwall to have been inadequately valued, and he increased the amount by £75,000, making a total surplus of above £100,000. Some discussion ensued with respect to the various items, and one, in connection with an establishment at Newmarket, elicited the reply that it had entirely passed out of Mr. Mare's hands. In referring to a resolution for the appointment of a committee, Mr. Lawrance stated that several instances had occurred in which creditors had been beneficially protected by the works of embarrassed firms being carried on under bankruptcy, and he specially alluded to the parties identified with the manufacture of wire-rope, in which the original patentee, after paying his creditors 20*s*. in the pound, was enabled advantageously to dispose of his interest. The whole of the explanations were favourably received, and Mr. Lawrance volunteered any further information, either on behalf of the official assignee, or of Mr. Mare, who himself was present. Mr. Murray, with one or two other professional persons, who represented claimants, having concurred in the desirableness of the course proposed, a committee of five creditors was appointed, including Mr. P. Rolt, who, possessing both a family and pecuniary interest, pledged himself to protect the interests of the general creditors. The following are the approximate statement and the resolution comprising the names of the committee :—

APPROXIMATE STATEMENT OF THE AFFAIRS, SEPTEMBER 20, 1855.

DEBTOR.

To creditors unsecured				£115,000	0	0
To creditors partially secured — amount of claims	£179,361	0	0			
Less estimated value of securities	137,900	0	0	41,461	0	0
To creditors fully secured—estimated value of securities	£115,746	0	0			
Less amount of claim	103,156	0	0			
Surplus, taken as an asset, *per contra*	£12,590	0	0	£156,461	0	0

CREDITOR.

By stock at Blackwall	£66,000	0	0			
Less *extent*	3,000	0	0	63,000	0	0
By Great Western steam-ship—cost of construction	£21,500	0	0			
Value of engines	17,500	0	0			
Proportion of expenses	3,900	0	0			
	£42,900	0	0			
Liens thereon	33,000	0	0	9,900	0	0
By Westminster Bridge Works—amount expended in plant, wages, and materials	£87,000	0	0			
Less received on account	50,000	0	0	37,000	0	0
By Saltash Bridge Works, sundry debtors, quarterly accounts ...				5,000	0	0
Works in progress—estimated amount due				12,000	0	0
Surplus securities in hands of creditors, *per contra*				12,590	0	0
Estimated profit to be received on the completion of unfinished contracts				46,600	0	0
				£186,090	0	0

MEMORANDUM.

The liabilities, after deducting assets at certain estimated values, amount to..........	£156,461	0	0
The assets, according to the statement, which is based upon rough estimates of immediate value, amount to	£186,090	0	0
Mr. Mare considers that the Blackwall property, consisting of the freeholds, leaseholds, fixed machinery, and plant, having cost £225,000 is greatly under-estimated in the statement, and that a deduction of £50,000 is ample to cover all possible deterioration, which enlarged estimate would effect an increase in the assets of..........	75,000	0	0
	£261,090	0	0

"Resolved—That a committee of five creditors be appointed, to confer and advise with the official assignee, on the affairs of this estate, during the interval before the choice of assignees, under the adjudication of bankruptcy, appointed to take place on the 12th of October. That the following gentlemen be appointed to act as the foregoing committee:—Mr. Peter Rolt, Mr. Jones, Mr. Cory, Mr. Marshall, and Mr. Hunter.'

THE ESTATE OF MESSRS. DE LISLE, JANVRIN, AND DE LISLE.

The suspension of this old-established firm was made public on the 3rd of October, and it excited very general regret. Occupying a high position as bankers and merchants, the event was little anticipated, but there was reason to believe that the interest of the creditors would be protected. Imprudent advances to a house in Quebec, whose affairs had become involved, necessitated this step, although the assets at the command of Messrs. De Lisle and Co. would, had they been so disposed, have enabled them to continue business without immediately sacrificing their credit. The house was amongst the oldest and most respectable in London, and the circumstances under which the stoppage was resolved upon appear to harmonize with the reputation for honour the firm have uniformly enjoyed. It appeared that they had a large balance at their banker's, as well as money at call in the hands of discount brokers, and in other available quarters, to the extent of about £100,000; but that, being involved in large advances in Canada, which assumed a more serious complexion on the arrival of the American mail, they determined at once to suspend, instead of risking the property of their creditors by any uncertain attempt to sustain themselves. The amount in question would, therefore, be almost immediately available for a dividend, while, with regard to the general assets, the statement was that, with the exception of the securities from Canada, they were of a favourable description, and would admit of a further considerable sum being realized during the present year. Many years back the firm were largely engaged in trade with Buenos Ayres, but their business was of a general description, and latterly they had become connected with an extensive firm at Quebec, who were involved in the American panic of 1837, and who again found themselves in difficulties at the commencement of the pressure in New York and Canada, about two years ago. On that occasion, their debt to De Lisle and Co. was enormously heavy, and was compromised upon conditions which have not been fulfilled. The amount remaining due under the compromise, coupled with fresh advances which De Lisle and Co. were induced to make by delusive representations, was understood to be equal to £200,000. The prospects from this source are most likely very doubtful, and it is upon the other assets only that reliance must be placed. Of Messrs. De Lisle's liabilities of £400,000, a portion was said to consist of acceptances for account of parties ultimately responsible, and who would meet them at maturity. In this way it was expected the total will be considerably reduced, and the prospects of the direct creditors proportionably improved.

Messrs. De Lisle and Co. were also connected with the trade of Guernsey and

Jersey, and were the correspondents of some of the principal banks in those islands, namely, the Guernsey Banking Company, the Guernsey Commercial Banking Company, the Jersey Commercial Bank (Janvrin and Co.), and the Jersey Old Bank (Godfray, Sons, and Co.) Messrs. Hankey and Co., the London bankers, interfered for all these institutions, and honoured the draughts drawn by them on Messrs. De Lisle and Co.

Annexed is the circular intimating the suspension :—

"16, *Devonshire Square, October* 3.

"It is with feelings of deep regret that we have deemed it necessary this day to suspend our payments. Large advances in Canada, and disappointment in receiving remittances, have led to this painful determination, and we feel convinced that by taking this course we shall best protect the interest of our creditors.

"We beg to assure those friends who have placed securities in our hands that they all remain intact, and are held at the disposal of the parties interested.

"We shall request a meeting of our creditors in a few days, of which you will be duly informed ; and, relying upon your sympathy and forbearance under these painful circumstances,

"We are, respectfully, your obedient, humble servants,

"De Lisle, Janvrin, and De Lisle."

FIRST MEETING OF CREDITORS.

A meeting of the creditors was held on the 8th of October, Mr. B. Dobree presiding. It was intimated that the proceedings were of a preliminary character, with the view of obtaining temporary forbearance on the part of the creditors, until some suggestions made by the friends of the house could be presented in a definite shape. Two proposals were thought of, and the course ultimately to be followed was left to the decision of the creditors. The first was the introduction of capital, if possible, through the assistance of friends, the payment of 20*s*. in the pound, and the immediate resumption of business. In the event of this plan failing, the next was a liquidation of the estate under inspection, and the distribution of the assets as they might be received. If the process of inspection be carried out, a new firm would then be formed to take up the connection, which is valuable, and could, without difficulty, be kept together. In this state of affairs an adjournment was moved until the following Saturday, when it was expected a final arrangement might be come to. Meanwhile the funds of the house were stated to be in perfect safety. There were no figures in relation to its liabilities or assets presented on this occasion, but it was mentioned that they were in an advanced state, and would be brought forward at the earliest moment. The proposal for an adjournment having met the unanimous approval of the creditors, a substantive motion to that effect was passed. On behalf of the Bank of British North America (represented by Mr. Gillespie and Mr. Brooking), it was announced that the directors had communicated with the agents in London of the Montreal and other banks connected with the dependencies of the Canadian firm in which Messrs. De Lisle were interested, and that instructions have been despatched abroad to ascertain that the securities were in order, and that the rights of all parties were properly preserved.

THE ADJOURNED MEETING.

The adjourned meeting of the creditors was held on the 13th of October, Mr. B. Dobree presiding, when the following statement was submitted by Mr. Coleman, the accountant :—

"LIABILITIES.

To creditors on open accounts ...		£198,500
To creditors on account, after giving them credit for our acceptances, which they will retire ...		65,600
The amount of drafts upon us at the date of suspension ...	£302,000	
Of which there were for account of other parties who will retire same, or for which credit has been given in the preceding item ...	182,000	
Amount expected to rank on our estate ...		120,000
		£384,100
Salaries, rent-charges, and sundry small accounts, to be paid in full ...		£1,800

"Beyond this, Mr. W. F. De Lisle, acting as our attorney in Canada, has guaranteed cargo bills, but which bills, we expect, will be duly met by the acceptors for whom such cargoes were shipped.

"He has also guaranteed the payment of local liabilities in Canada, for account of Messrs. Le Mesurier, Tilstone, and Co., and John Egan and Co., the exact particulars and amounts of which we are unable to state, but against which we were informed that there was a larger value in timber to cover the amount of such guarantees.

"ASSETS.

By cash balances at the banker's	£14,500	
By cash at call	57,300	
		£71,800
By bills receivable		79,700
By bonds, stock and shares, estimated to produce		12,500
By sundry debtors, estimated to produce		51,000
		£215,000
Amount of cash debt due from Le Mesurier, Tilstone, and Co.	£308,480	
Amount of acceptances included in our liabilities, as *per contra*	111,650	
	£420,130	

"Against the above, we hold the assignment of a debt due from John Egan and Co. to Le Mesurier, Tilstone, and Co., of £109,650, secured in part by a charge on large plots of land, mills, and timber, at Alymer, in Lower Canada; also a charge upon all the lands and properties of Messrs. Le Mesurier, Tilstone, and Co., in Lower Canada."

It was explained by the Chairman that, since the last meeting, the friends of the firm had taken into consideration their position, and that, upon mature reflection, looking at the extent of the Canadian dependencies, it was deemed advisable to recommend a liquidation under inspection. While it was proposed to distribute the assets with all possible expedition, it was felt that the partners should be allowed to carry on business, their character and credit remaining unimpeached. A dividend of 5*s*. in the pound was immediately declared; a further distribution was anticipated about January, and the whole of the £215,000 assets might, it was thought, be realized within six to eight months of that period. With regard to the large claim upon Messrs. Le Mesurier, Tilstone, and Co., it was stated that every endeavour would be made to place it in process of conversion, but that it was impossible, in the existing position of affairs, to estimate the probable return. An attempt was made by Mr. H. Godfray, of Jersey, to obtain a further adjournment, with the view of affording the creditors an opportunity for investigating the actual condition of the accounts; but representations from Mr. Hankey, Mr. Lavie, and others, respecting the impolicy of this course, induced the withdrawal of an amendment, and the annexed resolutions were unanimously agreed to. Mr. Brooking, as representing the Bank of British North America, intimated that every care would be taken to protect the interest of the estate in connection with the Canadian dependencies. In the course of the proceedings, sympathy was expressed for the position of the partners, whose future career, it is hoped, will be successful.

"At a meeting of the creditors of Messrs. De Lisle, Janvrin, and De Lisle, at their counting-house, Devonshire Square, on Saturday, the 13th of October, 1855, Mr. B. Dobree in the chair, a statement of the liabilities and assets was read. It was proposed and seconded, and resolved—

"First, that it is the opinion of the meeting that the affairs of the house should be liquidated under inspectorship, and that the following gentlemen be the inspectors:—Mr. T. H. Brooking, Mr. Bonamy Dobree, Mr. D. Meinhertzhagen, and Mr. J. Whateley.

"Secondly, that a dividend of 5*s*. in the pound be at once declared, and be payable on and after Saturday, the 27th of October.

"Thirdly, that a proper deed of inspectorship be prepared under the approval of the inspectors, and be executed by and on behalf of each creditor, on or before payment of the dividend.

"Fourthly, that such deeds shall contain covenants by the partners to liquidate the affairs of the house, according to the rules of administration adopted in bankruptcy, and covenants by the creditors not to sue, which shall operate as a release, upon the inspectors certifying that the liquidation has proceeded sufficiently, and upon the partners executing an assignment of any remaining assets to trustees for distribution among the creditors.

"Fifthly, that such deed shall be a deed of arrangement within the meaning of the 224th section of the Bankrupt Law Consolidation Act, and the 228th section shall be applicable thereto.

"Sixthly, that the inspectors shall have power to make the partners such allowances as they may think fit for their services.

"Seventhly, that the partners be at liberty to transact business on their own account on their covenanting, in the inspectorship-deed, not to use, either directly or indirectly, any of the existing assets of the firm, and to incur no new engagements which could, by any possibility, be thrown on the existing assets."

THE ESTATE OF MESSRS. STRAHAN, PAUL, AND BATES.

The following is the report prepared by Mr. Turquand, the accountant, of Old Jewry Chambers, on the affairs of this estate, together with a copy of the joint balance-sheet :—

"13, *Old Jewry Chambers, Dec.* 10.

"TO THE ASSIGNEES OF THE ESTATE OF MESSRS. STRAHAN, PAUL, AND BATES.

"Gentlemen,—I beg to make the following report of my investigation of the books and accounts of Messrs. Strahan, Paul, and Bates :—

"REPORT.

"The bank of Messrs. Strahan and Co. was one of the oldest on record, dating its origin from the early part of the reign of Charles II. At the time of the bankruptcy, the firm consisted of William Strahan, Sir John Dean Paul, and Robert Makin Bates. Sir John Dean Paul (then Mr. Paul) became a nominal partner in 1823, taking no share of the profits until the death of his father, the late Sir John Dean Paul, in January, 1852. The firm was composed of Robert Snow, Sir J. D. Paul, and J. D. Paul. William Strahan (who had changed his name from Snow on inheriting a very considerable property from an uncle in 1831) joined the above firm with his brother, Robert Snow, in 1832. Robert Snow, sen., died in 1835. Robert Snow, the younger, retired from the bank in 1841, and in January, 1842, Robert Makin Bates, who had for many years been a confidential clerk in the bank, became a nominal partner at a salary of £800 per annum, subsequently raised to £1,000 per annum, without any share in the profits. It is not necessary, for the purpose of elucidating the present position of the bank, to go further back than the partnership of Robert Snow, William Sandby, and John Dean Paul, formed in 1813. On the death of Mr. Sandby, in 1816, the partners were indebted to the bank in a sum of £29,000, which was apportioned in the following manner :—Robert Snow, £16,681 2*s.* 9*d.*; Wm. Sandby, £8,989 2*s.* 6*d.*; John Dean Paul, £3,329 14*s.* 9*d.* Total, £29,000.

"The debt of William Sandby was paid off by his executors by the end of the year 1826. At that period the debt due by the remaining partners had increased to £53,600, which was apportioned in the following manner:—Robert Snow, £36,319 10*s.* 9*d.* ; the late Sir J. D. Paul, £17,280 9*s.* 3*d.* Total, £53,600. This debt was represented by a joint note of the two. By an arrangement between themselves, and acquiesced in by succeeding partners, the amount was to be considered as a debt due to the bank, to be gradually liquidated by a certain portion being carried every year to profit and loss. By this means, at the death of the late Sir J. D. Paul, the amount had been reduced to £28,500, and at the date of the bankruptcy had been further reduced to the sum of £23,500.

"The balance-sheet now filed commenced on the 31st of December, 1851, showing a deficiency of £71,990 7*s.* 2*d.*

"Deducting the amount then standing to the credit of William Strahan's capital account, £10,330 6*s.* 1*d.*, less the amounts to the debit of the present Sir J. D. Paul,

£213 13*s*. 8*d*., and R. M. Bates (loan) £3,669 4*s*. 4*d*. (£3,882 18*s*.), £6,447 8*s*. 1*d*., left an actual deficiency between assets and liabilities of £65,542 19*s*. 1*d*.

"This deficiency appears to have been composed of the following items, viz.:—Balance due on joint note, £28,500; debt due by the late Sir J. D. Paul, £35,477 12*s*. 1*d*., less balance of subsequent receipts and payments to the credit of his account, £10,084 15*s*. 5*d*.; total, £25,392 16*s*. 8*d*.; bad and doubtful debts not written off, £2,446; bad and doubtful debts, Halford and Co., £15,705 12*s*. 8*d*.; estimated loss on valuation of bank assets, £4,369 1*s*. 6*d*.; total, £76,413 10*s*. 10*d*.

"Deduct—Amount standing to credit of balances written off as unclaimed, £4,073 4*s*. 1*d*.; Halford and Co., alleged surplus, as shown by books, £349 19*s*. 7*d*.; balance to credit of partners' accounts, £6,447 8*s*. 1*d*. (£10,870 11*s*. 9*d*.) Total, £65,542 19*s*. 1*d*.

"It should be remarked that there is one item included in the assets at its full amount, the actual value of which would materially affect the above position, and that is a sum appearing to the debit of Lord Mostyn, of £92,001 16*s*. 10*d*. Between the years 1848 and 1850, large advances had been made, and in January, 1850, a lease of the property known as the Mostyn Colliery was granted by Lord Mostyn to the bank, to secure a sum of £67,541 14*s*. 8*d*. then due by him.

"The colliery required very considerable outlay to bring it into a productive condition, and upwards of £45,000 was so expended by the bank; with this expenditure, and arrears of interest, the amount to the debit of the account at the time of bankruptcy was £134,940 17*s*. 1*d*. The colliery had thus been brought into a productive state, yielding, however, only sufficient to cover expenses, and to pay interest on a sum of £45,000, which had been borrowed by the bank on the security of the lease, to meet the necessary outlay. Taking the amount thus borrowed as an indication of the value of the asset, the deficiency of the bank would be increased to about £110,000. At this date, however, William Strahan was possessed of unencumbered private property to the extent of upwards of £100,000, Sir J. D. Paul about £30,000, and R. M. Bates about sufficient to cover his debt to the bank—say, £3000. Taking into consideration the security thus afforded to customers by the value of the private property, it might be inferred the bank would, with care and prudence, have recovered its position. The unfortunate connection, however, with Messrs. J. H. and E. F. Gandell, commencing in 1852, and resulting at the date of the bankruptcy in a debt to the bank of £269,382 3*s*. 5*d*., with liabilities in addition on their account to the extent of £103,870, coupled with the already heavy withdrawal of capital in respect of Lord Mostyn's debt and the Mostyn Colliery, may be fairly looked upon as the causes which brought about the disastrous failure of the bank.

"Messrs. Strahan and Co. were induced, by the representations made to them by Messrs. Gandell, to advance from time to time large sums of money for the purpose of enabling them to carry out certain contracts for the construction of railways in France and Italy, and for the drainage of the Lake Capestang, situated in the south of France. The profit to be derived by Messrs. Strahan in the transactions was five per cent. interest on money advanced, ½ per cent. commission on all payments made by them, and the payment of a debt of £1800, considered bad, due by J. H. Gandell to the bank in 1850.

"The debt to the bank (for which no tangible security was held) soon assumed such gigantic dimensions, and Messrs. Gandell's affairs were found to be in that condition, that Messrs. Strahan imagined they had no alternative left but to continue their advances for the purpose of maintaining Messrs. Gandell in a position to carry on the various contracts, the completion of which was looked to as the source whence the vast sums advanced were to be recovered.

"The time arrived when the resources of the bank were no longer able to meet this constant drain upon it. Acceptances were then given, and other heavy periodical liabilities incurred for the purpose of providing the required funds. These last acts appear to have led to the most distressing feature of this case; almost the whole property of the bank had been pledged, and Mr. Strahan's private estate was resorted to for the same purpose; the large sums so raised had disappeared, and there was no alternative but the payment of the acceptances and other liabilities as they fell due, or bankruptcy. In the vain hope that the anticipated funds would yet be forthcoming in time to avert the impending ruin, recourse was had to those means of raising funds, the consequences of which are now being visited on the bankrupts. The nature and extent of the securities held by the assignees in respect of the debt due by Messrs. Gandell are detailed in the balance-sheet.

"The present deficiency of the bank is as follows:—

Liabilities	£652,593	15	0
Estimated assets	127,670	16	7
Deficiency	£524,922	18	5

"It will be seen, by referring to the items composing this deficiency, as set forth in the balance-sheet, that the sums involved in the transactions with Messrs. Gandell and Co. and Lord Mostyn amount to upwards of £483,000 of the whole amount. Time has not permitted the completion of the balance-sheets of the separate estates, but they will be filed as expeditiously as possible. The position of the bankrupts' personal accounts with the bank at the time of its failure is, however, shown on the face of the joint balance-sheet now filed.

"I have the honour to be, gentlemen, your obedient servant,

"W. Turquand."

Balance-sheet of Messrs. Strahan, Paul, and Bates.

DEBTOR.

To creditors unsecured				£411,210	15	4
To creditors of Halford and Co.				26,800	16	4
Creditors for securities sold	£15,964	6	0			
Ditto pledged	94,911	7	8			
Ditto belonging to customers of Halford and Co.	12,802	17	0			
	123,678	10	8			
Deduct amounts for which certain of the above parties are debtors, or hold security	13,221	12	4			
				110,456	18	4
Creditors holding security fully covered, *per contra*	184,889	12	11			
Ditto partly covered	12,935	5	0			
Deduct estimated value of securities held	12,680	0	0			
				255	5	0
Creditors for advances on pledged securities *per contra*	100,109	17	4			
Creditors who issued extents, and were paid in full, deducted from assets, *per contra*	3,599	2	1			
Liabilities on account of Messrs. Gandell				103,870	0	0
Profit net				31,304	1	6
To W. Strahan—Balance to his credit after deducting amount drawn out by him for private expenditure				17,401	14	3
Sir J. D. Paul, ditto				772	12	3
W. Strahan—For separate property held by creditors of bank	70,457	0	0			
Amount required to cover their claims				32,673	4	0
				£734,745	7	3

CREDITOR.

By debtors—Strahan and Co., considered good							£76,536	14	1
Halford and Co., ditto							22,484	2	8
Doubtful and bad				£28,298	13	3			
Ditto, Gandell and Co.	£289,382	3	5						
Less purchase money of Capestang Lake	20,000	0	0						
				269,382	3	5			
Carried forward							£99,020	16	9

Brought forward			£99,020 16 4
Ditto, Lord Mostyn	£134,940 17 1		
Deduct proportion of debt mortgaged	44,967 3 4		
		89,973 13 9	
Property unencumbered, viz.:—Shares		£4,333 15 0	
Life policies		12,315 13 0	
Freehold and leasehold		2,500 0 0	
Exchequer bills		300 0 0	
			19,449 8 0
Capestang Lake, not carried out—the expenditure required to complete the drainage and other incidental circumstances rendering the value of this property uncertain		£20,000 0 0	
By cash in hand, 11th of June		1,767 2 5	
By bills receivable		9,949 3 6	
By cash at the Bank of England		83 8 0	
		11,799 13 11	
Deduct creditors paid in full *per contra*		3,599 11 8	
			8,200 12 3
By debtors, the securities representing which are held by creditors		£69,343 13 10	
By property of bank held by creditors		96,996 16 0	
Ditto of William Strahan, held by joint creditors	£70,457 0 0		
Amount required to cover their claims		32,673 4 0	
		£199,013 13 10	
Deduct amount held by creditors, partly covered		12,680 0 0	
Leaving do. held by do. fully do.		£186,333 13 10	
Deduct amount of claims of creditors fully covered		184,889 12 11	
Balance available for estate		£1,444 0 11	
Considered as a good asset in respect of one security to the extent of			1,000 0 0
The residue is not carried out, being subject to deduction for interest accruing due on creditors' claims.			
Total amount of assets considered good			127,670 16 7
By securities pledged, belonging to customers of the bank		£94,911 7 8	
Ditto, Halford and Co.		12,802 17 0	
		107,714 4 8	
Deduct amount so borrowed,	£100,109 17 4		
Less value of Exchequer-bills belonging to bank included	2,819 12 0		
		97,290 5 4	
		£10,423 19 4	
This balance is not carried out as an asset, being subject to the question as to the right of the assignees and of the parties to whom the securities belonged.			
By deficiency, December 31, 1851			71,991 7 2
By amounts not carried out above, viz.:— Doubtful and bad debts			28,298 18 3
Carried forward			£227,961

		£	s.	d.
Brought forward		£227,961	2	0
Gandell and Co.'s debts		269,382	3	5
Lord Mostyn's balance		89,973	13	9
Capestang Lake		20,000	0	0
Surplus value of property held by creditors as security	£1,444 0 11			
Considered a good asset only for	1,000 0 0			
		444	0	11
Surplus value of customers' securities pledged		10,422	19	4
By R. M. Bates's salary		5,750	0	0
Balance to his debit		4,381	7	10
Sir J. D. Paul (deceased), proportion of profit credited him in 1852, written back		4,560	0	0
Liabilities *per contra*		103,870	0	0
		£734,745	7	3

SEPARATE ESTATES OF MESSRS. STRAHAN, PAUL, AND BATES.

The Separate Estate of William Strahan, from January 1, 1852, to June 11, 1855.

DEBTOR.

		£	s.	d.
To creditors unsecured		£659	3	10
To creditors holding security		2,760	0	0
To liabilities on account of bank		134,008	11	8
To liabilities as co-trustee with Sir J. D. Paul		17,231	10	7
To income		15,458	13	4
To share of profits of bank passed to my credit in 1852	£5,700 0 0			
To ditto in 1853	7,700 0 0			
To ditto in 1854	7,000 0 0			
		20,400	0	0
Surplus assets on the 1st of January, 1852		128,048	2	0
		£318,566	1	5

CREDITOR.

		£	s.	d.
By debtors (good)		£1,471	6	0
By ditto (doubtful)		774	16	10
By property unencumbered:—				
Sundries		12,059	16	7
Shares		4,376	10	0
Reversionary interest		400	0	0
By property held as security by private creditors		2,760	0	0
By property, security to creditors of bank		70,457	0	0
By expenses, including improvements at Ashurst, and expenditure on farms:—				
Expenditure	£14,956 8 9			
Improvements, etc.	8,551 11 7			
		23,508	0	4
By annuity paid R. Snow		1,753	17	1
By losses		11,962	18	1
By Strahan and Co.:—				
As per joint balance-sheet	£17,401 14 3			
Add share of profits	20,400 0 0			
		37,801	14	3
By liabilities *per contra*		151,240	2	3
		£318,566	1	5

Separate Estate Balance-sheet of Sir John Dean Paul, Bart.

DEBTOR.

To creditors unsecured				£23,402	11	6
To creditors holding security fully covered	£14,500	0	0			
To creditors holding security partly covered	15,000	0	0			
Less value of security	1,188	9	0			
				13,811	11	0
To liabilities on account of bank				134,008	11	8
To liabilities as trustee, etc.				20,880	1	3
To income				3,546	9	2
To share of profits of bank passed to my credit in 1853	£3,300	0	0			
To ditto in 1854	3,000	0	0			
				6,300	0	0
To surplus of assets on the 1st of January, 1852				14,757	14	0
				£216,706	18	7

CREDITOR.

By debtors (good)				£2,241	4	0
By ditto (doubtful)				10	0	0
By property				9,306	19	0
By property held as security by creditors fully covered	£20,400	0	0			
Deduct amount of their claims	14,500	0	0			
				5,900	0	0
By expenses				17,175	11	9
By allowance to my son and other members of my family, etc.				11,726	2	9
By losses				8,395	16	1
By Strahan and Co., as per joint balance-sheet	£772	12	3			
For share of profit	6,300	0	0			
				7,072	12	3
By liabilities *per contra*				154,888	12	9
				£216,706	18	7

The Separate Estate of Robert Makin Bates.

DEBTOR.

To creditors	£85	3	7
To Strahan and Co.	4,381	7	10
To income	4,405	2	10
To surplus of assets over liabilities, January 1, 1852	729	10	8
To liabilities on account of bank	134,008	11	8
To ditto, as co-trustee with Sir J. D. Paul	4,336	12	1
	£147,946	8	8

CREDITOR.

By property	£1,903	3	7
By expenses	4,456	11	6
By losses	2,793	3	2
By interest to Strahan and Co.	448	6	8
By liabilities *per contra*	138,345	3	9
	£147,946	8	8

Report on Separate Balance-sheets.

"The separate balance-sheets commence on the 1st of January, 1852. In the report made on the 11th of December last, upon the joint balance-sheet, the surplus private property of the partners was estimated to have been as follows, on the 1st of January, 1852, viz.:—

W. Strahan	£100,000
Sir J. D. Paul	30,000
R. M. Bates	Nil.

"On making up the balance-sheets, the ascertained amounts are as follows:—

W. Strahan	£128,048	2	0
Sir J. D. Paul	14,757	14	2
R. M. Bates	72	1	8
At the date of the bankruptcy, the creditors of W. Strahan amounted to	3,419	3	10
The assets amounted to	92,299	9	5
Of which £70,457 are held as security by bank creditors.			
The creditors of Sir J. D. Paul, secured as well as unsecured, amounted to	52,902	11	6
The assets, including property held as security, to	31,958	3	0
The creditors of R. M. Bates, including debt due to the bank, amounted to	4,466	0	5
The assets amounted to	1,903	3	7

"It will be seen, on reference to the balance-sheets, that liabilities arising out of bank transactions, to the amount of upwards of £164,000, have ranked against the assets of the separate estates, independently of sundry liabilities in respect of trust funds.

"During the period embraced by the balance-sheets, the amounts paid into the bank by William Strahan and Sir J. D. Paul exceed the sums drawn out by them. In the case of William Strahan, such amount was supplied from his private estate; in the case of Sir John D. Paul, partly from the sale of his private property, and partly from moneys in respect of which parties are now creditors on his separate estate. "WM. TURQUAND."

SUSPENSIONS IN 1856.

Jan. Messrs. Palmer and Greene, Lichfield, bankers.
„ Messrs. G. Greig and Co., London, the Cape trade.
„ Mr. W. Schenck, London, Dutch and Newfoundland trade.
„ Messrs. Iliffe and Co., Birmingham, button manufacturers.
„ Messrs. Bruce and Kerr, Leith, corn factors.
„ Messrs. R. Clarke and Sons, Manchester, manufacturers.
Feb. Messrs. J. Scott Russell and Co., London, marine engineers.
„ Tipperary Joint-Stock Bank, Ireland, bankers.
„ Messrs. Chambers and Ellwood, London, wine trade.
„ Messrs. Cuylitz, Simond, and Co., London, general merchants.
„ Messrs. C. Leno and Co., London, corn trade.
„ Messrs. Schafer and Brown, London, general merchants.
„ Messrs. A. Jackson and Son, Glasgow, corn trade.
„ Messrs. R. Clarke and Sons, Manchester, manufacturers.
„ Messrs. Bacon, Price, and Co., Philadelphia, coal trade.
„ Messrs. Rockwood and Co., Springfield, dry goods trade.
„ Mr. H. W. Bigelow, Chicago, dry goods trade.
„ Messrs. Aspinall, Mackenzie, and Co., Shanghae, tea trade.
Mar. Messrs. Bohtlingk and Co., Liverpool, tallow trade.
„ Messrs. W. and S. Richardson, New York, hemp dealers, etc.
„ Messrs. H. M. Marley and Co., London, silk brokers.
April. Messrs. Syers, Walker, and Co., London, East India trade.
„ Messrs. W. O. Young and Co., London, ship owners and merchants.

April. Messrs. Saunders and Harrison, London, seed crushers and oil refiners.
" Mr. Thomas Harrison, London, ship owner.
" Messrs. Woollett and Nephew, London, ship brokers.
June. Messrs. Edward Bilton and Co., Newcastle, merchants.
" Messrs. Louis and Mier, Birmingham, American trade.
July. Messrs. Alie, Grand, and Co., Paris, bankers.
" Mr J. Nunns, Liverpool, cotton broker.
" Messrs. Marzetti and Sons, London, merchants.
" Messrs. Wright and Co., Stockport, manufacturers.
" Messrs. Lowe and Lawes, Manchester, manufacturers.
" Messrs. Baxter and Co., Manchester, manufacturers.
" Messrs. Reed and Sadler, Bristol, spirit merchants.
" Messrs. S. Adams and Co., Hertford, bankers.
Aug. Messrs. Pickford and Keene, London, American merchants.
" Messrs. Courtenay, Kingsford, and Co, London, American merchants.
" Messrs. Smith and Whittingstall, Hemel-Hempstead and Watford, bankers.
" The Newcastle-on-Tyne Commercial Bank.
Sept. Messrs. M'Larty and Lamont, Liverpool, Australian trade.
" The Royal British Bank, London.
" Mr. G. P. Simcox, Kidderminster, weaver.
" Mr. J. Shawcross, Manchester, yarn agent.
Oct. Messrs. T. and H. G. Gray and Co., London, colonial brokers.
" Messrs. Mason, Collins, and Co., New York, grain trade.
" Messrs. Ward, Badcock, and Riggs, New York, importers of dry goods.
" Messrs. Fox, Henderson, and Co., London and Birmingham, contractors.
Nov. Messrs. John Dick and Co., Glasgow, thread spinners.
" Mr. Jobson, Birmingham, iron founder.
" Mr. Dwight, jun., Boston (U. S.), cotton merchant.
" Messrs. Suydam, Reed, and Co., New York, flour trade.
" Mr. T. Parry, New York, flour trade.
" Messrs. Freeman and Bright, New York, dry goods importers.
" Messrs. Watson, Carter, and Co., New York, dry goods importers.
" Messrs. Bateman and Ruderrow, New York, turpentine trade.
Dec. Messrs. Kidd and Co., Hull, seed trade.
" Kidderminster Bank (Messrs. Gotch and Co.), Kidderminster, bankers.
" Mr. R. Johnson, London, warehouseman.
" Messrs. G. Ashworth and Co., Manchester, manufacturers.
" Mr. J. Little, New York, banker and stock operator.
" Messrs. Henslow and Sons, Boston, bankers and stock operators.
" Mr. — Dugard, Birmingham, carriage lamp manufacturer.
" Mr. T. A. Pervanoglu, London, Greek trade.

THE ESTATE OF MESSRS. J. SCOTT RUSSELL AND CO.

The suspension of Messrs. John Scott Russell and Co., the firm at whose works at Millwall the great ship for the Eastern Steam Navigation Company was constructed, was announced on the 4th of February, 1856. Their liabilities were stated to amount to £130,000, and the value of their assets depended much upon the success of arrangements in progress to prevent the abandonment of some of their principal contracts. The following circular was issued by the house, calling a meeting of their creditors, for the 15th of February :—

"*Millwall Works, Feb.* 4.

"SIR,—It is with deep regret that we have to inform you of our inability to fulfil our engagements as they severally become due, and have, therefore, to solicit the favour of your attendance at a meeting to be held on Tuesday, the 12th inst., at noon, punctually, at the Guildhall Coffee-house, King Street, Cheapside, to consider of and determine upon the most effectual and speedy mode of settlement; and we trust that the energies which will be exercised during the week will enable us to place before you a project for effecting the before-mentioned purposes, which will, under all circumstances, be considered satisfactory.

"We are, sir, yours obediently,

"J. SCOTT RUSSELL AND Co."

THE MEETING OF CREDITORS.

A meeting of the creditors of Messrs. J. Scott Russell and Co., of Millwall, who suspended on the 4th inst., was held on the 12th of February, Mr. S. Beale in the chair, when the following statement was presented by Mr. J. E. Coleman:—

STATEMENT OF THE AFFAIRS OF JOHN SCOTT RUSSELL AND CO., OF MILLWALL, LIMEHOUSE, AND GREAT GEORGE STREET, WESTMINSTER, MARINE STEAM ENGINEERS, ETC., FEBRUARY 4.

LIABILITIES.

To creditors on open balances and acceptances				£113,940	19	10
To creditors holding security	£15,000	0	0			
To estimated value of security held	6,000	0	0			
				9,000	0	0
				*£122,940	19	10

ASSETS.

By cash balance	£159	3	8
By book debts considered good	12,800	0	0
By assets, consisting of materials, engines, etc., at works	28,002	8	8
Ditto, Mr. Russell's interest in various ships	40,000	0	0
Ditto, contracts for vessels and machinery now constructing	19,392	0	0
	£100,353	12	4
By debts under £10	118	0	5
	†£100,235	11	11

With reference to the cause of suspension, it was explained by Mr. J. Freshfield that Mr. Russell was originally in partnership with Mr. Robinson, but that in 1852 the latter retired, and Mr. Russell assumed the whole control of the business. Being then supported by the late Mr. Geach, he considered himself justified in extending his operations, and, having taken the contract for building the large vessel for the Eastern Steam Navigation Company, he believed he was proceeding prosperously. A fire, however, occurred on the premises at Millwall in 1853, which destroyed a considerable amount of property, and impeded the works; but, although a consultation was then held with regard to the position of the firm, Mr. Geach appeared satisfied that Mr. Russell was justified in continuing operations. The capital being principally borrowed, and the premises heavily mortgaged, the Government contracts and other works, although profitable, required extended means; but, owing to the altered state of the money-market, it was impossible to provide sufficient to meet current engagements, and, under these circumstances, Mr. Russell thought it prudent to suspend, and place his affairs in the hands of his creditors. It was believed that, if an arrangement could be made for carrying on the Government contracts, and some of the other works, the mortgagees would allow the use of the premises; and the creditors, by adopting the mode of liquidating under inspection,

* The Eastern Steam Navigation Company reserve their right of any claim which they may establish on the estate for breach of contract, in case a satisfactory arrangement cannot be carried out with them.

† By shares in the Eastern Navigation Company, upon which calls have been paid, amounting to £22,410.

could thus avail themselves of advantages which would probably lead to an increased dividend, and prevent the sacrifice which otherwise must ensue. According to the statement of accounts, which had been carefully examined, the assets showed about 15*s.* in the pound, 10*s.* of which, it was believed, might be realized about June, leaving the balance open to contingencies, the result of which could not be safely estimated. Should the necessary preliminaries with the mortgagees and the Eastern Steam Company be completed, the maintenance of the establishment would prove more beneficial to the whole of the parties interested than any other course. Mr. Coleman confirmed the statement of Mr. Freshfield, in relation to a payment of 10*s.* at the period mentioned, and expressed an opinion that the Government contracts, and some of the other works in hand, could be steadily proceeded with. In answer to questions, it was mentioned that it is not proposed to continue the construction of the leviathan vessel of the Eastern Steam Company, the contract passing to the management of the directors. Up to the present time no loss had been sustained in connection with that steamer; but, if the work were continued, it would no doubt exhibit an unfavourable result. Looking at the whole of the property, although it now showed this deficiency, Mr. Russell had not, apparently, formed too sanguine an estimate of his position, since it was necessary to make allowance for the difference between the value of buildings, machinery, etc., when taken at cost price, and the amount likely to be obtained for them when brought into the public market. After some discussion, a series of resolutions, embodying the opinions of the meeting, was carried. Some of the creditors recommended that the entire management should be left to Mr. Scott Russell, but Mr. Freshfield said that, however encouraging such a compliment might appear, it would be necessary, for the general interests of the creditors, to follow out the scheme as presented. Annexed are the resolutions :—

"At a meeting of the creditors of Messrs. John Scott Russell and Co., at the Guildhall Coffee-house, on Tuesday, the 12th day of February, 1856—Mr. Samuel Beale in the chair—Mr. Coleman, the accountant, produced and read to the meeting a statement of the liabilities and assets.

"It was proposed and seconded, and resolved unanimously:—

"1. That it is the opinion of the meeting that the affairs of the house should be liquidated under inspectorship, and that the following gentlemen be the inspectors: —Mr. Samuel Beale, Mr. Thomas D. Grissell, and Mr. John Jones.

"2. That the inspectors be authorized to enter into arrangements for completing the pending contracts and works on hand, and to employ the funds of the estate for the purpose, and pledge the stock and assets of the estate, if necessary, to raise money for the purpose, but not so as to subject the creditors, individually, to liability; also to make arrangement with the mortgagees for the use of the premises and plant, and to employ Mr. Russell, if they shall think fit, in these matters.

"3. That a proper deed of inspectorship be prepared under the approval of the inspectors, and be executed by or on behalf of each creditor, as soon as prepared, of which due notice shall be given.

"4. That such deed shall contain covenants by Mr. Russell to devote his whole time and attention, under the direction of the inspectors, for so long as shall be necessary to complete the works and contracts in hand, to liquidate the affairs of the house according to the rules of administration adopted in bankruptcy, as if bankruptcy had taken place this day; and covenants by the creditors not to sue, which shall operate as a release upon the inspectors certifying that the liquidation has proceeded sufficiently, and upon Mr. Russell executing an assignment of any remaining assets to trustees for distribution among the creditors.

"5. That such deed shall be a deed of arrangement within the meaning of the 224th section of the Bankrupt Law Consolidation Act, and the 228th section shall be applicable thereto.

"6. That the inspectors shall have power to make Mr. Russell such allowance as they may think fit for his services.

"S. BEALE, Chairman."

THE ESTATE OF MESSRS. SYERS, WALKER, AND CO.

A meeting of the creditors of Messrs. Syers, Walker, and Co., largely engaged in the East India trade, who failed on the 4th April, 1856, was held 30th April, Mr. A. Warner presiding, when the following statement was submitted by Mr. J. E. Coleman, the accountant:—

DEBTOR.

	£	s.	d.	£	s.	d.
To creditors on open accounts				£12,737	0	0
To creditors on bills payable not secured	£56,450	0	0			
Less expected to be retired by the parties for whose account they were accepted	6,200	0	0			
				50,250	0	0
Creditors on bills payable partly secured	92,846	0	0			
Estimated value of goods hypothecated	65,647	0	0			
				27,199	0	0
Creditors on bills payable fully secured—						
Estimated value of goods hypothecated	55,273	0	0			
Amount of bills	50,748	0	0			
To *contra*	4,525	0	0			
Liabilities on bills receivable	199,192	0	0			
Of which it is expected there will be duly honoured	142,440	0	0			
	56,752	0	0			
Less cash balance in the hands of Roberts and Co.	619	0	0			
				56,133	0	0
				£146,319	0	0

CREDITOR.

	£	s.	d.
By cash balance in hand	£1,504	0	0
Debtors, good	7,262	0	0
Do., doubtful, £1,975			
Bills receivable on hand, office furniture, etc.	204	0	0
Estimated profits on charters	5,000	0	0
Bills of lading in hand, valued at	1,000	0	0
Balances on consignments, estimated at	20,273	0	0
Surplus from goods hypothecated against bills payable *per contra*, £4,525.			
	£35,243	0	0
Less amounts under £10, rents, rates, etc.	232	0	0
	£35,011	0	0

It was explained that, although this account can only be viewed as a preliminary statement, it discloses with sufficient accuracy the position of the estate. The principal losses have been incurred through the firm being large holders of produce, including rice, linseed, and jute; in addition to which there are bad debts and liabilities on bills. Allowing for the surplus of £4,500 from goods hypothecated against bills payable, the total indebtedness is about £141,000, to meet which the assets represent £35,000. Of this latter amount, upwards of £20,000 is receivable from two firms in India, whose means are stated to be unquestioned, while the remainder is believed to have been estimated with a due regard to contingencies. In November last the produce the house was interested in showed at the prices of the day a large surplus, but the subsequent decline has been ruinous. Before the extent of their misfortunes became apparent they obtained bank credits for a considerable sum, and sent out orders to India for indigo, cotton, and other articles, which were likely to be influenced favourably by the prospect of peace. The markets meanwhile advanced; and as their limits could not be fulfilled, the profit which they hoped to obtain from this source to meet their engagements or reduce their other losses was not realized. In answer to questions, it was mentioned that the trading of the firm commenced in 1851, Mr. M. Syers being joined by Mr. Walker, who brought in about £2,000. Mr. M. Syers possessed no capital beyond a claim on the residue of an estate in liquidation, the accounts of which have not yet been completed. Mr. D. Syers became connected with the firm in 1854, and introduced £4000 capital. The drawings of the partners have been moderate, but no balance has been regularly struck. Had the war continued, and produce maintained its price, they would have been enabled, it is alleged, to meet the whole of their engagements. The bill transactions with Mr. W. O. Young were originally based on mercantile transactions, but they latterly increased, and now show a large total. The great deficiency was a topic of comment, but it was stated that the books will exhibit the whole of the losses. With regard to the ultimate winding-up, it was intimated that Messrs. Syers and Co. are quite prepared to adopt any measures which may suit the convenience of the creditors, either through inspection, a composition, or, if necessary, immediate submission to bankruptcy. On this point there was a protracted discussion, and the feeling appeared to be in favour of a composition, if the payment could be in some measure secured. It was stated that, considering various contingencies, Messrs. Syers, with the assistance of friends here and in India, may be enabled to pay 4*s.* in the pound, in instalments extending over twelve months, and after some conversation, the following resolution was carried, appointing a committee to investigate the accounts, and to consider the proposal:—

"Resolved—That a committee of five be appointed for the purpose of investigating the affairs of Messrs. Syers, and of reporting the same to a subsequent meeting, the expediency of accepting the composition of 4*s.* in the pound, or to determine on the course to be taken. That the five following gentlemen be appointed:—Mr. Warner, Mr. J. Beatson, Mr. M'Kin, Mr. W. S. Grey, and Mr. R. M. Ebsworth."

At the adjourned meeting of the creditors of Messrs. Syers, Walker, and Co., held on the 16th May, Mr. C. Dearie presiding, a report from the committee appointed to examine the accounts, and the proposal of a composition of 4*s.* in the pound, was received. Four out of five of the committee have agreed to the following recommendation, but the dissentient, Mr. Warner, intimated his opinion that further investigation is requisite previously to any definitive conclusion:—

"Your committee, having obtained such information and explanations as they could collect, founded upon Mr. Coleman's *pro forma* statement of the affairs of Messrs. Syers, Walker, and Co. (which statement and investigation are by necessity imperfect, from the books not having been posted since July, 1855), have come to the conclusion that those gentlemen could not carry out their proposed proposition of 4*s.* in the pound. Messrs. Syers, Walker, and Co., still believing that with the assistance of their friends, and that influence which they could exercise over those persons having business transactions with them, by themselves administering the estate, offer to pay a composition of 3*s.* in the pound—say, 1*s.* on the signature of the deed by three-fifths of the creditors, 1*s.* in six months from that date, and 1*s.* in twelve months, the last instalment to be guaranteed. Your committee recommend that this offer should be accepted, or that the estate be wound-up under inspection, either mode promising a more favourable result than bankruptcy. Your committee are unanimous in this view as to the composition being more advantageous for the creditors, with the exception of their chairman, Mr. Warner, who advocates inspectorship as the most desirable course.

"R. M'Kin.
"R. M. Ebsworth.
"J. Beatson.
"W. S. Grey."

Mr. Warner explained his views with regard to the estate, and according to his estimate it may produce a larger dividend than the amount proposed under the revision of the committee. Instead of 3*s.*, or even 5*s.*, the sum shown by Mr. Coleman's *pro forma* statement, he thought, with management, 6*s.* or 7*s.* can be realized, if the produce be not forced to an immediate sale. He also believes it will be necessary to look further into the books and correspondence to ascertain the actual position of affairs, owing to the absence of due regularity by the partners in the arrangement of their transactions. A protracted discussion followed, in which it was stated that the firm were anxious to follow the wishes of the creditors in every respect, and that they have been induced to make this specific offer, believing it

was the wish of a majority to arrive at a direct settlement. After mutual explanations, a general desire was expressed to have the estate placed under inspection until the books shall have been made up and the correspondence examined, and the annexed resolution was eventually carried :—

"It is the opinion of this meeting that inspectors should be appointed for the period of four calendar months, to direct the making up of the books, and to superintend the realization of the assets, and to receive and examine the correspondence; and at the expiration of such time, or such earlier time as they may think fit, to report to the creditors the state and prospects of the estate. That the following gentlemen be requested to act as such inspectors, namely, Messrs. A. Warner, C. Dearie, and J. Beatson."

The estate was subsequently wound-up in bankruptcy.

THE ESTATE OF MESSRS. MARZETTI AND SONS.

A meeting was held, on the 12th of August, 1856, of the creditors of Messrs. J. G. Marzetti and Sons, merchants and ship insurance agents, whose failure was announced on the 18th of July. Mr. Cuthbert, of the firm of Cookson, Cuthbert, and Co., of Newcastle, presided, and, after some remarks from Mr. Upton, the solicitor of the firm, Messrs. Quilter, Ball, and Co. submitted the statement of affairs, dated the 16th of July, 1856, which was as follows :—

	£	s.	d.	£	s.	d.
Unsecured				£27,750	10	2
Partially secured	£43,153	16	6			
Less securities	34,214	16	11			
Deficiency				8,938	19	7
Fully secured	11,551	9	2			
Less securities	14,748	11	7			
Surplus	3,197	2	5			
Bad liabilities				2,379	13	7
Total estimated to prove				£39,069	3	4
Liabilities on bills receivable, current, considered good	£16,512	15	10			
The assets consisting of outstanding consignments	6,670	5	4			
Surplus securities in the hands of creditors	3,197	2	5			
Cash and goods in hand, and book debts	11,898	1	0			
	£21,765	8	9			
Deduct claims payable in full	700	0	0			
Total estimated assets				£21,065	8	9

A deficiency of £18,003 14*s.* 7*d.* was thus shown to exist, to which being added the capital at credit of the partners' accounts at their last balance, January 1, 1856, £11,761 10*s.* 9*d.*, and subsequent profits, £2,437 5*s.* 3*d.*, there resulted the sum of £32,202 10*s.* 7*d.* to be accounted for. This was done in a detailed statement, the principal items of which were losses on outstanding consignments to the Crimea and elsewhere, £14,255; bad debts and liabilities, £9,095; and outlay on premises in Vine Street, and North Street, £2,039. It was explained that the sudden cessation of the war with Russia had been one of the principal causes of loss, there being at that time in the East unsold consignments of provisions and wines and spirits of the cost value of about £10,000 to be sent home, and probably re-shipped on arrival, with a view to their realization at the least possible sacrifice. Much sympathy was expressed for the firm, which was an old and respectable one, and a resolution was unanimously passed, to the effect that the interest of the creditors would be best promoted by a liquidation under inspection, the parties named as inspectors being Messrs. Cuthbert, French, and Dawson.

THE ESTATE OF MR. W. O. YOUNG.

A meeting of the creditors of Mr. W. O. Young, ship and insurance broker, who failed on the 4th of April, 1856, was held on the 10th. Mr. D. Dunbar presided, when the following statement was submitted by Mr. J. E. Coleman, the accountant, showing assets valued at only £25,249 against debts for £65,750, exclusive of heavy liabilities on underwriting accounts :—

DEBTOR.

To creditors on shipbroking and other open accounts				£15,530	0	0
Creditors on underwriting accounts				2,560	0	0
Ditto, insurance accounts				3,100	0	0
Ditto, bills payable				40,160	0	0
Liabilities on bills receivable	£19,185	1	10			
Less bills expected to be duly honoured at maturity	17,265	1	10			
				1,920	0	0
Creditors holding security	4,000	0	0			
Security held	2,320	0	0			
				1,680	0	0
Liability in respect of Liverpool agency				800	0	0
				£65,750	0	0

CREDITOR.

By cash balance in hand	£25	5	10
Ship, and shares in ships, and profits of same, and sundry other assets	10,060	0	0
Debtors on shipbroking accounts	6,270	0	0
Ditto on underwriting accounts	5,000	0	0
Ditto on insurance accounts	634	0	0
Ditto on bills payable after the parties have taken up their draughts	2,535	0	0
Commissions not yet made up	1,500	0	0
Sundry shipments	277	0	0
Bills receivable	206	0	0
	£26,507	5	10
Less salaries, etc.	1,258	0	0
	£25,229	5	10

It was explained by the Chairman that he was consulted by Mr. Young immediately after the suspension of Syers, Walker, and Co. A glance at the state of his affairs showed that an immediate stoppage would be the only honourable course, and Mr. Young at once consented. With regard to the accounts, Mr. Coleman mentioned, that although they had been prepared at a very short notice, they were believed to approximate sufficiently to exhibit the condition of the estate. In the arrangement of the ordinary business of the firm, little difficulty is likely to be encountered, but the underwriting branch involves liabilities, the result of which cannot yet be ascertained. The balance due from Syers, Walker, and Co. is about £27,000. An old friendship having existed between Mr. D. Syers and Mr. Young previous to the failure of a former firm of Livingstone, Syers, and Co., in India, in 1848, it continued subsequently. When Syers and Co. started in business, Mr. Young became further connected with them. In this manner they were joint owners of ships, and were also interested in Manilla sugar and teak timber. Hence arose the bill transactions, which were in the first instance based on actual commercial operations, but which afterwards increased through the depreciation in produce and the great margins required on loans. Mr. Young hoped, up to a recent period, that these liabilities would run off; and Messrs. Syers and Co., it was alleged, considered in November last that their estate showed a surplus of £70,000. The great extent of their operations, however, caused the late fall in rice and other articles to result in their sudden ruin. But for the heavy liabilities associated with the underwriting account, Mr. Young, it was stated, would have been in a situation to make some definite offer on a realization of his assets. That course being impracticable, he was prepared to adopt any proceeding the majority of his creditors might advise. A question raised with respect to the exact relationship of the late Mr. Beckwith, of Newcastle, in connection with the underwriting carried on by Mr. Young, gave rise to considerable discussion, a difference of opinion prevailing as to the existence of a partnership. It was therefore agreed, before adopting any determination as to the way in which the estate should be wound up, to appoint Mr. Pitcairn and Mr. Stephens to investigate the books and documents, with the view of settling that question, and an adjournment for a week was immediately carried.

At the adjourned meeting on the 15th of April, it was agreed to take the necessary steps for placing the estate under the administration of bankruptcy—a course stated to be requisite in consequence of the liabilities arising from the underwriting branch, which, it was alleged, involved risks to the extent of nearly £2,000,000. The outstanding premiums of insurance averaged from £15,000 to £20,000, but allowance would have to be made for bad debts, while there was already from £3,000 to £4,000 owing on the account. Under these circumstances, no estimate could be given of the probable result, even presuming that the transactions of the late Mr. Beckwith, of Newcastle, may constitute a partnership, as indicated by the following communication received by the Chairman from Messrs. Pitcairn and Stephens, who were appointed to investigate the question :—

"SIR,—In conformity with the request made at the meeting of Mr. W. O. Young's creditors on the 10th inst., over which you presided, we have examined the letters relating to the alleged partnership between Mr. Young and the late Mr. Beckwith in the underwriting account, and we are of opinion, as mercantile men, that such partnership is established by Mr. Beckwith's letters of the 1st of April, 1853, 2nd of May, 1854, and 4th of May, 1854.

"We are, sir, your obedient servants,

"JOHN PITCAIRN,

"To Mr. D. Dunbar," etc. "THOMAS STEPHENS.

In the course of the day a petition in bankruptcy was issued against the estate, which was wound up before Mr. Commissioner Holroyd, Mr. Lee being the official assignee.

When the estate was brought under bankruptcy, the accounts showed the following results :—

BALANCE-SHEET FROM JUNE 24, 1849, TO APRIL 15, 1856.

DEBTOR.

	£	s.	d.
To surplus, June 24, 1849	£2,151	1	8
To creditors unsecured	38,072	7	6
Ditto fully secured	8,873	14	6
Ditto partially secured	2,121	1	0
Ditto, salaries, etc., payable in full, £380 18*s.* 2*d.*; liabilities on bills, etc., £58,465 4*s.* 8*d.*, of which £24,110 only are expected to be claimed against the estate	24,110	0	0
Risks on underwriting, about £62,000; in respect of which I estimate that about £10,000 losses may be proved	10,000	0	0
Brokerage profits	60,728	9	2
Premiums, etc., for underwriting	121,060	0	9
	£267,116	14	7

CREDITOR.

	£	s.	d.	£	s.	d.
By debtors, good				£22,538	6	0
Doubtful				11,672	19	8
Bad, carried to losses	£1,494	4	4			
Property given up to the official assignees	12,749	8	4			
Bills receivable	706	4	7			
	£13,455	12	11			
Less salaries payable in full				13,074	14	9
Consignments	£928	5	9			
My interest in an adventure in teak	2,000	0	0			
				2,928	5	9
Carried forward				£50,214	6	2

Brought forward		£50,214 6 2
Excepted articles		20 0 0
Liabilities *per contra*	£24,110 0 0	
Ditto ditto	10,000 0 0	
		34,110 0 0
Property held by creditors fully secured		9,294 15 4
Property held by creditors, partially secured		115 0 0
Trade expenses		17,570 3 8
Interest		3,428 13 5
Law charges		648 0 7
Domestic and personal expenses		9,927 14 3
Sundry losses	£28,329 10 3	
Ditto bad debts	1,494 4 4	
Ditto on underwriting	111,933 7 1	
		141,757 1 8
Difference		30 19 6
		£267,116 14 7

THE ESTATE OF MESSRS. FOX AND HENDERSON.

The failure was officially intimated on the 29th of October, 1856, although it was prematurely announced a day or two before in several quarters, owing to embarrassments which it was supposed could be adjusted. During the past year the firm have experienced occasional difficulties, which were greatly increased by the reckless circulation of reports affecting their credit. By great efforts they were enabled to maintain their position up to the present time; but the renewed pressure in the money-market, and the discovery of losses from heavy foreign contracts just finished, have now compelled them to call their creditors together. It is understood, that upon the completion of their annual stock-taking and balancing, the house have found that they have suffered to the extent of about £70,000, by the construction of the Zealand (Danish) Railway. This, combined with unprofitable results from some other works, and the impossibility of realizing or of obtaining sufficient advances upon the large amount of shares and debentures they hold in the various undertakings with which they have been connected, has left them in a position in which they could make no further sacrifices without jeopardizing the ultimate liquidation of their general liabilities. They have, therefore, taken the advice of the persons most largely interested, and have resolved to suspend. It appears, however, to be the general feeling that an extension should be granted them, and that they should be allowed at once to resume upon a full and satisfactory exposition of their engagements and assets being submitted. Their total debts are stated to be about £320,000, of which about half are unsecured.

"*London Works, Birmingham, Oct.* 29.

"SIR,—It is with the deepest regret we have to announce to you that we have been compelled to suspend our payments, and have, therefore, to solicit the favour of your attendance at a meeting to be held on Friday, the 7th of November, at one o'clock precisely, at Dee's Royal Hotel, in Birmingham, to consider and determine the most effectual and speedy mode for the adjustment of our affairs.

"A statement of our assets and liabilities will, in the meantime, be prepared for the consideration of the meeting, and we trust that the propositions then submitted will, under all the circumstances, be considered satisfactory.

"We remain, sir, your obedient servants,

"FOX, HENDERSON, AND CO."

The meeting of the creditors was held at Dee's Royal Hotel, Birmingham, on the 7th of November. About sixty persons were present, these being chiefly creditors resident in this district. Amongst the principal firms represented were the Patent Rivet Company; the London and North Western Railway Company; Tupper,

Carr, and Company; the Patent Tube Company; Messrs. Sims and Muntz; Mr. W. Williams, Mr. C. L. Browning, Mr. Samuel Blackwell, and other principal creditors of the firm in the neighbourhood, were also present. Mr. Murray, solicitor of London, attended on behalf of Messrs. Glyn and Co., and several of the London creditors; and for Messrs. Fox, Henderson, and Co., Messrs. Colmore and Beale were also present on behalf of those parties. Mr. Robinson, of the Ebbw Vale Ironworks, presided. The following statement was read by Mr. Coleman, the accountant:—

DEBTOR.

To creditors unsecured, whose claims are above £100	£77,920	7	9			
Ditto, whose claims range from £50 to £100	2,491	16	9			
Ditto, whose claims range from £10 to £50	1,462	16	0			
To creditors partly secured	30,355	1	0			
To estimated value of security	16,000	0	0			
				128,230	1	6
To creditors fully secured	155,689	13	0			
To liabilities on bills under discount expected to rank on this estate				12,948	0	0
Liabilities				£141,178	1	6

CREDITOR.

By valuation of tools and furniture				£26,000	0	0
By valuation of materials				27,000	0	0
By work in course of completion				20,000	0	0
By debtors				32,000	0	0
By estimated amount to be received in respect of the East Kent Railway Works	£30,000	0	0			
Less portion of same, mortgaged	12,500	0	0			
				17,500	0	0
By shares, etc., in hand, estimated value				1,200	0	0
				£123,700	0	0
Less creditors under £10, rent, taxes, salaries, etc., to be paidin full				2,000	0	0
				£121,700	0	0

Mr. Coleman mentioned that the estimate of the securities, put down at £16,000, was an exceedingly low valuation, inasmuch as that there could be no doubt whatever of its realizing the full amount; the securities to the creditors who held for the whole of their debts—in the gross of £155,690—had cost Messrs. Fox, Henderson, and Co., £215,000. The valuation of assets had been gone into very carefully by Mr. Henderson, and after making every allowance for depreciation, the computation of £105,000 might be regarded as a low figure. With reference to Messrs. Glyn and Co., they were in this position—they might dispose of the whole of their securities, and pay themselves at once; but with the liberality which invariably characterized the conduct of the firm in reference to such matters, it was not their intention to realize at present, but by holding them over for a few months, there would no doubt be sufficient realized to leave a surplus for distribution amongst the creditors. As to the tools and furniture, it had been the custom of the firm to take off their value a sufficient sum for depreciation from year to year, which enabled Mr. Henderson to take a pretty accurate estimate of what they were worth at this present moment; of course, if the concern was broken up, and these articles were brought to the hammer, as every one knew, that would be a very considerable loss. The materials had been valued at the market price. The amount due from debtors had been very carefully considered, and in his judgment the amount put down in the balance-sheet was a reliable one. In the month of June, 1855, Messrs. Fox and Henderson considered that they had a surplus of £115,000, and further profits had been made since that time; but the losses had been greatly in excess. There was a loss of £70,000 on the Danish (Zealand) Railway; this, with losses in other engagements, and on the securities, made a total of £160,000. Such was a general explanation of the affairs of the firm, and of the circumstances under which the present difficulty had been brought about. Beyond this, Mr. Coleman said he was ready to answer any question put to him, so that it could be done without prejudice to the estate. The meeting was next addressed by Mr. Murray, who stated that he represented both the secured and the unsecured creditors in London, all of whom were desirous of showing every consideration possible for the firm; in this spirit his clients were ready to do everything in their power to assist in realizing the assets as well as it was possible. Messrs. Glyn and Co. held securities, which they would take time to realize, so that as little loss as possible might ensue, which would be otherwise if there was a forced sale or bankruptcy. Other clients of his (Messrs. Hambro) were willing to wait; they also held securities, and would do anything rather than annihilate the estate, and send two such men again upon the world; rather than this they would waive their claim. There were three courses open to this meeting—the first was bankruptcy; the second to wind-up under inspection; and the third, to offer a composition. All present knew the amount of loss incident to a bankruptcy; there was not only the great cost, but the loss that would accrue from a forced sale of such stock as Messrs. Fox and Henderson possessed; and after much consideration, the creditors in London thought it best for the interest of all parties that the estate should be wound-up under inspection. The plan that he suggested was, that a deed of inspection should be prepared, and that five gentlemen should be selected from this meeting, with whom the creditors in London would co-operate; in this district the amount of debts did not exceed from £30,000 to £40,000, but as the creditors were numerous, it was thought better that the inspectors should be named here. The deed would of course contain all proper covenants, and power to make a considerable allowance to Messrs. Fox and Henderson; by this mode he thought the estate would be wound-up best for the interest of all parties.

Mr. J. Lord, merchant, made a few observations expressive of his regret that he had been the means of introducing the parties connected with the Danish (Zealand) Railway to Messrs. Fox and Henderson. At one period the affair promised to be extremely valuable to the contractors; but there was a dearth of labour, and some other unfavourable circumstances, which ultimately resulted in heavy loss; it was admitted, however, by all parties, that the work could not have been better done. As a small creditor, he was willing to accede to any proposition; and would gladly consent to leave the matter with Mr. Henderson.

The Chairman said he believed that all parties were willing to hold the securities, until the best that was possible could be made of them; it might be that ultimately they would realize more than the estimate.

Mr. Murray then suggested that the five gentlemen named below should act as inspectors, the debts of these in the aggregate amounting to £20,000.

A question was asked by Mr. Kempson as to whether the loss of £70,000 on the Danish Railway arose from an excess of work over the contract.

Mr. Coleman said it would not be prudent to reply more than generally. The fact was, that the number of yards to be excavated had been much under-stated; the Zealand Company were informed of the fact at the time, and the contractors allowed to go on. The work had been done, and satisfactorily done; and as representations had been made in the proper quarter, it was hoped that Messrs. Fox and Henderson would be to some extent reimbursed; but there was no legal claim; it was a debt of honour, and not regarded as anything in the assets.

Mr. W. Williams mentioned that he had had a conversation with a Dane as recently as yesterday, and he spoke of the excellent manner in which this line had been constructed.

On the motion of the Chairman, seconded by Mr. W. Williams, the following resolutions were passed, with one dissentient only:—

"That it is the opinion of this meeting that the affairs of Messrs. Fox, Henderson, and Co. should be liquidated under inspection, and that the following gentlemen be appointed inspectors, viz.:—Mr. Joseph Robinson, Mr. James Timmins Chance, Mr. Charles Lloyd Browning, Mr. Samuel Holden Blackwell, Mr. M'Gregor Laird. That a proper deed of inspectorship be prepared, under the direction of the inspectors, and that such deed shall contain covenants by Messrs. Fox and Henderson to arrange and liquidate their affairs according to the rules of administration adopted in bankruptcy, and as if bankruptcy had taken place on the 29th ult., and covenants by the creditors not to sue, which shall operate as a release upon the inspectors certifying that the liquidation had proceeded sufficiently, upon the partners executing an assignment of any remaining assets to trustees for distribution among the creditors, and such other terms and stipulations as may be requisite for the carrying out of such deed within the meaning of the Bankrupt Law Consolidation Act, 1849.

"That the inspectors shall have power to make Messrs. Fox and Henderson such allowance as they may think fit for their services.

"That these resolutions shall be, and be deemed to be, a memorandum of arrangement within the meaning of the 224th section of the Bankrupt Law Consolidation Act, 1849, and that the 228th section shall be applicable thereto.

"That as several copies of these resolutions may not be signed by all, or by the same creditors, the several parts shall constitute one memorandum of arrangement, within the meaning of the said 224th section of the above Act of Parliament."

In the course of the proceedings, it was mentioned that Sir Charles Fox and Mr. Henderson were in the adjoining room, but their attendance was not requested. The number of unsecured creditors is 150; of these, rather more than fifty have debts which do not amount to £100. It was stated that Government was a creditor to a considerable amount. This matter has been arranged, and consequently does not appear in the above balance-sheet. The firm was subsequently declared bankrupt, and the following is the balance-sheet:—

BALANCE-SHEET FROM JUNE 30, 1855, TO FEBRUARY 11, 1857.

DEBTOR.

To sundry creditors				£94,511	15	3
To creditors holding security on property	£131,454	0	6			
To liabilities	20,824	4	5			
Balance				35,243	7	1
				£129,755	2	4

CREDITOR.

By debtors, good				£28,233	13	7
By ditto, doubtful	£21,937	2	6			
By ditto bad	5,044	10	9			
	£26,981	13	3			
Taken at	6,500	0	0			
				6,500	0	0
By property to be taken by the assignee				40,770	14	5
By property on which creditors have security	£154,434	14	10			
Deduct amount due to creditors *per contra*	131,454	0	6			
				22,980	14	4
By special assets				31,270	0	0
				£129,755	2	4

The estate was eventually wound-up in bankruptcy.

THE ESTATE OF THE ROYAL BRITISH BANK.

At a meeting of the shareholders of the Royal British Bank, held on the 20th of September, 1856, a condensed statement of affairs was exhibited by Mr. J. E. Coleman, which set forth the liabilities at £539,131 12*s*. 9*d*., and the assets, exclusive of Welsh works, £288,644 8*s*. 11*d*. Subjoined, however, are the full details:—

Statement of the Affairs of the Royal British Bank, *3rd September,* 1856, *according to J. E. Coleman's Estimate.*

LIABILITIES.

	Head Office.	Strand.	Lambeth.	Islington.	Pimlico.	Borough.	Piccadilly.	Holborn.	Totals.
	£ s. d.	£ s. d.	£. s. d.	£ s. d.	£ s. d.	£ s. d.	£ s. d.	£ s. d.	£ s. d.
To sundry creditors	792 5 5	......	......	......	......	......	......	......	792 5 5
Do on drawing accounts	143,744 3 3	56,563 7 2	25,076 2 3	45,079 4 7	3,600 15 8	10,778 14 9	11,881 0 8	9,912 0 1	306,635 8 5
Do. on deposit accounts	122,425 4 5	57,906 18 5	18,802 8 1	30,359 10 2	6,386 13 2	6,687 11 9	4,232 16 7	6,194 15 7	252,995 18 2
Do. on cash credit accounts	834 19 6	136 13 1	309 18 1	32 8 9	......	......	......	15 8 11	1,329 8 4
Do. on promissory notes, with interest	5,000 0 0	......	......	......	......	......	......	......	5,000 0 0
Do. on do., without interest	2,589 11 0	......	......	......	......	......	......	......	2,589 11 0
Do. on bills in duplicate	161 0 0	......	......	......	......	......	......	......	161 0 0
Do. on circular notes	887 0 0	......	......	......	......	......	......	......	887 0 0
Do. on drafts on demand	567 10 10	46 0 0	......	18 19 9	......	......	120 4 0	2 4 2	754 18 9
Do. on unclaimed dividend	421 8 5	......	......	......	......	......	......	......	421 8 5
	277,423 2 10	114,652 18 8	44,188 8 5	75,490 3 3	9,987 8 10	17,446 6 6	16,234 1 3	16,124 8 9	571,566 18 6
Deduct amounts standing to credit of parties on their drawing or deposit accounts, who are the drawers, acceptors, or endorsers of bills under discount	30,731 7 10	4,462 3 2	871 7 11	1,893 5 8	155 9 9	2,130 7 10	1,346 17 8	749 5 11	42,340 5 9
	246,691 15 0	110,190 15 6	43,317 0 6	73,596 17 7	9,831 19 1	15,335 18 8	14,887 3 7	15,375 2 10	529,226 12 9
To further estimated liabilities	8,600 0 0	......	......	......	......	......	......	......	8,600 0 0
Liabilities on acceptances granted by the Bank for £5,433 5*s*. 1*d*., expected to be met by the drawers, who are indebted to the same amount	......	......	......	......	......	......	......	......	......
Liabilities on bills discounted, £103,335 15*s*. 2*d*., of which it is expected the amount of £102,030 15*s*. 2*d*. will be duly honoured	1,305 0 0	......	......	......	......	......	......	......	1,305 0 0
Liabilities	256,596 15 0	110,190 15 6	43,317 0 6	73,596 17 7	9,831 19 1	15,335 18 8	14,887 3 7	15,375 2 10	539,131 12 9

Statement of the Affairs of the Royal British Bank, *3rd September*, 1856, *according to J. E. Coleman's Estimate.*

ASSETS.

	Head Office.	Strand.	Lambeth.	Islington.	Pimlico.	Borough.	Piccadilly.	Holborn.	Totals.
	£ s. d.	£ s. d.	£ s. d.	£ s. d.	£ s. d.	£ s. d.	£ s. d.	£ s. d.	£ s. d.
By Cash at head offices and branches	......	......	......	......	......	......	......	......	48,528 1 3
Buildings and furniture at head offices and branches'...	......	......	......	......	......	......	......	......	25,730 0 11
Debtors on drawing accounts ...	2,895 5 1	125 6 8	272 10 11	92 15 1	8 12 10	1 17 0	59 17 6	24 18 4	3,481 3 5
Do. doubtful, estimated at ...	207 8 2	35 9 1	......	......	3 17 6	......	......	......	246 14 9
Do. on accounts, on which cash limits were allowed ...	52,236 4 7	12,957 13 3	1,928 17 4	1,361 6 2	270 11 9	221 12 8	......	4,025 12 8	73,001 18 5
Do., do., doubtful, estimated at	1,609 10 11	81 8 8	......	133 11 9	......	......	......	......	1,824 11 4
Advances on convertible securities	46,166 1 9	1,030 0 0	......	......	......	......	1,000 0 0	......	48,196 1 9
Do. doubtful, estimated at ...	500 0 0	......	......	......	......	......	......	......	500 0 0
Past-due loans, good	4,007 17 2	......	......	......	......	......	......	......	4,007 17 2
Do. doubtful, estimated at ...	75 0 0	......	......	......	......	......	......	......	75 0 0
Past-due bills, good	6,122 4 0	678 11 1	......	......	......	22 0 0	......	......	6,822 15 1
Do. doubtful, estimated at ...	3,823 14 11	108 0 10	......	......	......	26 9 6	62 10 0	......	4,020 15 3
London bills, discounted	77,213 12 11	17,366 7 2	4,626 12 6	3,252 13 10	603 12 6	7,110 9 10	5,898 7 10	3,203 4 8	119,275 1 3
Do., do., doubtful, estimated at	697 19 7	......	......	......	......	......	......	......	697 19 7
Country bills, discounted	2,364 4 4	1,052 3 0	......	......	......	......	......	......	3,416 7 4
Bills continued with collateral security, estimated at	100 0 0	......	......	......	......	......	......	......	100 0 0
Do., do., on account of W. Taite, £22,510 3*s.* estimated at..................................	1,300 0 0	......	......	......	......	......	......	......	1,300 0 0
Sundry debtors, good	695 14 8	......	......	3 2 2	......	......	......	......	698 16 10
Stamps..................................	345 0 4	......	......	......	......	......	......	......	345 0 4
	200,359 18 5	33,434 19 9	6,828 0 9	4,843 9 0	886 14 7	7,382 9 0	7,020 15 4	7,253 15 8	342,268 4 8
Less amounts against which parties have credit balances, contra	30,731 7 10	4,462 3 2	871 7 11	1,893 5 8	155 9 9	2,130 7 10	1,346 17 8	749 5 11	42,340 5 9
	169,628 10 7	28,972 16 7	5,956 12 10	2,950 3 4	731 4 10	5,252 1 2	5,673 17 8	6,504 9 9	299,927 18 11
Less allowance for contingencies, exclusive of any expenses, 5 per cent. on £225,669 16 9									11,283 10 0
Assets									£288,644 8 11

Welsh works (exclusive of interest) cost........................ ... £106,453 4 9

Statement of the Affairs of the Royal British Bank, *3rd September,* 1856, *according to the Books of the Bank.*

LIABILITIES.

	Head Office.	Strand.	Lambeth.	Islington.	Pimlico.	Borough.	Piccadilly.	Holborn.	Totals.
	£ s. d.	£ s. d.	£ s. d.	£ s. d.	£ s. d.	£ s. d.	£ s. d.	£ s. d.	£ s. d.
To sundry creditors	792 5 5	......	......	......	......	......	......	......	792 5 5
Creditors on drawing account	145,844 3 3	56,737 6 8	24,681 19 4	44,787 6 1	3,445 5 7	10,772 10 11	11,821 3 2	9,887 1 9	307,976 16 9
Do. on deposit account	168,845 13 1	59,570 15 3	18,802 8 1	30,537 14 9	6,518 0 7	6,687 11 9	4,232 16 7	6,194 15 7	301,389 15 8
Do. on cash credits	834 19 6	......	......	......	......	......	......	......	834 19 6
Do. on prom. notes with interest	5,000 0 0	......	......	......	......	......	......	......	5,000 0 0
Do. do. without interest	2,589 11 0	......	......	......	......	......	......	......	2,589 11 0
Do. on bills in duplicate	161 0 0	......	......	......	......	......	......	......	161 0 0
Do. on circular notes	887 0 0	......	......	......	......	......	......	......	887 0 0
Do. on drafts on demand	567 19 10	46 0 0	......	18 19 9	......	......	120 4 0	2 4 2	754 18 9
Do. on unclaimed dividends	421 8 5	......	......	......	......	......	......	......	421 8 5
Liabilities on acceptances granted by the Bank	5,433 5 1	......	......	......	......	......	......	......	5,433 5 1
Bills unaccounted for—contra	11,550 13 4	41,961 13 4	1,537 8 2	2,812 0 1	135 15 3	323 5 7	484 18 5	967 16 1	19,613 10 3
	342,927 9 11	113,115 15 3	45,021 15 7	78,156 0 8	10,099 1 5	17,783 8 3	16,659 2 2	17,051 17 7	645,854 10 10

	£ s. d.	£ s. d.	£ s. d.
To Capital	£150,000 0 0		
New share deposits	8,735 0 0		
		£158,735 0 0	
Bad debt fund		4,217 0 4	
Liquidation fund		528 18 5	
Premium		245 0 0	
Rent and taxes		112 17 8	
Income tax		2 0 0	
Branches		175,447 12 0	
		£339,288 8 5	
Reserve fund	£14,202 8 8		
Unappropriated balance, 30th June	1,060 2 7		
Interest received	1,877 12 7		
		17,140 3 10	
		£356,428 12 3	356,428 12 3
			£1,002,283 3 1

Statement of the Affairs of the Royal British Bank, *3rd September*, 1856, *according to the Books of the Bank.*

ASSETS.

	Head Office.	Strand.	Lambeth.	Islington.	Pimlico.	Borough.	Piccadilly.	Holborn.	Totals.
	£ s. d.	£ s. d.	£ s. d.	£ s. d.	£ s. d.	£ s. d.	£ s. d.	£ s. d.	£ s. d.
Cash at head office and branches	......	......	......	......	......	......	......	......	48,528 1 3
Buildings and furniture	......	......	......	......	......	......	......	......	35,252 17 1
Liabilities on acceptances to customers per contra	5,433 5 1	......	......	......	......	......	......	......	5,433 5 1
Debtors on drawing accounts ...	62,712 4 9	604 3 1	......	......	603 12 6	......	......	......	63,920 0 4
Debtors on account of which cash credits were allowed...	120,968 3 9	16,790 19 8	2,276 19 1	2,040 3 9	270 11 9	221 12 8	......	4,010 3 9	146,578 14 6
Debtors on convertible securities	130,661 16 9	1,030 0 0	......	......	......	......	1,000 0 0	......	132,691 16 9
Debtors on past-due loans	16,598 18 10	......	......	......	......	......	......	......	16,598 18 10
Debtors on past-due bills	74,410 5 4	5,958 18 1	677 7 0	......	......	513 1 6	250 0 0	......	81,809 11 11
Debtors on London bills discounted	94,707 4 9	17,366 7 2	4,626 12 6	3,252 13 10	......	7,110 9 10	5,598 7 10	3,203 4 8	136,165 0 7
Debtors on country ditto	5,864 4 4	1,052 3 0	......	......	......	......	......	......	6,916 7 4
Debtors on bills continued with collateral security	25,377 9 6	......	......	......	......	......	......	......	25,377 9 6
Sundry debtors	1,806 1 6	......	......	......	......	......	......	......	1,806 1 6
Stamps	345 0 4	......	......	......	......	......	......	......	345 0 4
Bills received for collection	10,682 6 0	1,801 13 4	1,537 8 2	2,812 0 1	125 15 3	323 5 7	406 7 5	967 16 1	18,666 11 1
By ditto remitted	868 7 4	......	......	......	......	......	78 11 0	......	946 18 4
	550,435 8 3	44,604 4 4	9,118 6 9	8,104 17 8	1,009 19 6	8,168 9 7	7,633 6 3	8,181 4 6	
Assets........................									721,036 15 2

By adjusting interest account........................		£22,356 10 6	
Branches........................		177,984 12 7	
Guarantee		33 5 0	
Clearing		4 13 0	
Suspense sundry accounts		60,916 14 2	
		£261,295 15 3	
Working expenses........................	£4,157 0 9		
Preliminary........................	15,793 11 11		
		19,950 12 8	
		£281,246 7 11	281,246 7 11
			£1,002,283 3 1

The estate was ultimately wound-up in bankruptcy, with a distribution equal to 14*s*. 6*d*. to 15*s*. in the pound.

THE ESTATE OF MESSRS. T. AND H. G. GRAY AND CO.

A meeting of the creditors of Messrs. T. and H. G. Gray and Co., colonial brokers, who failed on the 5th of October, 1856, was held on the 12th of November, when the statement regarding their affairs was presented by Mr. W. Quilter, the accountant.

STATEMENT OF AFFAIRS, OCTOBER 3, 1856.

DEBTOR.

To sundry creditors			£15,112 3 2
To liabilities on bills receivable		£9,911 4 2	
The holders of these bills have in hand, balance of Gray and Co.'s drawing account...	£90 11 1		
And also estimated surplus security beyond the amount advanced	862 9 10		
		953 0 11	
Drafts on us			9,346 0 9
To sundry creditors, old firm			1,742 6 6
			£26,200 10 5

CREDITOR.

By cash and bill in hand, viz. :—		
Cash at bankers'	£1,405 0 0	
Bill due 29th October	55 6 7	
		£1,460 6
By goods and office furniture in hand		317 0
By sundry debtors—		
Considered good		935 15 7
,, doubtful and bad, not estimated ...		0 0 0
By assets due to the late firm of Gray and Co. ...		1,254 1 6
		£3,967 3 8
Deduct—rent, salaries, expenses, etc.		600 0 0
		£3,367 3 8

It was explained, that although the balance-sheet shows only £3,300 to meet £26,000 debts and liabilites, the partners are desirous of effecting arrangements to ncrease the amount, and it is thought that, including the proceeds from an unsettled account with another firm, the assets may eventually reach about £7,000. Taking an estimate, it is believed a dividend of from 4*s.* to 5*s.* in the pound may be realized. A lengthened discussion took place with respect to some irregular bill transactions, and it was ultimately agreed that the estate should be wound-up under inspection, periodical distributions to be made as the assets are converted.

PART IV.

1857.

THE ESTATE OF W. FUSTANA AND CO., AND CALUTA BROTHERS.

A meeting of the creditors of Messrs. Fustana and Co., of Liverpool, and Messrs. Caluta Brothers, of Manchester, was held at the office of Messrs. Lowndes, Bateman, and Lowndes, the 11th of February, 1857; the representatives of thirty-three creditors being present, Mr. W. Courtnay Cruttenden in the chair, a summary of the accounts, prepared by Messrs. Bewley and Son, public accountants, was read, as follows:—

Statement of Affairs to February 11, 1857.

LIABILITIES.

Creditors on open accounts	£19,227	5	11
Creditors on Fustana and Co.'s acceptances	30,059	4	5
Creditors on Caluta Brothers' acceptances	18,579	0	4
Creditors on bills receivable	4,593	6	3
	£72,458	16	11

ASSETS.

Outstanding debts, good	£8,770	5	3
" doubtful and bad, worth	664	14	4
Bills and cash on hand	10,876	14	9
Goods on hand, and on the way	3,324	0	5
	£23,635	14	9
Amount at the debit of the foreign houses for consignments	49,749	13	5
	£73,385	8	2

From these accounts it appeared that there was every probability that, if time was given, all debts would be paid in full. Mr. Fustana and Mr. George Caluta were called in, and having given satisfactory answers to the questions submitted to them, they were asked what proposal they had to make, when they submitted the following, viz.:—To pay 4*s.* in the pound in August, 1857; 4*s.* in the pound in January, 1858; 4*s.* in the pound in July, 1858; 4*s.* in the pound in January, 1859; 4*s.* in the pound in July, 1859. It was resolved, that the above proposal be accepted. It was further resolved, that, in order to secure the funds of the concern, the same shall be placed, as received, in the Royal Bank, in the names of the Chairman and Mr. Stefano Franghiadi, but, in all other respects, Messrs. Fustana and Caluta be left to wind-up the business without inspection. That Mr. Bewley be authorized to proceed to London to explain to the Ottoman Bank, and other creditors there, the position of the affairs, and that he convey to them, and the other absent creditors, the importance of their acceding to the above resolutions.

(Signed) Wm. Courtnay Cruttenden, Chairman.

THE ESTATE OF MESSRS. C. FRANGHIADI SONS.

A meeting of the creditors of Messrs. C. Franghiadi Sons, Greek merchants, who failed on the 20th of February, 1857, was held on the 11th of March, Mr. Diggles presiding, when the following statement was presented by Mr. J. E. Coleman, the accountant:—

STATEMENT OF THE AFFAIRS OF C. FRANGHIADI SONS, OF GRESHAM HOUSE, OLD BROAD STREET, MERCHANTS, FEBRUARY 20, 1857.

DEBTOR.

	£ s. d.	£ s. d.
To creditors on open accounts		£12,754 14 8
To creditors on bills payable	£86,109 15 6	
Less debit balances against same	7,721 9 2	
		78,388 6 4
To creditors partially secured	£770 9 10	
Less security held	614 12 6	
		155 17 4
To creditors fully secured—		
Security held	£21,515 4 8	
Claims	17,874 19 7	
To *contra*	£3,640 5 1	
To liabilities on bills receivable, £114,536 8*s*. 9*d*., of which it is expected there will rank on the estate		24,760 0 0
		£116,058 18 4

CREDITOR.

	£ s. d.	£ s. d.
By cash		£573 8 5
By debtors (good)		1,578 3 7
By ditto (doubtful)	£1,517 14 2	
By ditto (bad)	6,065 17 6	
By bills receivable (good)		785 0 0
By ditto (doubtful)	3,800 0 0	
By assets, consisting of wheat, etc.		9,243 7 7
By surplus on securities held by creditors, *contra*		3,640 5 1
		£15,820 4 8
Less creditors under £10, and amounts to be paid in full		261 14 5
		£15,558 10 3
Amounts due by parties after they have provided for their draughts		5,639 19 7
Amounts due from Franghiadi and Valenti, of Alexandria		61,271 17 11
		£82,470 7 9
Balance in the hands of parties who are holders of bills receivable		817 0 8

It was explained that, although the balance-sheet showed liabilities to the extent of £116,000, with assets representing £82,000, these figures would have to be modified, owing to the position of the firms abroad. There were three houses in addition to the London establishment—one at Galatz (presumed to be solvent), a second at Alexandria, and a third at Trieste. The house at Trieste had stopped, and the liabilities would be increased through this event about £9,000, making a total of £125,000. Against this amount, the assets in London were taken at £15,550, and at Trieste at £3,500, while Mr. Coleman had thought it prudent, in consequence of the Trieste house being creditors for £32,000, to take the Alexandrian balance at £29,000, instead of £61,271, as it stood on the books. In this shape the accounts would show £48,050 to meet £125,000; and if the house of Franghiadi and Valenti, of Alexandria, was on a settlement, liable for, or in a position to pay, a greater amount, the excess would increase the general dividend. Since the London house stopped, £6,850 had been received, principally from Alexandria, and this, together with the £15,500, made £22,300, which it was thought might be safely realized, allowing a distribution, if the creditors concur, of 2*s*. in the pound at no distant period. As some of the claimants had commenced proceedings in bankruptcy, it was suggested that it would be desirable to ascertain what steps they intended to pursue, and Mr. Lavie, supported by the majority, explained that recourse to that tribunal would not improve the assets, or facilitate the administration of the estate. A long discussion ensued with regard to the manner in which one or two transactions have been conducted, and the parties opposing said they could not pledge themselves to accede to the mode of liquidation suggested. In answer to questions, it was stated that the firm commenced with a capital of £12,500, and that the losses and expenses had included from £50,000 to £60,000, £24,000 of which are through liabilities on bills. A balance was struck on the 1st of January, but several large sums were then kept under the head of suspense account instead of being carried to profit and loss. After some further conversation with regard to the expediency of despatching an agent to Alexandria, to ascertain the position of the house there, the annexed resolution was agreed to:—

"That the following gentlemen—Mr. Westmoreland, Mr. E. G. Franghiadi (Franghiadi and Rodocanachi), and Mr. Lazard—do form a committee, with authority either to select some competent person to go out to Alexandria or to forward instructions to some house at Alexandria with a view to ascertain the true position of the firm there, and to obtain the remittance of funds to the committee as quickly as possible." In the meantime the assets are to be realized, under the direction of the committee, with a view to an early dividend, and all funds to be deposited at the Bank of England, in the joint names of Mr. G. C. Franghiadi and Mr. Coleman.

The firm was subsequently made bankrupt, and the following is the balance-sheet:—

Balance-sheet from January, 1853, to March 14, 1857.

DEBTOR.

To creditors				£91,498	5	8
To creditors holding security				13,768	9	0
To capital				12,500	0	0
To profits				13,978	4	11
To liabilities on bills receivable	£82,939	16	7			
Of which it is expected there will rank on this estate				25,260	0	0
				£157,104	19	7

CREDITOR.

By debtors, good				£4,242	12	2
" doubtful	£4,939	0	4			
" bad	6,498	9	5			
	£11,437	9	9			
Estimated at 2*s.* 6*d.* in the pound				1,429	13	2
By property				14,748	3	8
By cash to official assignee				3,656	4	6
By property in the hands of creditors holding security				15,219	16	6
By amount due from Franghiadi and Válenti				23,245	14	6
By M. G. Valenti				3,412	4	4
By trade expenses				20,583	10	10
By losses				40,136	18	11
By excepted articles				20	0	0
By liabilities as *per contra*				25,260	0	0
By G. C. Franghiadi—drawing account				5,150	0	0
				£157,104	18	7

THE ESTATE OF P. SINANIDES AND CO.

A meeting of the creditors of P. Sinanides and Co., Greek merchants, who failed on the 21st of February, 1857, was held on the 10th of March, when an adjournment of three weeks was agreed to, with the view of receiving advices from Alexandria regarding the debt of P. H. Andrea and Son, amounting to £20,800.

The assets, irrespective of this sum, showed about 2*s*. in the pound, but it was hoped that they will be further increased. The creditors present expressed an opinion in favour of a liquidation under inspection in preference to an appeal to bankruptcy. Subjoined is the statement submitted by Mr. Quilter, the accountant:—

BALANCE-SHEET.

DEBTOR.

To creditors unsecured		£30,855 8 9
To creditors partly secured, viz.:—		
Claim	£5,125 17 11	
Less securities	2,623 16 8	
		2,502 1 3
To creditors fully secured, viz.:—		
Securities	£9,580 0 0	
Less claims	7,499 10 0	
Surplus taken as an asset *per contra*	£2,080 10 0	
To holders of our acceptances of P. H. Andrea and Son's drafts		3,183 17 3
To liabilities on bills receivable—		
Considered good	£21,431 17 6	
To liabilities on bills payable for our acceptances of H. C. H. Moyssi's drafts outstanding in the excess of the estimated balance owing to him, viz.:—		
Acceptances	£4.600 0 0	
Estimated balance	2,502 1 3	
	£2,097 18 9	
		£36,541 7 3

CREDITOR.

By cash and bills in hand, viz.:—		
At Bank of England	£9 18 1	
Bills receivable	1,139 16 0	
		£1,149 14 1
By debtors considered good		652 12 2
" considered bad	£10,979 15 2	
	£10,979 15 2	
By debtors in respect of bad liabilities	11,895 13 1	
By surplus securities in hands of creditors, *per contra*		2,080 10 0
By P. H. Andrea and Son:—		
Balance of account current, as per statement..	£17,709 11 3	
Bills accepted on their account	3,183 17 3	
		20,893 8 6
By shares—cost		300 0 0
By deficieney carried over		11,465 2 6
		£36,541 7 3

THE ESTATE OF MESSRS. SWAYNE AND BOVILL.

A meeting of the creditors of Messrs. Swayne and Bovill, merchants and engineers, who failed on the 3rd March, 1857, was held on the 9th, when it was agreed to wind-up the estate under inspection, the following gentlemen being selected to superintend the process:—Mr. Lockett, Mr. J. Cary, Mr. Rix, Mr. J. Freeman, and Mr. F. Bramwell. The whole of the discussion was in relation to the prospects of dividend, and the measures to be adopted for completing the works in progress, since the results of these, together with the proceeds of various patents, will greatly assist to increase the total assets. The annexed figures represent the principal items in the approximate statement submitted by Mr. Turquand, the accountant :—

To sundry creditors unsecured		£46,352 6 8
Ditto holding security as *per contra*	£80,304 9 0	
To liabilities		14,743 0 0
		£61,095 6 8
By cash balance		100 0 0
Debtors, good		10,648 7 5
Ditto, doubtful	£11,254 2 0 At 10*s.* per pound	5,627 1 0
Consignments of iron surplus over advances ...		3,700 0 0
Property unencumbered		19,147 5 6
		£39,222 13 11
Deduct estimated payments required to complete works		1,709 0 0
		£37,522 13 11
By property held as security *per contra*	£181,585 17 0	
By patents, etc., not included in the above assets, estimated probable value	90,000 0 0	

THE ESTATE OF MESSRS. COPLAND, BARNES, AND CO.

A meeting of the creditors of Messrs. Copland, Barnes, and Co., provision merchants, who failed on the 4th of March, 1857, was held on the 19th of March, Mr. Harker presiding, when the following statement was submitted by Mr. Pullein (of the firm of Harding and Pullein), the accountant :—

STATEMENT OF THE AFFAIRS OF COPLAND, BARNES, AND CO., OF BOTOLPH LANE, LONDON, PROVISION MERCHANTS, MARCH 4, 1857.

LIABILITIES.—DEBTOR.

To creditors unsecured		£25,298 19 11
Creditors partially secured	£524 11 11	
Less value of securities	408 0 0	
		116 11 11
To creditors fully secured	18,357 3 2	
Value of securities	21,231 8 11	
Surplus, see *contra*	2,874 5 9	
To creditors to be paid in full :—		
Rent and salaries	474 12 7	
Creditors under £10	181 10 7	
See *contra*	656 3 2	
To liabilities on bills receivable, etc.	9,610 10 9	
Less amount not expected to be claimed against the estate	8,157 14 1	
		1,452 16 8
		£26,868 8 6

ASSETS.—CREDITOR.

By debtors—considered good		£2,542 12 10
" doubtful and bad	£1,657 0 7	
By cash in hand		139 5 0
By stock in trade, etc.		4,322 15 8
Surplus securities held by creditors		2,874 5 9
		£9,878 19 3
Less creditors to be paid in full as *per contra*		656 3 2
		£9,222 16 1

It was explained that the assets show a dividend of about 6*s*. 8*d*. in the pound, and that the losses have been occasioned by the rise in the prices of provisions, the firm having a contract with the Royal Mail Steam Packet Company. The estimated distribution, it is thought, will be realized, since the company seem disposed to take the remaining stores, both at home and on the intercolonial stations, at a fair price. In answer to questions, it was mentioned that the private estates show no surplus, and that the amount of capital originally introduced has not been traced, the house having been upwards of thirty years in business. After hearing further details from Mr. Nicol (of the firm of Messrs. Allen and Nicol), relative to the cessation of the contract, and the manner in which it had been carried out, several creditors proposed a liquidation by inspection, but it having been suggested that an investigation is essential before coming to a final determination, the annexed resolution was ultimately agreed to:—

"That Messrs. Harker, Strong, Snelling, Thomas Pennick, Alexander A. Rattray, and Charles Phillips, be appointed a committee, three to be a quorum, to investigate the affairs of Messrs. Copland, Barnes, and Co., and report the same to a meeting of the creditors to be called for the purpose."

At an adjourned meeting, on the 27th of March, of the creditors of Messrs. Copland and Barnes, provision contractors, who failed on the 5th, it was agreed to wind-up the estate through a petition for private arrangement, under the control of the Bankruptcy Court, Messrs. Harker, Anderson, Ward, and C. J. Phillips being appointed inspectors and trustees. The loss by the Royal Mail Steam contract appears to have been £25,000, and the firm is stated three years ago to have been solvent. The firm was subsequently made bankrupt, and the following is the balance-sheet:—

BALANCE-SHEET, FROM DECEMBER 31, 1853, TO APRIL 3, 1857, THE DATE OF ADJUDICATION.

DEBTOR.

To creditors unsecured		£28,803 11 0
To creditors fully secured		144 6 10
To creditor partially secured (Royal Mail Steam Packet Company)		16,909 9 5
To creditors to be paid in full	£262 11 7	
To liabilities		1,452 16 8
To surplus on the 31st December, 1853		1,852 8 3
To profits		18,679 7 1
		£67,841 19 3

CREDITOR.

By debtors, good		£3,197 17 3
" doubtful		11 11 8
" bad (carried to losses below)	£510 2 5	
By cash paid to official assignee		439 2 5
By property realized	£2,910 5 2	
By ditto not realized	97 17 9	
Carried forward	3,008 2 11	3,648 11 4

Brought forward	£3,008 2 11	£3,648 11 4
Less—to be paid in full	262 11 7	
		2,745 11 4
Property held by creditor fully secured		190 5 8
Property held by Royal Mail Steam Packet Company		12,057 18 4
Liabilities as *per contra*		1,452 16 8
Trade expenses		10,805 8 6
Discounts		2,558 12 5
Law costs		138 0 6
Partners' drawings—Charles Copland	£2,815 17 1	
William George Barnes	2,963 0 4	
		5,778 17 5
Losses—On intercolonial contract	£27,178 9 10	
On realization of stock, etc.	777 4 10	
Bad debts	510 2 5	
		28,465 17 1
		£67,841 19 3

THE ESTATE OF MR. W. PITCHER.

A meeting of the creditors of Mr. William Pitcher, of Northfleet, shipowner, etc., who suspended shortly previous, was held on the 19th of May, 1857, Mr. White presiding, when the subjoined statement was presented by Mr. J. E. Coleman, the accountant:—

STATEMENT OF THE AFFAIRS OF WILLIAM PITCHER, OF NORTHFLEET, KENT, SHIPOWNER, ETC., MAY 9, 1857:—

DEBTOR.

To creditors unsecured		£60,803 9 6
Creditors partially secured	£93,185 6 9	
Estimated value of security held	59,010 0 0	
		34,175 6 9
Creditors fully secured—		
Estimated value of security held	10,950 0 0	
Amount of claims	6,000 0 0	
See *contra*	£4,950 0 0	
Liabilities on bills received	1,824 7 1	
Expected to be duly honoured at maturity.		
		£94,978 16 3

CREDITOR.

By cash balances		£81 7 2
West of Ireland Fishing and Fish Manure Company's shares, estimated at		100 0 0
Loose tools, stores, etc., estimated at		20,200 0 0
Debtors, good		765 0 0
Estimated amount to be received on completion of vessel now building for the Russian Government, say		1,400 0 0
Surplus security with creditors *contra*		4,950 0 0
		£27,496 7 2
Amounts under £10, and to be paid in full	£138 17 11	
Salaries, taxes, rates, and other charges	939 15 9	
		1,078 13 8
Household furniture, etc., stands at cost	1,023 3 2	
		£26,417 13 6

It was explained that the assets will realize about 5*s*. in the pound if the stock be sold by auction, and another valuation (could arrangements be made to secure an incoming purchaser) shows about 10*s*. In 1852 the estate exhibited a surplus of £13,000, but losses have since occurred, and these, with the interest on borrowed capital, private expenditure, and the difference in cost and estimated value of premises, have caused the deficiency. One creditor complained of the manner in which his debt had been contracted shortly before the suspension. A suggestion was made to wind-up by inspectorship, but several creditors thought that an offer of composition would be accepted. Mr. Coleman stated that if the creditors would agree to take 6*s*. 8*d*. in the pound, he would endeavour to make arrangements for the payment of that amount within two months. The majority supporting this proposition, a committee was appointed to carry it out, and a memorandum was drawn up, and signed by many of the parties present.

THE ESTATE OF MESSRS. GOTCH AND SONS.

The creditors of Messrs. Gotch and Sons, the bankers, of Kettering, agreed to wind-up that estate by inspection. By this course it was thought a dividend of 10*s*. in the pound would be realized. The debts and liabilities of the firm were stated to be £132,026, and the assets £82,003. As in many previous instances this suspension arose through advances made without adequate securities. The following are the statement and report prepared by the accountants, Messrs. Harding and Pullein:—

Statement of Affairs, June 9th, 1857.

DEBTOR.

To sundry creditors unsecured				£132,026	9	6
To sundry creditors holding security	£4,600	0	0			
To estimated value of securities held, as *per contra*	7,580	0	0			
Surplus, as *per contra*	£2,980	0	0			
To amount belonging to Kettering Savings' Bank payable in full and deducted from the assets, as *per contra*	£427	18	4			
To liabilities of bills rediscounted, but which are expected to be paid by the acceptors	£8,773	4	8			
				£132,026	9	6

CREDITOR.

By cash in hand	£880	0	0			
By bills receivable, considered good	1,548	6	5			
Less	2,428	7	1			
Amount immediately payable to Kettering Savings' Bank, as *per contra*	427	18	4			
				£2,000	8	9
Sundry debtors—						
Considered good				23,848	5	6
" doubtful	72,174	11	0			
Estimated to realize 5*s*. in the pound				18,043	12	9
Considered bad	21,010	5	9			
Property held by creditors—						
Estimated to realize	7,580	0	0			
Less amount of charges thereon	4,600	0	0			
Surplus				2,980	0	0
Carried forward				£46,872	7	0

Brought forward				£46,872	7	0
Property unencumbered	3,120	0	0			
Stock at tan works and factory.....................	20,960	17	5			
Goods consigned to Australia, etc.	3,000	0	0			
Stock and plant at Rowell Brewery..............	1,500	0	0			
				28,580	17	5
Estimated surplus from the separate estate of Mr. J. D. Gotch....................................	3,375	0	0			
Ditto of Mr. T. H. Gotch...........................	3,175	0	0			
				6,550	0	0
				£82,003	4	5

12*s.* 6*d.* in the pound will require £82,516 10*s.* 11*d.*

The accountants' report, of which the following is a copy, was presented by Mr. Harding:—

"To the Committee of Creditors of Messrs. J. D. and T. H. Gotch.

"Gentlemen,—In pursuance of your instructions, we have prepared a statement, showing the present position of the affairs of Messrs. J. D. and T. H. Gotch, of Kettering, by which it will be found that there is now due, in respect of deposits and balances upon drawing accounts with the bank, £110,306 19*s.* 2*d.* The amount of notes in circulation on the 8th instant, at the time of the suspension, was £9,805; of these there have been since tendered by debtors the sum of £2,215, which they claim to be entitled to set off against the balances due from them; upon these being allowed, the amount of notes out will be reduced to £7,590. The amount due to creditors at the tan works, manufactory, and brewery, is £14,129 10*s.* 4*d.*, making a total of £132,026 9*s.* 6*d.* due to unsecured creditors. The claims of creditors of the firm who are fully secured amount to £4,600, and there is a sum of £427 18*s.* 4*d.*, moneys belonging to the Kettering Savings' Bank, which, under the Savings' Bank Act (3rd William IV.), must be paid in full forthwith; that amount has, therefore, been deducted from the assets. The firm are also liable as the endorsees of bills and promissory notes not yet due, and which amount to £8,773 4*s.* 8*d.*; but it is not expected that any claim will arise from these bills against the estate. The assets consist of cash in hand and bills receivable, amounting together to £2,428 7*s.* 1*d.*, but from this sum we have deducted the amount payable to the trustees of the savings' bank, leaving a balance of £2,000 8*s.* 9*d.* There is due from sundry debtors (considered good), in respect of overdrawn accounts at the bank, and overdue bills and promissory notes, the sum of £17,531 17*s.* 11*d.*, and from debtors (also considered good), for goods supplied from the tan works and manufactory, £5,281 7*s.* 7*d.* The books of the brewery at Rowell being considerably in arrear, we have been compelled to estimate the good accounts due to that business at about £2,500, and we have no reason to suppose that, when the accounts have been completed, the debts will fall short of the estimate. The total amount of debts considered good is £23,845 5*s.* 6*d.* The doubtful debts amount to the sum of £72,174 11*s.*; and, after a careful inquiry into the circumstances of the debtors, we have every reason to believe that about £18,000 may be realized. Although each debt was considered separately, it will be seen that an average of 5*s.* in the pound will give a somewhat similar result. The bad debts amount to £21,010 5*s.* 9*d.* The property held by secured creditors is estimated to realize £7,580, and after payment of the mortgages, etc., thereon, is expected to leave a surplus of £2,980. The other property belonging to the firm consists of plant and stock in trade at the tan works, factory, and brewery, and freehold and copyhold property unencumbered; these, together, are estimated to realize the sum of £25,580 17*s.* 5*d.*, and there are goods consigned to Australia, Lima, and Natal, which are estimated to realize £3,000. In addition, the sum of £3,375 is expected to arise from the separate estate of Mr. J. D. Gotch; and the sum of £3,175 is expected to arise from the separate estate of Mr. T. H. Gotch, making a total of £82,003 4*s.* 5*d.* for assets available for distribution, which, if carefully realized, will be equivalent to 12*s.* 5*d.* in the pound. The separate estates comprise the freehold and copyhold farm land in Rothwell and Kettering, and some other copyholds together, of the estimated value of £18,750; but there are mortgages, etc., thereon, to the amount of £10,000, and the balance is subject to the payment of £3,000 to the Rev. F. W. Gotch, and for which he holds a joint and several bond—this bond was given about November, 1852, in pursuance of the terms of the will of the late Mr. Gotch, who bequeathed that amount to his son. The other portion of the separate estates consists of household furniture and effects, etc. The difference between the amount of the liabilities and of the assets will be found to amount to £50,023 5*s.* 1*d.*, and if to the amount of the bad debts (£21,010 5*s.* 9*d.*) be added the difference between the amount of the doubtful debts and the sum now expected to be realized therefrom (£54,130 18*s.* 3*d.*), it will be seen that the losses by debtors amount to £75,141 4*s.*, being £25,117 18*s.* 11*d.* beyond the present deficiency; this was about the amount appearing in the books as the aggregate capital of the partners in June, 1852. Although the debtors are, as may be expected, very numerous, we find that £46,417 2*s.* 4*d.* (almost the amount of the deficiency) is due from two persons only—one a well-known clergyman, who formerly lived near Kettering, the other a farmer and shoe-factor in the neighbourhood, also well known. In the case first mentioned, we find that, at the decease of the late Mr. Gotch, in May, 1852, there was due the sum of £2,852 11*s.* 6*d.* About December, 1852, he left this neighbourhood for a foreign country, where he has since resided; and, shortly after his arrival there, he appears to have engaged in patents, mines, and other speculative undertakings. He at first applied for small advances, holding out promises of immediate repayment, and that the old balance due from him would speedily be liquidated. These drafts continued to increase, without a single repayment, and, although no draft exceeded £200 in amount, yet the account has increased by this course to its present magnitude (£24,892 6*s.* 2*d.*). The only reason apparent for the continued payments appears to be the fear that any refusal would involve the loss of the whole debt, and the belief, arising from the individual's representations, that the enterprises in which he appeared to be engaged would enable him to repay every shilling in a very short time. We have perused the letters received from the person in question since January, 1853, and find that, on many occasions, he stated that his property was far more than sufficient to pay Messrs. Gotch; but it does not appear that on any occasion those statements were tested by investigation, or that any steps were taken to compel repayment of the amount drawn. Whether or not any considerable portion of the debt will be recovered, is still a matter of uncertainty. The security held will not produce more than £1,000. In the second case, the sum of £14,902 3*s.* 10*d.* was due at the death of Mr. Gotch, and must then have been very doubtful, yet the account has since been allowed to increase to £21,524 16*s.* 2*d.*, the amount now due. From this debt but a small dividend is anticipated. The assets, as before mentioned, show nearly 12*s.* 6*d.* in the pound, and that amount will be increased in the event of either of these large debts producing more than 2*s.* in the pound. We have carefully examined into the recent transactions of the firm, with the view of ascertaining whether or not any preferential payments have been made, and we have the satisfaction of being able to state that nothing of the kind has taken place; indeed, we have every reason to believe that the suspension of payment was not contemplated many hours before it actually took place, and was not determined upon until after four o'clock on the Monday afternoon (the 8th instant).

"Upon investigating the position of the affairs of the firm in May, 1852, the date of the death of the late Mr. Gotch, we find that a considerable portion of the debts now considered bad and doubtful were then standing in the books, yet we have every reason to suppose that, had the operations of the firm been then closed and wound-up, the estate would have been solvent, and that the real estate of the late Mr. Gotch would have remained available for the trusts of his will. The bequests amounted to £9,000, of which £2,500 only have been paid. The sum of £3,000 is still payable to the Rev. E. W. Gotch, and for which he holds a bond, as before mentioned. The balance passed upon the death of Mrs. Gotch and Miss Frances Gotch to J. D. Gotch and T. H. Gotch, and has remained in the estate. In conclusion, we think it due to the Messrs. Gotch to say, that all their books and papers have been placed at our disposal, that every facility has been afforded by them for the most searching investigation, and that they have most readily given the information we desired; in fact, that they have submitted themselves and their affairs to their creditors, at the request of the committee, as completely as if an adjudication in bankruptcy had taken place.

"We are, gentlemen, your obedient servants,

"HARDING AND PULLEIN.

The estate was ultimately wound-up in bankruptcy.

THE ESTATE OF MESSRS. EVANS, HOARE, AND CO.

A numerous meeting of the creditors of Messrs. Evans, Hoare, and Co., engaged in the Australian trade, who suspended business on the 9th of June, 1857, was held on the 3rd of July, Captain Denny in the chair, when the following statement, prepared by Messrs. Turquand and Young, the accountants, was exhibited. Mr. J. Linklater represented the firm ; and Mr. E. Lawrance attended on behalf of Mr. Hoare :—

STATEMENT OF AFFAIRS, 10TH JUNE, 1857.

DEBTOR.

To creditors unsecured						£59,252	2	4
To creditors partially secured						19,641	3	2
To creditors fully secured—Claims		£77,665	11	10				
To invoice value of security		107,073	1	10				
Per contra		£29,407	10	0				
To liabilities on bills—Payable		£13,954	16	5				
Of which may prove against the estate						3,500	0	0
To liabilities on bills—Receivable		72,124	16	3				
Considered good.								
To liabilities to creditors of the firm of E. D. Moore and Co., estimated at		500	0	0				
but which will be paid by that firm.								
To liabilities in respect of lease of premises at Southgate Street, held on lease for — years, annual rent		350	0	0				
						£82,393	5	6

CREDITOR.

By debtors						£2,572	1	2
Good		£822	1	2				
Doubtful and bad	£22,299 5 11							
Estimated to realize		1,750	0	0				
		£2,572	1	2				
By cash						84	17	2
By stock of wines, etc., as per statement						1,740	2	11
By stock in the hands of creditor								
By estimated surplus, as per statement						146	4	9
By consignments unencumbered at cost, as per statement						1,153	12	1
By office furniture and fittings, valued at						75	0	0
By surplus securities, *per contra*, being consignments at invoice price						29,407	10	0
By ditto at cost price		£—						
By value of our interest in the firm of E. D. Moore and Co., estimated, if sold, to realize						1,500	0	0
						£36,679	8	1
Deduct (to be paid in full)—								
Dock charges		£368	18	7				
Salaries, etc.		245	13	11				
						614	12	6
						£36,064	15	7

The creditors, at the commencement of the proceedings, complained of the delay in preparing the accounts, particularly as it was well ascertained that the assets are all of a doubtful nature. When the figures were explained their dissatisfaction increased, the prospects of the result of the realization of the consignments abroad being considered discouraging. It was intimated that Messrs. Evans, Hoare, and Co. are not in a position to make any proposal for an arrangement, and, consequently, their only course is to submit to whatever steps are dictated. Great sympathy was expressed for Mr. Hoare, who took into the firm £5,000, besides a credit for £6,000, which has been absorbed—his family likewise being claimants for goods supplied to the extent of £18,000. Mr. Lawrance placed these facts before the meeting on behalf of Mr. Hoare, and stated that his friends would have endeavoured to assist the firm in making a composition, but for the disastrous position in which the assets were placed. Some discussion occurred with regard to the probable returns from Australia, whither the principal consignments, beer, spirits, etc., have been sent, and a great difference of opinion prevailed. Several creditors alleged that goods had been obtained from them a few weeks before the suspension, and when it was evident the house would be compelled to pull up. In answer to questions, it was stated that the deficiency under the old estate of Mr. Evans now proved to be £16,000, but that he estimated his capital in December, 1855, taking the value of consignments, etc., into consideration, at £22,000. On an appeal to Mr. Linklater for his views, he said the only suggestion that could be made would be in favour of a liquidation by inspection ; but the great majority of the creditors dissented, urging that the estate should be administered in bankruptcy. A resolution to this effect was then passed, and the necessary measures for facilitating the proceedings will at once be adopted. Proceedings in bankruptcy were subsequently carried out.

THE ESTATE OF MESSRS. FORSTER, RUTTY, HALL, AND CO.

A meeting of the creditors of the above-named firm, who failed in the Scotch and Manchester trade, was held on the 14th of August, 1857, when the following statement of affairs was presented by Mr. Parrinton, of the firm of Parrinton, Ladbury, and Co., the accountants :—

DEBTOR.

	£	s.	d.	£	s.	d.
To creditors	£48,282	11	5			
Less securities held by them	3,154	7	2	£45,128	4	3
To liabilities on bills drawn upon and accepted by other persons for the accommodation of Messrs. Forster, Rutty, Hall, and Co.	10,000	17	6			
Less value to acceptors, who it is supposed will pay their acceptances	1,360	16	1	8,640	1	5
To liabilities on bills receivable, discounted, considered good	14,818	3	4			
To ditto ditto bad				1,114	0	2
To liability on promissory notes of the firm delivered to the acceptors of accommodation bills	2,817	4	11			
				£54,882	5	10

CREDITOR.

	£	s.	d.	£	s.	d.
By stock in trade at cost				£11,341	2	3
By book debts, good				6,283	15	8
Ditto doubtful	£98	19	1	49	9	6
Ditto bad	566	12	2			
By cash and bills in hand				8	16	3
By balance at bankers' (deducted from their claim)	69	18	0			
Carried forward				£17,683	3	8

	£	s.	d.	£	s.	d.
Brought forward				£17,683	3	8
By trade furniture, etc.				171	9	0
By bills received (dishonoured) amounting to £3,419 19*s.* 2*d.*, held by acceptors of accommodation bills as collateral security	£0	0	0			
By bills receivable, amounting to £1,906 2*s.* 6*d.*, estimated at	1,654	7	2			
Held by creditors as security upon advances of £1,350 :—						
By leases and fixtures of premises, Nos. 64 and 65, Friday Street, and No. 40, Cheapside, held for unexpired terms of 17, 12, and 12 years respectively, at a total rent of £930 per annum	1,500	0	0			
Mortgaged to Mr. J. Thomas, as security for his debt.						
Deducted *per contra*	£3,154	7	2			
				17,854	12	8
Less rent, taxes, salaries, etc.				902	13	10
				£16,951	18	10

It was explained that the estate shows about 6*s.* in the pound, but it is questioned whether this amount will be realized. The deficiency to be accounted for is £37,000, in addition to profits stated to have been made during the eight years' trading of £54,000. Mr. Parrington gave a most discouraging narrative of the condition of the firm, with regard to their mode of conducting business, the state of the books, and the drawings of the partners. No proper balance-sheet has ever been taken since the commencement of the trading. The cash account has been traced through the counterfoils of cheques ; and to bring the ledgers, etc., into a satisfactory condition, three months' labour must, it is alleged, be expended. The present statement must, therefore, be considered more as approximating to, than verifying, actual results. It is not presumed that any dishonesty or concealment has been practised, but the great recklessness exhibited will render it necessary that a strict investigation of the accounts shall be carried out. The loans and repayments have been of a most extensive nature, and the absence of good bookkeeping has, consequently, increased the confusion. The principal parties who are secured are the bankers, and a relative of Mr. Forster's, who has advanced about £8,000. It appears that the drawings of the partners gradually increased as their business extended, although it was conducted in such an unsatisfactory manner, and without prospect of ultimate amendment. The joint estate may be benefited to the extent of £300 by a surplus from the private estate of Mr. Forster, but those of Mr. Rutty and Mr. Hall are fully absorbed by claims upon them. A lengthened conversation then took place as to the best course to be adopted for effecting a liquidation. It was proposed, and eventually resolved, that the partners should be requested to sign a declaration of insolvency, that the estate should be wound-up by inspection, with the view of facilitating an early distribution of assets, and that recourse should be had to bankruptcy, if the accounts, on investigation, should exhibit features requiring the interference of that tribunal. The suggestion for an inspectorship emanated from Mr. J. Linklater, who considered it a more desirable proceeding than an ordinary assignment. The three partners were then introduced, and having signed the declaration of insolvency, some of the creditors interrogated them on various points. The most important facts elicited were the admission by Mr. Rutty that the firm had known for several years they were in a doubtful position, and that, according to a rough statement which he had prepared, the deficiency, etc., in the accounts would be found to consist of the expenses of business, interest, and discount, and discount allowed to customers on purchases, etc. Inspectors, with the general concurrence of the creditors, having been appointed, the meeting adjourned.

THE ESTATE OF MESSRS. CARR BROTHERS AND CO.

A general meeting of the creditors of Messrs. Carr Brothers, of Newcastle-upon-Tyne, coal owners, etc., was held on Tuesday, August 18th, 1857, Mr. Samuel George Smith, of the firm of Smith, Payne, and Smith, in the chair, when the following statement was presented by Mr. J. E. Coleman, the accountant. Of creditors amounting to £670,000, about £540,000 were represented, the principal parties being the Bank of England, the Northumberland and Durham District Bank, the Commercial Bank of Scotland, Messrs. Carr, Glyn, and Co., the National Discount Company, Messrs. Bailey, etc.

GENERAL STATEMENT OF THE LIABILITIES AND ASSETS OF MESSRS. CARR BROTHERS AND CO., OF NEWCASTLE-UPON-TYNE, AUGUST 1, 1857.

FULL CLAIMS AND ASSETS.

	£	s.	d.	£	s.	d.
To amount due to creditors fully secured				£301,549	0	0
Ditto, partially secured				202,091	0	0
Ditto, on acceptances discounted				149,072	0	0
Ditto, on acceptances for colliery stores and working purposes				13,395	0	0
To creditors on open accounts				12,500	0	0
				£678,607	0	0
By value of collieries	£439,315	0	0			
By value of shares, etc.	179,990	0	0			
By coal, stock, and sundries	10,953	0	0			
By sundry debtors	66,382	0	0			
By surplus from the separate estates of the partners after payment of claims thereon	20,000	0	0			
	£716,640	0	0			
Liabilities on bills discounted, the whole of which it is expected will be duly honoured at maturity				40,947	9	1
				£719,554	9	1

CLAIMS AND ASSETS AFTER DEDUCTING SECURITIES HELD BY CREDITORS.

	£	s.	d.	£	s.	d.
To creditors wholly secured:—						
Estimated value of securities	£468,640	0	0			
Amount of claim	301,549	0	9			
Surplus	£167,090	19	3			
To creditors partially secured:—						
Amount of claims	202,091	0	0			
Estimated value of securities	75,665	0	0			
				£126,426	0	0
To creditors on acceptances				149,072	0	0
To ditto for colliery purposes				13,395	0	0
To ditto on open accounts				12,500	0	0
				£301,393	0	0
By estimated surplus on English collieries and on shares				£167,091	0	0
By estimated value of Welsh mines				75,000	0	0
By coal, stock, and sundries				10,953	0	0
By sundry debtors estimated				66,382	0	0
Carried forward				£319,426	0	0

Brought forward	£319,426	0	0
By surplus from the separate estates of the partners, after payment of claims thereon	20,000	0	0
	£339,426	0	0
Liabilities on bills discounted, the whole of which it is expected will be duly honoured at maturity	£40,947	9	1

Mr. C. Freshfield stated the circumstances under which former consultations took place respecting the liquidation of the estate. When the bills of the firm were first dishonoured, Messrs. Smith, Payne, and Smith, as interested parties, proposed an investigation, and the Bank of England being also creditors, it was agreed that Mr. Coleman should be despatched to Newcastle to look into the accounts, and make a report. That object having been accomplished, the result was communicated to several of the principal creditors, and it was thought that it would be for the interests not only of Messrs. Carr Brothers, and the creditors, but of the locality itself, if an arrangement could be effected for a liquidation under inspection. Mr. Coleman, having fully investigated the estate, was now prepared to explain its position, in order that a conclusion might be arrived at. Mr. Coleman, in reply to the Chairman, announced that the estimate of the surplus of the property, which he before took at £40,000, will be now modified, and placed at £38,000. The principal assets consist of collieries in the North of England and in South Wales. The collieries in the North were valued in 1854, by Mr. Forster, at £466,000, but it is not considered prudent to take them at more than seven years' value of the produce; and, therefore, the amount is reduced to £364,315. With regard to the South Wales collieries, they had been valued at £116,000. Not being, however, in full operation, and the coal appearing less in demand than that obtained from the other collieries, the opinions of qualified persons have been accepted, who consider that they may safely be placed at £75,000. These two items constitute the total of £439,315 for the collieries. The shares described as valued at £179,990 consist of Blyth and Tyne Railway shares (the line having been originally the property of the firm), the preference descriptions paying ten per cent., and the ordinary descriptions six per cent. These have been taken at the market prices of the day, which, it is believed, form a fair representation of their worth, unless heavy sales take place, and they are forced to a depreciated point. The items of coal stock and sundry debtors will, it is expected, be realized, and the separate estates of Mr. Dryden and Mr. John Carr may yield £20,000. The debt of the Northumberland and Durham District Bank is about £200,000, for which they hold ample securities upon the collieries, but the directors are willing to remain in the same position in which they at present stand for twelve months, to see if the liquidation can be effected, reserving to themselves the right of receiving only the interest on the mortgage. Among the other secured creditors are the Edinburgh Insurance Company, the National Assurance Company, the North British Assurance Company, and Lord Ravensworth, most of whom hold Blyth and Tyne shares. The best mode of arranging the debts of the estate, it was contended, will be by inspection, an appeal to bankruptcy, under existing circumstances, being wholly out of the question, and the creditors were therefore urged to support this plan. In answer to questions, it was stated that the names of the partners in the collieries are Mr. John Carr, Mr. Charles Carr, Mr. W. Carr, Mr. Phillipson, Mr. Dryden, Mr. Pemberton, Mr. Burnett, and Mr. Roger, and that Mr Swan has not been admitted into the firm. A sum of £2,500 will be required to complete the workings of the South Wales Collieries; and the produce for the next year from the Northern collieries will, it is asserted, reach £45,000. A point was raised with respect to the acceptances of parties who had been connected with Messrs. Carr Brothers, and it was replied, that if not solvent, they will have to pay the penalty of their indiscretion. A long discussion ensued, relative to the responsibility of creditors in agreeing to a liquidation by inspection, if the property became deteriorated, or the results estimated were not realized; but it was asserted that, under the deed of inspection, the necessary precautions will be adopted to avoid any such contingency. As to general claims it was argued that one common course of action is desirable, the inspectors representing the entire body of creditors, a preponderance of whom were present on this occasion. It was then resolved that Mr. Bigg, of the Northumberland and Durham District Bank; Mr. Anderson, the manager of the branch of the Bank of England, at

Newcastle; and Mr. S. G. Smith, of the firm of Smith, Payne, and Smith, be appointed inspectors, with power to investigate and ascertain the position of the several parties who are liable upon bills, and to inform the holders of the same of the views they entertain of any propositions that may be made for a settlement, the holders to have seven days' option of assenting to such arrangements.

THE ESTATE OF MESSRS. THOMAS ASHMORE AND SONS.

A meeting of the creditors of Messrs. Thomas Ashmore and Sons, engaged in the drysaltery trade, who were compelled to suspend, was held on the 25th August, 1857, when the following statement was presented by Mr. H. Chatteris, the accountant :—

THOMAS ASHMORE AND SONS' STATEMENT OF AFFAIRS, AUGUST 7, 1857.

DEBTOR.

To creditors unsecured				£7,835	7	7
To creditors holding security	£26,719	6	10			
To value of securities in their hands	32,550	18	6			
Surplus carried to *contra*	£5,831	11	8			
To liability on dishonoured bills	£15,858	18	1			
To less dividends from other estates	12,869	2	2			
				2,989	15	11
To liability on bills which it is supposed will be duly honoured	£12,628	14	8			
				£10,825	3	6

CREDITOR.

By merchandise on hand	£235	7	11
By surplus property in the hands of creditors in excess of claims	5,831	11	8
By debtors	1,026	15	8
By cash	5	10	10
By office furniture	90	0	0
	£7,189	6	1
Less to be paid in full	66	2	9
	£7,123	3	4

It was explained that the accounts show a nominal dividend of about 13*s*. 4*d*., but that the amount which may be expected is 10*s*. in the pound. This difference will arise from losses on consignments, consisting of acids and argols, to America, where the operation of a new tariff has unfavourably influenced the value of these articles. The books of the firm, it is stated, have been well kept, and everything connected with their affairs has been conducted in an honest and straightforward manner. The house previousiy failed in 1846, when a composition of 12*s*. in the pound was accepted, and subsequently business was resumed with every appearance of success. In 1851, their accounts showed a surplus of £5,000, but it gradually decreased through bad debts and other unfavourable contingencies. At the commencement of the present year, the annual statement prepared by the partners exhibited a small balance in their favour, but it could not be considered accurate, as they had not allowed for bills upon which there were claims against them in association with other houses whose estates were winding-up. The creditors present having received these explanations as fully accounting for the position of the firm, a resolution was immediately passed in favour of an assignment, the trustees appointed being Mr. Crockat and Mr. H. Chatteris.

THE ESTATE OF MESSRS. MELROSE AND HUSSEY.

A meeting of the creditors of Messrs. Melrose and Hussey, ironfounders, whose failure was announced on the 6th, was held on the 26th August, 1857, when the following statement was submitted by Messrs. Thomas and Cates, the accountants:—

Statement of Affairs of Messrs. Melrose and Hussey, Phœnix Works, Tividale, near Dudley.

DEBTOR.

To creditors unsecured				£11,904	3	2
To creditors partially secured	£41,612	19	10			
Less value of security	22,808	5	10			
				18,804	14	0
To liabilities on bills receivable, considered good	£7,361	14	8			
				£30,708	17	2

CREDITOR.

By debtors, considered good				£2,593	10	6
Doubtful	£408	2	4			
By stock, plant, etc., realized under execution, say	12,006	0	0			
				£2,593	10	6

It appearing that the stock and plant have been seized by one of the principal creditors under execution, no compromise could be entertained, and it was resolved to wind-up the estate in the Court of Bankruptcy, which was eventually done.

THE ESTATE OF MESSRS. BRUFORD, DYER, AND CO., BRISTOL.

A meeting of the creditors of this firm was held on the 31st of August, 1857, at the banking-house of Messrs. Baillie, Baillie, and Co., Bristol, Mr. A. B. Savile, of the firm of Miles and Co., bankers, in the chair. Mr. Thomas, of the firm of Barnard, Thomas, and Co., accountants, reported the result of his investigation, which showed as follows:—

Total Liabilities.

DEBTOR.

	Secured and Unsecured.			Unsecured.		
On acceptances and promissory notes	£118,390	9	0	£118,390	9	0
Open account	19,637	17	4	19,637	17	4
Acceptances, receivable, considered doubtful and bad	14,119	15	0	14,119	15	0
Creditors holding security	18,425	0	0			
Bills under discount	87,299	11	7	20,000	0	0
	£257,872	12	11	£172,148	1	4

CREDITOR.

	Total Assets.			Available Assets.		
Available assets	£89,845	0	0	£89,845	0	0
Property held in mortgage	18,425	0	0			
	£108,270	0	0	£89,845	0	0

Bad debts and losses ascertained	£57,717	6	5
Estimated loss on bills	20,000	0	0
	£77,717	6	5

Mr. J. G. Shaw thereupon moved, and Mr. T. P. Jose, master of the Society of Merchant Venturers, seconded the resolution, requiring the members of the firm forthwith to execute a deed of conveyance and assignment of all their respective joint and several real and personal estate and effects to Messrs. H. H. Hareford (of Miles and Co.), G. O. Edwards (of Baillie and Co.), John Bates (of the West of England and South Wales District Banking Company), W. G. Coles (of Stuckey's Banking Company), John George Shaw, Thomas Chope, and John Wood, merchants, Messrs. Henry Brittan and Son, and Savery, Clark, and Fussell, were appointed solicitors, and Messrs. Barnard, Thomas, and Co., accountants to the estate. The resolution was signed by or for nearly fifty creditors, who are interested to the extent of upwards of £220,000. It is expected that with the amount to be contributed by Mr. Beeston, one of the members of the firm, the dividend will be at least 13*s.* 4*d.* in the pound.

THE ESTATE OF MESSRS. ROSS, MITCHELL, AND CO.

A meeting of the creditors of Messrs. Ross, Mitchell, and Co., in the Canadian trade, who failed on the 13th of October, 1857, took place on the 23rd, at the Guildhall Coffee-house, Mr. D. Price in the chair, when the following statement, showing the surplus of £57,000, was presented by Mr. Henry Chatteris, the accountant:—

STATEMENT OF THE AFFAIRS OF ROSS, MITCHELL, AND CO., IN LONDON, ON OCTOBER 13, 1857; AND IN TORONTO, AS ON JANUARY 31, 1857.

DEBTOR.

To creditors in Great Britain, unsecured							£101,267	7	3
To creditors in Great Britain, for advances on acceptances							99,617	0	5
To creditors in Great Britain holding acceptances in favour of Toronto house							25,703	0	6
To creditors in Great Britain holding security				£6,522	17	9			
Less estimated value of securities				9,250	0	0			
Surplus carried to *contra*				£2,727	2	3			
To creditors in Great Britain partially secured				£3,973	17	0			
Less estimated value of securities				2,400	0	0			
							1,573	17	0
Total creditors in Great Britain							£228,161	5	2
To creditors in Canada on open accounts				£41,871	3	9			
To creditors in Canada on bills payable				23,838	14	6			
To creditors in Canada for advances	£60,168	6	9						
Less securities held	13,608	0	0						
				46,560	6	9			
Total creditors in Canada							112,270	5	0
Total creditors, carried forward							£340,431	10	2

Brought forward			£340,431 10 2
To liability upon acceptances which are expected to be proved upon this estate		£41,790 12 10	
To liability upon acceptances which should, and it is believed will, be met by the drawers	£22,837 10 11		
To liability upon bills receivable under discount	106,778 15 4		
Out of which it is expected there will be claimed of this estate	14,394 5 9		
Less cash in the hands of bill-holders	333 11 6		
		14,060 14 3	
Total proveable liabilities			55,851 7 1
Total claims on the estate			396,282 17 3
Surplus			57,290 12 11
			£453,573 10 2

Note.—In addition to the above surplus, Ross, Mitchell, and Co. will have a claim upon the estates of the various parties for whom they have incurred the liability of £55,851 7*s.* 1*d.* From these estates they expect to realize a considerable sum. They also estimate their net profits in Toronto since February last at £15,000.

CREDITOR.

By freehold property in Renfrew and Castleton, cost	£12,300 0 0	
Mortgaged for unpaid balance of purchase-money	4,000 0 0	
		8,300 0 0
By ditto premises in Toronto, estimated at	£6,480 0 0	
Mortgaged for	4,000 0 0	
		2,480 0 0
By surplus securities in the hands of creditors		2,727 2 3
By 105 bonds of £100 each of the Ontario and Simcoe Railway, estimated to be of the value of		7,875 0 0
By Canadian debtors who remit direct to the London house		25,622 13 7
By debt due from J. Ross and Co., Melbourne		38,673 10 10
By stock of goods not yet shipped		6,340 13 4
By bills receivable on hand		12,804 11 10
By cash in hand		120 0 0
By office furniture		117 18 2
By debts due to Toronto house considered good		118,596 13 8
By ditto doubtful	£52,084 9 6	
Estimated to realize		34,622 7 1
By stock of merchandise		27,859 13 8
By goods shipped to Toronto since February 1, in excess of remittances		29,133 6 0
By bills receivable, good		52,370 6 9
By ditto, doubtful	£28,683 17 5	
Estimated to realize		17,603 3 5
By cash in hand		4,396 16 0
By debts not at present available, but for which securities are held		56,729 13 7
By Ross, Mitchell, and Co.'s interest in grocery store at Toronto		7,200 0 0
		£453,573 10 2

It appeared from the report of Mr. Chatteris that the cause of stoppage was the sudden discontinuance of accommodation between the firms of Ross, Mitchell, and Co., and other parties who have lately failed. The house took stock annually, and at their last balance in February of the present year, the joint capital was £134,063. About three years ago one of their buyers came to this country and purchased goods to the extent of £100,000 beyond the amount he was instructed, and more than their demands required. This circumstance is alleged to be the first cause of their difficulty, and of their obtaining advances upon bills to meet their payments. The balance-sheet shows that the house have raised nearly £100,000 by the discount of their acceptances through the agency of parties who have since suspended. Exclusive of this amount there is a further sum of £23,000 accommodation bills, for which Ross, Mitchell, and Co. have received no value whatever. Mr. Chatteris proceeds to say:—

"The claims against the estate being £396,000, I have directed my attention to the nature of the assets which are available for meeting this large sum. The assets amount in round numbers to £450,000, and they may be classified as follows:—

Stock, debts, and convertible securities, which Messrs. Ross, Mitchell, and Co. consider may be realized within eighteen months from the present time	£329,470	0	0
Debt due from J. Ross and Co., Melbourne, which they consider may be realized within three years	38,673	0	0
Debts for which security is held, and which may be converted during the next three years	56,729	0	0
	£424,872	0	0
Which would be more than sufficient to meet every demand, and would leave other securities amounting to	£28,700	0	0

Available for any contingencies, besides the dividends Ross, Mitchell, and Co. will be entitled to receive upon the indirect liabilities they will be called upon to pay.

"Assisted by a report of Mr. Fisken (the partner in Toronto) as to the character of every one of the Canadian debts, I have dissected and divided the assets as you see them in the balance-sheet, and have written off a very large sum from the amounts at which they were taken down in the balance. It appears to me, therefore, that you may very safely assume that the assets generally will realize the sums at which they are set down, always supposing that no hurried or forced realization be insisted upon, for in that case it is probable that a very large loss would arise."

In answer to questions, it was stated that some time since, when Mr. Ross proposed to retire, the surplus was about £130,000, but that circumstances which have lately occurred have reduced the total to the present amount. The books have been well kept, and show in every respect the transactions of the firm. One or two of the creditors complained of the accommodation bills, which were regarded as an unsatisfactory feature. Mr. Reed, of the firm of Reed, Langford, and Marsden, explained the proposal which Ross, Mitchell, and Co. were desirous of making to satisfy their creditors. It was their intention to pay 20*s.* in the pound, with 6 per cent. interest, by instalments of 5*s.* at six, twelve, eighteen, and twenty-four months, from the 1st of January, 1858, allowing at the same time the appointment of a committee to inspect their books as they progressed with the liquidation. He stated that the date is placed in January, 1858, because it will take some time to obtain the consent of the creditors, and to prepare the promissory notes, while the firm will depend upon the proceeds of the winter trade to pay the dividend at one period, and those of the fall trade to pay the dividend at the other. In this way they hope to preserve punctuality on the 4th of July and the 4th of January, until the whole amount is discharged. A conversation followed, during which it was suggested that it would be better to place the estate under inspection at once, but it was asserted that this would defeat the object of the firm, who propose to continue business and incur fresh liabilities, the larger creditors being favourable to this arrangement. The committee appointed to look into the bills will, it is said, have the power of preventing any renewal of accommodation engagements, which have produced this disastrous result, and the house will now have to concentrate business more within the sphere of their legitimate resources. After some discussion, during which the representatives of several Scotch houses accorded their assent previously to the meeting which is to take place at Glasgow on Monday, the annexed resolutions were agreed to:—

"At a meeting of the creditors of Messrs. Ross, Mitchell, and Co., of Gresham Street, London, and of Toronto, Canada West, merchants, holden at the Guildhall Coffee-house, King Street, London, on Friday, the 23rd October, 1857, Mr. David Price in the chair, Mr. Chatteris, accountant, having read a statement of Messrs. Ross, Mitchell, and Co.'s affairs, and given various explanations respecting them, and Mr. Reed, solicitor, having informed the meeting that Messrs. Ross, Mitchell, and Co. proposed to pay their creditors the full amount of their debts, with interest at six per cent. per annum, by instalments of 5*s.* in the pound, at six, twelve, eighteen, and twenty-four months, from the 1st January, 1858, and to deliver the promissory notes for such instalments to the creditors on or before that day, and also that for the satisfaction of their creditors Messrs. Ross, Mitchell, and Co., proposed to submit their books and affairs from time to time until the instalments are paid, to the inspection of a committee to be appointed by this meeting:

"It was Resolved—That Messrs. Ross, Mitchell, and Co.'s proposal be accepted, and that Mr. David Price (Price, Coker, and Co.), John Dillon (Morrison, Dillon, and Co.), and John Porter Foster (Foster, Porter, and Co.), all of London, and Messrs. W. Wingate (Wingate and Son and Co.), and James Arthur (Arthur and Fraser), of Glasgow, be appointed a committee for the purposes mentioned in such proposal; but the agreement of creditors to this resolution is to be without prejudice to their claims against third parties, or to any securities they may hold, and to be subject to the consent of such third parties (when necessary) being obtained."

An adjourned meeting of the creditors of Messrs. Ross, Mitchell, and Co. was held at the Guildhall Coffee-house, on the 29th June, 1858.

A strong feeling was generally expressed against the mode in which the house have conducted their affairs, especially since this was the second occasion on which they have been compelled to seek the indulgence of their creditors. In addition, also, to the doubtful nature of the assets, which it was proposed should be transferred for the benefit of British claimants, the opinions expressed with regard to the state of the Canadian law were of importance, as conveying to the traders in those provinces the sentiments of the manufacturing interest concerning the system of preferences so extensively followed out. The chairman was Mr. Price, of the firm of Messrs. Price, Coker, and Co.; and among those who took a very prominent part in the proceedings were Mr. Dillon, Mr. Morley, Mr. Foster, Mr. Greatorex, and the representatives of the Western Bank of Scotland and the City of Glasgow Bank. Several legal gentlemen were in attendance to advise with the creditors on the best course to be pursued, the principal being Mr. Reed, of Messrs. Reed, Langford, and Marsden, and Mr. Murray, of Messrs. Murray, Son, and Hutchings. In accordance with the arrangements previously made, the committee appointed to superintend the original liquidation prepared a report, which was now brought forward by Mr. H. Chatteris, the accountant. That document was as follows:—

"*London, June* 29, 1858.

"GENTLEMEN,—The committee nominated by the creditors to inspect the books and affairs of Ross, Mitchell, and Co., until they paid their creditors 20*s.* in the pound and interest, have called you together to make known to you that this firm informs them that it is no longer in a position to pay 20*s.* in the pound. The communication was made to them in a letter, dated Toronto, June 5, 1858, and the purport is an offer of 11*s.* 6*d.* in the pound, at one, two, three, and four years. The accounts upon which the offer of 11*s.* 6*d.* in the pound is based, are as follows:—

STATEMENT OF AFFAIRS IN LONDON, JUNE 22, AND IN TORONTO, AS ON MAY 15, 1858.

DEBTOR.

	£	s.	d.
To creditors in Great Britain, viz.:—			
On promissory notes	£320,846	13	8
On open accounts........................	12,113	5	2
Carried forward..............	£332,959	18	10

Brought forward	£332,959 18 10	
Less amount to be provided for by other parties	79,631 0 0	
		£253,328 18 10
To creditors in Canada unsecured		12,889 9 6
Ditto secured	£143,233 15 2	
Value of securities held	199,462 17 4	
Surplus to *contra*	£56,229 2 2	
To Alexander Mitchell, of Manchester	10,281 17 0	
Claims to have an assignment of Murray and Co.'s debt to the extent of	10,000 0 0	
		281 17 0
		£266,500 5 4

CREDITOR.

By freehold property in Renfrew and Castleton, cost	£12,300 0 0	
Mortgaged for	4,000 0 0	
Estimated of no value	£8,300 0 0	
By premises in Toronto—estimated value	£6,480 0 0	
Less mortgage thereon	4,000 0 0	
		£2,480 0 0
By John Ross and Co., Melbourne	£36,429 12 5	
Estimated to produce		6,000 0 0
By Whan, M'Lean, and Co.		2,607 18 9
By A. Murray and Co.	17,392 0 0	
Less claimed by A. Mitchell	10,000 0 0	
		7,392 0 0
By cash in hand		1,210 0 0
By bills receivable		5,638 18 11
By ditto L. Atterbury and Co.	£12,060 0 0	
Estimated at 1*s*. 3*d*. per pound		753 15 0

ASSETS IN CANADA.

By debts, mortgages, and bills receivable, the realization of which is undoubted	£90,149 16 0	
Ditto, which will be paid with time	156,131 3 6	
Ditto, which may be good after a long and doubtful period	52,168 15 9	
	£298,449 15 3	
Less in hands of secured creditors	199,462 17 4	
		98,986 17 11
Less stock in trade		29,361 16 10
Less cash in hand		232 11 10
Less surplus securities in hands of creditors		56,229 2 2
Or 15*s*. 10*d*. in the pound		210,893 1 5
Less deficiency		55,607 3 11
		£266,500 5 4

"It will be perceived by this statement that the assets are sufficient to pay 15*s*. 10*d*. in the pound, even assuming that the claim set up by Mr. Alexander Mitchell is a just one. If the asset to which he lays claim should be found available for distribution amongst the general body of creditors, it would make the assets sufficient to yield 16*s*. in the pound. The committee think it right to leave Messrs. Ross, Mitchell, and Co. to explain to their creditors the principle upon which they make the offer of 11*s*. 6*d*. in the pound. The committee have endeavoured, since the date of their appointment, to prevent any creditors obtaining preference over the general body; but so far as the Canadian creditors are concerned, their efforts have been of no avail. The statement now prepared shows that, with a trifling exception, the whole of the Canadian creditors are secured; and that, if no preferences existed, the assets would be sufficient to pay all the creditors a dividend of 17*s*. 4*d*. in the pound. Apart from the question as to whether this meeting shall accept the offer now made to them by Messrs. Ross, Mitchell, and Co., or any other offer of composition, the committee are of opinion that this meeting should express its disapproval of the state of the law which enables a trading firm in Canada to divide its creditors into classes, and to prefer out of the common assets any creditors whom it may choose to favour.

"Signed by order of the Committee,
"HENRY CHATTERIS, Accountant."

The discussion, which was of a diffuse and irregular character, tended at first in the direction of the causes for the depreciation of the assets. The two principal items were the debt due from Messrs. John Ross and Co., of Melbourne, and the large amount deducted for bad claims in Canada. The difference in the value of the freehold property in Renfrew and Castleton likewise excited comment, it having been before placed at a favourable price, while it is now stated to be unsaleable. These discrepancies, it was considered, required full explanation, apart from the objectionable nature of the proposal for composition, and the manner in which it was suggested it should be completed. With respect to the preferences to the Canadian creditors, that was a question which it was thought could be dealt with at a later stage of the proceedings, when explanations from the confidential *employé* of Messrs. Ross, Mitchell, and Co. were obtained. In reply to questions, Mr. Chatteris stated that the offer of composition was 11*s*. 6*d*. in the pound, leaving the firm to wind-up the estate by instalments of 2*s*. 6d. on the 1st February, 1859, and subsequent payments of 3*s*. at the same date in 1860, 1861, and 1862, the chief assets consisting of debts, mortgages, and bills. Not until after February in the present year were the committee aware that the creditors in Canada were preferred in any shape, and then only to a limited extent, the statement invariably being that ample assets existed to pay 20*s*. in the pound. The fact that claims to the extent of £200,000 had been so arranged did not transpire before the latest accounts were received. In the case of the consignments to Australia it appeared that a person had been sent out to look into the books, but he having died on his voyage home no accurate statement could be secured. There were some letters from him showing that the goods had been disposed of at depreciated prices, but it was supposed that he intended his explanations should be verbal. The details of the deficiency, placed at £112,000, including the surplus at first estimated of £57,000, and the adverse balance of £55,000, presented by the latest figures, was accounted for in the following manner:—£18,000 over-due interest on promissory notes; expenses and drawings of the partners since suspension, £3,280; Renfrew property, formerly placed at £8,300, now valueless; loss on the Melbourne account of Messrs. J. Ross and Co., £30,429; loss by L. Atterbury and Co., £11,307; bad debts written off, £42,900, etc. Although the account exhibited 15*s*. 10*d*. in the pound, it was considered essential to make deductions on some classes of the debts to the extent of 12½ and 25 per cent., which, being equal to £36,228, left a balance which would pay a dividend of 13*s*. 1¼*d*.; but the partners, presuming that a further allowance should be made for contingencies, proposed the 11*s*. 6*d*. as stated. It was intimated that there had been no reason to suspect that the accounts were wrong, because, after the figures presented in October, a balance-sheet received in January not only confirmed the original position, but increased the surplus to the extent of £1,500. Mr. Ross, it was stated, had gone out at the instance of the committee, with the view of preparing for the July payment; and the reason for Mr. Fisken, the partner, not returning to this country, but sending the principal clerk instead, was alleged to be fear of arrest. Mr. Mitchell remains in this country, and his attendance could be secured, if it were considered desirable; but it was believed that Mr. Snelling, the party deputed, and who was also a lawyer, could afford more information, particularly respecting the Canadian establishment. The arrangement concluded with the English creditors was for payment in two years, by four instalments, and that with the Canadian creditors in three years; the dates being April and October, whereas the others were July and January. Although instructions were sent out that the Canadian creditors were not to be paid unless the partners saw their way clear to satisfy the English dividend in July, the injunction received no attention, and the first instalment had been discharged, "and," thus observed one of the creditors, "we were let into a trap." After some conversation, it was agreed that Mr. Snelling should be introduced to tender an explanation of the serious difference in the appearance of the accounts, and state the ground on which the partners made the offer of 11*s.* 6*d.* in the pound, when, at the lowest computation, the assets showed 13*s.* 1*d.* Mr. Snelling, without any lengthened prelude, entered into an elaborate narrative of the position of the house. He asserted that the great discrepancy apparent arose from the London balance-sheet having been founded on a statement of the Toronto business in the early part of the year, unaccompanied by accurate knowledge of the prospective condition of business, which could only be ascertained when the new season commenced. When the roads became opened, and sleighing began, that was the period for the arrival of their customers, who then settled accounts, paid, or obtained an extension of time. On the late occasion, and after the first accounts had been prepared, it was discovered that the amount of indebtedness was enormous, without corresponding means to meet it. The traders were, consequently, obliged to announce their inability to pay, to give mortgages of their property, and to seek increased time, while, in many cases, the debts were altogether bad. These unfavourable circumstances had so altered the appearance of the assets, that, instead of the surplus anticipated, a deficiency was immediately manifested. The nature of the business of the provinces was such that it was impossible to arrive at accurate results until the season turned, when these accounts were adjusted. The prospects being discouraging, and doubts being still yet entertained of the ultimate realization of some of the accounts, the firm considered that they would not be justified in proposing more than 11*s.* 6*d.* in the pound, contingencies having to be provided for. The alleged preferences to creditors in Canada, he appeared to think, had constituted a subject of unnecessary complaint, inasmuch as he contended that the course pursued had in a measure protected the property for the parties interested in this country. The banks, having received from time to time collateral securities from Messrs. Ross, Mitchell and Co., could have taken proceedings against the customers, and thus effectually denuded the estate. The amount of £199,000, which was the aggregate secured to the Bank of Montreal, the City Bank of Montreal, and the Bank of Toronto, comprised the principal, and these arrangements had been progressing from about the 25th of October. The firm in Canada possessed no knowledge then that the house in London had stopped, but they had reason to believe that embarrassments existed, because remittances were not going forward in a regular manner. The effect of the crisis in New York had been experienced, and as it was speedily developed in the Canadas, the banks who held past-due notes would have proceeded, had not these arrangements been agreed to. He had advised the London firm, every mail, of the progress of the deposits made with these institutions; but it appeared that, although the resident partner received them, he neglected to inform the committee of their contents. Mr. Morley and some of the other creditors said that this was most disgraceful, and that it was evident, from these mortgages and other settlements, a great injustice had been committed against the English claimants. In defence of the conduct of the firm, Mr. Snelling gave a most graphic description of the excitement that ensued when the ravages of the panic became apparent. Great depression existed in January, but it was succeeded by an alarming sacrifice of property in February, March, and April. Horses, cattle, and land sold at a fearful depreciation, special descriptions not being able to find buyers; the local courts were inundated with cases, and although those which were litigated were numerous, a great number were settled with only a preliminary recourse to law. It was in consequence of this state of things that the general assets had become impoverished, and he regretted to learn that the proposal was not likely to be well received. From having ascertained the feeling of the creditors, he had since consulted with Mr. Mitchell, and he was authorized to submit amended terms —say, a total dividend of 12*s.* 6*d.* in the pound; the first payment to be 3*s.* 6*d.* in March, 1859, and three other payments of 3*s.* each in succeeding years, to make up the amount, the extension of a month being asked to allow for the benefit to be derived from the sleighing season to facilitate collections. Mr. Snelling having withdrawn, an animated debate followed; Mr. Morley, Mr. Dillon, and others, loudly censuring the conduct of the firm. Mr. Greatorex said that, although a considerable creditor, he was prepared to waive the question of dividend, if any recommendation could be made to stigmatize such a state of things. The conduct of Messrs. Ross, Mitchell, and Co. had been most discreditable, and their mode of transacting business in the Canadas was open to severe reprehension. Mr. Morley intimated that he was quite sure, from what he knew was passing in this particular branch of trade, that further losses would be experienced, unless great caution was exercised in giving credit. He was averse to individuals of this description receiving any countenance, the more so through the preferences given to the Canadian creditors. Mr. Reed and Mr. Murray both considered that the form of deed should be looked to which gave the powers asked, and that if the question of composition were entertained a committee should be appointed to investigate the security, etc. The representative of the liquidators of the Western Bank of Scotland, who are creditors to the extent of £100,000, while repudiating any notion of justifying the proceedings of Messrs. Ross, Mitchell, and Co., impressed upon the meeting the desirableness of accepting 12*s*. 6*d*. in the pound; the accountant of the bank, who had visited New York, had also gone to Toronto and investigated the books, and his outside estimate was 14*s*. in the pound. Bankruptcy would scatter the estate, and leave only a limited distribution. Mr. Morley questioned whether any dividend could be secured, and Mr. Foster said he was not inclined to accept less than the statement showed, viz., 13*s*. The representative of the City of Glasgow Bank and the Glasgow creditors, whose debts reached £70,000, supported the views urged by the liquidators of the Western Bank of Scotland. The question of the imperative necessity of an assignment was now mooted, and it was received with favour. Others advocated a composition of 13*s*. in the pound, in five half-yearly instalments, but the bulk of creditors, as represented by the Western Bank and the Bank of Glasgow, were inclined to adopt the proposal of 12*s*. 6*d*. Mr. Morley said little sympathy should be shown to the Western Bank of Scotland; the directors had encouraged over-trading to an alarming extent, and, mixed up with Messrs. Ross, Mitchell, and Co., were the notorious names of Macdonald and Co., Monteith and Co., and others, whose accommodation transactions had created great mischief in trade. Mr. Greatorex was against any plan which would suffer the firm to re-enter business. The present case was only a repetition of what formerly occurred in 1847; they then failed for a large amount, and promised a full dividend, which was not forthcoming; and now, after pretending to suspend with a surplus, a large deficiency was the actual result. The course they pursued in Canada was ruin to the trade of the country, and the sooner they were cleared from the field the better it would be for more scrupulous and disinterested merchants. The very fact of securing the banks was to get themselves into credit again. The Chairman advised circumspection; he should know what to do if they required credit of him; but he was not prepared to cast altogether aside the prospect of a dividend. After a long debate, it was agreed to accept a composition in the modified form proposed in the following resolution, and the opinion of the meeting was recorded against the practice of allowing preferences in Canada. A petition to the Legislative Assembly was signed in the room, to be forwarded to Toronto, requesting a repeal of the law which sanctioned such invidious distinctions between the creditors resident in England and Canada:—

"Resolved to accept a composition of 12*s*. 6*d*. in the pound, at dates to be agreed upon by a committee of five creditors, who are also empowered to decide as to the nature and the extent of the securities to be given for the payment of the composition."

THE ESTATE OF MR ALFRED HILL.

A meeting of the creditors of Mr. Alfred Hill, ship and insurance broker and commission merchant, who failed on the 13th of October, 1857, was held on the 2nd November, when the following statement was presented by Mr. H. Chatteris, the accountant:—

DEBTOR.

To creditors unsecured	£6,993 13 10	
To ditto for freight	979 9 11	
		£7,973 3 9
Liabilities on acceptances for the accommodation of the drawers, but which are expected to be proved on this estate		42,029 18 2
Carried forward		£50,003 1 11

Brought forward		£50,003 1 11
To liabilities on accommodation acceptances in favour of Ross and Mitchell, who are arranging to pay in full		8,447 17 0
To liabilities on acceptances which should, and it is believed will, be met by the parties on whose account they are accepted	£4,564 18 8	
To liabilities on bills receivable, negotiated, and supposed to be good	7,644 17 9	
	£12,209 16 5	
To liabilities on bills receivable, which are expected to be proved on this estate		2,817 10 0
		£61,268 8 11

CREDITOR.

By debtors, good		£8,828 1 8
By ditto, doubtful	£4,625 7 8	
Estimated to produce		2,613 16 3
By bills receivable, good		530 4 4
By cash at bank		12 9 3
By ditto in hand		190 14 3
By consignment to America, estimated to produce		100 0 0
By fixtures, estimated to be worth		70 0 0
By A. Hill's capital in the firm of A. Hill and Co., Liverpool, estimated on Dec. 30, 1856...		4,106 17 4
By A. Hill's capital in the firm of Hill, Anderson, and Co., London, exclusive of profits since June last		678 2 5
By household furniture, estimated to be worth		300 0 0
		£17,430 5 6

It was stated that the trade debts of Mr. Hill amounted to £7,973, and that the assets represented £17,430. Accommodation transactions, however, had been entered into with Messrs. Ross, Mitchell, and Co., Messrs. Macdonald, and other parties, for a sum of upwards of £50,000, and thus the liabilities had been increased until they reached a total of £61,268. This imprudent proceeding had been accompanied by the very questionable feature of charging "a commission" on such operations. The proposal made on behalf of Mr. Hill was, to pay to the general creditors a composition of 5*s*. 8*d*. in the pound by instalments, and to pay the trade creditors a further sum of 2*s*. 4*d*. in the pound from the contingent claims against the drawers of his accommodation acceptances. In the course of the discussion, it was mentioned that the estate of Messrs. Ross, Mitchell, and Co. will no doubt realize 20*s*. in the pound, with interest, and that other firms with whom Mr. Hill has been connected would prove equally solvent. With regard to Messrs. Macdonald and Co., it would be widely different, their accounts exhibiting an enormous deficiency. The following comprises the general terms of the resolution adopted :—

"At a meeting of the creditors of Mr. Alfred Hill, commission merchant and shipping agent, of No. 35, Milk Street, City, held at the Guildhall Coffee-house, King Street, Cheapside, on Monday, November 2, 1857, a statement of the affairs having been read by Mr. Chatteris, the accountant, Mr. Van Sandau, solicitor for Mr. Hill, proposed to pay the creditors a composition of 5*s*. 8*d*. in the pound, by three instalments, viz., 1*s*. 8*d*. in the pound at four months, 2*s*. in the pound at eight months, and 2*s*. in the pound at twelve months, from December 2, and to pay the trade creditors a further sum of 2*s*. 4*d*. in the pound, to be secured to them by an assignment to trustees of A. Hill's contingent claims against the drawers of his accommodation acceptances."

THE ESTATE OF MESSRS. NAYLOR, VICKERS, AND CO.

A meeting of the creditors of Messrs. Naylor, Vickers, and Co., steel manufacturers and iron merchants, whose suspension was announced on the 5th November, 1857, was held on the 24th, at Sheffield, Mr. J. P. Budd in the chair, when it was resolved to accept payment in full in four instalments, with interest, at the average Bank of England rate. The following are extracts from the report of Messrs. Harwood, Banner, and Son, accountants, of Liverpool, showing the full solvency of the firm, and the prospect of an immediate and favourable liquidation :—

"The partners constituting the houses in Sheffield, New York, Boston, and Liverpool,* are all identical, the various establishments being branches merely of the one concern centered in Sheffield, so that there is no distinction of accounts as regards the present adjustment, which contains one set of assets and one set of liabilities, amalgamated from the books of the concern in the four places. The statement of the affairs is as follows:—

Assets in England.			
Book debts and advances on consignments	£30,629 3 1		
Estimated surplus in the hands of creditors holding security	1,065 2 10		
Stock in trade, consisting of steel and iron, valued at ..	90,765 6 9		
Property, land, buildings, machinery, etc., after deducting mortgages as per valuation..	24,356 5 0		
Tools, implements, furniture, as per valuation	14,119 14 7		
Estimated value of unrealized consignments to New York..	19,496 0 5		
Shares in foreign steel works	300 0 0		
Cash balance	644 10 0		
Total assets in England....		£181,376 2 8	
Assets in America.			
Cash in hand	£5,564 0 0		
Metals { in warehouse £141,682; in transit to United States 40,864 }	182,546 0 0		
Advances on consignments secured by metals	30,825 0 0		
Property in New York and Boston, af.er deducting mortgages	26,294 0 0		
Bills receivable on hand	71,531 0 0		
Book debts	91,798 0 0		
Total assets in America..		408,558 0 0	
Total assets			£589,934 2 8
Liabilities.			
Acceptances in England	£119,265 17 8		
Ditto in America	22,828 0 0		
		£142,093 17 8	
Creditors on open accounts in England	73.712 4 2		
Ditto in America	134,218 0 0		
		207,930 4 2	
Creditors holding security, after deducting the value of securities held		14,262 18 11	
			364,287 0 9
Leaving a surplus of....................			£225,647 1 11

Which is at the credit of the partners in various proportions.

* Since this date, the firm have opened a house in London, for the conduct of a portion of their operations.

"The books balance exactly, and are most beautifully and admirably kept.

"In addition to this surplus, there will be the private estates of the partners, which, after satisfying private liabilities, will leave for the general estate a surplus of £30,000 to £40,000 in increase of the margin shown to the creditors to exist. It will be observed that the large portion of these assets is in America, and at first sight, looking to the present position of matters on that side, there might be thought to exist some uncertainty as to their value being realized; but, considering the large proportion that is in stock, the character of the debts, and the number of the parties (about 800) amongst whom they are distributed, bearing in mind, also, the fact of these being, in the majority of instances, old-established customers, and making the largest allowance for contingencies, we do not see any ground of apprehension as to the ultimate result to the creditors. The statements of the foreign houses are sent over monthly, and we have them complete by October 1st. The average sales there during the last five years have reached something like £570,000 per annum, while the bad debts have not exceeded £2,000 per annum, a very strong fact in confirmation of the excellence of the business there.

"Creditors are the best judges of their own interests, and it is scarcely for us to suggest the course which should be adopted in this matter, but the circumstances of the case are so peculiar, everything connected with it is so *bonâ fide*, and so perfectly free from anything speculative, and the character of the parties for exactness, ability, and integrity is so entirely beyond question, that we do not hesitate to say, that the wisest plan for all concerned is to leave in the hands of Messrs. Naylor, Vickers, and Co. the management of their own affairs, under such a deed, or letter of license, as will save them from molestation, fixed periods being allowed for the payment of their debts in full, by instalments of 5*s*. in the pound, the first maturing on the 15th July next (by which time the remittances from the spring sales will be coming forward), the remainder at successive intervals of three months each from that time; interest on the whole at the average Bank of England rate, to be paid in cash with the last instalment."

"*Sheffield, November* 24, 1857.

"At a meeting of the creditors of Messrs. Naylor, Vickers, and Co., held at their offices in Sheffield, on Tuesday, 24th November, 1857, J. P. Budd, Esq. (a principal creditor), in the chair, it being premised that the house stopped payment on the 4th November, inst., Mr. H. W. Banner, public accountant of Liverpool, read statements and balance-sheets of the house. When, after much discussion, the following resolutions were agreed to, moved by James Spence, Esq., seconded by Wm. S. Roden, Esq.: Resolved—

"'That the report read by Mr. Banner is approved and be adopted.

"'That, under the peculiar circumstances of this house and their estate, it is desirable, and will best promote the interest of the creditors of the house, that the partners should pay, and the creditors accept, their debts (that is, 20*s* in the pound), with interest at the average Bank of England rate, by four instalments of 5*s*. in the pound each; on the 15th July, 1858, 15th October, 15th January, and 15th April following, the interest on the whole to be paid in cash with the last instalment.

"'That new acceptances of the house be given to each creditor for the instalments.

"'That a deed be prepared by which the creditors on receipt of the acceptances will covenant not to sue the house, but give time until the acceptances are duly paid when and as they respectively fall due, and if and when they are all duly paid the house to stand released.'

"Carried with one dissentient voice.

(Signed) "J. P. Budd, *Chairman*."

The following is the circular announcing the resumption of payments in full:—

"*Sheffield and Liverpool, May* 17, 1858.

"Sir,—We beg leave to inform you that we have this day resumed payment in full, and every demand against us, matured or unmatured, in bills or in accounts, will be settled in cash on presentation at our respective offices. As the average of the Bank of England rate is below 5 per cent., interest will be regulated at 5 per cent. per annum.

"With sincerest thanks for your kind indulgence, with which the even to us unexpectedly rapid realization of our demands in America now enables us to dispense,

"We remain, sir, very respectfully,

"Naylor, Vickers, and Co."

THE ESTATE OF MESSRS. W. AND H. BRAND.

A meeting of the creditors of Messrs. W. and H. Brand, engaged in the West India and American trade, was held on the 11th November, 1857, when the following statement was presented by Mr. G. H. Jay, the accountant:—

STATEMENT OF AFFAIRS, OCTOBER 26TH, 1857.

DEBTOR.

To sundry creditors on open accounts	£5,920 12 3	
To rent and salaries	71 5 0	
		£5,991 17 3
To creditors on bills payable, as per statement...	164,669 9 3	
Less amount which it is expected will be retired by drawers	88,487 10 7	
		76,181 18 8
To creditors fully secured, as per statement	11,960 17 7	
To amount of securities held	20,788 15 9	
Balance—surplus carried to *contra*	8,827 18 2	
To liabilities on bills receivable	319,169 18 2	
To amount expected to be brought against this estate, inclusive of Dennistoun and Co.'s acceptances, amounting to	59,577 2 9	
To Western Bank of Scotland	67,000 0 0	
To City of Glasgow Bank	27,000 0 0	
		153,577 2 9
To liability to Rake, Kimber, and Co., Middlesborough-on-Tees, on a contract for building an iron paddle steamer, on account of G. P. Watson, of Demerara, amount of contract ...	8,750 0 0	
Less first instalment, paid on account	2,150 0 0	
Considered good	6,600 0 0	
		235,750 18 8
Deduct creditors under £20 each		226 6 6
		£235,524 12 2

CREDITOR.

By cash on hand, viz., at London Joint-Stock Bank	£58 18 2	
In the house	1 14 3	
		£60 12 5
By bills receivable, in hand	18,328 2 9	
Estimated to be of the value of		14,045 12 2
By sundry debtors, considered good		12,093 15 8
Considered bad	55,920 10 3	
By surplus security held by creditors	8,827 18 2	
Estimated to produce		3,000 0 0
By sundry property		95 19 7
By surplus of private estates of the partners		1,268 0 0
		£30,563 19 10
Deduct creditors under £20 each		226 6 6
Total		£30,337 13 4
Deficiency carried forward		205,186 18 10

Statement Accounting for Deficiency from 1st January, 1857, to 26th October, 1857.

DEBTOR.

	£ s. d.	£ s. d.	£ s. d.
To balance brought over, being deficiency brought from statement of affairs			£205,186 18 10
To partners' capital:—			
At credit of Wm. Brand		£4,587 10 4	
At credit of R. T. Brand		6,214 15 6	
			10,802 5 10
To profits, 1857, viz.:—			
Commissions		3,549 1 8	
Discounts		384 17 0	
Interest		458 10 10	
Sundries		175 0 3	
		£4,567 9 9	
Less:—			
Charges	£487 19 7		
Loss per "Tyne"	460 0 0		
Discount on bills	588 9 5		
Insurance	141 10 9		
Interest	220 12 11		
Law charges, say	50 0 0		
		£1,948 12 8	
			2,618 17 1
Total			£218,608 1 9

CREDITOR.

	£ s. d.
By balance at debit of J. G. Garner, but subject to the liquidation of his affairs	£55,903 11 7
By Anderson and Smith—bad debts	16 18 8
By bad liabilities, as per statement of affairs	153,577 2 9
By estimated depreciation in realizing securities held by creditors	5,827 18 2
By estimated depreciation in realizing bills receivable in hand ...	3,282 10 7
	£218,608 1 9

It was explained by Mr. Reece, the professional representative of the firm, that their embarrassments have arisen from a connection with Mr. Garner, of New York, who, although previously in very prosperous circumstances, has been brought down through the American convulsion. With respect to the actual condition of the estate, much will depend upon the liquidation of the affairs of the Western Bank of Scotland, the City of Glasgow Bank, and Messrs. Dennistoun and Co. The assets, under the most favourable view, show about 6*s.* 8*d.* in the pound, but should the liabilities not diminish, the dividend will scarcely exceed 2*s.* 6*d.* Of the drafts of Mr. Garner for about £110,000, about £52,000 will be retired, and with the exception of the facilities afforded that gentleman no irregular bill transactions whatever have occurred. The firm possess an interest, to the extent of £11,000, in the shares of the Demerara Railway, which are now marketable, and the principal assets consist of £30,000 exhibited on the face of the accounts. As the partners, owing to the suspension of Messrs. Dennistoun and the Western Bank of Scotland, are not in a position to make any proposal, they desired to place themselves entirely in the hands of their creditors, surrendering for their benefit every fraction of property. After some discussion, a proposal to wind-up under inspection was generally agreed to, a resolution to that effect being carried.

THE ESTATE OF MESSRS. J. HALY AND CO.

A meeting of the creditors of Messrs. J. Haly and Co., who had recently failed in the American trade, was held on the 12th November, 1857, when the following statement was presented by Mr. Quilter, of the firm of Quilter, Ball, and Co., the accountants:—

STATEMENT OF AFFAIRS, OCTOBER 28, 1857.

DEBTOR.

To sundry creditors		£30,264 13 2
To creditors partially secured—		
Amount of claim	£5,667 16 2	
Value of security	5,408 7 2	
Deficiency		259 9 0
To creditors fully secured—		
Estimated value of security	£6,300 0 0	
Amount of claim	2,962 9 7	
Surplus carried to credit	£3,337 10 5	
To liabilities on endorsements—		
Considered good	£20,974 12 6	
To liabilities on acceptances	43,233 4 11	
Considered bad—Jacot, Taylor, and Co.	6,678 4 4	
R. H. Brett	10,306 16 4	
		16,985 0 8
		£47,509 2 10

CREDITOR.

By cash balance in house		£216 1 3	
„ at Bank of England		33 14 8	
			249 15 11
By sundry debtors—			
Considered good			14,601 7 1
Considered bad		£7,906 18 5	
Considered doubtful		£3,028 7 0	
Estimated at			1,000 0 0
By bills receivable—			
Considered good			2,146 2 9
By consignment of railway iron to New York, in the hands of T. Dehon, of New York, and on joint account with him:— Our half net proceeds of 2,355 tons 7 cwt., remaining unsold, estimated at			9,000 0 0
By shares in vessels:—			
4-64ths of "Pride of Canada," cost, May, 1855	£1,167 15 3		
4-64ths of "Queen of the Lakes," cost, Sept., 1855	877 2 11		
Valued at		£2,000 0 0	
By mortgages held by T. Dehon, New York, for our account, with power of sale, viz.:—			
3-16ths of "Ellwood Walker," 12,000 dollars, supposed total value £11,000, say	£1,500 0 0		
28-64ths of "City of Manchester," supposed total value £9,000	2,500 0 0		
		4,000 0 0	
			6,000 0 0
Carried forward			£32,997 5 9

	£	s.	d.
Brought forward	£32,997	5	9
By counting-house furniture	100	6	0
By surplus securities in hands of creditors, as *per contra*	3,337	10	5
By dividends from the estates of Jacot, Taylor, and Co., and R. H. Brett. (Should the dividends to arise from our own estate and the estates of these two firms together exceed 20*s.* in the pound, a benefit will accrue to our estate equal to such excess.)			
	£36,435	2	2
Less sundry preferential payments, as rents, salaries, expenses, etc.	300	0	0
	£36,135	2	2
By deficiency carried forward	11,374	0	8
	£47,509	2	10

It was explained that the accounts show a deficiency of £11,374, and that the suspension has been occasioned through the failure of Mr. R. H. Brett, of Toronto, and Messrs. Jacot, Taylor, and Co., of Liverpool. Two years ago the firm possessed capital equal to £19,000, subject to bad debts of about £5,000. The estimate of Mr. Brett's estate, for present purposes, is taken at 6*s.* 8*d.* in the pound; but should a dividend of 13*s.* 4*d.* be realized, as anticipated, in Toronto, the Messrs. Haly and Co. will be solvent. Under any circumstances the assets, consisting of shares in vessels, iron, and book debts, may, it is thought, produce from 10*s.* to 15*s.* in the pound. A conversation took place with respect to the course proposed to be pursued. Some of the creditors appeared to think that an offer of composition would be accepted; but, as it was stated that one of the partners was abroad, a doubt was entertained of a result being immediately arrived at. The suggestion of Mr. J. Linklater, that a liquidation under inspection could at once be proceeded with, leaving the question of arrangement in abeyance until the views of the absent partner were ascertained, was eventually acted upon, and a resolution agreeing to this mode of proceeding was adopted.

THE ESTATE OF MESSRS. JAMES SCOTT AND CO.

A meeting of the creditors of Messrs. James Scott and Co., of Queenstown, shipowners, who suspended a short time previously, was held on Thursday, November, 19th, 1857. Mr. Edmund Burke having been called to the chair, Mr. Jameson, solicitor, laid before the meeting the following statement of the present financial circumstances of the firm.

The total liabilities amounted to £129,376, of which the secured debts were £47,679, leaving a balance of £81,697 unsecured. The available assets amounted to £109,225, which, deducting the secured debts, £47,679, left £61,546 to meet the unsecured debts amounting to £81,697. There thus appears to be a deficit of £20,151. The assets were made up as follows:—Ships, and freights due thereon, £53,000; houses, £44,600; other property, £11,625; total, £109,225. The house property is liable to two mortgages, for £10,000 to the Provincial Bank, and £4,000 to a firm in the City, with a contingent mortgage of £5,000, also to the bank. The ships are liable to mortgages of £8,000. The house property is let at an annual rental of £2,570, and includes some of the handsomest parts of Queenstown.

The failure of the firm was stated to have resulted from that of the Liverpool Borough Bank, together with the general pressure of the money-market, and the depreciation in the value of shipping.

Mr. Jameson then said he was authorized on the part of Messrs. Scott to submit the following proposition to the meeting:—"That the property should be placed in the hands of trustees, and that the creditors should take acceptances for a composition of 12*s.* 6*d.* in the pound, payable in three instalments, at intervals of six, twelve, and eighteen months."

The statement was received favourably by the creditors, several of whom declared their entire confidence in the honour and integrity of the firm, insomuch as one of them suggested that the Messrs. Scott should be permitted to work the concern themselves without the intervention of trustees; but this proposition was at once refused by the Messrs. Scott themselves. A committee of inspection was appointed on the suggestion of Mr. Jameson, with power to employ an accountant to examine the books of the firm and ascertain the correctness of the statement as to their liabilities and assets. Should they report favourably, the creditors expressed themselves ready to accede to the proposed arrangement.

THE ESTATE OF MESSRS. WILSON, MORGAN, AND CO.

A meeting of the creditors of Messrs. Wilson, Morgan, and Co., wholesale stationers, of Cheapside, who suspended on the 12th of November, was held on the 22nd of November, 1857, when the following statement of their affairs was presented by Mr. H. Chatteris, the accountant :—

STATEMENT OF AFFAIRS, NOVEMBER, 12, 1857.

DEBTOR.

	£	s.	d.	£	s.	d.
To creditors unsecured	£19,325	5	6			
Ditto under £10	195	6	10			
				£19,520	12	4
To Mr. Samuel Smith	4,520	12	5			
Less estimated value of security held by him	2,075	0	0			
				2,445	12	5
To creditors partially secured	1,041	10	5			
Less estimated value of securities	656	3	10			
				385	6	7
To liabilities on acceptances which should be provided for by the drawers	1,452	8	9			
Of which it is expected there will be proved on this estate				351	13	9
To liabilities on bills receivable under discount	20,984	1	3			
Supposed to be good	17,062	4	4			
Expected to be dishonoured	3,921	16	11			
Less cash in the hands of bill-holders	995	14	6			
				2,926	2	5
				£25,629	7	6

CREDITOR.

	£	s.	d.	£	s.	d.
By stock and utensils in trade, at cost				£5,005	7	6
By debtors—good				4,411	8	3
" doubtful	£229	14	4			
Taken at 10*s*. in the pound				114	17	2
Bad	2,045	13	4			
By debt due from Thomas Wilson—bad	6,777	0	9			
	£8,822	14	1			
By cash in hand				67	1	0
Estimated value of a printing plant assigned to us for a bad debt				300	0	0
				£9,898	13	11
Less rent, taxes, salaries, etc., to be paid in full				400	0	0
(About 7*s*. 4¾*d*. in the pound)				£9,498	13	11
Deficiency				16,130	13	7
				£25,629	7	6

The debts and liabilities amount to £25,629, and the assets to £9,499, showing a deficiency of £16,130. The affairs of the house appear to have been for some time in a doubtful state, owing to the involvement of the two original partners, Mr. William and Mr. Thomas Wilson. The first retired three or four years ago, in consequence of some liability incurred through an insurance company, and the second left the firm at a later period, having encountered responsibility in connection with the Royal British Bank. The firm were believed to be solvent in 1854, including a debt due from Australia, for consignments, which has since proved to be bad. When Thomas Wilson retired Morgan assumed that amount with the other assets, and it was agreed to allow Mr. Wilson an annuity, with permission, if he chose, at any time within a certain date, to re-enter as a partner. Mr. H. N. Smith having temporarily joined Mr. Morgan, his father advanced money for the business, and was now secured to the extent of £2,075. The proposal was to pay a composition of 7*s.* in the pound by instalments extending over a period of twelve months, but some of the creditors, including Mr. M'Murray and others, insisted that as the business was valuable a better arrangement should be attempted. Ten shillings, or a settlement under the arrangement clauses of the Bankruptcy Consolidation Act, was the point discussed at some length, and after a consultation among the parties, a kind of compromise in the following shape was agreed to :—

"It was proposed and resolved by the creditors present to accept a composition of 8*s.* 6*d.* in the pound, by three equal instalments at three, six, and nine months, from December 1, the last instalment to be secured to the satisfaction of Messrs. M,Murray, Templeton, and Pollock."

The estate was subsequently wound-up in bankruptcy.

THE ESTATE OF MESSRS. BARDGETT AND PICARD.

A meeting of the creditors of Messrs. Bardgett and Picard, who failed on the 16th November, 1857, in the corn trade, was held on the 9th December, when the following statement of affairs was presented by Mr. Turquand, of the firm of Turquand, Youngs, and Co., the accountants :—

STATEMENT OF AFFAIRS, NOV. 14, 1857.

DEBTOR.

To creditors unsecured			£33,596 0 8
To creditors partially secured		£38,890 12 9	
To estimated value of securities		33,687 14 8	
			5,202 18 1
To creditors fully secured		17,556 0 3	
To estimated value of securities		18,981 10 8	
		£1,425 10 5	
To liabilities on bills of exchange		112,652 16 0	
Considered good	£65,023 3 8		
" bad	47,629 12 4	47,629 12 4	
Deduct cash *per contra* in the hands of bill-holders...		1,285 14 5	
			46,343 17 11
			£85,142 16 8

CREDITOR.

By cash in hands of bankers and bill-holders...		£1,285 14 5	
Deducted from liabilities *per contra*—			
By debtors, good		3,623 3 2	
By ditto, doubtful and bad...	£7,481 12 6		
Estimated to realize		935 3 3	
Carried forward			£4,558 6 5

Brought forward		£4,558 6 5
By debtors in America	£38,646 0 1	
Estimated to realize		10,000 0 0
By property		1,924 6 0
		£16,482 12 5
Deduct claims payable in full		376 0 0
		£16,106 12 5

By doubtful assets—	
Property	£6,824 0 0
Surplus securities *per contra*	1,425 10 5
	£8,249 10 5

It was stated that the assets show about 5*s.* in the pound, but that there is not the prospect of the whole amount being realized, in consequence of the large proportion of debts due from America. The operations of the firm have been on an extensive scale, and the difficulties are attributed to losses, bad debts, and the depreciation in the value of grain. In 1856 the accounts exhibited a surplus of £24,314, but an accurate estimate of the debts had not then been made. The house not being in a position to make any distinct offer, the partners are prepared to adopt any course which may be suggested for the arrangement of their affairs. After some discussion an adjournment of a week was agreed to, three creditors being meanwhile appointed to look into the books, with the view of recommending the measures that should be resorted to. At an adjourned meeting it was determined to appoint a committee to consider the propriety of accepting 2*s.* 6*d.* in the pound.

The estate was subsequently wound-up in bankruptcy.

THE ESTATE OF MESSRS. BROCKLESBY AND WESSELS.

A meeting of the creditors of Messrs. Brocklesby and Wessels, who failed on the 18th November, 1857, in the grain trade, was held on the 3rd December, when the subjoined statement of their affairs was presented by Mr. G. H. Jay, the accountant:—

Brocklesby and Wessels' Statement of Affairs, November 18, 1857.

DEBTOR.

To sundry creditors unsecured	£31,793 4 2	
To rent, taxes, etc.	392 18 6	
		£32,186 2 8
To sundry creditors fully secured—		
Amount of claims	£18,354 13 11	
Amount of security	27,357 16 6	
Balance—surplus to *contra*	£9,003 2 7	
To sundry creditors, partially secured—		
Amount of claims	50,570 4 11	
Security held	42,269 11 4	
		8,300 13 7
To liabilities on acceptances for sundry cargoes, of which we are relieved by the consignees suspending the delivery	2,804 17 7	
To liabilities on bills receivable per list—Considered good	24,401 8 5	
To liabilities on guarantee of Levy and Arndt's drafts, not expected to become claims	4,319 0 5	
		£40,486 16 3

CREDITOR.

By cash at bankers'		£1,695 11 11	
Retained by them against bills *per contra*—			
In the house		19 13 11	
At the National Bank		92 18 2	
			£112 12 1
By bills receivable in hand, as per statement, considered good			5,609 15 0
By sundry debtors, as per statement—			
Considered good		£6,795 18 0	
Doubtful	£190 0 0 say	170 0 0	
Bad	2,538 3 4		
			6,965 18 0
By stock in hand			1,200 0 0
By surplus security held by creditors, as *per contra*			9,003 2 7
By counting-house furniture			40 0 0
By debts due to consigners, not collected		956 8 3	
			£22,931 7 8

It was explained that the estate showed a dividend of about 11*s*. 3*d*. in the pound, the assets consisting of grain taken at a fair valuation, and which can be steadily realized. The whole of the transactions of the firm have, it is stated, been characterized by regularity, and the bankers being satisfied with the conduct of Messrs. Brocklesby and Wessels, have agreed to abide by the general decision of the creditors. Mr. M'Leod, the professional representative of the firm, intimated that if a liquidation under inspectorship were adopted, a first dividend of 5*s*. could be speedily declared, and a second of 2*s*. 6*d*. within two months, leaving the further distribution in the hands of the gentlemen appointed to act under the arrangement. General testimony having been borne to the satisfactory mode of transacting business by the house, a resolution was passed authorizing a liquidation by inspection.

THE ESTATE OF MESSRS. J. AND A. DENNISTOUN AND CO.

A meeting of creditors of Messrs. J. and A. Dennistoun and Co., of Glasgow, with branch houses at London, Liverpool, New York, and New Orleans, whose suspension was announced on the 7th, was held on the 26th November, 1857, when the following statement of affairs was submitted by Mr. Coleman, the accountant:—

BALANCE STATEMENT OF AFFAIRS TO NOVEMBER 7, 1857.

DEBTOR.

To creditors on open accounts		£91,947 10 8
Ditto, holding security		217,276 1 9
Ditto, on bills payable (this amount includes our acceptances (£296,687 17*s*. 5*d*.) given for the account of Dennistoun Brothers and Co., of Australia)		1,833,478 3 1
Liabilities		£2,142,701 15 6
To amount of capital standing to the credit of the several partners	£560,897 2 9	
Ditto, Liverpool Borough Bank, shares cost	208,873 2 11	
		769,770 5 8
Balance of profit and loss account, etc.		23,520 7 5
		£2,935,992 8 7

CREDITOR.

By cash in hand		£9,038 10 0
By counting-house furniture, etc.		2,500 0 0
By produce on hand		26,360 0 0
By bills receivable on hand		111,786 9 0
By debtors in Great Britain, etc.	£213,010 7 11	
Ditto and property in America	1,584,215 14 9	
		1,797,226 2 8
By securities with creditors		187,080 12 10
By amount to debit of Dennistoun Brothers and Co., Australia		354,345 14 6
Inclusive of acceptances amounting to £296,687 17*s.* 5*d.*		2,488,337 9 0
By further separate estates of the partners available		190,379 16 8
By debts due to separate partners in America		48,392 0 0
By Liverpool Borough Bank shares, *contra*		208,873 2 11
		£2,935,982 8 7

Claims and Assets after Deducting Contra Entries.

DEBTOR.

To creditors on open accounts		£91,947 10 8
Ditto on bills payable	£1,833,478 3 1	
Less balances to the debit of parties for whose account the bills are accepted, the principal part of which are considered good	1,214,414 5 3	
		619,063 17 10
To creditors partially secured	177,142 14 11	
To security held	142,531 15 0	
		34,610 19 11
Creditors fully secured—		
Security held	44,548 17 10	
Claims	40,133 6 10	
	4,415 11 0	
Liabilities on bills discounted, £514,769 10*s.* 4*d.*, of which it is estimated there may rank on this estate		50,000 0 0
		£795,622 8 5

To liabilities in respect of letters of credit.
Ditto of partners on Liverpool Borough Bank shares or any other company.

CREDITOR.

By cash in hand, etc. (subject to the due payment of bills discounted by Commercial Bank of Scotland)	£9,038 10 9
By counting-houses, furniture, etc.	2,500 0 0
By produce on hand	26,360 0 0
By bills receivable on hand	111,786 9 3
By debtors and property	582,814 17 5
By surplus from securities, *contra*	4,415 11 0
By amount to debit of Dennistoun Brothers and Co., Australia (inclusive of acceptances amounting to £296,687 17*s.* 5*d.*)	354,345 13 6
By further separate estates of the partners available	190,379 16 8
By debts due to the separate partners in America	48,399 0 0
	£1,330,039 18 7
By Liverpool Borough Bank shares, held by the several partners	£208,873 12 11

Mr. Coleman explained that in the above account only one item is estimated arbitrarily—viz., the profit and loss; and that the value set on the produce is calculated from the price between the 7th and 14th of November. The debts and property in America may be thus classified :—£695,000 is represented mostly by goods and bills receivable of customers; £325,200 consists of other bills receivable; £140,000 guaranteed securities; £151,000 open accounts, and the remainder is composed of railway bonds, mortgages, and various securities of that character, but of the best description. The house of Dennistoun and Co., of Australia, being a distinct firm, the creditors in connection with that business will be in a somewhat better position than the ordinary creditors, as the returns from Australia will accrue more quickly than the proceeds to be expected from the realization of the American assets. Messrs. Dennistoun, as large shareholders in the Liverpool Borough Bank, are necessarily subject to a heavy liability on account of the failure of that institution; but it is believed, not only that the £5 call already made will be sufficient to clear off all claims, but that there is every probability this sum will be returned. All the transactions of the firm have been on a sound and legitimate basis, and no discreditable act can in any way be imputed to them. Mr. Murray, their London solicitor, then made the following proposition on the part of Messrs. Dennistoun, namely, that the creditors of the houses in Great Britain and the United States be paid in six instalments, namely, 3*s.* in the pound on or before the 30th of January, 1858; 2*s.* on the 31st of July; 3*s.* on the 31st of December; 4*s.* on the 30th of June, 1859; 4*s.* on the 31st of December; and 4*s.* on the 30th of June, 1860; with 5 per cent. interest on each instalment; and that the creditors through the Australian house should receive 6*s.* 8*d.* in the pound on the 30th of January, 1858; 3*s.* 4*d.* on the 30th of June; 5*s.* on the 31st of January, 1859; and 5*s.* on the 31st of July, also with interest at the rate of 5 per cent. In answer to a question, Mr. Murray stated that when the first payment is made, promissory notes for the remaining instalments will be given and the present acceptances cancelled. He hoped the proposal would be accepted, as such a course would prove of the greatest assistance, both morally and legally, in the realization of the assets and the enforcement of the claims in America. It was further stated that the accounts from the United States were encouraging, and that a possibility exists of the instalments being paid before the dates fixed, although, on considering the whole circumstances of the case, Messrs. Dennistoun could not feel certain of being able to carry out better terms. After a short discussion, a resolution accepting the proposition was passed unanimously.

The estate has paid 20*s.* in the pound with interest, having anticipated some of the instalments.

THE ESTATE OF MESSRS. BENNOCH, TWENTYMAN, AND RIGG.

A numerous meeting of creditors of Messrs. Bennoch, Twentyman, and Rigg, extensively engaged as manufacturers in London and Manchester, was held on the 27th November, 1857, Mr. Foster presiding, when the subjoined statement, prepared by Mr. Parrinton, of the firm of Messrs. Parrinton, Ladbury, and Co., was submitted. Mr. Reed represented the firm; Mr. J. Linklater, Mr. Bower, Mr. Crowther, and Mr. Abrahams, various creditors.

DEBTOR.

To creditors as per list	£167,552	5	10			
Less securities held by them	6,722	5	3			
				£160,830	0	7
(In the above claims, £50,950 11*s.* 9*d.* is included, depreciation in the value of silk in the hands of the brokers.)						
Bills accepted by consigners	£99,663	18	9			
Less bills fully or partly secured by goods on hand	62,680	13	6			
				36,983	5	3
To liabilities on bills receivable, discounted, considered good	127,511	14	10			
Carried forward				£197,813	5	10

	£	s.	d.	£	s.	d.
Brought forward				£197,813	5	10
Ditto, ditto, considered bad	£23,547	6	11			
Less security, held on account thereof	550	0	0	22,997	6	11
To liabilities on guarantees, in respect of which it is considered there will be a claim of				3,000	0	0
To creditors of Messrs. Lovatt and Gould, whose debts Messrs. Bennoch, Twentyman, and Rigg have undertaken to pay				33,884	3	2
				£257,694	15	11

CREDITOR.

	£	s.	d.	£	s.	d.
By stock in trade in London and Manchester, cost	£65,768	5	8			
Estimated at				£39,195	18	4
By book debts, good	41,347	5	2			
Less debts on consigners' account	10,191	0	1	31,156	5	1
By book debts, doubtful and bad	16,267	18	1			
Estimated at				5,057	15	4
By bills receivable on hand	7,248	7	0			
By less portion due to consigners	435	13	9	6,812	13	3
By cash in hand and at banker's				874	13	4
By cash at banker's	2,028	16	4			
Deducted from their claim of	5,000	0	0			
By fixtures at Manchester				154	0	0
By furniture, etc., at Wood Street				300	0	0
By shipments to Melbourne, Sydney, New York, etc., amounting to	41,021	5	11			
Subject to advances and claims thereon	20,163	10	1			
	£20,857	15	10			
Estimated at				12,794		
By value of shares				100		
By securities deposited with creditors, and deducted *per contra*	£6,722	5	3			
By leases and fixtures of premises, 77 and 78, Wood Street, and land in Silver Street				1,400	0	
By stock, book debts, utensils in trade, furniture, etc., of Messrs. Lovatt and Gould, of Leek				4,382	11	
By equity of redemption in freehold land and mill of Messrs. Lovatt and Gould, subject to mortgage of £10,000.						
				£102,228	5	2
Less rent, salaries, etc.				781	11	8
				£101,446	13	6
Deficiency				£156,248	2	5

It was explained that the accounts showed a deficiency of £156,248, the difference between £257,694 debts and liabilities, and £101,446 assets. The dividend, therefore, taking the balance-sheet as it stands, would be from 7*s.* 10*d.* to 7*s.* 11*d.* in the pound; but as a portion of the assets consists of silk, they are still liable to some fluctuation. The firm in 1848 commenced with a capital of £9,410, which subsequently increased until it was taken in December, 1856, at £38,247. This, together with the profits in trade, by silk, commissions, and the amount of deficiency, constituted a total of which the firm had to discharge themselves. It could, however, be readily done; for although some of the items presented unfavourable results, they accounted for the several amounts. The loss on the depreciation of silk, now held by themselves and the brokers, was £50,900; the depreciation of other stock, £26,573; loss by the firm of Messrs. Lovatt and Gould, of Leek, £41,728; loss by Messrs. T. S. Reed and Co., of Derby, £16,692; loss by Mr. John Taylor, of

Coventry, £17,426; and these, with other amounts, including the drawings of the firm, £5,838; loss on bills receivable, £23,547; on shipments, £7,100; trade expenses, £8,213; and bad debts, £14,510, represented the aggregate. The assumption of the debts of Messrs. Lovatt and Gould, amounting to £33,800, arose from a joint connection, by which Messrs. Bennoch and Co. became liable with them in silk purchases from Messrs. Durant, and eventually they took the property of Lovatt and Gould, and arranged to discharge their engagements. In the case of Messrs. Reed and Co., of Derby, the loss accrued through depreciation, the same being the result with the firm of Messrs. Taylor and Co., of Coventry. Being consignees to Messrs. Bennoch and Co., they had overdrawn to an amount in excess of the value of the goods as now estimated, Mr. Bennoch and Mr. Twentyman being at the same time partners in the firm of Messrs. Reed and Co. Mr. Reed, as the professional adviser of the firm, showed that the chief injury to the estate had ensued through depreciation, the further sacrifice, ever since the suspension, having been at least 10 per cent. In a period of depression like the present, it was impossible to predict what might be the result; but it was hoped that a favourable reaction might ensue and increase the value of the assets. While explaining the minutiæ of the business with Messrs. Lovatt and Gould, Messrs. Reed and Co., and Mr. J. Taylor, he showed that they originated under proper influences, and that no bill transactions of a doubtful or irregular character had occurred. The proposal to be placed before the creditors did not include any specific offer, because, looking at the nature of their estate, Messrs. Bennoch and Co. felt that it would be impossible for them to make an arrangement which would be satisfactory. They had therefore determined to yield to any course of action that might be suggested with the view of liquidating their estate in the best possible manner. It was not presumed that bankruptcy would be proposed, but it was believed that, through the process of inspection, affairs might be wound-up without detriment to existing interests. The separate estates of Mr. Bennoch and Mr. Twentyman each presented a small surplus, which, if required, could be made available for the benefit of the joint estate, but that of Mr. Rigg exhibited a deficiency. After a lengthened conversation, in which it was elicited that the operations of the firm in silk had been based on *bonâ fide* purchases, they having acted in the character of silk dealers as well as manufacturers, it was agreed that a liquidation under inspection should be effected. In the course of the proceedings, Mr. Reed intimated that it would in all probability be found advisable to give the inspectors power, after a dividend to the amount of 6*s*. had been declared, to deal either with the members of the firm, or with other parties, for a final arrangement, and, this view being agreed to, the suggestion was incorporated in the following resolution, which was immediately adopted:—

"At a meeting of the creditors of Messrs. Bennoch, Twentyman, and Rigg, of Wood Street, London, and of Manchester, merchants, holden at Guildhall Coffee-house, King Street, Cheapside, on Friday, the 27th of November, 1857, Mr. Foster in the chair, Mr. Parrinton, accountant, read a statement of Messrs Bennoch and Co.'s affairs, and gave various explanations with reference to their estate. Mr. Reed, solicitor, stated that, in consequence of the serious reduction in the present value of the assets, Messrs. Bennoch, Twentyman, and Rigg were unable to propose any offer of composition which would be safe for themselves, and, at the same time, satisfactory to their creditors, and, from the magnitude of the liabilities, it was doubtful whether they could find an adequate guarantee, even if the creditors assented to a composition; that, with the hope of securing to their creditors the present value of their assets, at all events, and a greater dividend than is at present shown, should a better state of things ensue, Messrs. Bennoch and Co. propose to realize their estate under inspection, in the usual way, but that the creditors shall give power to the inspectors, when the estate shall have been brought within smaller compass by inspection, and the payment of a certain amount of dividend (say 6*s*. in the pound), to dispose of the remainder of the estate to Messrs. Bennoch, Twentyman, and Rigg, or one or more of them, or to any other person, at such price and on such terms as the inspectors may think proper. Mr. Reed also stated that Messrs. Bennoch and Co. placed themselves unreservedly in the hands of their creditors, and would adopt any course that the latter might consider most conducive to their interests.

"Resolved—That Messrs. Bennoch, Twentyman, and Rigg's estate be realized under inspection, in accordance with their proposal; that inspectors be appointed, and that it be left to those gentlemen to decide (on behalf of the creditors) upon the terms of the deed of inspection, including those upon which Messrs. Bennoch, Twentyman, and Rigg are to obtain their release. The assent of creditors to this resolution is to be without prejudice to their claims against other parties, or to any securities they hold, and to be subject to the consent of third parties (when necessary) being obtained."

A dividend of 6*s*. 6*d*. in the pound paid, and a release given to the partners.

THE ESTATE OF MESSRS. BROADWOOD AND BARCLAY.

A meeting of the creditors of Messrs. Broadwood and Barclay, engaged in the West India trade, who failed on the 9th November, 1857, was held on the 10th December, when the following statement was presented by Mr. J. Weise, of the firm of Messrs. J. E. Coleman and Co., the accountants:—

Statement of Affairs, November 9, 1857.

DEBTOR.

	£	s.	d.	£	s.	d.
To creditors on open accounts				£959	0	0
To ditto on bills payable	£259,547	0	0			
Of which it is expected there will be provided for by the parties for whose account the bills were accepted	133,410	0	0			
Leaving to rank on this estate				126,137	0	0
To creditors after they have retired their drafts				10,865	0	0
To creditors on underwriting accounts				1,829	0	0
To ditto, fully secured	£11,300	0	0			
To ditto, security held	14,300	0	0			
Surplus *contra*	3,000	0	0			
To creditors, partially secured	66,500	0	0			
To ditto, security held	44,500	0	0			
Deficiency				22,000	0	0
To liabilities on bills receivable	117,917	4	3			
Of which it is anticipated there will rank on this estate				50,230	0	0
				£212,020	0	0

Liability on Ships Chartered.

CREDITOR.

	£	s.	d.
By cash on hand	£89	0	0
By bills receivable on hand (good)	660	0	0
By assets	2,360	0	0
By debtors (good)	23,982	0	0
By ditto, doubtful (estimated)	1,000	0	0
By surplus from creditors fully secured (*contra*)	3,000	0	0
	£31,091	0	0
Less sundry small debts, salaries, etc., to be paid in full	400	0	0
	£30,691	0	0
By Mayaguez refinery, stands at debit £81,097 12*s*. 3*d*.	25,000	0	0
By Decastro, Lindegren, and Co., stands at debit £61,550 11*s*. 4*d*.	10,000	0	0
	£65,691	0	0

It was explained that the accounts showed a dividend of more than 5*s*. in the pound, the assets being taken at the lowest valuation. The large amount of bills accepted, and the capital sunk in the sugar refinery at Porto Rico, represent the principal causes of the suspension, which, together with the fall in produce, exhibit the total deficiency. It appears that Messrs. Decastro, Lindegren, and Co., of Porto

Rico, have authorized the payment of a composition of 10*s.* in the pound, by three instalments, to the holders of their drafts; and it was mentioned that the necessary precautions have been adopted to secure the sugar refinery for the benefit of the creditors. As this item represents so large an asset, it is proposed to continue the operations for some period in order to effect a sale of the stock, and also, if possible, of the property. The capital absorbed since 1849 has been £70,000. A lengthened discussion took place respecting the general condition of the estate; and eventually resolutions were passed, authorizing a liquidation by inspection, Mr. Kirkman Hodgson and Mr. A. Klockmann being appointed to act for the creditors.

THE ESTATE OF MESSRS. J. H. BAIRD AND CO.

A meeting of the creditors of Messrs. J. H. Baird and Co., in the Australian trade, was held on the 10th December, 1857, when the following statement of affairs was presented by Mr. C. F. Kemp, the accountant:—

LIABILITIES.

DEBTOR.

	£	s.	d.	£	s.	d.
To creditors on open accounts	£1,813	14	4			
To ditto on bills payable	13,661	9	3			
Unsecured				£15,475	3	7
To ditto on bills payable in respect of consignments secured				3,729	6	9
To liabilities on bills receivable	7,609	0	6			
Of which it is expected will be paid in full	5,038	9	10			
Estimated dividend	1,606	13	1			
Leaving	953	17	7			
Estimated further loss on bills held as collateral, security and bills over-due	1,100	0	0			
				£2,053	17	7
				£21,258	7	11

ASSETS.

CREDITOR.

	£	s.	d.	£	s.	d.
By cash in hand				£45	0	0
By debtors, good				34	0	0
„ doubtful	£404	3	3			
„ estimated				269	8	10
By bills receivable, held as collateral security by the City Bank	3,960	12	8			
Less overdrawn account	140	0	0			
				3,820	12	8
By cash advances on consignments	1,082	18	6			
By advances by bills *per contra*	3,729	6	9			
				4,812	5	3
By bills of lading held by City Bank	2,670	10	6			
By brandy certificates	300	0	0			
				2,970	10	6
By commission on consignments secured	7,395	12	1	300	0	0
By shipments to C. and L. Wharton and Co., on joint account	37,579	0	0			
Less drafts, two-thirds	25,052	13	0			
By remittances on account	3,500	0	0			
Carried forward	£28,552	13	0	£12,251	17	3

Brought forward	£28,552	13	0	£12,251	17	3
By insurance and freight	159	17	7			
By cash paid to Mr. Baird	36	5	11			
	£28,748	16	6			
By balance due to J. H. Baird, or respective of profit on the above				8,831	3	6
				£21,083	0	9

The immediate cause of the suspension was stated to be the discontinuance of remittances from Messrs. Wharton and Co, of Melbourne, with whom Messrs. Baird had been connected in business. It was finally resolved that Mr. Baird should proceed to Australia to make the best arrangements in his power, and Mr. Kemp was instructed to act in his absence.

THE ESTATE OF MESSRS. SANDERSON, SANDEMAN, AND CO.

A meeting of the creditors of Messrs. Sanderson, Sandeman, and Co., the bill-brokers, who suspended on the 11th November, 1857, took place on the 10th December, Mr. W. Prescott presiding, when the following statement of accounts was presented :—

MESSRS. SANDERSON, SANDEMAN, AND CO.'S STATEMENT OF AFFAIRS, NOVEMBER 11, 1857.

DEBTOR.

To creditors unsecured									£378,827	14	10
To creditors partially secured					£237,864	5	4				
Less bills of exchange held as security					230,205	9	5		7,658	15	11
This balance will be increased by amount of such bills as may ultimately prove bad.											
To creditors secured					£2,810,008	5	3				
To bills of exchange held as security	£2,959,468	9	6								
To Government stock	45,497	10	0								
					3,004,965	19	6				
Surplus *per contra*					£194,957	14	3				
To liabilities on bills re-discounted					2,015,585	2	11				
Of this amount it is impossible at present to form an estimate of what may prove bad and come against this estate.											
									£386,486	10	9

Interest to date of this statement has not been calculated on debit and credit balances.

CREDITOR.

By cash and bank shares				£1,725	10	9
By bills of exchange on hand, viz. :—						
On houses considered good	£40,891	0	0			
On houses which have suspended payment	52,847	13	10			
				93,738	13	10
It cannot at present be estimated what this item may ultimately produce.						
Carried forward				£95,464	4	7

Brought forward		£95,464 4 7
By loans and debts for which securities are held	£79,701 14 11	
Estimated to realize		73,500 0 0
By sundry debtors, viz. :—		
Good	1,544 19 9	
Debit balances, £206,315 7*s*. 6*d*., estimated to realize	86,980 18 5	
		88,525 18 2
By partners' private estates, about		40,000 0 0
		£297,490 2 9
By surplus securities, as *per contra*		194,957 14 3
This item will be reduced by amount of such bills as may ultimately prove bad.		
		£492,447 17 0

It appeared from the statement made by Mr. Turquand, the accountant, that of the £5,298,990, the total liabilities for which the firm were responsible at the date of the failure, about £2,013,600 have run off. The bills returned in this sum represent about £91,000, of which £49,000 it is estimated will ultimately be paid. Valuing the remainder at about 10*s*. in the pound, the loss in this respect will not exceed £20,000. It is estimated that there will be a further sacrifice of about £30 000 on the £3,000,000 of paper which has yet to arrive at maturity, notwithstanding it is considered to be of a higher character. The firm have discounted £132,000 for the Liverpool Borough Bank, but an amount of about £60,000 has run off, and the remainder is asserted to be secured by promissory notes. The assets are now in course of collection, about £19,000 having already been paid into the Bank of England, the proceeds being invested from time to time in Exchequer bills. The capital originally was about £34,000, but at the date of the failure it had augmented through the accumulation of profits to £83,820. A proposal for winding-up the estate under inspection was unanimously agreed to, and although no doubt was expressed with regard to the payment of 20*s*. in the pound, nothing transpired to indicate whether a reorganization of business will be attempted. The liquidation may, it is thought, last a year ; but all expedition, with due regard to the interests of the creditors, will be used to bring affairs to a close as soon as possible. The principal discussion was with regard to the release to be given to the surviving partner, the power being left in the hands of the inspectors. The resolutions in favour of the liquidation were carried unanimously, Mr. W. G. Prescott, Mr. L. Lloyd, jun., and Mr. Robertson being selected to superintend the administration of the estate.

At present dividends to the extent of 11*s*. in the pound have been paid, and others are expected.

THE ESTATE OF MESSRS. HOARE, BUXTON, AND CO.

A numerous meeting of the creditors of Messrs. Hoare, Buxton, and Co., lately engaged in the Swedish trade, whose suspension was announced on the 17th November, 1857, took place on the 14th December, Mr. Hawkins presiding, when the following statement, prepared by Mr. J. E. Coleman, the accountant, was presented :—

STATEMENT, DECEMBER 14, 1857.

DEBTOR.

To creditors on open accounts		£16,986 1 5
Bills payable		337,121 12 2
To creditors holding security :—		
Security held	£122,545 3 9	
Claims	105,739 18 7	
Surplus to *contra*	16,805 5 2	
Carried forward		£354,107 13 7

Brought forward				£354,107	13	7
To liabilities on bills receivable	£341,495	8	6			
Of which there will not be met at maturity	112,493	7	2	112,493	7	2
Leaving the amount expected to be duly honoured at maturity	£229,002	1	4			
				466,601	0	9
To capital	£38,192	12	8			
To profit and loss, Jan. 1, to Nov. 16, 1857	15,939	4	8			
				54,131	17	4
				£520,732	18	1

CREDITOR.

By cash in hand and at bankers'	£7,118	6	8			
Less amount retained by bankers against loans on security	5,727	18	1			
				£1,390	8	7
By bills receivable on hand				12,604	19	1
By debtors				345,794	14	0
By merchandise				24,578	0	0
By office furniture and sundry assets				802	1	3
				£385,170	2	11
By surplus from creditors holding security				16,805	5	2
				£401,975	8	1
By partners' drawings				6,264	2	10
By liabilities on bills receivable *per contra*				112,493	7	2
				£520,732	18	1

ESTIMATED LIABILITIES AND ASSETS.

DEBTOR.

To creditors on open accounts				£16,986	0	0
To bills payable	£337,121	0	0			
Of which it is expected there will be provided for by the parties for whose account the bills were accepted	172,138	0	0			
Leaving to rank on this estate				164,983	0	0
To creditors holding security :—						
Security held	£122,545	0	0			
Claims	105,740	0	0			
Surplus	£16,805	0	0			
To liabilities on bills receivable	£341,495	0	0			
Of which it is known there will not be met at maturity	£112,493	0	0			
Amount expected to rank on this estate in respect of the last amount	75,000	0	0			
				75,000	0	0
Leaving	£37,493	0	0			
				£256,969	0	0
Which will be taken up by the drawers, etc.						

CREDITOR.

By cash balance				£1,390	0	0
By bills receivable on hand	£12,605	0	0			
Carried forward	£12,605	0	0	£1,390	0	0

Brought forward	£12,605 0 0	£1,390 0 0
Of which there are considered good	6,977 0 0	6,977 0 0
Leaving	£5,628 0 0	
By debtors considered good		10,327 0 0
By debtors considered doubtful	144,071 0 0	
	£149,699 0 0	
Estimated at 7*s*. 6*d*. in the pound		56,137 0 0
By amount for which parties will be debtors after retiring the bills accepted for their account		19,258 0 0
By merchandise		24,578 0 0
By office furniture and sundry assets		802 0 0
		£119,469 0 0

These accounts exhibit a dividend of about 9*s*. in the pound, but the assets are still liable to some fluctuation. The liabilities on the underwriting account are taken at £5,000, and the separate estates of the partners at £20,000; and it is, consequently, hoped that the general amount of distribution may be increased, and not diminished. The two items on the credit side, which will mainly affect the dividend, are the debts, estimated to produce £56,137, and the amount, £19,258, for which parties will be debtors after retiring the bills accepted for their account. It was stated by Mr. Coleman that he is sanguine of a favourable result being arrived at, and that the estimate will be fully realized. At the date of the suspension the accounts showed a surplus of £54,130, and it was anticipated that arrangements could have been effected which would have enabled the partners to resume business. The subsequent failure of Messrs. Carr, Josling, and Co., with the other disasters in this particular branch of business, rendered the attempt nugatory, and hence the depreciation of assets and the present large deficiency. The whole of the operations of the house have been legitimate, and conducted on a purely mercantile basis, the capital absorbed, including the separate estates, being upwards of £72,000. Had it not been for the effects of the crisis in Sweden, there would have been ample to pay the English bill-holders 20*s*. in the pound. As it is, the assets first to be dealt with may, in the course of the next two months—that is to say, in February—produce about 3*s*. 4*d*. in the pound; and when this shall have been accomplished a better estimate of what may be the prospect of further payment can be formed. A liquidation under inspection was the course proposed, but before the resolutions affirming that determination were agreed to, some discussion took place with regard to the position of Mr. Kleman in connection with the firm. It transpired that that gentleman was a partner, but, in addition, transacted business on his private account. His affairs are in a very involved condition, and his property, including landed and mineral estates, being situate in Sweden, it is expected he will have to avail himself of the process of bankruptcy in that country. In the books of Messrs. Hoare, Buxton, and Co., there stands to his debit a sum of £31,000. Great sympathy, in the course of the proceedings, was expressed at the position in which the house is placed, and as it is considered desirable to take immediate action in winding-up, three representatives of the creditors (who, it is suggested, shall have the assistance of two of the partners) were appointed for the purpose.

Dividends to the amount of 10*s*. in the pound have been paid, and others are expected.

THE ESTATE OF MR. G. H. T. HICKS.

At a meeting on the 14th December, 1857, of the creditors of Mr. G. H. T. Hicks, described as an East India merchant, who suspended on the 19th November, the following statement, prepared by Messrs. Turquand, Youngs, and Co., showing very discouraging results, was presented. The assets exhibit a dividend of not more than 3*s*. in the pound, and the debts and liabilities reach the large total of £151,900. Affairs were arranged to be wound-up under inspection.

Statement, December 14, 1857.

DEBTOR.

	£	s.	d.	£	s.	d.
To creditors unsecured				£67,488	3	9
To creditors holding security partly covered	£91,653	3	9			
Deduct estimated value of security	39,700	5	8			
				51,952	18	1
To creditors holding security fully covered	£67,385	1	2			
To liabilities on account of Birrell and Co.:—						
Partly covered	£7,462	6	6			
Deduct securities held	5,500	0	0			
				1,962	6	6
Fully covered	3,745	1	8			
To liabilities on bills receivable, likely to be claimed against estate				28,697	19	6
To liabilities on bills, payable over and above amount included in creditors				1,798	16	3
				£151,900	4	1

CREDITOR.

	£	s.	d.	£	s.	d.
By property held as security by creditors fully covered, estimated at	£83,652	9	11			
Deduct amount of their claims	67,385	1	2			
				£16,267	8	9
By debtors				1,723	3	3
By cash in hand				57	2	6
By homeward freight per "Queen of Ava"	£2,000	0	0			
By ship "Kossuth"	£5,000	0	0			
Average recoverable	2,000	0	0			
	£7,000	0	0			
Deduct debts secured on ship	3,745	1	8			
				3,254	18	4
By ship "Tudor," 4-64, about				650	0	0
By ship "Carnatic," 4-64, about				400	0	0
By iron for Hastings				1,345	3	7
By office furniture, about				50	0	0
By bills of lading of goods shipped to Rangoon	£750	0	0			
				£23,747	16	5
By Birrell and Co., balance of account current, about	£30,000	0	0			
By ships "Arabella," and three-fourths of "Royal Albert," assigned to creditors of Birrell and Co.	£10,000	0	0			

THE ESTATE OF MESSRS. MENDES DA COSTA AND CO.

A meeting of the creditors of Messrs. Mendes Da Costa and Co., West India merchants, whose suspension was announced on the 2nd, was held on the 15th December, 1857, when the following statement was presented by Mr. J. E. Coleman, the accountant:—

Statement, December 15, 1857.

DEBTOR.

	£	s.	d.	£	s.	d.
To creditors on open accounts				£38,117	0	0
To creditors on bills payable	£217,689	0	0			
Carried forward	£217,689	0	0	£38,117	0	0

Brought forward..............	£217,689	0	0	£38,117	0	0
Of which it is expected there will be provided for by the parties for whose account the bills are given	46,833	0	0			
Leaving to rank on this estate				170,856	0	0
To liabilities on bills discounted	£94,204	3	2			
Of which it is expected there will be duly met at maturity	69,004	3	2			
	£25,200	0	0			
Less security held	2,500	0	0			
				22,700	0	0
By bills purchased, negotiated, etc., £36,475 16*s.* 7*d.*, expected to be duly met at maturity.						
				£231,673	0	0

CREDITOR.

By cash at bankers'		£208	0	0
By railway shares and shares in the bank of St. Thomas		1,830	0	0
By bills receivable on hand (good)		21,599	0	0
Ditto (bad)	£8,022 15 0			
By debtors (good)		7,574	0	0
Ditto (doubtful)	£47,425 18 8			
Ditto (bad)	42,592 11 6			
Amounts due by parties after they have retired the bills accepted for their account..		12,520	0	0
By property, consisting of houses and stores at St. Thomas, belonging to Mr. Da Costa, sen., cost £15,000, estimated to realize		10,000	0	0
By cargo of sugar, wood, etc., per "Lotus," on hand, estimated at..		2,000	0	0
		£55,731	0	0

Amount due from Messrs. Da Costa and Co., of St. Thomas......	£112,311	1	10
Ditto, old firm of Charles Da Costa and Co.	2,558	7	1
Amount estimated to be realized from doubtful debts and bills on hand	£30,000	0	0

It was explained that this house, like many others, scarcely expected to find themselves in their present position a month ago, and that when, about the middle of November last, the partners sought the assistance of their bankers, they thought they would have been able to proceed with their business as usual. The devastation which has since occurred among mercantile firms, through the money pressure, has, however, involved them in the general ruin; and early in December they were compelled to succumb. Between the partnership in the London firm and that in St. Thomas there is a distinction, Mr. J. Mendes Da Costa being alone interested in the foreign establishment. The liabilities of the London firm are £231,673, and the assets £55,731. The amount due from Messrs. Da Costa and Co., of St. Thomas, is £112,311, arising out of purely mercantile transactions, and fully represented on the other side. According to the accounts of the 30th of August, the assets were then in due order, and it is expected that the result will be as follows:—It is estimated that about £42,000 of bills will go back upon the house at St. Thomas; that the firm are liable for and owe about £10,000; and that the damages and interest on bills will reach a further sum of £10,000, making a total of £62,000, and leaving a balance of £50,000. If forbearance is exercised, and assistance afforded, it is reasonably hoped that £30,000 of the amount may be recovered. The general statement will then stand in this position:—The assets of £55,000 will be increased, by the addition of the £30,000, to £85,700; and the liabilities decreased from the £231,000, by the £31,000 bills paid at St. Thomas, to £200,000, showing a dividend of about 8*s.* 9*d.* in the pound. The capital absorbed, after allowing for the drawings of the partners, has amounted to £50,000. The only discussion was in relation to the liquidation of the estate, the process of inspectorship being agreed to. It was stated at the close of the proceedings that the partners are denuded of everything, and that consequently, if Mr. Da Costa, jun , proceeds to Porto Rico to protect the interests of the creditors, it will be necessary to make him an allowance.

A dividend of 7*s*. in the pound paid, and a release given to the partners.

THE ESTATE OF MESSRS. GORRISSEN, HÜFFEL, AND CO.

At the meeting held on the 17th December, 1857, of the creditors of Messrs. Gorrissen, Hüffel, and Co., engaged in business as banking and exchange agents, whose suspension was announced on the 18th October, Mr. E. Cohen presiding, the subjoined statement was presented by Mr. Quilter, of the firm of Quilter, Ball, and Co., the accountants :—

STATEMENT, DECEMBER 17, 1857.

DEBTOR.

To sundry creditors, as per statement			£84,134 12 3
To creditor holding security		£7,757 0 0	
Estimated value of security		7,000 0 0	
			757 0 0
To liabilities on acceptances, viz. :—			
Provided for since the suspension		£255,898 7 3	
Not yet provided for, but which it is expected the drawers will meet		196,683 12 11	
Assumed claim on the estate		19,332 2 6	19,332 2 6
		£471,914 2 8	
To liabilities on bills receivable, as per statement		£337,296 16 6	
Of which those considered likely to prove claims amount to		21,643 11 1	
Less cash and surplus securities retained by bankers		2,157 1 7	
			19,486 9 6
To liabilities on charter parties			1,600 0 0
			£125,310 4 3

CREDITOR.

By bills receivable on hand, considered good, as per list			£5,331 15 5
Ditto, belonging to other parties		£2,601 18 0	
By sundry debtors, considered good		£61,179 1 6	
Ditto, doubtful	£15,346 17 2		
Estimated to produce		6,400 0 0	
Ditto, bad	£3,235 13 6		
			67,579 1 6
By sundry merchandise on hand, as per statement estimated at			11,070 0 0
			£83,980 16 11

It was stated that the debts in America, amounting to £33,000, may be considered good, and that, consequently, the estimate of £67,579 under the head of assets, in which that sum is included, may be realized. A large amount of liabilities on acceptances has been provided for, so that the general total since the date of the suspension has been greatly reduced. According to the figures exhibited the dividend will be about 13*s*. 4*d*. in the pound, and at the suggestion of Mr. Hollams, the solicitor, it was agreed to follow out a liquidation by inspection, Mr. E. Cohen, Mr. J. H. Schroeder, and Mr. Melville, being appointed to act as the committee. It transpired in the explanation rendered concerning the deficiency that the losses have been £58,000, and that the capital at the commencement of the year was about £6,000.

Dividends to the extent of 4*s*. in the pound have been paid.

THE ESTATE OF MESSRS. ALEXANDER HINTZ AND CO.

At a meeting on the 18th December, 1857, of the creditors of Messrs. Alexander Hintz and Co., merchants, whose suspension took place on the 24th November, Mr. Wulff presiding, the annexed statements were exhibited by Mr. Quilter, of the firm of Messrs. Quilter, Ball, and Co. :—

STATEMENT, DECEMBER 18, 1857.

DEBTOR.

To sundry creditors unsecured, as per statement				£31,899	10	7
To creditors partially secured—						
Amount of claims	£37,550	4	10			
Less estimated value of securities	26,225	3	11			
				£11,325	0	11
To creditors fully secured—						
Estimated value of securities	£17,703	10	3			
Amount of claims	15,078	3	7			
Surplus	£2,625	6	8			
To liabilities on acceptances—						
For account of other parties	£169,118	7	8			
Assumed claims on the estate				28,192	5	9
To liabilities on bills receivable, as per statement	171,039	6	4			
Of which it is estimated the claims on the estate will amount to				26,522	13	11
To liabilities for goods purchased principally on account of continental houses, and which should be paid for by them, assumed claim on this estate				3,500	0	0
				£101,439	11	2

CREDITOR.

By cash at bankers'—						
Subject to bills current, but considered good				£5,620	11	0
By bills receivable on hand—						
Considered good				12,284	0	11
Ditto, doubtful	£1,388	6	11			
Estimated to realize				462	15	8
By sundry debtors—						
Considered good				9,165	14	11
Ditto, doubtful	14,480	17	10			
Estimated to produce				5,000	0	0
Considered bad	2,085	1	8			
By sundry merchandise on hand, as per statement, estimated at				6,187	16	2
By surplus securities, in hands of creditors				2,625	6	8
By office furniture				200	0	0
				£41,546	5	4

STATEMENT ACCOUNTING FOR THE DEFICIENCY EXHIBITED BY THE STATEMENT OF THEIR AFFAIRS, NOVEMBER 25, 1857.

DEBTOR.

To deficiency, November 25, 1857				£59,893	5	10
To capital, January 1, 1857	£16,537	14	5			
Add subsequent profits	11,166	14	1			
				27,704	8	6
Carried forward				£87,597	14	4

Brought forward			£87,579 14 4
Note.—It appears from these figures that, at the date of the suspension, the capital account stood thus :—			
At credit, as above		£27,704 8 6	
At debit, as *per contra*		6,004 14 8	
Balance at credit		£21,699 13 10	
			£87,597 14 4

CREDITOR.

By expenditure and losses incurred and ascertained up to the date of the suspension, November 25, 1857, viz. :—			
Trade charges		£779 6 0	
Bad debts and sundries		2,543 5 8	
Partners' drawings		2,682 3 0	
			£6,004 14 8
By losses accruing from and since the date of the suspension, viz. :—			
Estimated loss on bills receivable		£925 11 3	
Ditto on doubtful debts		9,480 17 10	
By depreciation in merchandise on hand thus—			
Cost	£46,355 14 6		
Value	33,384 3 7		
		12,971 10 11	
			23,378 0 0
By liabilities in respect of bills which should be provided by other parties, but which, it is assumed, will be claimed on this estate, viz. :—			
On our acceptances		£28,192 5 5	
On bills receivable		26,522 13 11	
			54,714 19 4
By liabilities in respect of unfulfilled contracts...			3,500 0 0
			£87,597 14 0

The assets, it was explained, show a dividend of about 8*s.* in the pound, but if the liabilities are increased through fresh failures, this amount will not be realized. The suspension of the firm at the period indicated has, it appears, proved beneficial, and the balance at the bankers', it is considered, may be relied on, should an important house at Hamburg be able to struggle through present difficulties. Mr. Hollams, the solicitor, intimated that the course of arrangement suggested was the process of inspectorship, and that the creditors proposed to select Mr. Brown, Mr. Cæsar, and Mr. Abegg to watch over their interests. Much sympathy was expressed for the position of the firm, and resolutions assenting to this mode of liquidation were passed.

The estate wound-up under inspection, and dividends to the extent of 5*s.* already paid ; others anticipated.

THE ESTATE OF MESSRS. ALLEN AND SMITH.

At a meeting, on the 18th December, 1857, of the creditors of Messrs. Allen and Smith, whose suspension was announced on the 18th of November, the subjoined statement was presented by Mr. Ball, of the firm of Messrs. Quilter, Ball, and Co., showing, on a trading of eight months, a deficiency of £15,762. This was principally created by the failure of houses at home and abroad, connected with the Swedish trade :—

Statement, December 18, 1857.

DEBTOR.

To sundry creditors			£1,032 4 2
To liabilities on acceptances not expected to be proved against this estate, viz.:—			
Drafts from Stockholm against securities deposited there		£13,653 10 4	
Drafts on account of persons who will probably provide for them		9,958 14 6	
To sundry other acceptances		984 5 11	
		£24,596 10 9	
To liabilities on acceptances on account of parties considered bad			9,825 8 5
To liabilities on bills receivable—			
Considered good		£2,485 7 2	
Considered bad		9,620 17 2	
Less securities held		2,472 0 6	
			7,148 16 8
To liabilities in respect of estimated amount of loss on sundry unfulfilled contracts for saltpetre			2,300 0 0
			£20,306 9 3

CREDITOR.

By cash in hand			£17 0 11
By bills on hand		£1,000 0 0	
Estimated to realize			750 0 0
By sundry debtors—			
Considered good		280 10 3	
Ditto doubtful	£1,250 16 10		
Estimated to realize		230 0 0	
Estimated bad	7,579 4 11		510 10 3
By consignments outstanding, estimated to realize			1,000 0 0
By one-third share of the ship "G. B. Carr," of the estimated value £5,000			1,666 13 4
By timber on hand, estimated to realize			480 0 0
By office furniture, estimated value			120 0 0
			£5,544 4 6
Balance, being deficiency carried forward			14,762 4 9
Note.—This deficiency is liable to increase to the extent of any amount of the above liabilities, estimated as good proving bad.			
			£20,306 9 3

The feature in this case, according to the statement of Mr .M'Leod, was the position in which the senior partner was placed. Having joined Mr. Smith, they commenced with a capital belonging to Mr. Allen, of £5,000, which was subsequently increased to £7,000, and, after trading for the short period stated, they were compelled to suspend. The assets of the joint estate do not exceed £4,544, and they may be increased by the separate estate of Smith to the amount of £500. The separate property of Mr. Allen is, however, considerable, and there appears no doubt but that the whole of the creditors will receive 20*s.* in the pound. Mr. Murray, who represented large and influential claimants, agreed to the course proposed, but some legal arrangements will have to be effected before the entire separate estate of Mr. Allen can be made available for the joint creditors. An immediate dividend of about 5s. in the pound will be declared, pending the proceedings for the realization of the other assets. In answer to a question, it was mentioned that the suspension was necessary to prevent the sacrifice of the whole of the property of Mr. Allen. Resolutions were then passed for a liquidation under inspection, and Mr. Ball was appointed official inspector to superintend the winding-up of the estate.

THE ESTATE OF MESSRS. BARBER, ROSENAUER, AND CO.

The attendance of creditors, on the 18th December, 1857, at the meeting of Messrs. Barber, Rosenauer, and Co., whose suspension was announced on the 3rd, was very numerous, and the proceedings were of a rather protracted nature. The chair was occupied by Mr. J. W. Mitchell, and the statement presented by Mr. Henry Chatteris exhibited the subjoined unsatisfactory results:—

STATEMENT, DECEMBER 18, 1857.

DEBTOR.

To creditors unsecured			£17,771 16 6
To ditto, partially secured		£33,091 13 7	
Value of securities held		25,476 9 1	
			7,615 4 6
To creditors fully secured		398 13 6	
Value of securities held		1,009 10 3	
Surplus *contra*		£610 16 9	
To liability on bills payable which should be retired by drawers		2,768 10 5	
Expected to be claimed on this estate			1,509 15 6
To liability on bills receivable negotiated		55,451 16 10	
Expected to be dishonoured		£6,419 7 1	
Less balance due to acceptors included with creditors	£735 11 11		
Less cash at bankers	92 4 11		
		827 16 10	
			5,591 10 3
			£32,488 6 9

CREDITOR.

By produce abroad expected to realize	£631 6 7	
Less due to consignee	58 9 10	
		£572 16 9
By wine on hand		35 18 9
By debtors considered good		1,993 1 11
By ditto, doubtful	£916 14 2	
Estimated to realize		136 0 0
By disputed debts	£657 5 11	
Estimated to realize		220 0 0
By bad debts	£3,202 2 3	
By surplus securities held by creditors		610 16 9
By cash in hand		127 16 9
By office furniture		100 0 0
		£3,796 10 11
Less rent and salaries, to be paid in full		46 6 1
Carried forward		£3,750 4 10

Brought forward	£3,750	4	10
By deficiency..	28,738	1	11
	£32,488	6	9

The circumstances elicited in relation to this estate were of the most discouraging description. Having traded between two and three years to an enormous extent upon a limited capital, there is, upon debts and liabilities amounting to £32,488, a deficiency of £28,738. The remaining assets are of a very doubtful nature, and there seems to be little expectation of a dividend. The firm commenced in 1854, with a capital of about £4,000, and at the end of 1855, the profits reached £1,048, but the partners drew to the amount of £1,139. They continued to operate without the least regard to the probabilities of success, and in the end the net loss at the date of the present year when they suspended was £30,101, included in which were bad debts for £10,103, and depreciation of merchandise £13,967. Not confining their transactions to any particular branch, they traded in everything, but loss was invariably the consequence; and to illustrate the amount of their business, it was stated that the value of merchandise which passed through their hands in 1856, was £213,000, while this year it represented £187,000. Complaints were made of the conduct of the partners in endeavouring to lead individuals to believe that their estate was better than it really appeared to be, especially as it transpired there was no separate property, and that Mr. Barber had settled his furniture, and a policy of insurance for £2,000, upon his wife. To a firm trading desperately like this, the certainty of a suspension must be apparent; and, as a creditor very sagaciously remarked, it would be sure to follow whether "the rate of money ruled at 2 or 10 per cent." The discussion which subsequently occurred, was directed to the measures that should be adopted to wind-up the estate, and the great majority being in favour of bankruptcy, the annexed resolution was passed:—

"The creditors, after hearing a statement of the affairs, resolved—That the estate be wound-up under bankruptcy, and that Messrs. Barber, Rosenauer, and Co. be requested to sign a declaration of insolvency."

This estate was wound-up in bankruptcy.

THE ESTATE OF MESSRS. KRELL AND COHN.

A meeting of the creditors of Messrs. Krell and Cohn, commission agents, who failed on the 7th, was held on the 21st December, 1857, when the following statement was presented by Mr. Collison, of the firm of Messrs. Parrinton, Ladbury, and Co., the accountants:—

STATEMENT, DECEMBER 21, 1857.

DEBTOR.

To creditors as per list	£13,517	4	5			
Less amounts secured by goods in their hands ...	104	11	6			
				£13,412	12	11
To creditors in respect of bills drawn or accepted by Messrs. Krell and Cohn, and discounted, for which they have had value in full				6,187	10	9
To liabilities on bills accepted for persons who have failed, and which will be proved against the estate				2,290	4	2
To liabilities on bills receivable discounted, considered good	£20,406	7	4			
To liabilities which will be proved on this estate	2,181	9	4			
Less cash in the hands of bankers, who hold the bills	554	17	0			
				1,626	12	4
				£23,517	0	2

CREDITOR.

By stock in London and Lyons, at cost		£1,547 5 9
By stock with agents at Berlin, Vienna, Milan, Hamburg, Rotterdam, etc.	£2,698 10 6	
Subject to claims thereon	104 11 6	
		2,593 19 0
By book debts, good		8,551 7 3
By book debts, doubtful, bad, and disputed	£4,746 10 6	
Estimated at		1,216 6 1
By cash on hand		625 1 4
By bills receivable on hand, good		2,358 10 11
By bills receivable on hand, doubtful	£836 12 9	
Estimated at		568 3 10
By cash at bankers'		113 16 7
By cash with bankers (held against bills discounted)	£554 17 0	
By fixtures in London and Lyons		60 0 0
		£17,634 10 9
Less rent, salaries, etc.		96 15 0
		£17,537 15 9

It was explained that the assets, according to the statement, show a dividend of about 14*s.* 11*d.* in the pound, but that since the accounts were made up bad debts have increased. The firm possessed establishments both in London and Lyons. As commission agents, purchasing exclusively on orders, they should hold no stock, and that which is in their hands has been returned through the unsettled state of affairs on the Continent, buyers not having completed their transactions. They commenced business without capital, and last year their balance-sheet showed a surplus of about £1,700. Their difficulties are wholly attributed to the progress of the panic throughout Germany, which has involved them in a large amount of loss and bad debt. Mr. Reed, their professional adviser, detailed at length the circumstances under which the firm suspended, and the prospects of the creditors. Messrs. Krell and Cohn were, he stated, ready either to facilitate a liquidation through bankruptcy, to make an assignment, or to comply with any arrangement under inspection. With regard to a composition, they felt themselves placed in a doubtful situation, the state of business abroad and the difficulty of collecting debts not permitting them to make so good an offer as they should wish. Although the accounts exhibited 14*s.* 11*d.* in the pound, Messrs. Krell and Cohn were not prepared to propose a greater amount than 10*s.*; and if the creditors appeared inclined to accept it, they will endeavour to obtain security for the last instalment. Several of the creditors considered that the margin allowed between the assets represented, and the sum required to pay 10*s.* in the pound was too large; but Mr. Reed and Mr. Parrinton both intimated that, owing to the situation of affairs in Hamburg, Berlin, and other places, the debtors felt that they could not hope to realize a larger amount, though they stated their readiness, if an excess were obtained, to make it available for the general benefit of the creditors. After some discussion, in which the agent of the foreign bankers bore testimony to the honest nature of the transactions of Messrs. Krell and Cohn, the following resolution was agreed to:—

"Resolved—That it be referred to a committee, consisting of the Chairman, Mr. Dunn, Mr. Kohnstammer, and Mr. John Scott, to consider and decide on the propriety of accepting the offer of composition made by Messrs. Krell and Cohn, or any larger amount of composition, and what security should be required for the last instalment, with liberty to require a variation in the times of payment."

THE ESTATE OF MESSRS. REHDER AND BOLDEMANN.

The creditors of Messrs. Rehder and Boldemann, who failed on the 25th of November, 1857, in the German trade, held a meeting on the 21st December, when the following statement was presented by Mr. Ball, of the firm of Messrs. Quilter, Ball, and Co., the accountants :—

Statement, December 21, 1857.

DEBTOR.

To sundry creditors unsecured		£39,952 3 2
To creditors partially secured		
Amount of claims	£9,498 9 8	
Estimated value of securities	3,874 4 0	
		5,624 5 8
To creditors fully secured—		
Estimated value of securities	£20,528 8 7	
Amount of claims	15,984 17 0	
Surplus	4,543 11 7	
To liabilities on acceptances	114,348 14 5	
Secured to the extent of the value of consignments on hand of the estimated value of	32,922 2 6	
Assumed to be claims on this estate		2,070 11 9
To liabilities on bills receivable as per statement £136,294 19 6		
Of which those considered likely to prove claims amount to	10,208 19 2	
Less cash balance retained by bankers	982 16 6	
		9,226 2 8
To estimated claim in respect of goods bought and not delivered		1,000 0 0
		£57,873 3 3

CREDITOR.

By cash and bills receivable on hand, considered good, as per statement		£4,312 12 2
By cash retained by bankers, *per contra*	£982 16 6	
By sundry debtors—		
Considered good	£5,474 10 10	
Ditto, doubtful £5,199 9 11		
Estimated to produce	1,949 16 3	
		7,424 7 1
Ditto, bad £642 17 11		
By sundry merchandise on hand		
At home	£18,090 1 1	
Abroad	306 0 0	
		18,396 1 1
By sundry property		550 0 0
By surplus securities held by creditors, *per contra*		4,543 11 7
By property deposited as security for certain returned bills, which it is considered will be paid		2,273 0 0
		£37,499 11 11

Statement accounting for the Deficiency.

DEBTOR.

To deficiency, November 25, 1857		£20,373 11 4
To partners' capital—		
January 1, 1857	£6,991 1 9	
Add subsequent profits	3,698 18 3	
		10,690 0 0
		£31,063 11 4

CREDITOR.

By trade charges		£1,152 16 8
By losses on merchandise realized, estimated depreciation in value of unrealized stock, now valued at £43,437 9*s*. 8*d*., and sundries		15,128 6 0
By bad and doubtful debts, viz.:—		
Estimated loss on doubtful	£3,249 13 8	
Bad	642 17 11	
		3,892 11 7
By partners' drawings, from January 1, 1857		2,096 14 2
By liabilities, viz.:—		
On acceptances	£2,070 11 9	
On bills receivable	10,208 19 2	
On goods bought, but not delivered	1,000 0 0	
	£13,279 10 11	
Less produce held by creditors	4,486 8 0	
		8,793 2 11
		£31,063 11 4

These figures exhibit a dividend of about 12*s*. 6*d*. in the pound, but it is not certain that this amount will be realized. The firm commenced business on the 1st of January in the present year with a capital of £6,991, and the failure, it appears, has been caused by the depreciation of produce and the failure of other houses. In answer to a question, it was stated that there had been no accommodation transactions, but blank credits have been opened on foreign account to the extent of £114,348. On these, however, only a small amount of claim is anticipated. Some of the creditors appeared to believe that there is little difference between the terms "blank credits," and "accommodation transactions." Mr. Hollams, who attended for Messrs. Rehder and Boldemann, intimated that these gentlemen are not in a position to make any proposal, and that it has been suggested the most prudent course would be, to effect a liquidation through inspection. The deficiency between the assets and liabilities is £20,000, but the depreciation on produce has been upwards of £23,000. The produce in which the firm is interested consists of tea, cotton, and other articles, and consequently the estate could be readily put in process of liquidation. It was eventually agreed to wind-up under inspection, the following creditors being appointed to superintend:—Mr. Maurencrether, Mr. Alvers, and Mr. J. Trueman.

This estate has already paid dividends to the extent of 6*s*. in the pound.

THE ESTATE OF MESSRS. HERMAN SILLEM AND CO.

A meeting of the creditors of Messrs. Herman Sillem, Son, and Co., in the Hamburg and German trade, was held on the 22nd December, 1857, Mr. J. H. W. Schroeder in the chair, when the following statement was presented by Mr. Quilter, of the firm of Messrs. Quilter, Ball, and Co., the accountants:—

STATEMENT, DECEMBER 22, 1857.

DEBTOR.

To sundry creditors unsecured, as per statement		£61,084 6 7
To creditors in respect of bills receivable, to be given up	33,033 11 5	
To liabilities on acceptances	£344,851 19 10	
Actually cancelled	£67,934 13 5	
Interfered for, referred, or guaranteed	153,431 12 10	
Carried forward	£221,366 6 3	£61,084 6 7

Brought forward				£221,366	6	3	£61,084	6	7
Not interfered for at present, but believed to be good				101,388	6	10			
Estimated to be claimed on this estate				22,097	6	9	22,097	6	9
				£344,851	19	10			
To liabilities on bills receivable and bills negotiated, as per statement, £171,274 18*s.* 9*d.*, of which those considered likely to prove claims amount to				£10,889	8	5			
Less secured by balance at bankers', and an amount due on bills negotiated				1,023	2	9			
							9,866	5	8
							£93,047	19	0
CREDITOR.									
By cash at bankers'				£9,281	7	0			
Less liable to pay bills receivable, considered bad *per contra*				517	17	10			
							£8,763	9	2
By bills receivable on hand at the date of the suspension				£66,512	2	5			
Less bills to be given up				33,033	11	5			
							33,478	11	0
By sundry debtors—									
Considered good				£42,444	18	11			
Ditto doubtful	£6,736	6	1						
Estimated to produce				2,455	5	1			
Ditto bad	£1,037	15	1						
							44,900	4	0
By lease of house, No. 2, Crosby Square							317	10	0
Surplus of the private estate of the partners, estimated at							2,000	0	0
							£89,459	14	2

This estate showed nearly 20*s.* in the pound, and the expectation was, that a dividend of 6*s.* 8*d.* might be paid at the end of February, when the position of the bills not interfered for shall have been ascertained. There was cash at the bankers' bearing interest, £25,000; balance at the bankers', £5,000; cash at the Bank of England, not bearing interest, £7,000; with good bills on hand, £8,000, making a total of £45,000. The amount of the capital of the firm was £32,000, and this, together with a further slight sum, has been wholly absorbe dthrough the disasters which have taken place in Hamburg and Germany. Previous to these events, the house, as proved by the accounts, was solvent. Although the liabilities on acceptances have reached the large total of £344,851, the estimate of the amount to be claimed on the estate was £22,097. The sum not interfered for was £101,388, but the character of the houses they represent led to the expectation that they would be paid. A distribution cannot be made before the end of February or the commencement of March, as some of the bills will not run off before that date. In answer to questions, it was stated that the open credits have been granted to first-class houses in the regular course of trade, against consignments, etc., and that it testifies greatly to their stability to find that the estimated loss, so far as can at present be ascertained, will be only £22,000. The proposal for a liquidation under inspection was moved by the Chairman, and immediately seconded, the gentlemen appointed to act being Mr. T. Baring, M.P., Mr. J. H. W. Schroeder, and Mr. L. Miéville.

Dividends to the amount of 20*s.* in the pound have been paid, with interest.

THE ESTATE OF MESSRS. DRAPER, PIETRONI, AND CO.

The creditors of Messrs. Draper, Pietroni, and Co., in the Mediterranean trade, assembled on the 22nd December, 1857, Mr. Fitzpatrick, of the firm of Maudsley, Field, and Co., presiding; when the subjoined statement was presented by Mr. J. E. Coleman, the accountant:—

STATEMENT, DECEMBER 22, 1857.

DEBTOR.

To creditors on open accounts		£40,467 14 7
To bills payable	£189,369 18 4	
Of which it is expected there will be provided for by the parties for whose account the bills are accepted	134,284 5 10	
Leaving to rank on this estate		£55,085 12 6
To creditors partly secured—		
Claims	£19,639 0 1	
Security held	18,895 8 3	
Deficiency		£743 11 10
To creditors fully secured—		
Security held	£120,022 3 8	
Claims	106,440 8 6	
Surplus to *contra*	£13,581 15 2	
		96,296 18 11
To liabilities on bills receivable	£175,237 16 11	
Of which it is expected there will be duly honoured at maturity	171,984 14 9	
Leaving to rank on this estate		3,253 2 2
		£99,550 1 1

Further liabilities on account of the Transatlantic Steam Navigation Company, in respect of credits opened for their account against bottomry bonds on the vessels of the company, £26,440; of which amount the company have accepted bills to the extent of £8,460.

CREDITOR.

By cash at bankers', subjected to bills discounted now running, considered good		£3,465 4 2
By bills receivable on hand		5,325 9 6
By debtors, good		11,734 12 11
By amounts for which parties will be debtors after taking up the bills accepted for their account		10,598 13 11
By debtors doubtful, £18,385 11*s.* 9*d.*, estimated at 3*s.* 6*d.* in the pound		3,217 9 4
By debtors, bad	£22,897 18 5	
By silk on hand, etc.		3,456 11 0
By surplus from creditors holding security, *contra*		13,581 15 2
By shares and sundry assets		2,477 0 0
		£53,856 16 0
Less creditors to be paid in full		294 19 0
		£53,561 17 0

Amount claimed by Messrs. Draper, Pietroni, and Co., of the Transatlantic Steam Navigation Company, exclusive of all liabilities, £33,000, subject to the delivery of 1,285 shares to the company.

From the explanations afforded by Mr. Coleman, it appeared that there was a prospect of a dividend of from 10*s*. to 12*s*. 6*d*. in the pound, and that 10*s*. at least might be considered certain. The solidity of the connections of the house may, it was thought, be estimated by the fact that, of the large amount of the bills, only an amount equal to two per cent. of the total is likely to come against the estate. The creditors on open account included £15,000 on behalf of the Transatlantic Steam Company. With respect to the assets, the cash and bills were considered good, the debts stand secured, and the surplus from silk would, it was expected, be realized. The connection of Messrs. Draper and Pietroni with the Transatlantic Company was entered into at length, showing how their claim against the directors arose, and it was distinctly alleged that the neglect of the company to provide remittances had necessitated the suspension of the firm. It appears that Messrs. Draper and Pietroni entered into an arrangement with the Transatlantic Company to provide one-fourth of the capital. A contract for the construction of seven vessels was then entered into. Four vessels were built at a cost of £260,525, the contract being for £400,000, but, since it has not been carried out, instead of taking the 3,000 shares, as originally intended, they accepted 1,715 shares, and claim £32,000, the difference between that number and the quantity for which they at first subscribed. These and other points would be submitted to the consideration of parties who might have to determine the question of account, Messrs. Draper and Pietroni being quite ready to meet the Transatlantic Company on fair business ground. It was stated that the company were about to settle with the contractors an amount of £13,000, and if such was the case, the liabilities would be diminished to that extent. The deficiency shown by the statement was £42,000, which was fully accounted for through bad debts and liabilities, the capital of the firm having been about £45,000. No definite proposal could be made, but a liquidation through the process of inspectorship would permit facilities for arrangement. A considerable dividend, it was believed, may be paid in February, and then the partners would be in a situation to see what further can be done in settling fixed terms by their promissory notes for future payments. The assets in hand were about £12,500, and the collection of everything would proceed as expeditiously as possible. After some conversation, it was agreed that the estate should be wound-up under a deed of inspection, and the following representatives of creditors were selected to act:—Mr. Rennie, Mr. Yames, Mr. Fitzpatrick, and Mr. Price.

A dividend of 3*s*. in the pound has been already paid, and further distributions are expected.

THE ESTATE OF MR. THOMAS MELLADEW.

A meeting of the creditors of Mr. Thomas Melladew, commission agent, who had recently failed, was held on the 22nd December, 1857, when the following unsatisfactory statement of the debtor's position was presented by Messrs. Turquand, Youngs, and Co. A proposal was made to pay 3*s*. in the pound, cash down, and a committee of two creditors was appointed to investigate and to report to an adjourned meeting:—

STATEMENT, DECEMBER 22, 1857.

DEBTOR.

To creditors unsecured		£19,544 6 2
To creditors holding security, partly covered ...	£714 8 1	
Deduct estimated value of security	555 2 0	
		159 6 1
To creditors holding security, fully covered ... See *contra*.	£17,295 18 3	
To liabilities on bills payable, over and above amount included in creditors	7,370 18 8	
Deduct amount considered likely to be retired by parties for whom same were accepted...	914 15 3	
		6,456
To liablities on bills receivable....................	£10,011 8 7	
(None of which it is expected will be claimed against this estate.)		
		£26,159 15 8

CREDITOR.

By debtors, good		£1,857 0 11
By ditto, doubtful	£1,238 3 7	
Considered worth about		200 0 0
By ditto, bad	7,896 4 2	
By property unencumbered		1,785 0 0
By bills receivable in hand		54 3 5
By property held by creditors—		
Estimated value....................................	£17,598 15 8	
Deduct creditors' claims	17,295 18 3	
		302 17 5
		£4,199 1 9
Deduct creditors to be paid in full		321 19 7
		£3,877 2 2

At the adjourned meeting on the 5th January, a composition of 3*s.* was accepted.

THE ESTATE OF MESSRS. E. SIEVEKING AND SON.

A meeting of the creditors of Messrs. E. Sieveking and Son, connected with the Swedish trade, who failed in December, 1857, took place on the 23rd of the same month, Mr. Meinertzhagen in the chair, when the subjoined statement was presented by Mr. Quilter, of the firm of Quilter, Ball, and Co., accountants:—

STATEMENT, DECEMBER 23, 1857.

DEBTOR.

To sundry creditors, as per statement		£77,229 3 10
To sundry creditors partially secured, as per statement	£18,000 0 0	
Less estimated value of security	15,866 0 11	
		2,133 19 1
To sundry creditors fully secured—		
Amount of claims	£4,970 9 2	
Value of security	7,000 0 0	
Surplus to *contra*	£2,029 10 10	
To liabilities on acceptances, as per list	£454,690 2 4	
Of which it is expected claims will arise on...		125,014 1 4
To liabilities on bills receivable, as per statement	£348,956 12 0	
Of which those reckoned as bad and claimable on us amount to	£57,110 4 8	
Deduct securities in the hands of holders, as per statement................................	3,122 3 7	
		53,988 1 1
		£258,365 5 4

CREDITOR.

By bills receivable on hand, as per list—			
Considered good		£26,191 15 2	
Ditto doubtful	£9,041 3 6		
Estimated at		2,000 0 0	
Carried forward..............			28,191 15 2

Brought forward			£28,191 15 2
By sundry debtors, as per list—			
Considered good		£56,900 13 1	
Ditto, being represented by merchandise on hand, as per statement		45,195 4 0	
Ditto, doubtful	£24,810 1 3		
Estimated at		10,800 0 0	
Ditto, bad	7,677 2 7		112,895 17 1
By surplus securities in the hands of creditors			2,029 10 10
			£143,117 3 1

This estate shows a dividend of from 10*s.* to 11*s.* in the pound, and it was immediately agreed that the most satisfactory mode of proceeding would be a liquidation under inspection. The deficiency which the accounts exhibits is fully explained by the losses incurred through failures since October, amounting to £206,603; the capital of £72,000, with a small balance of £2,000 for commissions and interest, being likewise swept away. Of the £182,000 represented by bills, £125,000 have reference to Sweden, £18,000 to Hamburg, and £39,000 to the West Indies and other places. There is no separate property of the partners except furniture, and this, it is believed, will be little more than sufficient to satisfy private claims. With respect to four of the largest items of liability connected with establishments at Stockholm and Gottenburg, amounting to £36,555, it was mentioned that the partners take a more favourable view than the accountants of the probable result. Much sympathy was expressed for the position of the firm, and the representatives of Messrs. Hambro and other influential houses intimated a desire that the furniture and plate belonging to the partners should not be interfered with. The inspectors appointed to superintend the liquidation of the estate were Mr. H. Huth, of the firm of Messrs. F. Huth and Co., and Mr. W. H. Göschen, of the firm of Messrs. Fruhling and Göschen.

The estate has already paid 7*s.* in the pound, and further dividends are anticipated.

THE ESTATE OF MESSRS. HIRSCH, STROTHER, AND BRISI.

A meeting of the creditors of Messrs. Hirsch, Strother, and Brisi, who suspended in the early part of the month, was held on the 23rd of December, 1857, when the following balance-sheet was presented by Mr. Maynard, the accountant :—

STATEMENT, DECEMBER 23, 1857.

DEBTOR.

To creditors unsecured			£12,198 15 1
Ditto, partly secured		£10,472 6 3	
Securities valued at		7,383 10 9	
			3,088 15 6
To creditors secured		£9,416 14 9	
Securities valued at		10,048 18 11	
Surplus carried to *contra*		£632 4 2	£15,287 10 7

LIABILITIES.

On bills payable		£25,481 0 2	
Whereof there has been or is expected to be provided for by other obligations	£3,706 10 6		
Included in unsecured creditors	7,612 19 10		
		11,319 10 4	
			£14,161 9 10
Carried forward			£29,449 0 5

Brought forward			£29,449 0 5
On bills receivable		£80,502 17 9	
Whereof there has been provided for by other obligations	£26,697 4 9		
Expected further to be provided for by others	37,918 17 5		
		64,616 2 2	
			15,886 15 7
On rape oil		1,000 0 0	
Bought abroad for delivery in London contingent on the vendor's enforcing delivery.			
On contracts for forward delivery of tallow ...		1,000 0 0	
			2,000 0 0
On contingencies		£5,000 0 0	
			£47,335 16 0

Note.—It is probable that other contingencies will arise, out of which claims may be made against the estate to the extent of (say) this amount.

	Ledger Balances.	Estimd. Value.	
By oil and seed on hand......	£677 13 10	£677 13 10	
By cash and bills receivable on hand	£1,775 3 3	1,775 3 3	
Ditto at bankers' retained against advance	264 15 2		
By contracts.....................	£5,171 18 8	4,000 0 0	
The realization contingent on the business being continued and our firm being able to make deliveries in conformity therewith, and of the parties being able to pay differences.			
By debtors, good..............	£4,210 0 7	3,000 0 0	
Ditto, doubtful	£7,541 18 11	1,000 0 0	
By bills receivable	£269 16 0		
Belonging to other parties.			
By surplus securities	632 4 2		£10,452 17 1
Held by creditors.			
	£902 0 2		
Deduct salaries, rent, and sundries, W. Culverwell, one year's salary to 31st December, 1857	£80 0 0		
Less paid on account	78 19 5		
		1 0 7	
S. W. Moens, one year's salary to 31st December, 1857	250 0 0		
Less paid on account, per ledger	240 18 6		
		9 1 6	
Carried forward		£10 2 1	£10,452 17 1

Brought forward				£10	2	1	£10,452	17	1
Thomas Reynolds, Great St. Helen's, one quarter's rent to 31st December, 1857, at £75 per annum				18	15	0			
Queen's taxes (say)				30	0	0			
Creditors, under £10	66	4	0						
Sundries	150	0	0	216	4	0	275	1	1
							10,177	16	0
Deficiency................................							37,158	0	0
							£47,335	16	0

The operations of the firm were considered in some respects to have been of an irregular character, but still it was proposed to wind-up through the process of inspection. A proposal to make a composition of 3*s.* 6*d.* in the pound (2*s.* 6*d.* down, and 1*s.* in six months), a release to be given on the payment of the last instalment, was well received, but one creditor objected, and it remains to be seen whether it can be carried out.

THE ESTATE OF MESSRS. HADLAND, SHILLINGFORD, AND CO.

At a meeting of the creditors of Messrs. Hadland, Shillingford, and Co., in the Manchester trade, held on the 23rd December, 1857, the appended statement was presented by Mr. C. F. Kemp, the accountant. After some discussion, a resolution, agreeing to accept a composition of 16*s.* in the pound by three instalments of 6*s.*, 6*s.*, and 4*s.*, at three, six, and nine months' date, the last secured to the satisfaction of three of the principal creditors, was adopted. Annexed is a copy of the balance-sheet presented to the meeting :—

STATEMENT, DECEMBER 23, 1857.

DEBTOR.

To creditors, as per list.............................	£29,063	17	6			
Ditto under £10	184	18	1			
				£29,248	15	7
To liability on bills receivable, £7,300 5*s.* 8*d.*, of which it is estimated will come against this estate	£1,968	6	8			
less balance held by Fuller and Co.	13	10	6			
				1,954	16	2
Surplus				639	0	10
				£31,842	12	7

CREDITOR.

By debtors, good				£10,473	4	3
Ditto, doubtful and bad	£4,334	14	2			
Estimated to realize				941	11	8
By cash on hand				1,120	0	0
By bills receivable, on hand	3,930	1	8			
Estimated to realize..				3,537	1	6
By stock on hand, at cost				16,108	5	9
By fixtures				200	0	0
Carried forward.................				£32,380	3	2

Brought forward		£32,380 3 2
Less salaries, taxes, and rent, to be paid in full		537 10 7
		£31,842 12 7
By surplus		£639 0 10
And Mr. Hadland's private estate.		

THE ESTATE OF MESSRS. LICHTENSTEIN AND CO.

A meeting of the creditors of Messrs. Lichtenstein and Co., who failed in the German trade, was held on the 23rd of December, 1857, Mr. Brown presiding, when the following statement, prepared by Mr. J. E. Coleman, the accountant, was submitted :—

STATEMENT, DECEMBER 23, 1857.

DEBTOR.

To creditors on open accounts		£4,967 15 10
To creditors on bills payable........................	£14,597 13 6	
Less debit balances due by the parties for whose accounts the bills were accepted	12,071 12 8	2,526 0 10
To creditors partially secured	£5,540 6 5	
To security held....................................	3,621 0 0	1,919 6 5
To creditors fully secured—		
Security held	£4,470 10 11	
To claims	4,330 2 7	
Contra..	£140 8 4	
Direct liabilities		£9,413 3 1
To liabilities on bills discounted, negotiated, etc.	£47,680 4 8	
Of which it is expected there will be met at maturity	32,601 17 11	15,078 6 9
To liabilities on open contracts		1,010 0 0
		£25,501 9 10

CREDITOR.

By cash balances		£2,845 14 7
By debtors, good		3,498 0 6
By debtors, doubtful	£121 12 4	
Estimated to produce 10*s.* in the pound		60 16 2
By goods, etc., in hand		3,960 7 3
By bills receivable in hand..		484 2 0
By surplus securities with creditors, *per contra*		140 8 4
By amounts due from parties after taking up the drafts for their account		155 9 5
		£11,144 18 3
Less amounts to be paid in full		119 18 3
In respect of the bills, amounting to £15,078 6*s.* 9*d.*, which will at present rank on this estate, it is expected that reclamations will be made from the acceptors to some extent, in which case such reclamations will be available for division amongst the general creditors.		£11,025 0 0

Policy for £2,000 on the life of S. Lichtenstein, effected in 1853, in the Scottish Widows' Fund Assurance.

It was explained that the difficulties of the house had wholly arisen from the failure of individuals abroad, and that a dividend of about 10*s.* in the pound was anticipated. The cash balance of goods in hand represent together £6,805; consequently a distribution of 5*s.* in the pound might be made by the 1st February. At the commencement of the year the capital was £8,000, but the disasters in Hamburg have entirely absorbed it, and left them with an additional large amount of responsibility. Both Mr. Coleman and Mr. Nicholson, the professional representative of the house, bore testimony to the legitimate character of the operations, and it was finally resolved to carry out a liquidation under inspectorship, Mr. Carey and Mr. Schroeder being appointed to act for the creditors.

Dividends to the extent of 5*s.* 6*d.* in the pound have been paid, and further distributions are anticipated.

THE ESTATE OF MESSRS. CHARLES NICHOLSON AND CO.

A meeting of the creditors of Messrs. Charles Nicholson and Co., of Cannon Street and St. Paul's Churchyard, who failed in the Manchester trade, was held on the 24th December, 1857, when the following statements of affairs were presented by Mr. Parrinton, of the firm of Parrinton, Ladbury, and Co., the accountants :—

Statement, December 24, 1857.

DEBTOR.

	£	s.	d.	£	s.	d.
To creditors, as per list				£47,317	10	7
To liabilities on customers' bills discounted, which will be claimed on this estate				4,482	5	2
To liabilities on bills for which value has been received by Messrs. C. Nicholson and Co.				4,314	15	9
To liability on joint promissory note				200	0	0
To liabilities on customers' bills discounted, considered good	£20,637	11	11			
To liabilities on bills to be provided for by other persons	1,337	13	0			
				£56,314	11	6

CREDITOR.

	£	s.	d.	£	s.	d.
By stock in trade at Cannon Street, cost	£17,267	13	9			
By stock in trade at Angel Street, cost	6,270	10	11			
				£23,538	4	8
By book debts :—						
Good				13,414	19	5
Doubtful	£1,138	7	2			
Estimated to produce				569	3	7
Bad	3,594	13	6			
Estimated to produce				150	0	0
By cash and bills on hand				973	6	2
By lease of premises, Nos. 39 and 40, Cannon Street, West, and No. 20, Friday Street, held for a term of about 90 years, from Christmas, 1856, at the annual rent of £1,200, cost with fixtures, etc.				8,903	9	0
Held by Messrs. Candy and Co., as a collateral security for customers' bills discounted.						
By lease, fixtures, and utensils at Angel Street, cost				582	0	0
By ten shares in the Unity Bank, on which £50 per share has been paid, estimated at				150	0	0
By surplus expected from customers' bills, claimed in the sum of £4,482 5*s.* 2*d.*, *per contra*				1,944	12	5
Carried forward				£50,225	15	3

Brought forward			£50,225 15 3
Less rent, taxes, gas, and salaries			1,515 11 5
			£48,710 3 10

By debt due from Messrs. Charles Nicholson and Co., St. Paul's Churchyard	£13,387 5 6

DEBTOR.

To creditors as per list	£15,428 5 9	
Less security held by creditor on this estate	4,140 0 0	£11,288 5 9
To liability on bills receivable—		
Discounted, considered good	£601 18 6	
Ditto, ditto, bad		282 7 1
		£11,570 12 10

To debt due by this estate to C. Nicholson and Co., Cannon Street	£13,387 5 6

CREDITOR.

By stock in trade, at cost			£14,482 18 3
By book debts—			
Good			2,620 19 1
Doubtful		£660 18 0	330 9 0
Bad		771 4 2	
By lease of No. 61, St. Paul's Churchyard, held for an unexpired term of about eight years at £350 per annum.			
By lease of No. 62, St. Paul's Churchyard, held for an unexpired term of about thirteen years, and of Nos. 58 and 59, Paternoster Row, held for an unexpired term of about five years. The whole at a rental of £405 per annum, estimated with fixtures in the last balance-sheet at		£6,962 0 0	
Deduct—			
By mortgage debt due to Captain Bague's executors, owing by C. Nicholson and Co., Cannon Street	£1,815 0 0		
And mortgage debt and interest owing to Messrs. C. Candy and Co. by C. Nicholson and Co., St. Paul's Churchyard	4,140 0 0		
		5,955 0 0	
			1,007 0 0
			£18,441 6 4
Less rent, taxes, gas, and salaries			823 15 0
			£17,617 11 4

Although there are two estates, it was proposed to treat them as one, with the view of obviating difficulties in an arrangement. Mr. Nicholson has a partner in the Cannon Street business, and also one in the London Mantle Company in St. Paul's Churchyard, and the accounts have been in a great measure kept distinct. The assets exhibit nearly 19*s.* in the pound, but the valuation of the stock, as made by Mr. Parrinton, could not, Mr. Nicholson asserted, be realized; and, therefore, he considered himself only in a position to make an offer of 12*s.* 6*d.* in the pound, payable by instalments, extending over a period of sixteen months. The stock, which was stated to be extensive and of a peculiar description, especially in St. Paul's Churchyard, would require to be cautiously dealt with, and could not be for a lengthened period disposed of. The time consequently asked for the ultimate payment was stated to be not too long, considering the difficulties Mr. Nicholson will have to contend with in completing this composition. The failure is attributable to over-bought stock, and trading beyond due limits, his capital in his early career having been very limited. Considerable discussion took place respecting the proposal, and it was eventually accepted by the great majority—the order of payment being 3*s*. 6*d*. in the pound at four months, 3*s*. at eight months, 3*s*. at twelve months, and 3*s*. at sixteen months, the last payment to be secured to the satisfaction of the three principal creditors, the estate to go to bankruptcy, if the arrangement be not concluded within a specified period. The necessity of co-operation among the creditors was strongly urged to prevent litigation and the certain sacrifice of assets.

THE ESTATE OF MESSRS. SVENDSEN AND JOHNSON.

A meeting of the creditors of Messrs. Svendsen and Johnson, coal and iron merchants at Newcastle-on-Tyne, and general commission merchants in London, was held on the 30th December, 1857, when the following statement was presented by Mr. Ball, of the firm of Messrs. Quilter, Ball, and Co., the accountants :—

STATEMENT, DECEMBER 30, 1857.

DEBTOR.

To creditors unsecured				£17,745	14	7
To creditors secured—						
Estimated value of securities	£2,258	6	4			
Amount of claims	2,119	3	4			
Surplus, *per contra*	£139	3	0			
To liabilities on acceptances, £78,828 10*s*. 4*d*. Of which it is expected claims will arise on				34,286	2	4
To liabilities on bills receivable, £74,389 1*s*. 6*d*. Of which it is estimated will prove claims on this estate				40,993	16	2
				£93,025	13	1

CREDITOR.

By bills receivable on hand				£1,417	18	0
By debtors, considered good	£7,939	4	6			
Doubtful, £5,748 10*s*. 10*d*., estimated at	1,437	2	8			
Bad, £3,737 16*s*. 10*d*.	0	0	0			
				9,376	7	2
By share of ship, estimated value				1,000	0	0
By merchandise and sundries on hand, estimated at				231	14	2
By surplus securities in hands of creditors				139	3	0
				£12,165	2	4

It appeared from the general explanations that the firm commenced business in Newcastle about September, 1854, with £381. A branch was subsequently opened in London, and business extended, but the system of blank credits had again proved unfortunate, involving the partners in heavy engagements. They had also been connected in bill transactions with Messrs. Hoare, Buxton, and Co., and Messrs. Rew, Prescott, and Co.; and the amount of loss which had thus altogether accrued was upwards of £75,000, out of a total deficiency, as shown by the statement, of £80,860. At the commencement of the year the accounts of the London house showed an adverse balance of £3,000, but those of the Newcastle establishment presented a surplus of £3,000. The assets consist of a share in a ship, and good debts in Sweden and Norway. To collect the latter one of the partners proposed to visit those countries, and it was thought that he might be successful in realizing a considerable proportion. The books of the firm had not been well kept, but it appeared to be considered that they pretty accurately represented the general results. Some of the creditors were desirous that the debtors should make an offer of composition, and it was suggested that 2*s.* 6*d.* in the pound would probably be accepted. It seemed, however, that they were not prepared to make any definite proposal, owing to the uncertainty of the realization of some of the estates with which they have had transactions. After general discussion and further explanation with respect to matters of detail, it was resolved to wind-up under inspection, Mr. W. Geipel, of Newcastle-on-Tyne, and Mr. Quilter, the accountant, being appointed inspectors.

This firm eventually wound-up with a composition of 1*s.* 6*d.* in the pound.

THE ESTATE OF MESSRS. W. J. POWELL AND SON.

A meeting of the creditors of Messrs. W. J. Powell and Son, engaged in the Manchester trade, was held on the 31st December, 1857, when the following statement was presented by Messrs. Parrinton, Ladbury, and Co.:—

STATEMENT, DECEMBER 31, 1857.

DEBTOR.

To creditors, as per list	£43,932	10	2			
Less amount secured	1,365	9	9			
				£42,567	0	5
To liability on bills discounted, which will be claimed on the estate	£2,186	4	0			
Less security held by creditors	1,056	3	0			
				1,130	1	0
To liabilities on bills discounted, considered good	25,052	19	9			
To liability on acceptance due March 4, secured by consignment to the full amount	1,000	0	0			
				£43,697	1	5

CREDITOR.

By stock in trade at cost				£18,563	16	10
By book debts—						
Good				20,542	1	6
Doubtful	£3,181	19	8	918	18	7
Bad	962	16	2			
By cash and bills in hand				806	2	10
By cash and bills in the hands of Messrs. Barclay and Co., and Messrs. Weston and Laurie, which they hold against bills discounted	1,629	4	9			
Less claims thereon	1,056	3	0			
				573	1	9
By lease of business premises and fixtures Nos. 10 and 11, Friday Street, held for a term of 21 years, from Midsummer, 1852, at a rental of £300 per annum	2,450	0	0			
Held by Messrs. Stagg and Co., as security for their debt of	1,365	9	9			
				1,084	10	3
By lease of business premises and fixtures, No. 9, Friday Street, held for a term of 17 years, from Michaelmas, 1857, at the rental of £500 per annum				150	0	0
Carried forward				£42,638	11	9

Brought forward		£42,638 11 9
Less rent, taxes, gas, and salaries		609 7 7
		£42,029 4 2
By amount of the private estate of Mr. W. J. Powell	£2,065 0 0	

The position of this estate, it appeared, was rather anomalous, and it will be perceived by the figures that a dividend of nearly 20*s.* in the pound is presented. The trading had been of a satisfactory character, the returns being from £160,000 to £170,000 per annum. The capital, when the firm commenced business in 1852, was £23,000, brought from another large establishment, and the cause of suspension is entirely attributed to the drawings of the senior partner, which have been at the rate of £3,000 to £4,000 per annum. His private property shows a surplus of £2,065, including a considerable quantity of wine, but the creditors expressed considerable dissatisfaction at the want of prudence so painfully exhibited. The situation of the son, who has not drawn more than £150 per annum, was the subject of deep commiseration, and it was proposed that he shall in future carry on the business. The offer made to the creditors was 17*s.* in the pound, but some discussion arose with respect to the period of payment. It was ultimately agreed to arrange the instalments at three, six, and nine months, from the 15th January; and a committee was appointed to determine the security.

THE ESTATE OF MESSRS. CARR, JOSLING, AND CO.

There was a numerous meeting, on the 7th January, 1858, of the creditors of Messrs. Carr, Josling, and Co., lately engaged in the trade of the north of Europe, who failed on the 23rd November previous, when the following statement was presented by Mr. J. E. Coleman, the accountant:—

STATEMENT, JANUARY 7, 1858.

DEBTOR.

To creditors on open accounts		£50,686 0 2
To bills payable	£164,384 2 2	
Of which it is expected there will be provided for by the parties for whose account the bills were accepted	110,571 14 4	
Leaving to rank on this estate		53,812 7 10
To creditors fully secured—		
Security held	£15,004 7 0	
Claims	12,500 0 0	
Surplus to *contra*	£2,540 7 0	
To liabilities on acceptances for account of other parties to whom Carr, Josling, and Co. had granted credits, but which parties have since suspended payment, or from other causes are unable to meet their engagements...		72,442 5 2
To liabilities on bill receivable, discounted	£159,390 19 3	
Of which it is expected there will be duly honoured at maturity, or taken up by the drawers	121,334 19 4	
Leaving to rank on this estate		38,055 19 11
Liabilities		£214,996 13 1

CREDITOR.

	£	s.	d.	£	s.	d.
By cash at bankers'	£2,091	16	9			
Less amount which will be retained against bills discounted	151	17	1			
				£1,939	19	8
By bills receivable on hand, good				2,122	16	0
By bills doubtful	£4,083	17	9			
Estimated at 7*s.* 6*d.* in the pound				1,531	9	0
By shares and sundry assets				7,210	11	11
By debtors, good				69,551	8	11
By debtors, doubtful	55,418	7	2			
Estimated at				14,440	0	0
By consignments of goods				10,831	9	10
Surplus from creditors holding security, *per contra*				2,540	7	0
				£110,168	2	4
Less creditors to be paid in full				250	18	7
				£109,917	3	9
By surplus from separate estate of G. B. Carr (In respect of the bills amounting to £38,055 19*s.* 11*d.*, which will at present rank on this estate, it is expected that reclamation will be made from the acceptors to some extent, in which case such reclamations will be available for division amongst the general creditors.)				18,833	7	4
Assets				£128,750	11	1

General Balance Statement, from July 1, 1855, to November 23, 1857.

DEBTOR.

	£	s.	d.
To liabilities as above	£214,996	13	1
To capital (including separate estates)	76,878	17	8
To profits	44,208	2	5
	£336,083	13	2

CREDITOR.

	£	s.	d.
By assets, as above	£128,750	11	1
By cash retained by bankers	151	17	1
Loss by bad and doubtful debts and bills, amounting to	69,730	11	2
By other losses	21,036	18	7
By liabilities on bills discounted, *per contra*	38,055	19	11
By ditto, on acceptances	72,442	5	2
By personal expenses of partners	5,915	10	2
	£336,083	13	2

The above figures show a dividend of about 12*s.* in the pound, and Mr. Coleman estimated that an amount of 10*s.* might at least be realized. It was expected that the assets would be increased through the reclamation of bills in connection with estates in Sweden, the Government of which country is making great exertions to assist its commercial interests. No exaggerated value had been placed upon the debts, but the amounts at which they had been taken would, it was believed, be secured. The surplus of the private estate, £18,900, and the property in shares, will assist to increase the general assets available by the creditors. It was intimated that, in addition to the general statement, a balance statement had been prepared in order to show the capital and progress of the firm, and to combat an opinion expressed, that it was the practice of accountants, at these meetings, to avoid entering into sufficient details. Although Mr. Carr, who commenced business in the year 1829 with a capital of £2,050, exclusive of personal bequests since received of £12,500, has, during the latter part of his career, been making £4,000 a year, his private expenditure has not exceeded £900 a year. His operations in the north of Europe, where he has had extensive connections, have occasioned his failure. As a circumstance illustrating his position and credit, it was mentioned that, of the total of the open accounts included in the statement, £38,000 is represented by money deposited in his hands at interest. Mr. Lawrance, of the firm of Lawrance, Plews, and Boyer, the professional adviser of the firm, stated the course proposed to be adopted. In the first instance, Messrs. Carr and Josling thought of offering a composition, but subsequently it was found that this could not be satisfactorily accomplished, and, therefore, it was proposed to liquidate under inspection. The proceedings taken in connection with the private arrangement clauses of the Bankruptcy Act had been only to protect the interests of the general creditors against any special creditor who might have been disposed to seek an undue advantage; and in the succeeding steps for winding-up the estate, care would be taken to obtain the affidavits of claims in proper form. The inspectors appointed, in accordance with the resolutions adopted, are Mr. Stephen Cave, Mr. Méivelle, and Mr. Bowness.

The amount already distributed has been 4*s*. in the pound.

THE ESTATE OF MESSRS. C. A. JONAS AND CO.

The meeting of the creditors of Messrs. C. A. Jonas and Co., engaged in the Baltic trade, who failed on the 4th December, 1857, was held on the 7th January, 1858, when the following statement was submitted by Mr. Ball, of the firm of Messrs. Quilter, Ball, and Co., the accountants :—

Statement, January 7, 1858.

DEBTOR.

	£	s.	d.	£	s.	d.
To sundry creditors unsecured				£19,705	5	10
To creditors partially secured—						
Amount of claims	£9,753	14	4			
Less estimated value of security	6,596	0	2			
				3,157	14	2
To creditors fully secured—						
Estimated value of security	£4,457	0	2			
Less amount of claims	3,922	17	5			
Surplus	£534	2	9			
To liabilities on bills receivable—						
Considered good	£17,136	16	2			
Considered bad	£4,161	0	0			
To less estimated value of security	3,470	0	4			
				690	19	8
To liabilities on acceptances—						
Provided for	£16,863	9	7			
Secured	12,410	15	9			
Outstanding	22,852	12	9			
	£52,126	18	1			
Of which it is estimated will prove on this estate				8,591	5	0
				£32,145	4	8

CREDITOR.

By cash on hand		£277 10 9
By cash detained at bankers'	£1,010 0 4	
By bills receivable on hand—		
Considered good	2,922 13 10	
Considered doubtful, £1,500. Estimated at 7*s.* 6*d.* in the pound	575 0 0	
		3,497 13 10
By sundry debtors—		
Considered good		11,420 4 4
Considered bad	596 10 2	
By office furniture, estimated value		100 0 0
By surplus securities, *per contra*		534 2 9
		£15,829 11 8
Less rent, salaries, etc., payable in full		100 0 0
		15,729 11 8
By deficiency		16,415 13 0
		£32,145 4 8

According to the explanations afforded, the estate showed a dividend of about 10*s.* in the pound, and of the deficiency of £16,415 presented by the statement, special losses to the extent of £14,273 had occurred from the failure of other houses during the recent crisis. On the 1st January, 1857, the apparent deficiency was £2,307, but the father of one of the partners advanced £6,000 to carry out operations, for which he is now an unsecured creditor. The profits of the year exceeded the ordinary losses, private drawings, etc., and the trading, it was stated, had been perfectly legitimate. Another feature was, that the books had been well kept. The whole of the explanations being regarded as satisfactory, the resolutions prepared by Mr. Hollams, the solicitor, for a winding-up under inspection, were agreed to.

THE ESTATE OF MESSRS. WIENHOLT, WEHNER, AND CO.

A meeting of the creditors of Messrs. Wienholt, Wehner, and Co., in the East Indian and Australian trade, who failed on the 11th December, 1857, was held on the 7th January, 1858, when the following statement, prepared by Mr. Clarke, the accountant, was presented:—

STATEMENT, JANUARY 7, 1858.

DEBTOR.

To creditors unsecured, as per statement, viz.:—		
Trade	£29,758 7 9	
Permanent family loans	17,597 7 6	
		£47,355 15 3
To creditors partially secured—		
Amount of claims	£52,706 4 7	
Less estimated value of securities	43,975 0 0	
		8,731 4 7
To creditors fully secured—		
Estimated value of securities	£46,855 0 8	
Amount of claims	41,757 12 7	
Surplus, *per contra*	£5,097 8 1	
To liabilities on acceptances	£146,907 1 5	
Carried forward		£56,086 19 10

Brought forward			£56,086 19 10
Viz. :—			
Protected by other parties		£46,304 11 7	
Secured by the documents of produce invoiced at	£82,786 10 0		
To amount of bills	75,675 10 0	75,675 10 0	
To surplus *per contra*	£7,111 0 0		
Estimated to be claimed on the estate		24,926 19 10	24,926 19 10
		£146,907 1 5	
To liabilities on bills receivable, as per statement		£152,196 0 7	
Of which it is estimated will be proved on the estate			12,500 2 0
To creditors to be paid in full		550 0 0	
			£93,514 1 8
Capital on January 1st, 1856			£30,758 7 7
Balance of profit and loss from January 1st, 1856, to December 11th, 1857			35,000 0 0
			£65,758 7 7
And permanent family loans			£17,597 7 6

CREDITOR.

By cash and surplus securities in the hands of bankers			£4,624 2 10
By debtors—			
Good		£21,270 18 11	
Doubtful	£13,874 1 1		
Estimated to realize		7,000 0 0	
Bad	506 15 4		
		£28,270 18 11	
Less creditors to be paid in full, *per contra*		550 0 0	
			27,720 18 11
By surplus securities in the hands of creditors			5,097 8 1
By surplus on produce			7,111 0 0
By merchandise and other securities, as per statement			13,321 15 9
By office furniture			150 4 0
			£58,025 9 6
By outstanding debts and assets in India		£66,871 3 8	
Which at only 7*s.* 6*d.* in the pound, will yield			25,071 13 6
			£83,097 3 0

The estate, it will be noticed, showed about 17*s.* 9*d.* in the pound; but as the assets are principally Indian, the actual dividend could not be accurately estimated. They included a small indigo factory, and consignments of goods, which it was hoped might be realized without any great sacrifice. With the view of conducting a liquidation which should be satisfactory to the creditors, it was proposed to adopt the process of inspectorship, and three gentlemen were appointed to act with the debtors. In answer to questions, it was stated that the failure had been solely occasioned through the crisis, and that the item of permanent family loans had not been diminished. After the ordinary resolutions introduced to the notice of the meeting by Mr. Peachey, of the firm of Messrs. Oliverson, Peachey, and Co., were passed, one expressing sympathy at the difficulties of Messrs. Wienholt, Wehner, and Co., was brought forward and carried. It was then announced that although the permanent family loans were unsecured, further assistance would have been rendered had the partners entertained an impression that they could have surmounted their embarrassments. Several creditors testified to the honourable conduct of Messrs. Wienholt, Wehner, and Co., in their various business engagements.

THE ESTATE OF MESSRS. LUPTON, HOOTON, AND CO.

A meeting of the creditors of Messrs. Lupton, Hooton, and Co., in the Manchester trade, who failed on the 1st, was held on the 11th January, 1858, when it was agreed to accept a composition of 17*s.* in the pound, payable by instalments at three, six, and nine months' date. It will be noticed from the following statement, prepared by Messrs. Parrinton and Ladbury, that the estate shows a small surplus. Some surprise was, therefore, created when it was announced that an offer of 16*s.* was about to be made. After a short discussion, it was resolved to take the amended proposal, a committee having been appointed to superintend the arrangement. Should the estate when investigated not be able to pay the composition named, it is then to be wound-up under inspection. The cause of suspension was the withdrawal of capital which had been left in the firm by a previous partner:—

Statement, January 11, 1858.

DEBTOR.

	£	s.	d.	£	s.	d.
To creditors, as per list	£29,786	14	5			
Less security held by creditors	2,705	8	10			
				£27,081	5	7
To liability on bills receivable discounted, and which will be proved on this estate				515	6	0
Ditto, ditto, considered good	£7,597	16	0			
To liability on acceptance, for which the firm has received a collateral security, but the acceptance will be proved on this estate				1,400	0	0
To surplus (which would be increased by any value that may arise from the collateral security above-mentioned)				913	16	0
				£29,910	7	7

CREDITOR.

	£	s.	d.	£	s.	d.
By stock in trade, at cost				£12,783	13	3
By book debts—						
Good				10,399	17	10
Doubtful	£950	5	3	633	10	2
Bad	1,007	11	7			
Estimated to produce				136	1	0
By estimated surplus to arise from acceptances charged *per contra*				255	6	0
By cash in hand and at bankers'	359	16	1			
Less—retained by bankers, against loss	83	18	6			
				275	17	7
By bills in hand	£5,865	12	5			
Estimated to produce				5,718	19	11
By lease of premises and fixtures, Nos. 53 and 54, Bread Street, No. 6, Star Court, and Nos. 79 and 79½, Watling Street, held for the term of 21 years from Christmas, 1857, at the annual rent of £400, taken in balance-sheet, June, 1857	1,700	0	0			
Expended thereon since that time	1,005	8	10			
	£2,705	8	10	£30,203	5	9
Held by cash creditors as security for their debts, and deducted, *per contra*				£30,203	5	9
Less rent, taxes, gas, and salaries				292	18	2
				£29,910	7	7

THE ESTATE OF MESSRS. SABEL AND CO.

A meeting of the creditors of Messrs. Sabel and Co. was held on the 11th January, 1858, when the following statement was submitted by Mr. Saffery, of the firm of Palmer and Saffery, accountants:—

Statement, January 11, 1858.

CLAIMS.

	£	s.	d.	£	s.	d.
To sundry creditors unsecured, viz.:—						
On acceptances	£15,396	14	0			
On open accounts	8,432	6	3			
				£23,829	0	3
To creditors partially secured	£917	12	6			
Less securities held	850	10	6			
				67	2	0
To creditors for advances on 300 shares in the London General Paper Company, as *per contra*	£700	0	0			
Less value of life policy also held	50	0	0			
				650	0	0
To liabilities on acceptances on account of other parties	£5,856	1	4			
Of which it is expected claims will arise on				2,466	5	3
To liabilities on bills receivable	£12,323	9	4			
Of which it is expected claims will arise on	£2,424	12	2			
Less cash at bankers	200	15	0			
				2,223	17	2
				£29,236	4	8
To balance deficiency				£14,556	12	5
To outstanding liability of the late firm of Sabel and Sabel	£12,755	17	6			

ASSETS.

	£	s.	d.	£	s.	d.
By cash in hand				£323	7	9
By bills receivable on hand, considered good				1,423	10	3
By ditto, doubtful	£563	3	3			
Estimated to realize 5*s.* per pound				140	15	10
By sundry debtors, good				7,051	15	11
By consignments, balances outstanding				315	17	11
By property—						
Stock in trade, cost	£5,415	9	9			
Office furniture, estimated to realize	200	0	0			
Household furniture, estimated to realize	400	0	0			
				6,015	9	9
				£15,270	17	5
Deduct claims to be paid in full				591	5	2
				£14,679	12	3
By balance "deficiency"				14,556	12	5
				£29,236	4	8

STATEMENT ACCOUNTING FOR DEFICIENCY.

DEBTOR.

To deficiency as above				£14,556	12	5
To capital, July 1st, 1856				730	17	11
To subsequent profits				5,763	4	7
				£21,050	14	11

CREDITOR.

By cost of shares—						
In London General Paper Company	£1,500	0	0			
(Held by creditors, *per contra.*)						
In Caillaud's Patent Tanning Company	750	0	0			
				£2,250	0	0
By trade charges, from June, 1856, to December, 1857, viz. :—						
Interest, discount, commission, etc.	£2,357	13	5			
Salaries, rent, travelling expenses, etc.	4,703	9	2			
				7,061	2	7
By E. Sabel, drawing account				1,936	7	1
By losses prior to stoppage, viz. :—						
Doubtful and bad debts	£3,697	0	5			
Bills receivable (doubtful)	422	7	5			
Consignments	113	0	0			
				4,232	7	10
By losses on realization, viz. :—						
Household furniture	£380	0	0			
Office furniture	300	0	0			
				680	0	0
By losses in respect of bills, which should be provided for by other parties, as *per contra*, viz. :—						
On our acceptances	£2,466	5	3			
On bills receivable	2,424	12	2			
				4,890	17	5
				£21,050	14	11

Mr. Lawrance appeared for the principal creditors, and after some discussion the meeting was adjourned to allow Messrs. Sabel and Co. time to consider the possibility of making a definite offer of composition of 9*s.* in the pound to the creditors.

THE ESTATE OF MR. EDWARD SMITH.

A meeting of the creditors of Mr. Edward Smith, woolstapler, who failed on the 15th December, 1857, took place on the 13th January, 1858, at which the following statement was presented :—The liabilities on bills appear very large, and an amount of £109,397 is expected to be proved against the estate. It was proposed to pay 3*s.* in the pound, secured at three months; but subsequently the creditors, without accepting or rejecting the offer, appointed a committee to carry out a further investigation, and to report to an adjourned meeting.

STATEMENT, JANUARY 13, 1858.

DEBTOR.

To creditors unsecured				£68,967	12	1
To creditors holding security	£133,874	17	10			
Less value of security held by them	109,274	11	11			
				24,600	5	11
To creditors fully secured	£1,630	1	4			
To value of security held by W. C. Haigh	9,219	12	8			
Surplus available as an asset *contra*	£7,589	11	4			
Carried forward				£93,567	18	0

Brought forward		£93,567 18 0
To liabilities on endorsements of bills receivable	£201,724 10 8	
Of which there are expected to be proved upon this estate	£109,397 6 9	109,397 6 9
		£202,965 4 9

CREDITOR.

By stock on hand, estimated at		£25,163 17 4
By book debts		285 15 6
By cash .... (Retained by bankers to meet liabilities.)	£146 1 11	
By cash with country agents		169 6 5
By surplus stock in hands of creditors		7,589 11 4
		£33,208 10 7
Less rent, taxes, and small debts proposed to be paid in full		551 0 10
		£32,657 9 9
By value of private estate, estimated at		756 11 9
		£33,414 1 6

At the adjourned meeting the estate was placed under bankruptcy.

THE ESTATE OF MESSRS. S. BEGG AND CO.

A meeting of the creditors of Messrs. S. Begg and Co., engaged in the Australian trade, who failed on the 7th December, 1857, was held on the 13th January, 1858, when the subjoined statement, prepared by Mr. Tappling, the accountant, was presented. The debts represent £18,769, and the assets £10,239, showing a dividend of about 12*s.* in the pound; but as some of the latter were considered to have been over-estimated, the nearer amount was regarded as 10*s.* A proposal to pay 8*s.* by instalments, the last one secured to the satisfaction of a committee, was, after a general discussion, adopted. The professional representatives of the firm were Messrs. Hudson and Francis.

STATEMENT, JANUARY 13, 1858.

DEBTOR.

To creditors, viz.:—		
Of S. Begg and Co.	£4,387 10 5	
Of S. Begg	12,146 6 3	
In Melbourne, September 30th, 1857	1,574 11 5	£18,108 8 1
To creditors secured	3,792 1 11	
To estimated value of property in their hands	3,906 2 6	
Surplus, see *contra*	£114 0 7	
To liabilities, viz.:—		
On bills receivable	12,440 0 0	
On consignments	1,110 6 6	
	£13,550 6 6	
Of which sum is considered bad		661 2 2
		£18,769 10 3

CREDITOR.

	£	s.	d.	£	s.	d.	£	s.	d.
By debtors, viz. :—									
Good				£1,000	0	0			
Doubtful	£678	14	0						
Estimated to produce				339	7	0			
							£1,339	7	0
Bad	1,120	6	9						
By debts in Melbourne on Sept. 30th, 1857, estimated at							1,700	0	0
By estimated value of property in the hands of creditors over and above the advances thereon							114	0	7
By property, viz. :—									
In Melbourne				£3,170	2	7			
By shipments in or on the way to Melbourne				3,250	0	0			
By life policies				250	0	0			
							6,670	2	7
By balance of cash, viz. :—									
At London and Westminster Bank				3	18	9			
In hand				386	13	8			
By bill receivable				24	14	10			
							415	7	3
By deficiency							8,530	12	10
							£18,769	10	3

THE ESTATE OF MESSRS. SEWELLS AND NECK.

A numerous meeting of the creditors of Messrs. Sewells and Neck, lately engaged in the trade with Norway, who failed on the 5th December, 1857, was held on the 14th January, 1858, at the London Tavern, when the following statement, showing a surplus of £57,581, was presented by Mr. W. Turquand, of Messrs. Turquand, Youngs, and Co., the accountants:—

STATEMENT, JANUARY 14, 1858.

DEBTOR.

	£	s.	d.	£	s.	d.	£	s.	d.
To creditors unsecured—									
On open accounts				£40,632	1	2			
On bills payable				60,790	5	0			
On family deposits				22,656	7	4			
							£124,078	13	6
To creditors holding security—									
Securities held				42,809	19	0			
Claims				41,389	8	3			
Surplus subject to interest				£1,420	10	9			
To liabilities on bills payable accepted on account of the drawers, who are returned as creditors for certain amounts, subject to their retiring our acceptances, as per statement page	£97,618	5	4						
Of which it is estimated will be claimed against this estate ...				10,825	16	3			
Carried forward	£97,618	5	4	£10,825	16	3	£124,078	13	6

Brought forward	£97,618 5 4	£10,825 16 3	£124,078 13 6
Accepted on account of drawers who are debtors after retiring our acceptances, as per statement page	104,802 4 10		
Of which it is estimated will not be retired by the drawers......		9,742 0 0	
			20,567 16 3
	£202,420 10 2		
To liabilities on bills receivable—			
Amount run off and paid......	£108,322 0 1		
Amount run off, not met by acceptors, but expected to be retired by drawers or endorsers	29,766 0 1		
Expected to become claims ...	4,118 5 10	4,118 5 10	
To amount still running, expected to be met by acceptors, drawers, or endorsers	137,972 0 1		
Expected to become claims	1,911 0 10	1,911 0 10	
			6,029 6 8
	£282,089 6 11		
			£150,675 16 5
To balance, being surplus of joint estate			37,581 3 10
			£188,257 0 3

CREDITOR.

By balances at bankers', subject to repayment of loan and to liabilities on bills under discount	£5,755 17 10	
By bills receivable on hand		£34,820 17 10
By debtors—		
Considered good		86,619 9 0
Considered doubtful	1,634 7 7	
Considered bad..................................	11,394 16 7	
	£13,029 4 2	
By property, estimated value		18,990 0 0
By Borregaard estate, estimated value, after deducting mortgages and other claims		30,000 0 0
By bills of exchange and cash deposited as collateral security for payment of bills discounted and negotiated		14,826 13 5
		£188,257 0 3

By balance, being surplus of joint estate brought down	£37,581 3 10	
By partners' private estates, estimated to realize	20,000 0 0	
		£57,581 3 10

STATEMENT SHOWING THE POSITION OF THE FIRM ON 31ST DECEMBER, 1856, AND ACCOUNTING FOR PRESENT SURPLUS.

DEBTOR.

To capital, December 31st, 1856		£72,[illegible] 10 0	£72,834 10 0
To Mr. Joseph Sewell's deposit account	£5,700 0 0		
Carried forward	£5,700 0 0	£72,834 10 0	£72,834 10 0

Brought forward	£5,700 0 0	£72,834 10 0	£72,834 10 0
To family deposits	22,656 7 4		
		28,356 7 4	
		101,190 17 4	
To profits from Dec. 31st, 1856, to Dec. 5th, 1857, viz. :—			
To interest account balance		£7,656 6 10	
By commission		13,217 14 6	
By insurance		825 10 0	
By brokerage, postages, etc.		462 14 3	
		22,162 5 7	
Less expenses		2,071 14 2	
			20,090 11 5
			£92,925 1 5

CREDITOR.

By partners' accounts, from December 31st, 1856, to December 5th, 1857—			
By drawings of Mr. C. Sewell and Mr. Neck ...		£2,916 4 6	
By Thomas Sewell, balance at his debit in general ledger		2,210 15 8	
		5,127 0 2	
Deduct—Joseph Sewell, amount to his credit since December 31st, 1856		1,903 7 2	
			£3,223 13 0
By losses, viz. :—			
By sundries—balance of profit and loss account		4,348 0 10	
On estimated realization of debtors	£13,029 4 2		
By property	795 1 8		
By Borregaard estate	600 4 4		
By securities held by creditors	1,420 10 9		
By Holmsund stock............	2,000 0 0		
By shipping in Norway	3,329 19 11		
		21,175 0 10	
			25,523 1 8
By liabilities—			
On bills payable		20,567 16 3	
On bills receivable		6,029 6 8	
			26,597 2 11
By surplus, as per statement of affairs			37,581 3 10
			£92,925 1 5

The announcement that the estate showed 20*s.* in the pound, leaving a large surplus, was received with much satisfaction, particularly when it was ascertained that of this sum £20,000 is derived from the separate properties of the partners. The Borregaard estate, valued at £30,000, after deducting mortgages and other claims, was originally purchased for £43,000 ; and the remaining assets can be steadily realized. It was explained by Mr. Hollams, the professional representative of the firm, that the suspension was entirely voluntary, Messrs. Sewells and Neck, notwithstanding the offer of assistance from the Bank of England and private friends, having declined to proceed owing to the uncertainty which prevailed a few weeks since with regard to the trade of the north of Europe. They were not then aware that Norway could have so well withstood the effects of the crisis as has now proved to be the case, and they consequently hesitated to jeopardize the funds which would have been supplied for their use. The local government having stepped forward to the assistance of the mercantile community, they have been enabled to contend with their difficulties, and in credit they now present a strong contrast to other neighbouring countries. The assets being principally in Norway, the only favour Messrs. Sewells and Neck ask at the hands of their creditors is an extension of time, and it is proposed to pay 20*s.* in the pound with 5 per cent. interest, by instalments, at six, twelve, eighteen, and twenty-four months. The resolution containing the proposal included a suggestion for the appointment of inspectors; but two of the creditors, Mr. Boyson and Mr. Churchill, intimated that there was no necessity for this provision, it being the interest of Messrs. Sewells and Neck to effect a liquidation to the best possible advantage. Great sympathy was expressed during the discussion for the position to which the firm had been reduced through the panic of November and December; and it was unanimously decided that it was not desirable to enforce any supervision.

The arrangements for 20*s.* in the pound were carried out.

THE ESTATE OF MESSRS. REW, PRESCOTT, AND CO.

A numerous meeting of the creditors of Messrs. Rew, Prescott, and Co., engaged in the Swedish trade, who suspended on the 14th December, 1857, took place on the 15th January, 1858, at the Guildhall Coffee-house, Mr. Prescott, the banker, in the chair, when the following favourable statement was presented by Mr. J. E. Coleman, the accountant:—

STATEMENT, JANUARY 15, 1858.

DEBTOR.

	£ s. d.	£ s. d.
To creditors on open accounts		£33,008 10 11
To ditto on bills payable	£81,625 1 8	
Of which it is expected there will be provided for by the parties for whose account the bills were accepted	52,708 18 11	
Leaving to rank on this estate		28,916 2 9
To creditors fully secured—		
Security held	£3,004 18 2	
Claims	935 8 0	
Surplus to *contra*	£2,069 10 2	
To amount due to the executors of the late H. J. Prescott		25,963 0 8
To liabilities on bills receivable	69,057 9 11	
Of which it is expected there will be taken up by the drawers or endorsers	61,241 10 1	
Leaving to rank on this estate		7,815 19 10
Liabilities		£95,703 14 2

CREDITOR.

	£ s. d.	£ s. d.
By cash and short bills in hands of bankers		£5,108 11 8
By cash at Bank of England, etc.		58 3 6
By bills receivable on hand, expected to be duly honoured at maturity		10,682 16 4
By bills receivable on hand, the payment of which may be deferred, but which are expected ultimately to be paid in full...		5,778 16 11
By debtors, considered good		73,403 18 1
By ditto, doubtful	£14,979 3 9	
By bills receivable on hand, doubtful	2,680 16 6	
	£17,660 0 3	
Estimated at		6,623 10 0
Carried forward		£101,655 16 6

Brought forward	£101,655	16	6
Surplus from creditors, *contra*	2,069	10	2
	£103,725	6	8
Less creditors under £10, and to be paid in full	274	3	5
Assets	£103,451	3	3

The cash and short bills with the bankers will be retained by them until the bills which they have discounted arrive at maturity.

PRO-FORMA BALANCE STATEMENT FROM JANUARY, 1847, TO DECEMBER 14, 1857.

DEBTOR.

To liabilities as above							£95,703	14	2
To capital at commencement							48,295	18	10
To profits				£128,574	3	6			
Less charges, etc.	£26,740	8	1						
Less losses, etc.	19,061	5	7						
				45,801	13	8			
							82,772	9	10
							£226,772	2	10

CREDITOR.

By assets, as above				£103,451	3	3
By amounts drawn out by the partners				45,817	3	11
By portion of the late Mr. Prescott's capital paid to his executors	£30,000	0	0			
By ditto, now due	25,963	0	8			
				55,963	0	8
By estimated losses in 1857, by doubtful and bad debts, and bills receivable				13,724	15	2
By liabilities on bills that have been dishonoured which ought to have been provided for by other parties				7,815	19	10
				£226,772	2	10

The estate, it will be noticed, exhibits 20*s*. in the pound, and a surplus of about £8,000. The assets and liabilities have been carefully examined, and Mr. Coleman asserts that there is every prospect of these results being realized. The operations of the house, which is of old standing and great respectability, have been conducted with prudence, and have not been based upon irregular bill transactions or open credits, which have unfortunately distinguished many other recent cases. The debts due in Sweden are connected with firms of the highest character, the trading relations of Messrs. Rew, Prescott, and Co. abroad being of the most undoubted description. The profits, during their trading of eleven years, have amounted to £82,000, or at the rate of £7,500 a year, and the drawings of the partners have been £45,000, or at the rate of £4,150 a year, leaving £37,000 to be carried to capital. The losses during the last year, including liabilities on bills, amount to £21,540. It is proposed that Messrs. Rew, Prescott, and Co. shall continue business, taking the management of their affairs entirely under their own control, and that they shall pay 20*s*. in the pound, with interest at the rate of five per cent., to the whole of their creditors. The first payment is to be 5*s*. in cash, on or before the 31st of March, 1858, with subsequent distributions, at the rate of 2*s*. in the pound, at intervals as speedy as the realization of the assets will permit. Great sympathy was expressed for the position of the firm, and the resolutions approving of the plan of liquidation were moved by Mr. Burmester, and seconded by Mr. Tottie.

The estate was finally liquidated by payment in full with interest.

THE ESTATE OF MESSRS. ALBERT PELLY AND CO.

A meeting of the creditors of Messrs. Albert Pelly and Co., engaged in the Norwegian trade, who failed on the 7th December, 1857, was held on the 16th January, 1858, Mr. W. Tottie in the chair, when the following satisfactory statement was presented by Mr. G. H. Jay, the accountant:—

STATEMENT, JANUARY 16, 1858.

DEBTOR.

	£ s. d.	£ s. d.	£ s. d.
To sundry creditors—			
On open accounts		£11,945 11 6	
On our acceptances.........	£2,286 14 1	2,286 14 1	14,232 5 7
To creditor for cash loan secured by Mr. A. Pelly ...		£12,751 6 0	
To liabilities on our acceptances, given on account of the drawers, as per statement	84,311 15 6		
Who, it is expected, will be unable to retire them to the extent of...		10,000 0 0	
To creditors for amount due to them after retiring our acceptances in their favour, amounting to	20,659 16 1	7,084 7 6	17,084 7 6
Total acceptances	£107,258 5 8		
To liabilities on bills receivable	£78,929 14 6		
Of which it is estimated will prove claims			5,000 0 0
Total claims			£36,316 13 1
To surplus			49,425 5 10
			£85,741 18 11

CREDITOR.

	£ s. d.	£ s. d.	£ s. d.
By cash at bankers, viz.:—			
Bank of England		£34 10 9	
Barclay and Co.		1,433 10 1	£1,468 0 10
By bills receivable in hand		£6,818 8 4	
Considered good		£5,444 13 3	
Doubtful, £1,373 9*s.* 1*d.*, estimated at 10*s.* in the pound		687 0 0	£6,131 13 3
By sundry debtors—			
Secured	£14,921 17 8		
Unsecured	34,178 0 0		
	£49,099 17 8		
Considered good		34,687 4 0	
Doubtful, £14,412 13*s.* 8*d.*, estimated at 10*s.* in the pound		7,206 6 10	41,893 10 10
Carried forward			£49,493 4 11

Brought forward		£49,493 4 11
By private estates of the partners for estimated surplus	£49,000 0 0	
Thereof less creditor secured thereon, *per contra*	12,751 6 0	
		36,248 14 0
Total assets		£85,741 18 11

STATEMENT SHOWING THE ESTIMATED RESULT OF THE FIRM'S OPERATIONS FROM JANUARY 1 TO DECEMBER 7, 1857.

DEBTOR.

To capital, January 1, 1857	£17,127 14 4
To balance of profit and loss to December 7, 1857	6,190 13 5
To private estates of partners included in statement of assets ...	49,000 0 0
	£72,318 7 9

CREDITOR.

By claims anticipated to be made on our acceptances in favour of drawers unable to retire them	£10,000 0 0
By claims anticipated in respect of bills receivable considered bad	5,000 0 0
By loss anticipated in realizing the bills receivable in hand	686 15 1
By loss anticipated in realizing the debts due	7,206 6 10
By partners' drawings, nil	0 0 0
By balance being present surplus as per statement of affairs	49,425 5 10
	£72,318 7 9

The estate thus exhibited debts and liabilities of £36,316, and assets £85,741, or 20*s*. in the pound, with a surplus of £49,425. It was explained that the account had been drawn out with great care, and that there was every expectation of the assets producing the amounts respectively set forth. The suspension was solely occasioned by the crisis, the capital on the 1st of January, 1857, having been £17,127, and the private property of the partners £49,000. It will be noticed that this latter item now represents £86,248, and that it constitutes the principal proportion of the surplus. Very little discussion was entered into, but great sympathy was expressed for the position of the firm, who, like Messrs. Sewells and Neck, and Messrs. Rew, Prescott, and Co., have suddenly been brought to the ground. Mr. Teesdale, the legal representative of the house, announced the proposal for the payment of the creditors. It was satisfaction in full for all demands, with interest at the rate of 5 per cent. per annum, by four equal instalments of 5*s*. in the pound, at six, twelve, eighteen, and twenty-four months, from the 25th day of February next; the extentension of this date being necessary to obtain the consent of the Norwegian claimants. Messrs. Albert Pelly and Co. were immediately to resume and carry on their business, and retain uncontrolled possession of their affairs, the creditors declining to avail themselves of the privilege of appointing inspectors. The resolutions empowering the fulfilment of this arrangement were proposed by Mr. Boyson, and seconded by Mr. Fenning.

The payment of 20*s*. in the pound was, under this agreement, effected.

THE ESTATE OF MESSRS. T. H. ELMENHORST AND CO.

The meeting of the creditors of Messrs. T. H. Elmenhorst and Co., who suspended on the 11th December, 1857, took place on the 16th January, 1858, when the annexed statement was presented by Mr. Quilter, of the firm of Messrs. Quilter, Ball, and Co., the accountants. It showed debts and liabilities equal to £11,167, and assets £14,281, nominally sufficient to pay 20*s.* in the pound, and to leave a surplus of £3,114. The house was engaged in the Swedish and Norwegian trade, and the failure was caused by the suspension of other large establishments. The partners not being in a position to make a definitive offer, and an amount of bills having still to mature, it was agreed that the best course to adopt would be a liquidation under inspection:—

STATEMENT, JANUARY 16, 1858.

DEBTOR.

To sundry creditors, as per statement				£516	15	2
To creditors fully secured, as per statement, viz.:—						
Estimated value of security	£1,000	0	0			
Amount of claim	208	6	7			
Surplus, *per contra*	791	13	5			
To liabilities on acceptances, as per statement	71,486	14	6			
Of which claims are expected to arise on, say				8,816	9	8
To liabilities on bills receivable, as per statement	47,537	17	1			
Of which, considered bad	2,830	0	0			
Less balances in the hands of bankers, held against liability	995	12	5			
				1,834	7	7
				£11,167	12	5

CREDITOR.

By bills receivable on hand—						
As per statement	£6,359	14	5			
Considered good	5,200	12	5			
Doubtful £1,159 2 0						
Estimated at	579	11	0			
				£5,780	3	5
By sundry debtors, as per statement—						
Considered good				4,209	16	10
Considered bad	£317	10	5			
By surplus securities in the hands of creditor, *per contra*				791	13	5
				£10,781	13	8
By dividends on the liabilities *per contra*, treated as claims, say 6*s.* 8*d.* in the pound thereon, viz.:—						
On acceptances	8,816	9	8			
On bills receivable	2,830	0	0			
	£11,646	9	8			
Say				3,500	0	0
				£14,281	13	8

The estate was realized, and produced 20*s.* in the pound.

THE ESTATE OF MESSRS. WILLIAM DRAY AND CO.

A meeting of the creditors of Messrs. William Dray and Co., agricultural implement makers, who failed on the 19th Dec., 1857, was held on the 19th Jan., 1858, when the following statement, prepared by Messrs. Turquand, Youngs, and Co., was presented. On the part of the debtor an offer of 10*s.* in the pound was made, and a committee of creditors was appointed to investigate and report to an adjourned meeting.

STATEMENT, JANUARY 19, 1858.

DEBTOR.

To creditors unsecured		£41,546 5 8
To creditors holding security partly covered	£1,887 17 7	
Deduct estimated value of security	1,350 0 0	
		537 17 7
To creditors fully covered (See *contra.*)	4,392 4 2	
To liabilities on bills receivable, expected to become claims against the estate		547 3 0
To liabilities to bankers on overdrawn accounts, and on bills receivable, expected to become claims against the estate	£5,515 1 2	
Less estimated value of security	3,652 13 9	
		1,862 7 5
To liabilities to Deane, Dray, and Co., uncertain as to amount until the partnership accounts are closed		
Total liabilities		£44,493 13 8

CREDITOR.

By debtors, good	£8,605 11 2	
Less 10 per cent.	860 11 0	
		7,745 0 2
By creditors, doubtful	£914 17 5	
Estimated to realize		250 0 0
By ditto, bad	1,518 17 8	
By distillery accounts	4,355 2 1	
Estimated to realize		2,000 0 0
By cash in hand		180 2 0
By stock at Swan Lane and various warehouses		5,431 19 10
By stock and debts at Paris		2,460 12 10
Ditto at Vienna		350 0 0
Ditto at Farningham		2,148 17 0
By farming stock at Farningham		4,053 0 0
By horses and carts in London		501 7 8
By shares in Oxford Distillery		300 0 0
By consignments, estimated at		370 0 0
By property held by creditors	4,550 0 0	
Deduct amount of their claims	4,392 4 2	
		157 15 10
By lease of farm at Farningham		
By lease of warehouse, Swan Lane		
By licenses of distillery patent		
		£25,948 15 4
Deduct creditors to be paid in full amount, under £5, rent, taxes, etc.		793 1 5
Total assets		£25,155 13 11

THE ESTATE OF MESSRS. POWLES BROTHERS AND CO.

A meeting of creditors of Messrs. Powles Brothers and Co., South American merchants, whose suspension was announced on the 4th of November, 1857, was held on the 19th of January, 1858, when the following balance-sheet was submitted by Mr. Moates, the accountant:—

STATEMENT, JANUARY 19, 1858.

DEBTOR.

To creditors unsecured		£154,441 6 7
To creditors secured—		
Claims	£18,384 6 1	
Value of security	15,267 10 0	
		3,116 16 1
To liability on bills bearing our endorsement, which will be paid by the acceptor at maturity	14,823 18 5	
To liability on bills which are expected to be proved on this estate		2,811 9 10
		£160,369 12 6

CREDITOR.

By cash in hand		£13 14 11
By bills receivable in hand, accepted		2,902 4 5
,, unaccepted	£6,485 0 0	
By shares		908 5 0
By book debts		62,787 12 10
By capital in firms in New Granada		20,738 18 2
		£87,350 15 4
Less rent, taxes, and small debts, proposed to be paid in full...		100 0 0
		£87,250 15 4
By value of private estate, estimated at		1,000 0 0
		£88,250 15 4

The dividend will thus be rather more than 10*s.* in the pound, but will depend upon the fair realization of the assets in South America, which form a large portion of the estate. Liabilities may also accrue from Mr. Powles being a shareholder in several companies which are now in process of liquidation. It was resolved to wind-up the concern under inspection.

THE ESTATE OF MESSRS. KIESER AND CO.

A meeting of the creditors of Messrs. Kieser and Co., lately engaged in the continental trade, who failed on the 3rd December, 1857, was held on the 20th January, 1858, Mr. Valentine Corrie in the chair, when the following statement, prepared by Mr. G. H. Jay, the accountant, was presented :—

STATEMENT, JANUARY 20, 1858.

DEBTOR.

To sundry creditors—		
Open account	£1,435 17 1	
Brokerage commission and exchange	650 0 0	
Grunelius and Co., subject to their retiring our drafts on them	2,410 5 8	
		£4,496 2 9
To sundry creditors partially secured—		
Amount of claim	£2,523 8 3	
Securities estimated at	2,341 16 2	
		181 12 1
Carried forward		£4,677 14 10

	£	s.	d.	£	s.	d.
Brought forward				£4,677	14	10
To sundry creditors fully secured—						
Amount of claim	£2,600	0	0			
Securities estimated at	2,839	14	5			
Surplus carried to *contra*	£239	14	5			
To liabilities (our acceptances)	£18,660	7	6			
Less amount accepted for, and since retired by drawers	£8,656	6	0			
Less amount accepted for, and will be retired by drawers at maturity	1,400	0	0			
Less amount accepted against securities lodged in the hands of Jacob and Son, and which are reported of sufficient value	4,500	0	0			
Leaving to rank upon the estate	4,104	1	6	4,104	1	6
	£18,660	7	6			
To liabilities on bills receivable—						
Foreign drafts	£22,201	16	7			
Considered good	12,259	15	9			
	£9,942	0	10			
Less securities held against Messrs. Dependorff's drafts for	2,356	3	1			
				7,585	17	9
Less inland bills, considered good	£4,334	9	5			
Note.—Cash held by Messrs. Cohen, against £3,964 3*s.* 5*d.* of the above, carried to *contra*	1,200	0	0			
To liabilities on sundry open contracts, estimated at				900	0	0
				£17,267	14	1

CREDITOR.

	£	s.	d.	£	s.	d.	£	s.	d.
By cash at bankers' and in the house							£2,160	2	3
By bills receivable, considered good							2,309	2	8
By sundry merchandise—									
In hand, at market value				£5,584	15	4			
Sundries, estimated at				100	0	0			
							5,684	15	4
By sundry debtors—									
Considered good				£3,485	16	11			
Doubtful	£392	15	9						
Estimated at				200	0	0			
Bad	260	5	3						
							3,685	16	11
By surplus value of securities held by creditors, *per contra*							239	14	5
By surplus to be received of Messrs. Cohen, *per contra*							1,200	0	0
							£15,279	11	7

Note.—Further assets may be expected to arise from reclamations to be made on other parties through whose suspension claims have arisen *per contra*, and the amount, as far as at present can be ascertained, will enable Messrs. Kieser and Co. to pay 20*s.* in the pound.

Although the liabilities of the firm represent about £50,000, the amount which will be proved against the estate is not likely to exceed £17,267. The assets are expected to realize 20*s.* in the pound; but the creditors appear inclined to refuse the acceptance of more than 17*s.* High testimony was afforded to the conduct of Mr. Kieser, whose transactions have been regulated by a scrupulous regard to honour. The failure has been caused by the Hamburg disasters; but notwithstanding the firm could have proceeded for some months longer, it was considered essential for the interests of the body of the creditors to suspend. At the termination of the proceedings most favourable resolutions were passed, of which the subjoined is an abstract:—

"That the partnership now subsisting between Messrs. Kieser and Jacob being dissolved, Mr. Kieser be at liberty to resume and carry on his business; and that the creditors, notwithstanding the desire of Mr. Kieser to undertake the payment of 20*s.* in the pound, accept payment of their demand by three instalments, amounting together to 17*s.* in the pound; the first 10*s.* in the pound, on the sanction of the creditors being obtained to this arrangement; the second of 3*s.* 6*d.* in the pound at four calendar months; and the third and last instalment of 3*s.* 6*d.* at eight calendar months from the date thereof. That the creditors beg to express their sympathy with Mr. Kieser under his present position, and their entire approval of his conduct throughout."

An eventual settlement of 12*s.* 8*d.* in the pound was accomplished.

THE ESTATE OF MESSRS. HERMANN, COX, AND CO.

A meeting of the creditors of Messrs. Hermann, Cox, and Co., engaged in business as general merchants, who suspended on the 1st of December, 1857, was held on the 21st of January, 1858, when the following statements were presented by Mr. Ball, of the firm of Messrs. Quilter, Ball, and Co., the accountants:—

STATEMENT, JANUARY 21, 1858.

DEBTOR.

	£	s.	d.	£	s.	d.
To sundry creditors unsecured				£88,531	17	2
To creditors partially secured—						
Amount of claims	£52,546	7	5			
Less estimated value of security	39,852	17	0			
				12,693	10	5
To creditors fully secured—						
Estimated value of security	£7,300	18	1			
Less amount of claims	6,506	10	6			
Surplus	£794	7	7			
To liabilities on bills receivable—						
Outstanding amount	£244,224	0	5			
Estimated to prove claims on the estate	£42,302	19	1			
Less cash and securities deposited	2,080	19	5			
				40,221	19	6
To liabilities on acceptances outstanding	£64,536	2	2			
Estimated to prove claims on the estate for				4,464	6	2
Estimated claims in respect of goods purchased				9,000	0	0
				£154,911	13	3

CREDITOR.

	£	s.	d.	£	s.	d.
By cash and good bills on hand				£14,950	2	6
By sundry debtors—						
Considered good				38,549	3	0
Considered doubtful	£52,095	13	6			
Estimated at 6*s.* 8*d.* in the pound				17,365	4	6
Estimated bad	£17,168	7	7			
By sundry merchandise on hand				13,320	11	2
Carried forward				£84,185	1	2

Brought forward		£84,185 1 2
By surplus securities, *per contra*		794 7 7
By consignment to Sydney		400 0 0
		£85,379 8 9
Less rent, salaries, etc., payable in full		250 0 0
		£85,129 8 9
Deficiency		69,782 4 6
		£154,911 13 3

Note.—Since the suspension, the cash and short bills in hand have been increased by merchandise sold and debts collected to £31,037 2 10

Statement accounting for the Deficiency.

DEBTOR.

To deficiency 1st December, 1857, as per statement of affairs		£69,782 4 6
To capital, 1st January, 1857		40,410 3 11
Profits, viz. :—		
Commission	£7,900 19 11	
Cotton	280 10 2	
Insurance	94 9 7	
Interest	425 11 6	
Sundries	5,087 10 9	
		13,789 1 11
		£123,981 10 4

CREDITOR.

By trade charges—		
London and Liverpool	£5,691 3 2	
H. Cox's drawings	2,067 6 11	
		7,758 10 1
By sundry losses on produce		8,556 18 5
By bad and doubtful debts, viz. :—		
Estimated loss on doubtful	£34,730 9 0	
Bad	17,168 7 7	
		51,898 16 7
By liabilities proved bad		47,767 5 3
By estimated claims in respect of goods purchased		8,000 0 0
		£123,981 10 4

The estate, it was explained, showed about 11*s*. in the pound, but there was every reasonable prospect of 10*s*. being secured. The assets already in hand were sufficient to pay a dividend of 4*s*., but a distribution could not be made until the whole of the liabilities shall have run off. March was the period when the result in this respect would be ascertained, though the total was diminished by £160,000. The estimate of the doubtful debts was taken on a fair average; and the assets since the failure had increased, through the sale of merchandise and the collection of debts. The cause of the suspension was heavy losses through the depreciation of produce, etc., the capital on the 1st of January, 1857, having been £40,000. The proposal made by Mr. Hollams, the professional representative of the firm, was a liquidation, through the process of inspection, which was generally assented to, and some of the creditors were appointed to act on behalf of the general body.

A dividend of 4*s*. in the pound has been distributed.

THE ESTATE OF MESSRS. BISCHOFF, BEER, AND CO.

A meeting of the creditors of Messrs. Bischoff, Beer, and Co., engaged as merchants in London, Switzerland, and Calcutta, who failed in November, 1857, was held on the 27th January, 1858, when the following statement was presented by Mr. Weise, of the firm of Messrs. J. E. Coleman and Co., the accountants :—

Statement, January 27, 1858.

DEBTOR.

	£ s. d.	£ s. d.
To creditors on open accounts		£7,358 9 3
To creditors on bills payable	£15,038 2 1	
Amount expected to be taken up by other parties	1,030 2 7	
Amount expected to rank on this estate		14,007 19 6
To creditors holding security—		
Securities held	2,951 8 11	
Amount of claim	2,574 1 5	
Surplus to *contra*	£377 7 6	
To liabilities on bills—		
Receivable discounted	£29,042 8 1	
Of which it is expected there will be paid at maturity	15,694 0 2	
Leaving to rank on this estate		13,348 7 11
Liabilities		£34,714 16 8

CREDITOR.

	£ s. d.	£ s. d.
By assets, consisting of goods in London office, furniture, etc.		£87 0 0
By debtors, good		2,291 10 5
By ditto, doubtful, estimated to produce		146 12 9
By surplus securities with creditors, *per contra*		377 7 6
By consignments, etc.		1,928 18 9
By amount due from Bischoff and Co., of Calcutta	£11,658 9 11	
Less amount due to Bischoff and Co., of Teufen, Switzerland	1,953 9 11	
		9,705 0 0
		£14,536 9 5
Less amounts under £10, and to be paid in full		127 12 11
Assets		£14,408 16 6

General Balance Statement from January 1855, to November 26, 1857.

DEBTOR.

	£ s. d.
To liabilities as above	£34,714 16 8
To amount standing to the credit of F. Beer	229 7 0
	£34,944 3 8

CREDITOR.

	£ s. d.
By assets as above	£14,408 16 6
By losses by doubtful and bad debts	1,480 10 2
By profit and loss balance	5,706 9 1
By liabilities on bills receivable discounted	13,348 7 11
	£34,944 3 8

It was stated that, although a longer time had elapsed than usual in convening a meeting, it had not arisen from any irregularity in the accounts, but solely from a desire to ascertain the position of the houses at Calcutta and Teufen, in order that the creditors of the London establishment might have the benefit of any assets to be received thence. It appeared that Mr. Bischoff had since arrived from Teufen, and had explained that if time was allowed the Calcutta house, there was every prospect of realizing nearly the entire amount which that firm was indebted to the London partners. Mr. Weise having entered into the full details of the statement, intimated that Messrs. Bischoff, Beer, and Co. of course placed themselves unreservedly in the hands of their creditors, either to liquidate under inspection, or through any other medium which might be thought advantageous. Looking, however, to the nature of the assets, the greater part depending on the Calcutta house, which, if at all pressed or interfered with, would probably not show favourable results, it was considered advisable to request Messrs. Bischoff, Beer, and Co. to be prepared with an offer to pay a composition, and to retain the management of the estate themselves. The proposal was a payment of 7*s.* 6*d.* in the pound, by instalments of 2*s.* at two months, 2*s.* at five months, 2*s.* at eight months, and 1*s.* 6*d.* at ten months, the last to be secured, which was ultimately accepted. In answer to questions, it was stated by Mr. Nicholson, the legal representative of the firm, that the establishments in Calcutta and Switzerland had suspended, and that the security for the last instalment would be approved to the satisfaction of the creditors. The house in London commenced without any capital, and it stood in the relation of an agency between the firm in Teufen (who were manufacturers of Swiss goods, and who exported at the rate of from £30,000 to £40,000 a year) and the firm in India. At the conclusion of the proceedings, two inspectors were appointed until the payment of the first instalment, and the approval of the security for the last.

The estate eventually paid 7*s.* 6*d.* in the pound.

THE ESTATE OF MR. PETER BROWN.

At a meeting, on the 2nd February, 1858, of the creditors of Mr. Peter Brown, engaged in business as a carpet warehouseman, who failed on the 21st January, the following satisfactory statement, showing 20*s.* in the pound, with a surplus, was presented by Mr. Turquand, of the firm of Messrs. Turquand, Youngs, and Co., the accountants:—

STATEMENT, FEBRUARY 2, 1858.

DEBTOR.

	£	s.	d.	£	s.	d.
To creditors unsecured				£29,630	5	5
To creditors holding security (see *contra*)	£1,181	4	4			
To liabilities on bills payable over and above amounts included in creditors				445	15	8
To liabilities on bills receivable	16,148	10	0			
Considered good	15,691	16	4			
Expected to be claimed against the estate				456	13	8
				£30,532	14	9
Surplus carried down				7,719	0	5
				£38,251	15	2

CREDITOR.

	£	s.	d.	£	s.	d.
By debtors—						
Good				£7,966	15	10
Doubtful	638	0	11			
Estimated to produce				159	10	0
Bad	1,091	18	10			
By cash in hand				25	0	0
By bills receivable in hand				157	12	10
Carried forward				£8,308	18	8

Brought forward				£8,308	18	8
By bills deposited with the bankers—						
Considered good	5,037	18	1			
Less amount advanced thereon	2,690	8	4			
				2,347	9	9
By stock and fixtures				23,040	1	10
By property				1,330	0	0
By ditto, in the hands of creditors	£3,825	9	11			
Less amount of their claims	1,181	4	4			
				2,644	5	7
By bills taken as liabilities, *per contra*	445	15	8			
	456	13	8			
	£902	9	4			
Amount likely to be recovered				580	19	4
				£38,251	15	2
Surplus				£7,719	0	5

It was explained that the position of the debtor had been principally produced by the pressure of one creditor, and that, although the estate exhibited such a favourable prospect, there was no other resource than to adopt the present proceedings. Mr. Brown, it appeared, commenced business in 1851 with £300, and had since gradually extended his connection until the present period, when his accounts showed 20*s*. in the pound, with a surplus of £7,700. Universal testimony was offered to his straightforward conduct and economical expenditure, his drawings having only been £250 per annum. Mr. J. Linklater, as his legal representative, stated that it was at first thought that a proposal for the payment of 20*s*. in the pound, by instalments of 5*s*. in three, six, nine, and twelve months, with interest at the rate of five per cent., should be made, but that Mr. Brown considered payments at four, eight, twelve, and sixteen months would better meet his convenience, if the creditors had no objection. A liquidation under this arrangement, was then unanimously agreed to, and full confidence being reposed in Mr. Brown, it was considered unnecessary to appoint inspectors.

Payment of all debts was eventually effected at the rate of 20*s*. in the pound.

THE ESTATE OF MESSRS. HENRY HOFFMANN AND CO.

A meeting of the creditors of Messrs. Henry Hoffmann and Co., engaged in business as general merchants, who failed on the 30th November, 1857, was held on the 3rd of February, 1858, when the following statement was presented by Mr. J. E. Coleman, of the firm of Messrs. Coleman and Co., the accountants :—

STATEMENT, FEBRUARY 3, 1858.

DEBTOR.

To creditors on open accounts				£12,155	0	3
To ditto, on bills payable	£108,639	13	9			
Of which it is expected there will be provided for by the parties for whose account the bills were accepted	49,632	12	1			
Leaving to rank on this estate				59,007	1	8
To creditors fully secured—Security held	£13,436	12	1			
Claims	12,018	18	0			
Surplus, which will be retained against bills discounted	£1,417	14	1			
Carried forward				£71 162	2	4

Brought forward				£71,162	2	4
To creditors partially secured—Claims	£60,916	5	2			
Securities held	44,661	16	2			
Deficiency				16,254	9	0
To liabilities on bills receivable, discounted	£221,905	6	7			
Of which it is expected there will be duly honoured, or taken up by the drawers	185,321	10	10			
Leaving to rank on this estate				36,583	15	9
				£124,000	7	1

Liability as shareholder in the Australian Auxiliary Steam Clipper Company (Limited.)

CREDITOR.

By cash and bills on hand	£215	5	4
By sundry assets, consisting of consignments, etc.	2,044	3	6
By debtors, good	19,115	11	3
Estimated amount due to H. Hoffmann, as partner in the firms of Kirchner, Sharp, and Co., of Melbourne, and Kirchner and Co., of Sydney, accruing from his share of profits	30,000	0	0
Amount for which parties will be debtors after they retire the drafts accepted for their accounts	858	1	1
	£52,233	1	2
Less creditors under £10, and amounts to be paid in full	243	9	6
	£51,989	11	8

60 shares in the Australian Auxiliary Steam Clipper Company (Limited), £10 paid	£6,000	0	0
300 ditto, ditto, £15 paid	4,500	0	0
	£10,500	0	0

General Balance from January 1, 1850, to November 30, 1857.

DEBTOR.

To liabilities as above				£124,000	7	1
To capital in the firm of H. Hoffmann and Co.	£1,152	16	0			
To ditto, in the firms of Kirchner and Co., of Sydney, and Kirchner, Sharp, and Co., of Melbourne	30,000	0	0			
				31,152	16	0
To profits				68,645	13	5
				£223,798	16	6

CREDITOR.

By assets, as above				£51,989	11	8
By shares in the Australian Auxiliary Steam Clipper Company	£10,500	0	0			
By ditto, Australian, London, and Emigration Company	2,500	0	0			
				13,000	0	0
By sundry bills upon failed houses				4,040	2	10
By surplus securities in the hands of creditors held against bills discounted				1,417	14	1
By personal expenses for eight years				11,910	10	3
By trade ditto for eight years				17,997	8	10
By losses				58,372	2	0
By estimated losses from bad and doubtful debts on Nov. 30, 1857				28,487	11	1
By liabilities on bills receivable discounted, *contra*				36,583	15	9
				£223,798	16	6

It was stated by Mr. Coleman that the cause of the failure in this case was the pressure of the period, and the suspension of other houses with which the firm was connected. The amount of bills payable likely to rank upon the estate was £59,000, and it principally arose from the adoption of the system which had hitherto proved so fatal, viz., that of open credits. The connection of Messrs. Hoffmann and Co. with Messrs. Rugin and Hoffmann, of Zurich, who had since failed, would still leave the latter creditors on this estate, through the account current, and bills for which they were liable, for upwards of £50,000, after deducting the total of the securities, was then valued. The assets, consisting of consignments, might be realized, and it was hoped that the amount of good debts would be secured, but the greater part of these had to be collected from abroad. The principal amount of £30,000, which was exhibited on the creditor side of the account, was the estimated sum due to Mr. H. Hoffmann, as partner in the firms of Messrs. Kirchner, Sharp, and Co., of Melbourne, and Messrs. Kirchner and Co., of Sidney. It was stated that there was no distinct articles of partnership existing between these respective parties, but that the correspondence showed the actual position in which they stood in relation to each other. Although no regular accounts had been furnished, some statements were in existence which led to the presumption that Mr. Hoffmann was entitled to about £40,000, and, making allowance for bad debts, which might have since been incurred, the outstanding balance was taken at £30,000. It was alleged that the object of the debtor had, for several years past, been to concentrate his connections in Australia and California, believing that from those resources he would make a favourable position. He was also a shareholder to the extent of £10,500 in the Australian Auxiliary Steam Clipper Company and, notwithstanding assurances had been given that the liquidation would be satisfactory, it had not been considered advisable to place them at any value in the usual category of assets. As four-fifths of the amount estimated to be realized depended upon the Australian partnership, it was thought that the presence of Mr. Hoffmann in Port Philip and New South Wales might expedite remittances, and since Mr. Kirchner, who was then staying in Germany, was about shortly to return to Australia, it was suggested that Mr. Hoffmann should also proceed thither. The firms in Australia were understood to be thoroughly solvent. In answer to a question, Mr. Nicholson, the legal representative of the debtor, intimated that there was no question but that the partnership was valid. After some general conversation, in which the opinion of the creditors appeared to be in favour of Mr. Hoffmann proceeding abroad to arrange with the Sydney and Port Philip firms, for the transmission of his share of profits, resolutions were passed agreeing to a liquidation under inspection, and Mr Cahlmann and Mr. Dreutler were appointed to act on behalf of the general body.

THE ESTATE OF MR. CHARLES FAUNTLEROY.

A meeting of the creditors of Mr. Charles Fauntleroy, engaged in the wool trade, who failed on the 31st December, 1857, was held on the 5th February, 1858, Mr. Barber in the chair, when the following statement was presented by Mr. Quilter, of the firm of Messrs. Quilter, Ball, and Co., the accountants:—

STATEMENT, FEBRUARY 5, 1858.

DEBTOR.

To creditors unsecured		£19,242 18 10
To creditors partially secured—		
Amount of claims	£95,408 0 2	
Less estimated value of securities	44,611 3 8	
		50,796 16 6
To creditors fully secured—		
Estimated value of securities	£14,371 17 7	
Less amount of claims	12,972 12 8	
Surplus	£1,399 4 11	
Carried forward		£70,039 15 4

Brought forward		£70,039 15 4
To liabilities on acceptances, as per list	5,409 14 5	
Estimated claims in respect thereof		2,693 10 4
To liabilities on bills receivable, as per list	£13,635 15 2	
Estimated claims in respect thereof		3,609 14 0
		£76,342 19 8
CREDITOR.		
By cash in hand		£41 19 1
By bills receivable on hand, considered good		484 0 8
By ditto, bad	£430 1 1	
By sundry debtors, considered good		2,464 6 7
By ditto, doubtful	1,136 19 5	
Estimated to produce		284 4 10
By ditto, bad	404 10 2	
By stock of wool in hand		6,646 13 4
By mining shares, estimated value		1,429 3 9
By house and office furniture, estimated value		900 0 0
By surplus securities in hands of creditors, *per contra*		1,399 4 11
By plant and utensils at Market Harborough, Coleman Street, and Bermondsey		2,000 0 0
		£15,649 13 2
Less creditors to be paid in full		157 16 9
		£15,491 16 5

The debts and liabilities being £76,342, and the assets £15,491, a dividend of about 4*s*. in the pound was exhibited, which, it was thought, with careful management, might be realized. The connection of the insolvent with Messrs. Haigh and Messrs. W. Cheesbrough and Son, of course raised suspicions with respect to his trading; but, although the debt due by him in the one case is £20,000, and in the other, £28,000, the amount of accommodation paper, properly so called, has been limited. About eighteen months since, the insolvent possessed a capital of £35,000; but the depreciation in wool, and extensive operations in English mining shares, have absorbed the whole, leaving, in addition, a serious deficiency. The books throughout have been very badly kept; and, notwithstanding several of the creditors appear to consider that the insolvent has not been actuated by improper motives, his conduct in this respect was severely censured. The question of an administration under bankruptcy was raised, but it was stated that the present system would involve too great an outlay. A suggestion was offered with a view of promoting reform in this direction; but, after some discussion, it was agreed to effect a liquidation under inspection.

THE ESTATE OF MESSRS. B. C. T. GRAY AND SONS.

A meeting of the creditors of Messrs. B. C. T. Gray and Sons, Canadian merchants, who failed on the 13th January, 1858, was held on the 10th February, Mr. H. Lancaster in the chair, when the following statement was presented by Mr. Quilter, of the firm of Messrs. Quilter, Ball, and Co., the accountants :—

STATEMENT, FEBRUARY 10, 1858.

DEBTOR.		
To creditors unsecured	£27,773 19 3	
Deduct estimated amount of interest, commissions, and charges to be placed to sundry accounts	800 0 0	
		£26,973 19 3
To creditors partially secured—		
Amount of claims	£5,929 6 4	
Estimated value of security	4,750 0 0	
Deficiency		1,179 6 4
Carried forward		£28,153 5 7

	Brought forward		£28,153 5 7
To creditors fully secured—			
Estimated value of security		£8,660 4 1	
Amount of claims		7,965 9 11	
Surplus, *per contra*		£694 14 2	
To liabilities on acceptances running		£72,722 19 4	
Of which it is estimated will prove claims on this estate			9,215 0 0
To liabilities on bills receivable, considered good		£48,056 7 8	
			£37,368 5 7

CREDITOR.

By cash in hand		£282 8 5	
By cash at bankers'		329 13 7	
			£612 2 0
By bills receivable—			
Considered good		£1,023 3 0	
Ditto, doubtful	£412 10 0		
Estimated to produce		60 0 0	
			1,083 3 0
By debtors—			
Considered good		£16,221 8 5	
Ditto, doubtful and bad	£17,246 13 3		
Estimated to produce		1,263 0 8	
			17,484 9 1
By merchandise on hand, estimated value			1,204 15 6
By surplus securities in hands of creditors, *per contra*			694 14 2
By 4-64ths of ship "Queen of the Lakes," estimated value			320 0 0
By life policies, estimated value			400 0 0
By sea insurance policies			120 16 7
By land in Australia and New Zealand, value cannot be estimated			0 0 0
			£21,920 0 4
By deficiency			15,448 5 3
			£37,368 5 7

The debts and liabilities, according to this account, stood at £37,368, and the assets at £21,920, showing a dividend of about 12*s*. in the pound. It was, however, not expected that this amount would be fully realized, a considerable portion of the assets being due from mercantile houses in St. Vincent's and Bermuda. The cause of suspension was alleged to have been the absence of remittances from Halifax. The capital on the 1st of January, 1857, was £14,434, and the transactions of the firm had, on the whole, been regular. Some open credits had existed in favour of establishments abroad, but produce had generally been forwarded to meet the drafts. Among the secured creditors were personal friends, who advanced £6,000 shortly before the pressure of the period compelled the firm to suspend. The debtors not being in a position to make an offer, left themselves entirely in the hands of the creditors, and, through Mr. Ellis, their solicitor, expressed their readiness either to submit to bankruptcy or to agree to a liquidation by the process of inspection. The majority of the creditors intimated that the latter would be the proper course to pursue, and, in accordance with this opinion, resolutions were passed, authorizing the arrangement, and appointing two creditors to represent the general body.

THE ESTATE OF MESSRS. COTTON AND TRUEMAN.

A meeting of the creditors of Messrs. Cotton and Trueman, metal brokers, who failed on the 20th November, 1857, was held on the 15th February, 1858, when the following statement was presented by Mr. G. H. Jay, the accountant :—

STATEMENT, FEBRUARY 15, 1858.

DEBTOR.

To sundry creditors—		
Unsecured, per list	£1,000 12 5	
Underwriting accounts	903 9 5	
		£1,904 1 10
To creditors holding securities, as per statement, claims	£37,949 8 2	
Deduct estimated value of metals and shares held as security	35,470 11 8	
		2,478 16 6
To liabilities on bills payable, per list, amounting to	£71,290 8 8	
To be met by the drawers, who will then have claims amounting to		9,219 6 8
To liability on bills payable, accepted for the Mexican and South American Company, per statement	16,000 0 0	
Which it is expected this estate will be relieved from by the Company liquidating in full.		
To liabilities receivable, as per list	41,080 17 9	
Of which will be proved		2,954 0 0
To liability on underwriting		
Estimated risks	300,000 0 0	
To liability on joint and several promissory notes for £3,000, estimated share of liability in respect thereof		500 0 0
		£17,056 5 0
Less creditors to be paid in full		596 7 6
		£16,459 17 6

CREDITOR.

By cash at bankers'			£196 18 0
By bills receivable—			
In hand, considered good			459 2 11
By promissory notes of the Mexican and South American Company		£13,932 14 6	
To be given up on payment of bills accepted for £16,000, and balance of account current for £4,041 2*s.* 7*d.*			
By sundry debtors, per list—			
Considered good	£3,753 15 6		
Allowed 10 per cent.	375 2 6		
		3,378 13 0	
Mexican and South American Company, balance due		4,041 2 7	
			7,419 15 7
By stock of copper in hand, estimated at			5,300 0 0
By counting-house furniture, estimated at			100 0 0
By private property, considered worth			1,680 0 0
By 4,800 Wheal Vor shares, cost £7 per share		£33,600 0 0	
By premiums due on underwriting account		4,363 12 5	
			£15,155 16 6
Less creditors and expenses to be paid in full			1,096 7 6
			£14,059 9 0

It was explained that the accounts showed about 17*s.* in the pound, but that there was the expectation of 20*s.* being obtained. The failure was caused by the breaking-up of the Mexican and South American Association, to whom the debtors were brokers, and for whom they had come under engagements to the extent of £16,000. The firm originally possessed a capital of £33,600, which was invested in the Wheal Vor Mines. The assets would be increased by any rise which might take place in the price of metals, and the balance due from the Mexican and South American Company would be recovered, as, under the operation of the Winding-up Act, the contributors must provide for the outstanding claims. After some discussion, it was agreed to place the estate under the process of inspection, and sympathy was expressed for the position in which the firm was placed.

THE ESTATE OF MESSRS. HEINE, SEMON, AND CO.

A meeting of the creditors of Messrs. Heine, Semon, and Co., bankers and exchange brokers, who suspended on the 10th December, 1857, was held on the 17th February, 1858, Mr. J. H. W. Schroeder in the chair, when the following satisfactory statement, exhibiting a surplus of £52,832, was presented by Mr. Turquand, of the firm of Messrs. Turquand, Youngs, and Co.:—

STATEMENT, FEBRUARY 17, 1858.

DEBTOR.

To creditors unsecured—			
Open accounts		£15,115 3 9	
Bills payable		42,353 14 6	
			£57,468 18 3
To creditors partially secured—			
Amount of claims		£9,316 0 6	
Value of securities		3,913 10 0	
			5,402 10 6
To creditors fully secured—			
Value of securities		£33,263 2 4	
Amount of claims		28,734 15 1	
Surplus taken as an asset, *per contra*		£4,528 7 3	
			£62,871 8 9
To liabilities on bills payable	£517,709 19 10		
Accepted for account of parties who are creditors, subject to their retiring such acceptances		290,157 6 8	
Returned and cancelled	£198,301 11 11		
Guaranteed	10,401 4 2		
		208,702 16 1	
Leaving outstanding		£81,454 10 7	
Which will all be retired by the drawers.			
Accepted on account of parties who are debtors, after retiring such acceptances		227,552 13 2	
Returned and cancelled	£187,845 16 9		
Guaranteed	300 14 6		
		188,146 11 3	
Leaving outstanding		£39,406 1 11	
Carried forward			£62,871 8 9

Brought forward			£62,871 8 9
Of which it is expected there will be claimed against the estate			8,780 0 0
To liabilities on bills receivable, discounted	£508,514 15 9		
To ditto on foreign bills negotiated........................	371,104 11 8		
	£879,619 7 5		
Run off, viz. :—			
Bills receivable...........		£487,221 13 8	
Foreign bills..............		352,077 19 0	
Of which there have been returned unpaid, and will be claimed against the estate, viz. :—			
Bills receivable...............	£132 17 0		
Foreign bills.................	21,300 0 0		21,432 17 0
To amount still running, viz. :—			
Bills receivable..............	£22,093 2 1		
Foreign bills	18,226 12 8	40,319 14 9	
		£879,619 7 5	
			£93,084 5 9
In respect of which it is anticipated there will not be any claim against the estate.			
To balance, being surplus of joint estate, carried down...........			37,832 11 4
			£130,916 17 1

CREDITOR.

By cash at bankers' and in hand			£4,043 7 1
Bills receivable in hand..........................		£49,382 19 10	
By sundry foreign bills receivable	£14,571 1 11		
Estimated to realize ...	12,000 0 0	12,000 0 0	61,382 19 10
Carried to losses	£2,571 1 11		
By sundry debtors—			
Considered good			36,000 0 0
Doubtful..		£1,582 11 6	
Bad		971 10 7	
		£2,554 9 1	
Estimated to realize		1,004 7 0	
Carried to losses..................................		£1,554 2 1	
By property		£8,685 8 8	
Estimated to realize		4,932 2 11	4,932 2 11
Carried to losses..............................		£3,753 5 9	
By property held by creditors fully covered, surplus as *per contra*			4,528 7 3
Carried forward..............			£110,886 17 1

	Brought forward	£110,886 17 1
By amount estimated to be recovered upon bills treated as liabilities, viz. :—		20,030 0 0
Bills payable	£8,780 0 0	
Bills receivable	21,432 17 0	
	£30,212 17 0 (As *per contra.*)	
		£130,916 17 1
By balance, being surplus of joint estate brought down	£37,832 11 4	
By partners' private estates estimated to realize—		
Mr. R. Heine's	1,500 0 0	
Mr. C. Semon's	13,500 0 0	
		£52,832 11 4

STATEMENT SHOWING THE POSITION OF THE FIRM ON THE 31ST OF DECEMBER, 1856, AND ACCOUNTING FOR PRESENT SURPLUS.

DEBTOR.

To capital at credit of partners, Dec. 31, 1856		£48,293 10 10
Profit from Dec. 31, 1856, to Dec. 10, 1857, viz. :—		
To profit and loss account, balance, sundries	£1,332 12 6	
To interest, commission, etc.	£12,472 8 6	
	£13,805 1 0	
Less expenses	1,554 13 8	
		12,250 7 4
To partners' private estates, as per statement of affairs, viz. :—		
Mr. R. Heine	£1,500 0 0	
Mr. C. Semon	13,500 0 0	
		15,000 0 0
		£75,543 18 2

CREDITOR.

By partners' drawings—		
R. Heine		£1,034 9 4
C. Semon	£1,372 10 9	
Ditto, furniture	2,243 0 0	
		3,615 10 9
		£4,650 0 1
By losses on estimated realization of assets, as per statement of affairs, viz. :—		
On foreign bills on hand	£2,571 1 11	
On debtors	1,554 2 1	
On property	3,753 5 9	
		7,878 9 9
By liabilities, viz. :—		
On bills payable	£8,780 0 0	
On bills receivable	21,432 17 0	
	£30,212 17 0	
Deduct amount estimated to be recovered	20,030 0 0	
		10,182 17 0
By surplus, as per statement of affairs		52,832 11 4
		£75,543 18 2

It was explained by Mr. Turquand that the reason of the meeting having been so long deferred had arisen from the desire of Messrs. Heine, Semon, and Co. to ascertain the result of the bills running, before presenting a proposal for the arrangement of their estate. Being now in a condition to do so, they were able to exhibit very favourable accounts. As proving not only the care with which operations have been conducted, but as also showing the sound and satisfactory position of the business relations of the house, attention was directed to the debtor side of the statement. The total liabilities which were running at the date of the suspension were, on bills payable, upwards of £517,000, and on bills receivable £508,000, in addition to foreign bills negotiated, £371,000. The total amount of bills not met, or which it was anticipated might not be met at maturity, in respect of these three items, was £30,312, of which, out of the bills receivable, amounting to no less a sum than £371,000, the loss figures only for the very small amount of £132; so that of liabilities amounting to nearly £1,500,000, the comparatively limited amount of £30,312 alone would rank against the estate, of which it was anticipated £20,000 would be recovered. Besides this amount of £30,312, Messrs. Heine and Semon had returned to them prior to their suspension foreign bills receivable, included among the assets, to the amount of £14,571. With respect to this total a recovery of £12,000 was also anticipated, which would leave the entire loss on liabilities at only about £12,571. There was also another fact which it was thought right should be mentioned, viz., that the great bulk of the bills which had not been met, or were known to Messrs. Heine and Semon, would not be met, at maturity, at or before the date of their suspension. Since their stoppage on the 10th December, bills bearing the endorsement of Messrs. Heine and Semon had only been returned to the amount of £8,000, and bills receivable only to the amount of £2,900, for which ample security was held to cover them. Mr. Nicholson, the legal representative of the firm, in alluding to the causes which led to their suspension, stated that the decision had not been come to without mature reflection, although the partners had received offers of assistance both from personal friends and the Bank of England. Painful as the alternative was, the house, having incurred large liabilities on bills and possessing extensive business relations with banks in Germany, considered that it would not be prudent to accept loans from friends, which might be jeopardized by the current of events abroad. Although at that date they had every expectation of being enabled to exhibit as favourable a balance-sheet as now presented, still they could not but regard some circumstances with suspicion, especially when large special remittances were intercepted, and funds which they had a right to rely on failed to arrive in due course. Notwithstanding the partners had refrained from availing themselves of the assistance offered, they nevertheless now desired to return their best thanks to those friends and to the Directors of the Bank of England for the overtures made at the time they found it necessary to cease payments. The proposal made by Messrs. Heine, Semon, and Co. was to pay 20*s.* in the pound, with interest at the rate of 5 per cent., on the 25th March; but if certain arrangements then in progress could not be carried out, then the firm would on that date pay 10*s.* in the pound in cash, and give their acceptances or promissory notes, payable at three months' date from the 25th March next, for the balance, with interest. It was, however, considered that the full payment would be made in the manner and at the period described. Mr. Melville moved, and Mr. Abegg seconded, the annexed resolution, which was unanimously passed, great sympathy being expressed at the unfortunate situation in which the firm has been temporarily placed:—

"Resolved unanimously,—That the proposal now made by Messrs. R. Heine, Semon, and Co. be accepted, and that the time necessary to carry the same into effect be granted. And the creditors now present do for themselves give, and recommend to absent creditors to give, to Messrs. R. Heine, Semon, and Co. free leave and license in as full and ample a manner as if a formal deed had been executed."

Payment in full, with interest, as proposed was effected.

THE ESTATE OF MESSRS. ROCHUSSEN AND CO.

A meeting of the creditors of Messrs. Rochussen and Co., who failed in the Mogadore trade, was held on the 26th February, 1858, Mr. F. J. Price presiding, when the following statement was presented by Mr. A. Young, of the firm of Messrs. Turquand, Youngs, and Co., the accountants :—

STATEMENT, FEBRUARY 26, 1858.

DEBTOR.

To creditors on open accounts		£4,661 16 4	
" on bills payable		11,465 5 4	
		£16,127 1 8	
To creditors secured—			
Value of security	£800 0 0		
Claims thereon..............	750 0 0		
Surplus, *per contra* ...	£50 0 0		
To liabilities on bills receivable........................	£2,623 7 7		
Of which considered bad		482 8 2	
			£16,609 9 10
To liabilities in respect of our acceptance for balance of purchase money of "Paulina" steamer	£7,575 0 0		
Secured by mortgage of the steamer, which cost £9,150, but, if realized by bill-holders, may not produce..........................	5,000 0 0		
	£2,575 0 0		£16,609 9 10

CREDITOR.

By cash in hand	£30 0 0		
By bills receivable on hand, good	191 9 10		
		£221 9 10	
By doubtful	£1,583 11 0		
Estimated to realize...		227 17 0	
By bad	222 18 4		
By sundry debtors—			
Considered good		542 19 11	
Doubtful....................	1,832 10 5		
Estimated to realize..........................		508 0 3	
Bad	37 3 4		
By sundry debtors on current accounts .... Holding goods on consignment, their accounts being debited with the prime cost of such goods.		3,745 2 10	
By consignments not yet realized, at cost price		1,590 0 0	
By stock at wharves, etc.		349 0 0	
By office furniture		50 0 0	
By surplus, *per contra*		50 0 0	
By amounts in hands of creditors in anticipation of dividend		1,159 10 0	
		£8,443 19 10	
Deduct creditors to be paid in full.........		184 0 7	
			8,259 19 3
By vendors of the "Paulina"...... 4,056 4 0			
By amount of our claim for breach of contract.			£8,259 19 3

The debts and liabilities being £16,609, and the assets £8,259, a dividend of about 10*s*. was presented, but as an individual creditor had commenced proceedings against the insolvent, it was agreed that he should make an assignment to trustees, and that the estate should be wound-up in bankruptcy under the Private Arrangement Act. The causes of his suspension were stated to be the absence of remittances from Mogadore, and his connection with the houses of Messrs. Leopold Sampson and Messrs. Ward, of Liverpool, both of which have failed. A claim exists against the vendors of the "Paulina" steamer of £4,056 for a breach of contract. The estates of the Liverpool firms would, it was believed, pay satisfactory dividends, and in this case the creditors of Messrs. Rochussen and Co. were likely to obtain some benefit. Through an attempted arrangement, before the notice in bankruptcy was served upon the insolvent, he had endeavoured to effect a composition, and some parties therefore received small sums in anticipation of dividend. The feeling of the creditors present was favourable to the proposed proceeding, and it was agreed that Mr. Price should act as trustee, in conjunction with Mr. Graham, the official assignee.

THE ESTATE OF MESSRS. CHARLES WALTON AND SONS.

A meeting of the creditors of Messrs. Charles Walton and Sons, engaged in business as shipbrokers, who failed on the 19th January, 1858, was held on the 15th March, Mr. Johnson presiding, when the following statement was presented by Mr. J. E. Coleman:—

STATEMENT, MARCH 15, 1858.

DEBTOR.

To creditors on open accounts	£9,259	15	7			
Ditto underwriting ditto	19,458	8	11	£28,718	4	6
To creditors on bills payable	£19,689	13	5			
Less amount expected to be provided for by parties for whose account the bills were accepted	11,221	4	3			
Leaving to rank on the estate				8,468	9	2
To creditors fully secured—						
Securities held	£23,640	0	0			
Claims	18,438	6	8			
Surplus to *contra*	£5,201	13	4			
To creditors partially secured—						
Claims	£14,634	3	5			
Securities held	11,300	0	0	3,334	3	5
To liabilities on bills receivable, discounted	£29,012	18	2			
Of which it is expected there will be duly met at maturity	28,685	17	9			
Leaving to rank on the estate				327	0	5
				£40,847	17	6

CREDITOR.

By assets, consisting of cash in hand, office furniture, etc.	£186	8	10
By debtors, good	3,139	10	4
By surplus securities with creditors, *per contra*	5,201	13	1
Carried forward	£8,527	12	3

Brought forward	£8,527	12	3
Less creditors under £10, and salaries, etc., to be paid in full	628	19	7
	£7,898	12	8
By amount due from the executors of C. Walton, deceased	8,420	19	4
	£16,319	12	0

GENERAL BALANCE-SHEET FROM APRIL 15, 1856, TO JANUARY 18, 1858.

DEBTOR.

To liabilities as above				£40,847	17	6
To profits				10,932	17	1
To amount due to the separate estate of C. Walton	£3,866	6	8			
Less amount standing to his debit	3,243	14	8			
	£622	12	0			
To amount due to the separate estate of W. Walton	1,451	0	6			
				2,073	12	6
				£53,854	7	1

CREDITOR.

By amount as above	£16,319	12	0
By suspense account (amount paid into Court on account of ship "Hope")	311	9	9
By partners' drawings	3,907	2	5
By amount paid for furniture, etc.	669	7	0
By charges and expenses	5,881	4	3
By estimated loss on bad debts (new accounts)	4,063	4	9
By two-thirds estimated loss on bad debts (old accounts)	4,794	17	4
By losses on shipping, etc.	17,580	9	2
By liabilities on bills receivable discounted	327	0	5
	£53,854	7	1

The accounts, it will be noticed, showed debts and liabilities to the amount of £40,847, while the assets were not placed at a greater sum than £7,898. It was explained that the present was the residue of an old estate, for, notwithstanding the sons have been identified with the firm, the chief of the indebtedness arose through the transactions of the father. At his death it was supposed he was a person of considerable wealth, and probate was taken out for £50,000; but it ultimately appeared that his affairs were in an embarrassed condition, large losses having been incurred through a Mr. Cochrane, in addition to the depreciation in the value of vessels. The dividend, according to the statement, was about 3*s.* 10*d.* in the pound, but it was believed 3*s.* 6*d.* might be realized if a liquidation under inspection be resorted to. Both Charles and William Walton carried on business on separate account as underwriters, and these estates would have to be wound-up. No difficulty was likely to be experienced in the case of Charles Walton, a committee of creditors at Lloyd's having arranged to adopt the liabilities in consideration of being allowed to deal with the assets, and a negotiation was pending for a similar settlement, if possible, with regard to the affairs of William Walton. The deficiency presented was thoroughly entered into, and it was shown to arise from bad debts, losses on vessels, and shipments to Australia. Mr. Ellis, the legal representative of the firm, intimated that Messrs. Walton were quite prepared to abide by any decision arrived at, and would, if necessary to facilitate proceedings in bankruptcy, sign a declaration of insolvency. At the same time, if it were wished, a liquidation by inspectorship could be pursued, and they would endeavour to realize at least to the extent of 3*s.* 6*d.* in the pound, procuring security for the last payment. Several creditors and their representatives, while concurring in the proposal for a winding-up in the manner described, objected to a release on the distribution of a fixed amount, and a variety of questions were put, with the view of eliciting information respecting matters of account, etc. A long conversation eventually took place in relation to the sale of the "Sarah Sands," a vessel in which the debtors possessed an interest. Although the explanations appeared to be complete, an investigation was desired previously to adopting resolutions in favour of a liquidation, and an adjournment was consequently agreed to. The feeling of the meeting evidently tended towards a winding-up, as proposed, it being considered advisable to avoid the expense of an appeal to bankruptcy.

The separate estate of Charles Walton showed debts and liabilities, principally on underwriting account, of £11,074, and assets £8,656. The separate estate of William Walton includes debts and liabilities, £4,776, and assets, £2,993. In this latter case the risks, it was understood, were not of so favourable a character.

The estate was subsequently wound-up in the Bankruptcy Court.

THE ESTATE OF MR. H. P. MAPLES.

A meeting of the creditors of Mr. H. P. Maples, insurance broker and Custom-house agent, who suspended on the 24th, was held on the 31st March, 1858, Mr. Gore presiding, when the following statement of affairs was submitted by Mr. J. E. Coleman, the accountant :—

STATEMENT, MARCH 31, 1858.

DEBTOR.

To amount of general creditors, above £10				£8,230	0	0
To amount of creditors on insurance accounts above £10				5,695	0	0
				£13,925	0	0
To amount due to the London and Brighton and the French railways for advances on five steam-boats	£38,100	0	0			
To cost of these five boats	60,000	0	0			
To estimated value of the five steam-boats, as ships at work	54,300	0	0			
To amount due to other creditors, for advances on steam-boat "Brighton"	5,600	0	0			
To estimated value of steam-boat	5,600	0	0	13,925	0	0

To underwriting liabilities.

CREDITOR.

By amount of debtors				£3,050	0	0
By sundry assets				520	0	0
By shipping cost	£6,102	0	0			
By estimated value				3,000	0	0
				£6,570	0	0
By difference between estimated value of steam-boats as working vessels	£54,300	0	0			
By charges on same	38,100	0	0	16,200	0	0
				£22,770	0	0

Premiums due on underwriting account.

It was explained that the difficulties of the debtor had arisen from his having entered into an arrangement with the Brighton and Western of France Railway Companies to develop the traffic by steam-boats between Shoreham and Jersey, and

Newhaven and Dieppe. The trade was a lucrative one, but its enlargement had been too rapid for the means of Mr. Maples, and he had been brought to a stand, not being fully able to carry out the arrangements with those companies. The accounts, although the books were not yet balanced, presented nearly an accurate statement, or at least sufficient to show what was the condition of his affairs. In his character as insurance broker, etc., his liabilities were about £13,925, and the assets to meet them £6,570. With respect to the agreement with the London and Brighton and the Western of France Railways, the value of the fleet, as working vessels, was placed at £54,300 ; while the advances made had been £38,100, leaving a surplus of £16,200. This, added to the assets, constituted a total of £22,770 to meet £13,925. Without relying on such a favourable issue, it was hoped that the creditors might receive 20*s.* in the pound, but it was considered that the first step to be adopted was the appointment of a committee, who could assist the debtor and his legal adviser, Mr. Pearce, in opening a communication with the companies, in order to arrive at a settlement. In the course of a month the committee would be able to report to a future meeting, and as the pressure of the unfavourable season had passed, and the traffic was becoming remunerative, a satisfactory arrangement was deemed probable. So late as January the estimate of the value of the boats was formed, and hence it was presumed that the companies would not object to come to terms on a fair and honourable basis. Preparations for the adjustment of the underwriting account were in progress, similar to those effected in the case of Mr. C. Walton. At the conclusion of the meeting the committee appointed to act were Mr. Wheeler, Mr. F. Morris, and Mr. Brett, who were likewise empowered to effect arrangements for the sale of the steam-boat "Brighton," which was mortgaged for £5,600, with the view of satisfying the claim upon the vessel. An adjournment then took place, as recommended, for one month.

At an adjourned meeting, held on the 28th May, it was agreed to accept a composition of 7*s.* 6*d.* in the pound.

THE ESTATE OF MESSRS. MAITLAND, EWING, AND CO.

At a meeting held on the 21st April, 1858, of the creditors of Messrs. Maitland, Ewing, and Co., in the China trade, who failed on the 3rd, the following statement, showing a surplus of £28,650, was presented by Mr. J. E. Coleman, of the firm of Messrs. Coleman, Turquand, Youngs, and Co. :—

STATEMENT, APRIL 21, 1858.

DEBTOR.

To creditors unsecured—			
On open accounts		£5,452 13 10	
On bills payable		26,725 18 2	
			£32,178 12 0
To creditors partially secured—			
Amount of claims		67,909 11 9	
Estimated value of securities held		52,228 5 11	
			15,681 5 10
To creditors fully secured—			
Estimated value of securities held		52,866 0 0	
Amount of claims		39,514 5 6	
Surplus taken as an asset, *per contra*		£13,351 14 6	
To liabilities on bills payable, etc.—			
For account of Moncrieff, Grove, and Co.	£105,067 19 8		
Ditto, ditto, part of letter of credit for £10,000 ...	8,346 18 6		
		113,414 18 2*	
Carried forward		£113,414 18 2	£47,859 17 10

* The whole of the bills drawn by Messrs. Moncrieff, Grove, and Co., are in our opinion fully represented by produce and other properties in their hands ; and we have every reason to believe that Messrs. Moncrieff, Grove, and Co., not knowing the depreciation of prices in England, considered that the produce and remittances already forwarded by them were sufficient to meet

Brought forward		£113,414 18 2	£47,859 17 10
To liabilities for account of sundries	16,268 17 6		
Of which parties for whose account they were drawn will provide to the extent of	13,144 13 1	3,124 4 5	
			116,539 2 7
To liabilities on bills receivable, as per statement		£51,715 2 9	
Of this amount it is not expected anything will come against the estate.			
			£164,399 0 5
To surplus			28,650 16 9
			£193,049 17 2

CREDITOR.

By cash and bills receivable on hand		£5,237 2 11
By debtors—		
Considered good	£11,543 2 3	
Secured	1,321 11 10	
		12,864 14 1
Doubtful	32,239 6 3	
Estimated to realize	5,000 0 0	
		5,000 0 0
Carried to losses	£27,239 6 3	
By Moncrieff, Grove, and Co.—		
Balance to their debit, after crediting the estimated value of produce unsold consigned to us by them	£10,584 9 4	
Amount of bills accepted for their account, *per contra*	113,414 18 2	
		123,999 7 6
By produce on hand—		
On our own account—estimated value	1,444 0 0	
Consigned by Moncrieff, Grove, and Co.—Estimated value	23,590 0 7	
		25,034 0 7
By adventures outstanding—		
Balance in ledgers	11,043 4 10	
Estimated value	4,600 0 0	
		4,600 0 0
Carried to losses	£6,443 4 10	
By shares in ships—		
Balance in ledgers	6,257 13 0	
Estimated value	3,000 0 0	
		3,000 0 0
Carried to losses	£3,257 13 0	
By estimated surplus from property in hands of creditors		13,351 14 6
Lease and improvements, furniture, etc., premises, 21, Birchin Lane, estimated value		500 0 0
		£193,586 19 7
Deduct sundries to be paid in full		537 2 5
		£193,049 17 2

the bills that have matured. In consequence of the altered state of matters, we have in our late advices strongly urged the necessity of speedy remittances, and to which no doubt they will readily respond.

STATEMENT SHOWING POSITION OF THE FIRM AT DECEMBER 31, 1856, AND ACCOUNTING FOR PRESENT SURPLUS.

DEBTOR.

To capital, December 31, 1856	£50,086 14 1	
Brought in since, less drawings	4,073 2 10	
		£54,159 16 11
To profits, from December 31, 1856, to April 3, 1858 :—		
Adventures	5,451 13 5	
Commission, interest, insurance, etc.	9,103 14 11	
		14,555 8 4
		£68,715 5 3

CREDITOR.

By estimated losses on realization of debtors	£27,239 6 3	
By adventures	6,443 4 10	
By shares in ships	3,257 13 0	
		£36,940 4 1
By liabilities on bills payable		3,124 4 5
Statement of affairs—surplus		28,650 16 9
		£68,715 5 3

The whole of the explanations were received as satisfactory, the circumstances attending the suspension of the firm, and the connection between Messrs. Maitland, Ewing, and Co., and Messrs. Moncrieff, Grove, and Co., being fully entered into. The last advices from Messrs. Moncrieff and Co., together with the personal testimony of one of the partners, Mr. King, led to the impression that that house was perfectly solvent, and able to meet the whole of its engagements. The position of the bill-holders created the only difficulty, and these parties, represented by the London Joint-stock Bank, the Mercantile Bank of India, and the Agra and United Service Bank, requiring time for intelligence from China, a short adjournment was suggested. It was, however, intimated that such a proceeding could not be adopted without danger, since, as bills continue to become due, some of the small creditors might throw the estate into bankruptcy, which would seriously depreciate the assets. Besides, the situation of the bill-holders had already been in a measure secured, by £10,000 having been paid in to the Bank of England on their account. In answer to questions, it was stated that Messrs. Moncrieff, Grove, and Co. had no important connections with any other house except Messrs. Dennistoun, Cross, and Co., of Melbourne, and that the whole of their transactions with Messrs. Maitland, Ewing, and Co. had been on the most sound basis. The original proposition that the estate should be wound-up under inspection, Mr. Hugh Mattheson, Mr. Alexander Anderson, and Mr. John Scott, the chairman, acting for the creditors, was then agreed to, and it was thought that a first payment of 20 or 25 per cent. might be made within a month, and others at a correspondingly early period. Every confidence was expressed in the position of Messrs. Maitland, Ewing, and Co., and also in that of the house of Messrs. Moncrieff, Grove, and Co., great sympathy being exhibited at the slightest interruption to the business of the former. With respect to the books of Messrs. Maitland, it was stated that they were in perfect order, a trial balance having been produced immediately after the announcement of suspension.

The estate has paid 20*s*. in the pound with interest.

THE ESTATE OF MESSRS. JAMES HOSKING AND CO.

At a meeting of the creditors of the estate of Messrs. James Hosking and Co., who suspended at the end of March, 1858, held at the office of Messrs. Lawrence, Plews, and Co., on the 29th April, the following balance-sheet was presented by Mr. Turquand, of the firm of Coleman, Turquand, Youngs, and Co., accountants:—

Statement, April 29, 1858.

DEBTOR.

To creditors unsecured			£3,917 11 6
Ditto, partially secured			883 6 10
To liabilities on bills payable, viz.:—Good		£1,575 3 11	
Bad		1,647 1 3	1,647 1 3
		£3,222 5 2	
To liabilities on bills receivable, viz., good		2,890 16 3	
As per statement		£6,113 1 5	
To liabilities on account of goods supplied to ship "Lyme Regis," for which we guaranteed the payment, as per statement		£442 6 9	
To liability to owners of sundry vessels for freight on 22 vessels chartered to load with guano at the Kooria Mooria Islands, but which, from unforeseen difficulties, were obliged to leave the islands without obtaining cargoes			64,926 19 2
Note.—These vessels have proceeded to various ports to obtain freights, the amount of which would go in reduction of this liability.			
To liability to owners of sundry vessels for freight on four vessels chartered for a similar purpose			9,431 4 8
Note.—These vessels are advised to have commenced loading with guano, and if successful this liability for freight would be discharged.			
To liability to lessees under Government of the Kooria Mooria Islands for claim for royalty and license fees on an engagement to import 30,000 tons of guano			
This liability is disputed.			
			£80,806 3 5

CREDITOR.

By cash balance in hand, 18th March, 1858		£150 0 1	
Deduct since paid in full for salaries, etc., as per statement		70 0 0	
			£80 0 1
By debtors:—Good			171 9 1
Doubtful		£569 8 3	
Estimated at 10*s.*		284 14 1	284 14 1
		£284 14 2	
Bad, as per statement		195 9 6	
		480 3 8	
By property, as per statement			368 5 7
			£904 8 10
Deduct rent for one quarter to March 25, 1858, payable in full			35 0 0
			£869 8 10
By nine cargoes of guano *in transitu* from Kooria Mooria Islands, carrying about 4,440 tons, estimated at £9 per ton		£39,960 0 0	
Subject to claims of owners for freight, dead freight, labour, and demurrage	£28,181 17 6		
Carried forward	£28,181 17 6	£39,960 0 0	

Brought forward	£28,181 17 6	£39,960 0 0	
And license fee and royalty on three cargoes	1,496 0 0		
		29,677 17 6	£10,282 2 6
And subject to claims of parties who have entered into agreements to participate in profits..........		2,458 7 1	
As per statement		£7,823 15 5	

After some discussion, it was decided to leave the management of the estate in the hands of Mr. Turquand, and for him to call another meeting of the creditors when circumstances should render such necessary.

THE ESTATE OF MESSRS. CHRISTODULO AND SUGDURY.

A meeting of the creditors of Messrs. Christodulo and Sugdury, engaged in the Levant trade, who suspended some months previous, was held on the 4th May, 1858, Mr. G. Lascaridi in the chair, when the following statement, prepared by Mr. Weise, of the firm of Messrs. Coleman, Turquand, and Co., was presented. Although the estate appears small, the interests involved were important, the firm having been connected with other establishments at Constantinople, Marseilles, and Odessa. Attempts were made by the partners to arrange their own affairs, and previous meetings had been held on the accounts, but with the exception of appointing a committee of examination nothing was done, and it was at length determined to call in professional aid.

STATEMENT OF THE AFFAIRS OF MESSRS. CHRISTODULO AND SUGDURY, OF GRESHAM HOUSE, OLD BROAD STREET, MERCHANTS, EXCLUSIVE OF THE LIABILITIES AND ASSETS OF THE OTHER HOUSES IN WHICH E. CHRISTODULO IS A PARTNER, MARCH 31, 1858.

DEBTOR.

To creditors on open accounts		£8,344 4 3
Ditto, bills payable	£6,362 15 6	
Less amount standing to the debit of parties for whose account the bills were accepted	26 17 0	
		6,335 18 6
To creditors fully secured—		
Security held	£1,384 2 4	
Claims..........	1,100 0 0	
Surplus to *contra*	£284 2 4	
To creditors partially secured—		
Claims..........	4,800 0 0	
Security held	3,800 0 0	
		1,000 0 0
To liabilities on bills discounted, the whole of which are expected to be met at maturity ...	£7,725 2 0	
Liabilities		£15,680 2 9

CREDITOR.

By assets, consisting of cash at bankers' and office furniture				£1,901	12	2
By debtors, good				882	5	9
By surplus securities with creditors, *per contra*				284	2	4
Assets				£3,068	0	3
By Sugdury, Son, and Co., Odessa				5,975	0	1
By E. Sugdury, Constantinople				47	2	0
By Emanuel Christodulo, Marseilles	£5,737	14	3			
Less capital standing to his credit	4,000	0	0			
				1,737	14	3
				£10,827	16	7

General Balance-sheet, from December 1, 1857, to March 31, 1858.

DEBTOR.

To liabilities, as above				£15,680	2	9
To profits	£15,733	15	6			
To three-quarter share, credited to Emanuel Christodulo, of Marseilles	11,799	17	1			
				3,933	18	5
				£19,614	1	2

Messrs. Christodulo and Sugdury will also be liable for about £100 on the safe arrival of the ship "Adele" at Ibraila, to which place she is now on her way.

CREDITOR.

By assets, etc., as above				£10,827	16	7
By loss on bad debts	574	12	9			
By loss on consignment, etc.	16,656	7	2			
By charges and expenses	3,675	13	1			
	£20,906	13	0			
By three-quarter share debited to Emanuel Christodulo	£15,680	0	6			
				5,226	12	6
By G. Sugdury's drawings				3,559	12	1
				£19,614	1	2

There is a sum of £160 to be recovered on the policy of the ship "Arpadina," but it is very doubtful, on account of various claims, whether anything will be available for the estate.

It was explained that the estate was in an involved position, owing to its relations with other houses abroad; but it was at the same time stated that the books were well kept, affairs having apparently been conducted in a straightforward manner. It was also intimated that the creditors of the head house, viz., E. Christodulo, of Marseilles, had agreed that all the firms in which Mr. E. Christodulo was interested should be considered as one, and that the proposal of settlement made, 25 per cent., had been accepted by a large majority. After a lengthened discussion, it was agreed to wind-up the estate through the process of inspection, the creditors appointed being Mr. Lascaridi and Mr. Zarifi. The individuals present recommended that when the creditors of the three other houses, viz., Messrs. E. Christodulo, of Marseilles; Messrs. Sugdury and Sons, of Odessa; and Mr. E. Sugdury, of Constantinople, had agreed to accept the offer that Mr. E. Christodulo, of Marseilles, made, that the English creditors should agree to such offer, and for that purpose hand over the English assets to a general fund for the payment of the said 25 per cent.

THE ESTATE OF MESSRS. FELIX CALVERT AND CO.

The investigation into the affairs of Messrs. Felix Calvert and Co., the brewers, of Upper Thames Street, having been completed by Messrs. Quilter, Ball, and Co., the annexed statements were issued on the 4th May, 1858, for the information of the public, with the accompanying memorandum:—"We are authorized to state that a full investigation of the affairs of Messrs. Felix Calvert and Co. has now been gone into, under the superintendence of Messrs. Quilter, Ball, and Co., and the following is a copy of the statements prepared. From these it would appear that, if the assets could be realized at their estimated value, the joint estate would show a surplus of £242,513 3*s.* 2*d.*, a sum more than sufficient to cover the estimated deficiency shown on the private estates, and it is therefore hoped that, if the settlement can be effected, as proposed, under a deed of inspectorship, all parties will be paid in full, with interest. But it appears, in the event of unfriendly proceedings, and a consequent hostile liquidation, so many questions will arise among the different classes of creditors, that it will be very doubtful whether all the creditors will be paid in full, besides that great delay and expense will necessarily result from the litigation that will ensue." This memorandum emanated direct from the representatives of the firm, and it may therefore be viewed as the commentary of the partners on their own position. Looking at the figures themselves, it appeared that the amount owing to depositors and customers was £471,000, and that the loss by the Westminster Brewery was between £50,000 and £60,000. Although it was attempted to be assumed that the Westminster Brewery was a distinct concern, from what is publicly known, there was little doubt of the validity of the partnership. To meet this £471,000 liabilities, the assets were taken at £713,450, and stood represented by a variety of important items. The brew-house estate and public-house property, although valued on the same principle as the formation of the existing firm in 1854, is probably over-estimated, and that to a considerable amount, because those acquainted with the trade entertained an impression that, in the interim, the actual value of this class of property has declined.

STATEMENT OF THE AFFAIRS OF MESSRS. FELIX CALVERT AND CO., APRIL 10, 1858.

DEBTOR.

To sundry creditors, viz. :—						
Depositors	£326,435	16	2			
Trade accounts	144,501	16	5			
				470,937	12	7
To creditors fully secured, viz. :—						
Estimated value of property	£739,705	13	7			
Estimated amount of claims	265,131	0	6			
Surplus, *per contra*	£474,574	13	1			
To liabilities, viz. :—						
On account of the Westminster Brewery	£19,110	14	11			
On account of debts secured by, and expected to be paid out of, the private estates of the partners	40,000	0	0			
	£59,110	14	11			
Note.—This statement is made out on the assumption that the Westminster Brewery is a distinct concern, and that the arrangements now in progress for carrying it on irrespective of Messrs. Calvert will be completed.						
				£470,937	12	7

CREDITOR.

By cash at bankers'	£4,138	16	8
By sundry debtors	125,550	10	6
Carried forward	£129,689	7	2

		Brought forward	£129,689 7 2
By brewery estate, plant, utensils, etc., ...	£261,865 0 0		
Deduct mortgages	128,540 7 5		
		£133,324 12 7	
By leases and freeholds of public-houses, etc., estimated at what they would realize if sold to a tenant	£468,692 15 8		
Deduct claims secured by deposit of deeds	127,442 15 2		
		341,250 0 6	
Being surplus, *per contra*			474,574 13 1
By malt, hops, etc.			11,622 10 6
By ale, stout, and porter			17,942 0 0
By casks			13,122 5 0
By wine establishment (debts and stock)			34,000 0 0
By surplus of private estates, viz.:—			
John Johnson's, estimated at.....................		£30,000 0 0	
N. Calvert's reversionary interest in Irish property at the decease of his mother, estimated at		2,500 0 0	
			32,500 0 0
By Westminster Brewery		£40,000 0 0	
			£713,450 15 9

Note.—The brew-house estate and the public-house property have been valued on precisely the same principle as was adopted at the time of the formation of the present firm in 1854.

STATEMENT OF THE AFFAIRS OF THE PRIVATE ESTATES OF EDMOND CALVERT AND FELIX LADBROKE, APRIL 10, 1858.

DEBTOR.

To creditors holding bonds and notes		£127,464 0 0
To creditors unsecured (private deposit accounts)		203,324 18 5
To creditors fully secured—		
Estimated value of property	£630,853 6 8	
Amount of claims	362,107 10 0	
Surplus, *per contra*............................	£268,745 16 8	
		£330,788 18 5

CREDITOR.

By private estate of E. Calvert	£290,200 0 0	
Deduct mortgages thereon	182,307 10 0	
		£107,892 10 0
By private estate of F. Ladbroke	340,653 6 8	
Deduct mortgages thereon	179,800 0 0	
		160,853 6 8
		£268,745 16 8
By deficiency..		62,043 1 9
		£330,788 18 5

A very numerously-attended and highly influential meeting of ladies and gentlemen, who had deposited money with the firm, was held at the London Tavern on the 12th of May, 1858, for the purpose of agreeing that the business should be worked for two years under inspectors, in order to avoid the ruinous litigation and disastrous losses which would ensue if the business were broken up, and the affairs of the company placed in the hands of the gentlemen of the long robe, for the purpose of being carried through the Bankruptcy Court. Mr. William Smalley, the Secretary of the Incorporated Society of Licensed Victuallers, and a depositor to a considerable amount, was unanimously requested to preside.

The CHAIRMAN commenced his observations by requesting any person who might not happen to be depositors to withdraw, inasmuch as it was only that class of creditors who were concerned in the proceedings which were about to take place; and, in reply to a question, he stated that persons representing societies would be considered depositors, but would only be entitled to one vote. He had attended at the brewery on the previous Saturday, in company with persons representing £15,000 or £20,000. They heard the deed read, and were requested to sign it. By the desire of some of his friends, he asked a few questions of the partners and other persons present. They thought that as they were very large creditors—the depositors being in fact the largest creditors—they were entitled to some explanation as to the means the firm had of paying the principal and 4 per cent. interest in two years. The first thing they had to look at was the magnitude of the sums in the hands of Calvert and Co., which, according to the balance-sheet of Messrs. Quilter and Ball, amounted to the sum of £326,435 16*s.* 2*d.* The amount of claims (advances by way of mortgage) were £265,131 0*s.* 6*d.*, and the total amount on which interest would have to be paid was £591,566 16*s.* 8*d.* If they looked at the foot of Messrs. Quilter and Ball's balance-sheet, they would see that the private estates were completely exhausted. They were valued at £630,853. There had been issued on bonds and notes £127,464; private deposit accounts, £203,324; issued by way of mortgage, £362,107. All this bore interest, and he was not aware at what rate; but as the private estate was a separate account, that did not concern them, and so much the better.—(Hear, hear.) Supposing it realized all that it was valued at—which was very seldom the case—taking that into consideration, if Messrs. Calvert and Co. paid interest at the rate of 4 per cent. on these deposits and mortgages on the partnership estate, it would amount to £23,662 per annum.—(Hear, hear, and Oh.) He did not learn on Saturday morning, but he had been since informed by Mr. Morse, that the trade debts—that was, sums due to hop-merchants and others—were entitled to receive 4 per cent. interest, because they had agreed to wait for their money. The trade debts amounted to £144,501, and the interest on that sum would be £5,780. The total interest on money payable at this time was £29,442, and interest receivable £3,590, making the annual charge for interest £25,852. There must be a large business done to pay off that sum in two or three years.—(Hear, hear.) Next, he had asked the partners in what way they thought they could realize this money. They said first, that they could save in the brewery so much per quarter on so many thousand quarters of malt. He found that to be a very large sum, and they did not quite agree with him as to the figures. The total sum proposed to be saved was £20,000 per annum; of this, £6,000 a-year would be on malt. Whether this was to be saved by the skill exercised in buying it, or the difference of the markets, or by paying cash instead of taking long credit, he did not know. Stables, £2,500; annuities, £1,000; public-house repairs, £7,000; salaries, £2,500; total, £20,000. This was the only direct information he could get. He suggested to them that they should not simply call their creditors together to read a long document which no one but a lawyer could understand, and, even when they had read an abstract, some could hardly tell whether they ought to sign it or not. When he went to the brewery some weeks since, Mr. Dewen told him that one of the gentlemen connected with the Royal Exchange Assurance Company was engaged to be one of the inspectors. This was not now the case. He was also given to understand that the private estates would realize £300,000. The deposits amounted to about that. The object of these proceedings was, of course, to try to avoid bankruptcy, if possible—(Hear, hear, and loud applause.) He was a depositor, and he had also a near relative who was a depositor, and he could tell them that there was a likelihood of getting their money, if they would have patience and forbearance one with another—not one man saying, I will have my money, thinking he will be paid.—(Hear, hear, and applause.) Let them share and share alike.—(Hear.) For his part, he would rather take a small dividend than any one else should have the preference over him and his friends—(Hear, hear.) He would now read the accounts to them. He had conversed with Mr. Quilter on the subject, and Mr. Quilter seemed to have gone into the accounts in a very careful manner, but he had arranged the items in a somewhat different way from that gentleman. The first item of the debts due by Messrs. Calvert and Co. was as follows;—To sundry creditors, £326,435 16*s.* 2*d.* That formed 44⅓ per cent. on the whole amount of their liabilities, so that they (the depositors) were the largest claimants in the whole matter. The second item was, trade accounts, £144,501 16*s.* 5*d.* That was as near 20 per cent. as possible (19⅔). The creditors fully secured were on property valued at £739,705 13*s.* 7*d.*, that was on the brewery, plant, leases, freeholds, and public-houses (359 in number), on which these mortgages were fully secured. He had arranged these accounts in a different way from Quilter and Ball, putting all the liabilities together, as being better adapted for a mixed audience—(Hear, hear) —in order to show the exact position of the debts and assets. The total sum on which they would have to pay interest was £736,068 13*s.* 1*d.* Under any circumstances the mortgages must be paid, if they wound-up under an inspectorship, or if they went through bankruptcy, and whatever loss there was to be borne would fall twice as much on them as on the trade creditors, and therefore it was right for them to consider how much they would be likely to realize when this £265,131 for mortgages was paid off. Though the debts were large, there were not many items. They lay in a small compass, and could easily be analyzed. Messrs. Calvert and Co., when consulted about that meeting, advised them to take their own course, and if the meeting wanted any question answered Mr. Quilter was at hand, and would give the required information.—(Hear, hear.) Now, what were the assets of the firm? The first item was, cash at bankers', £4,138 16*s.* 8*d.* This was an easily realized and available item. Sundry debtors, £125,550. Of this £79,000 was for loans to different licensed victuallers, and the rest was for rent and for beer. Then they came to the two principal items, which were the brewery estate, plant, utensils, etc. If they took them together, they would be represented by the sum of £730,557 15*s.* 8*d.* He said, if they were "taken together," because of what use would the public-houses be without the brewery, or the brewery without the public-houses?—(Hear, hear.) Without the public-houses the brewery would be simply a freehold, with so much frontage and fixtures. If sold off, it would be at an enormous loss! It would now perhaps not more than pay the mortgage of 49 or nearly 50 per cent. The public-houses were mortgaged to the extent of 26 per cent. These public-houses were valued, not exactly at the valuation of the present day, but as they were valued in 1854, when the new firm was formed by the admission of the new partners, Messrs. Johnson and Phelips. On paper these figures showed a surplus of nearly a quarter of a million, but what would there be if sold off? That was the point for them to consider. £700,000 had been invested in the brewery, in public-houses, and in stock of one kind or another. He could not, of course, tell what the profits of the brewery were; still it appeared to him that £700,000 sunk in a business ought to yield a large profit—a profit which, with the saving they could effect, would go far to pay the interest and a part of the principal, till by gradually diminishing the principal they would get into such a position as that capitalists would be found to come in, pay off the liabilities, and carry on the business. The mortgages on the brewery were only held by two firms, and, being fully secured, he should say they would have no objection to wait for their money. The mortgages on the public-houses were held by only seven firms. He was telling them this in order to show them how few hands the thing was in. The trade creditors were not more than a dozen, except under the item "tradesmen's bills, about £3,000." About £141,000 out of £144,000 was held by twelve parties. Therefore the only difficulty was the depositors, who were 369 in number altogether, including forty-four benefit societies. They had all these conflicting interests to reconcile. The total amount of the claims of those parties was £11,627 13*s.* 5*d.* That was not a great sum, but under this deed he feared it would be impossible to pay any of them out at present, though he did hope that some mode might have been devised of doing it, as 4*d.* in the pound on the whole liabilities would have sufficed for the purpose, and besides gratifying so many persons, it would have made the firm popular.—(Hear.) Now, under the act in conformity with which this deed had been prepared, he found they must have six-sevenths of the creditors for number and value to make it legal. No doubt they could induce creditors to sign in the required proportion of the value, but the difficulty would be with regard to number, for it was generally small creditors who were the most troublesome. Their interest, however, was at stake equally with that of others who were creditors for larger amounts, and if they did not come in he feared the result would be a bankruptcy. There would then be a depreciation in the value of the property of more than 50 per cent.; but if it once found its way into the Court of Bankruptcy there was no knowing where it would end.—(Hear, hear.) They knew what lawyers were, and what saturnalia the bankruptcy of Calvert and Company would afford them to revel in. He understood that there was one gentleman, a creditor for a large amount, who was not disposed to come in, no doubt from the expectation that he would get his money ; but he could assure him that he would not ; and if any one was to have it before another, he should much prefer that it should be one who needed it, rather than one who had his thousands.—(Cheers.) It was a hardship for all of them, no doubt, for the richer among them might want the money to settle on their wives, and they ought all, therefore, to learn to suffer together, and assist each other in their common object.—(Hear, hear.) He had inquired that morning—for he knew that in the history of public companies it had been seen to be a favourite scheme to bring in more capital to go on with—he had inquired whether it would be possible for any party, during the two years that this deed of inspectorship would be in operation, to bring more capital into the concern. It certainly would not be ; for the profits belonged to them, the depositors, and if the business would not suffice to pay them, they could not want money for extending the brewery or building more public-houses. On the other hand, if it could not pay interest, and a part of the principal of the debt, it could not be in a healthy condition.—(Hear, hear.) The number of public-houses they had was 359. This was a large number, and the fault of the firm appeared to have been that they were too anxious to extend their operations, and had gone into the wine trade, when the brewery was enough. Their speculations had, consequently, been very heavy, far exceeding the fair limits of trade, and they had used their deposits as if they were never to be asked for them. There had, no doubt, been great extravagance and mismanagement, but up to the year 1846, there were no mortgages on the property at all. In that year, however, the mortgages commenced, but he was assured that the money which was produced by the mortgages was laid out on the brewery, and in the business generally, and not diverted to other speculations. That being so, the value of the plant and house ought to be much greater than it was some years ago ; but he had not had access to the books to investigate how that was, and could only glean his inferences from the statement of Messrs. Quilter and Ball. What a time would they be in getting a dividend if they went into bankruptcy ; and then, after paying off the mortgages, it would be very doubtful whether they would realize a large dividend.—(Hear, hear.) Having now laid the facts of the case before them, he had to express his strong hope that all parties who had to address the meeting would do so in a calm and temperate spirit, and that they would not suffer their feelings to oversway their judgment.—(Hear, hear.) They were all subject to reverses, and when they did fall on us it was unwise and unmanly not to grapple with them in the way that promised the best and speediest deliverance from them.—(Hear, hear.) He could conscientiously say that he had endeavoured to advise them for their benefit, and hoped they would unanimously agree to accept the view of the case which he had offered them. Little had he ever thought the day would arrive when he should see the firm of Calvert and Company in such a position. It was indeed painful to reflect upon ; but, bad as it was, he did believe that by judicious and careful management, their prospects would be retrieved. If they had confidence in the inspectors, if they believed them to be men of integrity of purpose, ready and anxious to carry out their wishes, and to act generally for their benefit ; if they believed that, they might confidently expect much good to be done.—(Hear, hear.) A great hardship had been inflicted on the benefit societies, and it was enough almost to make men declare that they would not be provident any more. Out of evil, however, might come good, if they would only look the matter calmly in the face, and not be led away by any rash desire of driving that great concern into bankruptcy.—(Hear, hear.) There was one suggestion he had to make to the meeting, and it was one to which even the firm itself, as he apprehended, would not object. He did not think the number of inspectors was sufficient. The inspectors were Mr. H. Kingscote, of Eaton Square ; Mr. W. R. White, of the firm of Wigan and White, hop merchants, Hibernia Chambers, London Bridge ; and Mr. Morse. The latter gentleman was connected with the brewery, but he did not want his name struck out. If one or two more gentlemen were added to the list, he thought the arrangement would work better.—(Hear.) He should also propose that the inspectors report to them from time to time, say at the end of six, nine, or twelve months, in order that they might then be sooner able to judge with more correctness of their situation. It was possible, though he believed the contrary would be the result, that much mischief might be done before a year had elapsed. The parties who already were to act were undoubtedly highly respectable, but he imagined that if one or two more were added to their number, their responsibility to the depositors would be increased, and the depositors be more likely to get a good dividend. His own idea was, that the liabilities of the firm might be ultimately so much reduced as to render it worth while for wealthy parties to come in and get rid of the debt altogether.—(Hear, hear.) He had now communicated to them all that was necessary or interesting that they should know, and in doing so, he hoped he had not detained them too long, or indulged in too sanguine expectations.—(Hear, hear.) He had simply endeavoured to explain the accounts to them; he believed the views he had expressed concerning them to be well founded, and he hoped that in the course they were now about to take, they would show that they were influenced by judgment and discretion, and not led away by the rash promptings of passion.—(Cheers.)

Mr. Funnell, referring to the Chairman's last suggestion, which he thought a very valuable one, proposed that they should ask any member of the firm who might be present, whether there would be any objection to allow one or two more gentlemen to act with the inspectors already named, on behalf of the meeting.

Mr. Morse said that question could only be answered by a member of the firm, and there was none present.

The Chairman stated that he had been assured, in the presence of two of the partners, that they would not oppose such a proposition.

A conference here took place between the Chairman, Mr. Paine (of Chartham, Canterbury), Mr. Foster, and some other gentlemen, after which

Mr. Paine stated that they had considered the point, and found it would be impossible to effect the necessary alterations in the deed, and obtain over again the signatures in sufficient time, namely, by Thursday next. They were, therefore, of opinion that they had better let things stand as they were, and, for one, he was perfectly satisfied with the arrangement.

Mr. Thomas Jones said that the deed would require to be signed over again, and it would be impossible to get through that by Thursday. Without signing the deed again nothing they might endorse on the deed would have the necessary force.

Mr. Funnell said he had no doubt that, at the request of the meeting, Mr. Smalley would at any time be allowed to look at the accounts. If they could not appoint him an inspector, they could at any rate pass a resolution, requesting Mr. Smalley to inspect the accounts, and report upon them to the depositors. He should, therefore, propose that that be done.

The Chairman here read an abstract of the deed of inspection, stating that it was framed under the 224th section of the Bankrupt Law Consolidation Act of 1849.

Mr. L. Paine, a creditor to the amount of £4,000, eventually proposed the following resolution, drawn up by Mr. T. Jones:—"That this meeting strongly recommends that the proposed three inspectors do allow Mr. Smalley, the Chairman of this meeting, to act on the part of the depositors, and to be allowed to inspect the books of the firm that now exist, and which may be in future used; and the three inspectors be requested to supply Mr Smalley with every information that he may require on the part of the depositors; and that Mr. Smalley do report to a future meeting of the depositors to be convened when and for such purposes as he may see fit."

Mr. Wood, of Chelsea, seconded it.

On the question being then put, the resolution was carried unanimously.

Mr. Paine next moved the following resolution, which was also drawn up by Mr. Jones, expressing his confidence that it laid down the true line of policy for the depositors to follow:—"That this meeting is of opinion that it would be most advisable, for the interests of all parties concerned, that the business of Messrs. Felix Calvert and Co. should be carried on under the inspection proposed, and pledges itself to use its utmost endeavours to carry out that arrangement, and that the depositors present do now sign the deed of inspection."

Mr. Wood seconded the resolution.

Mr. Jones, in support of it, stated that he had been consulted with a view to proceedings in bankruptcy, but so strong was the feeling with which he received the proposition, that he replied that, rather than have his name tarnished by such a proceeding as striking a docket against the firm, he would forego all future connection with his client. (Hear, hear.) Such a course on his part would have been nothing short of murderous as regarded the interests of others, and he told him so, adding that, were he to follow it out, he would be pointed at in the streets. Let them imagine the consequences of 359 houses being sent into the market, as they then would be, and their property would be sacrificed to the extent of 10, 15, or 20 per cent. during the next four years. A greater act of imprudence could not possibly be committed, and, after such a warning, it would have been premeditated murder to persist in it. (Hear, hear.)

The resolution was then put, and unanimously agreed to, in the midst of loud and general cheering.

Mr. Funnell next moved a vote of thanks to the Chairman for the lucid and able statement he had made, the pains he had taken to investigate the matter, and the judicious advice he had given.

Mr. W. Harrison, of Thames Street, seconded the resolution, which having been unanimously agreed to,

Mr. Smalley, in reply, said he had felt great anxiety in undertaking to preside over that meeting, and it was their kindness, rather than his own ability, that had led to so good a result. It was seldom, indeed, that a meeting of creditors so numerous, and having such different interests, were seen to come together in such a spirit of mutual sacrifice, desirous only of their success as a body. He thanked them most sincerely for the gratifying mark of their confidence, and felt highly gratified at having endeavoured, with so much success, to save that great concern from falling into bankruptcy. (Hear, hear.) In conclusion, he would urge all present to sign the deed at once, lest, from the want of only a few names, the scale might be turned against them in number. Let them sign immediately, and depart with the perfect assurance that the best would be done for them. (Cheers.)

The business of executing the deed then went rapidly and orderly forward, and among the large number of depositors present there was scarcely one who did not sign.

A second meeting of the deposit creditors of Messrs. F. Calvert and Co., the brewers, took place on the 27th June, 1859, Mr. Smalley in the chair, to consider a proposal of the inspectors under liquidation to conduct the business by means of a joint-stock company. It is intended that the creditors shall have the option of converting their claims either into ordinary shares of the undertaking or of taking debentures at the rate of 15*s*. in the pound, redeemable in seven years. The last proviso is introduced to meet the convenience of trustees or representatives of friendly societies, clubs, etc., who may be legally precluded from becoming stockholders. The present plan is suggested to prevent a piecemeal realization of the estate, which might result in a disposal of all the public-houses, so as to depreciate the value of the brewery itself and render it nearly unsaleable. A suggestion had been made for the disposal of the business in one lot at a reduction of 22 per cent. upon the valuation, but the plan was found unsuccessful. During the past year the profits of the brewery have been £44,990, and the beer supplied has been of equal if not better quality than formerly. It is therefore estimated that with a continuance of the present returns, of which there appears every probability, a sufficient sum may be realized not only to pay a 5 per cent. dividend, but to provide a fund applicable for the redemption of the debentures. After a long discussion it was resolved on a division that the proposed company be formed, only three or four creditors voting against it. In the course of the proceedings it was stated that the private estates of one of the partners have not realized the original valuation, while in another case there has been a considerable excess. 120 public-houses have been sold for about £95,000, or about 9 per cent. below the nominal estimate. This result is considered favourable, as it was believed that the loss would have been as much as 25 per cent. The proceeds have been applied to the liquidation of secured claims. The debts provable on the estate represent a total of £726,000, of which £315,000 is for deposits, £252,000 private debts, and £159,000 trade debts.

THE ESTATE OF MESSRS. ROBERT BROWNE AND CO.

A meeting of the creditors of Messrs. Robert Browne and Co., East Indian and Australian merchants, who suspended on the 15th April, was held on the 20th May, Mr. C. G. Grainger in the chair, when the following statement was presented by Mr. Glegg, of the firm of Messrs. Quilter, Ball, and Co.:—

Statement, May 20, 1858.

DEBTOR.

To sundry creditors unsecured		£577 15 8
To creditors fully secured, estimated value of securities	£6,082 7 8	
Deduct amount of claims	5,690 9 6	
Surplus to *contra*	£391 18 2	
Carried forward		£577 15 8

Brought forward..............		£577 15 8
To liabilities on drafts accepted or guaranteed by us, viz.:—		
Secured by produce, amount of acceptances	£27,352 0 4	
Deduct estimated value of securities ...	25,832 9 2	
		1,519 11 2
Unsecured		30,391 8 7
Note.—These liabilities are subject to reduction by the value of any securities that may have been lodged against the drafts on Ceylon.		
		£32,488 15 5

CREDITOR.

By cash and bills on hand		£1,263 2 11
By sundry debtors—		
Considered good		137 16 8
Ditto bad	21 7 9	
By sundry property		2,119 5 7
By surplus securities held by creditors, *per contra*..		391 18 2
By debt at Melbourne, £2,887 4*s*. 11*d*. estimated to produce		2,000 0 0
		£5,912 3 4
By J. Swan and Co., Colombo	£1,320 12 10	
By Melbourne firm, in liquidation	952 0 3	
By deficiency carried forward		26,576 12 1
		£32,488 15 5

It was explained that the operations of the house had not been conducted on a very extensive scale, and that the connection with Messrs. J. Swan and Co., of Ceylon, had produced their embarrassments. Bills to the extent of £31,000 had been accepted on account of that firm, and the absence of remittances to meet them had necessitated the suspension. The total liabilities were estimated at £32,488, and the assets are £5,912. The principal creditors were the Oriental Bank and the Mercantile Bank of India, the ordinary claimants representing only about £1,500. After some discussion it was resolved that a composition of 3*s*. 4*d*. in the pound should be paid to the general creditors, exclusive of the Mercantile Bank of India and the Oriental Bank, the period fixed to be within one month if the banks agreed to this liquidation. It was also arranged that the creditors should not be prejudiced or affected by these proceedings as to their rights or remedies against other persons than the said Robert Browne.

THE ESTATE OF MESSRS. R. BAINBRIDGE AND CO.

An adjourned meeting of creditors of Messrs. R. Bainbridge and Co., of London and New York, who suspended during the late crisis, took place on the 4th of June, 1858. From the report of a committee appointed on the 5th of January, it appeared that a confidential agent was despatched to New York. After considerable difficulties, which led to the waste of much valuable time, he was permitted by Mr. Bainbridge to examine the books and accounts, and found them to be in such utter disorder, that it was impossible readily to discover the actual position of affairs. Many of the balances were thus supplied by Mr. Bainbridge himself, and not from the books, particularly the estimates of the stock in trade and real estate. These were taken too low; and, again, the claims of the English creditors were calculated at an unduly high rate of exchange. The result was, that there was a difference in the valuation of assets by the two parties, Mr. Bainbridge and the creditors' agent, of £12,000. Supposing the statement of the latter to be correct, there was sufficient to pay in full. No legal means, however, existed to enforce the claims of the creditors, except by an expensive and doubtful course, there being no Court of Bankruptcy in New York. In the meantime the usual preferential payments had been made, the American creditors receiving the whole of their claims, which had the following effect:—The total net liabilities assented to by the insolvent were £47,000, with admitted assets of £37,000, or about 15*s.* 9*d.* in the pound. Providing for the American claims, however, the matter stands thus:—Assets, £17,000, to meet English debts of £27,000, or about 12*s.* 6*d.* in the pound. This calculation was irrespective of the amount in dispute as understated. Mr. Bainbridge, without attempting to controvert the positions of the agent of the creditors, had already determined on his course of action, and offered a composition of 12*s.*, boldly stating the estate would realize no more. After considerable delay and long negotiations, an agreement was at length come to to pay all claims under £10 in full in two months from the 17th of March, 12*s.* 6*d.* on claims between £10 and £100 in six months from the 20th of March, and 13s. 4*d.* in four instalments extending over two years from the 10th of February on debts of £100 and upwards. A guarantee was at first positively refused, and it was only after repeated pressure that certain securities were obtained, worth about £10,000, or nearly 9*s.* 6*d.* in the pound of the composition. The report concluded with a recommendation that the terms be accepted, and accordingly a resolution to that effect was passed unanimously.

THE ESTATE OF MESSRS. RAWSON SONS AND CO.

A meeting of the creditors of Messrs. Rawson Sons and Co., engaged in the East India and China trade, whose suspension took place on the 2nd June, 1858, was held on the 8th, when the following satisfactory statement, showing a surplus of £48,000, was presented by Mr. J. E. Coleman, of the firm of Messrs. Coleman, Turquand, Youngs, and Co.:—

STATEMENT, JUNE 8, 1858.

DEBTOR.

To creditors on open accounts				£39,344	0	0
To creditors on bills payable	£489,634	0	0			
Less, expected to be provided for by parties for whose account the bills were accepted	125,038	0	0			
Amount to rank on this estate				364,596	0	0
To creditors partially secured	156,731	0	0			
To securities held	138,520	0	0			
				18,211	0	0
To creditors fully secured—						
Securities held	359,891	0	0			
Claims	274,923	0	0			
Contra	£84,968	0	0			
To liabilities on bills receivable, £297,310 12*s.* 3*d.*, the whole of which will be duly honoured at maturity.						
				£422,151	0	0

CREDITOR.

By cash balance	£2,613	0	0
By bills receivable on hand	22,733	0	0
By sundry assets, consisting of shares, etc.	650	0	0
By debtors, good	36,506	0	0
By debtors after taking up the bills accepted for their account	13,067	0	0
By produce on hand	12,580	0	0
By surplus from creditors holding securities, *contra*	84,968	0	0
By amount of various exports estimated at	125,000	0	0
By amount standing to debit of Blenkin, Rawson, and Co., of China	32,902	0	0
Carried forward	£331,019	0	0

Carried forward			£331,019 0 0
By amount standing to debit of Ker, Rawson, and Co., Singapore			7,365 0 0
Ditto, Leach, Rawson, and Co., Calcutta			42,940 0 0
By capital of T. S. Rawson and S. Rawson, in the firms of Blenkin, Rawson, and Co., and Ker, Rawson, and Co.			53,560 0 0
By estimated proportion of commission of profits since April 30, 1857			22,515 0 0
By sundries from the separate estates of the partners.........	£12,595 0 0		
Creditors to be paid in full		£522 0 0	
			£457,399 0 0

It was explained by Mr. Coleman, that although the foregoing was only a *pro forma* statement, it showed as nearly as possible the correct position of the firm, and that in consequence of the admirable manner in which the books and accounts had been kept he had been enabled to lay the results thus early before the creditors, so that they might be communicated by the mail which left on the 9th. With respect to the aggregate of liabilities, £125,000 of the bills would be taken up, leaving only £364,000 as chargeable against the estate. Of the bills receivable, representing £300,000, the whole would, it was thought, be met, and they were not likely to constitute the least liability. The assets were of a favourable character, and the cash and bills might be considered a very respectable item. The good debts consisted of balances due from correspondents, brokers, and prompts, all of which would be speedily available. It appeared that the produce in hand, and the surplus produced after the payments of the loans, amounting to £97,548, had been estimated at the value on the date of the suspension, and that also a portion had been taken at prices at which offers had been actually made, while some descriptions had slightly advanced in value since the estimate was completed. In connection with the exports calculated to produce £125,000, it was believed that they would before that have been sold, and the proceeds would be in course of remittance. About two-thirds of the amount would, according to the examination, be receivable by the shippers, and must go in reduction of acceptances to a similar extent, and the remaining third would be appropriated to the general creditors. It was stated that the balances due from the China, Singapore, and Calcutta houses, amounting to £83,216, the £33,560 of the capital of T. S. Rawson and S. Rawson, and the £22,500 the proportion of commission and profits, were all ordered, in December and January last, to be forwarded to this country, and that an acknowledgment of the request had been received. It was, therefore, anticipated that the bills to represent the greater amount were also in course of transmission. No doubt was entertained of the value of the India and China assets, since they had been thoroughly investigated and reported upon by a gentleman of experience from Manchester. In answer to questions it was intimated that Mr. T. S. Rawson and Mr. S. Rawson were directly interested in the houses in India, Singapore, and China, and that Mr. C. Rawson possessed an indirect interest through the father. The cause of suspension had been the absence of remittances, and the indisposition to force £400,000 produce in the hands of creditors as security suddenly upon the markets. The reason of the private estates appearing for so small a total was that the senior partner had not long since brought in £83,000, independently of other contributions previously, to meet the recent losses, which were taken in round numbers at £125,000. These explanations having been received as satisfactory, Mr. Charles Freshfield submitted the propriety of effecting a liquidation under inspection. The application of bankruptcy to an estate of this description would be highly prejudicial, and not warranted by the circumstances of the case. The assets showed 20*s.* in the pound, with a large surplus, including the private property of the partners; and if the usual process were adopted, the house could recommence business without interruption. The opinion of the creditors being unanimously in favour of this proceeding, after a short conversation, during which it was stated that no bills could come back from India or China unless they were document bills, resolutions were at once passed agreeing to the proposition. At the close of the discussion it was distinctly understood that the houses abroad should be advised by the morrow's mail, that without they felt themselves perfectly competent to meet all maturing engagements it would be their duty immediately to suspend, and not to allow any priority or preference. Before the meeting finally separated a strong feeling was expressed with regard to the honourable conduct exhibited by the partners, and to the satisfactory position in which the accounts were found, confidence at the same time being entertained in the result of the liquidation. Annexed are the resolutions, passed in official form:—

At a meeting of the creditors of Messrs. Rawson, Sons, and Co., held at their offices, No. 62, Moorgate Street, on Tuesday, the 8th day of June, 1858, Mr. Richard Durant in the chair, Mr. Coleman, the accountant, produced and read to the meeting a statement of the liabilities and assets.

It was proposed and seconded, and resolved unanimously—

"1. That it is the opinion of the meeting that the affairs of the house should be liquidated under inspectorship, and that the following gentlemen be the inspectors—George Dewhurst, Esq., Alexander Mackenzie, Esq., and W. Lyon, Esq. 2. That a proper deed of inspectorship be prepared under the approval of the inspectors, and be executed by or on behalf of each creditor, on or before payment of the first dividend. 3. That such deed shall contain covenants by the partners to liquidate the affairs of the house according to the rules of administration adopted in bankruptcy, and covenant by the creditors not to sue, which shall operate as a release upon the inspectors certifying that the liquidation has proceeded sufficiently, and upon the partners executing an assignment of any remaining assets to trustees for distribution among the creditors. 4. That such deed shall be a deed of arrangement, within the meaning of the 224th section of the Bankrupt Law Consolidation Act, and the 228th section shall be applicable thereto, and the creditors executing it shall not be prejudiced as to any securities or lien they may be entitled to, or as to their rights against third parties. 5. That the private property of the partners, after payment of their separate liabilities, shall be applied in the payment of the debts of the firm, according to the rules of distribution in bankruptcy. 6. That the inspectors shall have power to make to the partners such allowance as they may think fit for their services. 7. That instructions be sent out by the mail on the morrow, to the several firms in India and China, to adopt all necessary measures for protection of consigned goods and their proceeds, and to send home statements of all goods on hand, and account of sales of such as have been sold, with remittances direct to the respective consignees where no advances have been made. 8. That the partners be at liberty to transact business on their own account, on their covenanting in the inspectorship deed not to use, either directly or indirectly, any of the existing assets of the firm, and to incur no new engagements which could by any possibility be thrown on the existing assets.

(Signed) "R. Durant, *Chairman*."

THE ESTATE OF MESSRS. BRISTOW, WARREN, AND HARRISON.

A meeting of the creditors of Messrs. Bristow, Warren, and Harrison, who failed in the wholesale grocery trade at the end of June, 1858, was held on the 2nd of July. The subjoined accounts, prepared by Messrs. Coleman, Turquand, Youngs, and Co., exhibit a dividend of about 15*s.* in the pound. A heavy depreciation had occurred in the stock of sugar, and other losses had been considerable; but the partners, it transpired, commenced with a capital of upwards of £3,000. At the close of the meeting, the firm leaving themselves entirely in the hands of their creditors, a committee of six, including Mr. C. Coles, of Messrs. B. and J. Coles; Mr. R. A. Boyd, of Messrs. Goodby and Co.; Mr. N. Martineau, of Messrs. Martineau and Co.; Mr. Harrison, of Messrs. Harrison and Wilson; and Mr. J. Redpath, of Messrs. Wackerbath and Co., was appointed to look into the estate, and report, on a future occasion, the best course that can be recommended for carrying out a liquidation.

Statement of Affairs of Messrs. Bristow, Warren, and Harrison, June 26, 1858.

DEBTOR.

To creditors on open accounts—						
Trade	£29,334	0	5			
Cash	1,387	0	0			
To ditto on bills payable	14,161	6	5			
Carried forward				£44,882	6	10

Brought forward			£44,882 6 10
To creditors partially secured—			
Claims		£14,264 14 3	
Security held		9,949 0 10	
			4,315 13 5
To liability on contracts for purchase of goods of which we have not taken delivery, estimated at			365 17 6
To liabliities on bills payable, which will be retired by the drawers		£590 8 6	
To liabilities on bills receivable, the whole of which are expected to be duly met at maturity		23,500 12 8	
			£49,563 17 9

CREDITOR.

By cash balance at bankers'			£796 9 6
By bill receivable on hand			173 6 10
By debtors, good		£32,809 12 0	
Doubtful	£3,421 14 4		
Estimated at 10*s.* in the pound	1,710 17 2		
		1,710 17 2	
	1,710 17 2		34,520 9 2
By ditto, bad	243 13 0		
Carried to losses	£1,954 10 2		
By stock, viz. :—			
In warehouse		£1,233 2 11	
Docks, wharves, etc.		2,787 0 5	
			4,020 3 4
By lease of premises in Rood Lane, cost			300 0 0
By office furniture and trade utensils, estimated at			50 0 0
			£39,860 8 10
Deduct sundries to be paid in full		£214 8 8	
Estimated expenses of liquidation and allowance for realization of debts		1,500 0 0	
			1,714 8 8
			£38,146 0 2

The liquidation of this estate was favourable, and a dividend of 16*s.* or 17*s.* was realized.

THE ESTATE OF MESSRS. SKEEN AND FREEMAN.

The failure of Messrs. Skeen and Freeman, timber brokers, which took place on the 30th June, 1858, proves to have been more disastrous in its results than was generally anticipated. Not only had their suspension shown that their own resources had become completely exhausted, but it was quite clear that, through their bill transactions, they had involved other individuals. Among several who found themselves placed in difficulty through this state of things were Mr. E. A. Skeen and Mr. C. Snewin, both of whom were seriously compromised, and had to surrender their estates to their creditors. The circumstances of the stoppage indicated a reckless course of trading, but the principal responsibility rested upon the junior partner, who, in these days of competition, was entrusted with the finance department, and had committed a variety of irregularities with respect to warrants, cash advances, etc., illustrating the desperate straits to which he was driven to support the position of the firm. The senior partner had, it was represented, not exercised sufficient control, and, having been engaged at the waterside in the practical pursuit of the business, was not fully acquainted with the whole proceedings. So discreditable were some of the transactions that, under the estate of Messrs. J. Lilley and Co., of the East and West India Docks, who were recently made bankrupts, the funds which arose from the sale of stock, and which, in regular course, should have passed to the official assignee, were retained and absorbed by Messrs. Skeen and Freeman. Other operations of an equally discouraging character were disclosed, and the attendant risk was proportionate. A meeting of the creditors took place on the 2nd July, at which the following statement was presented, exhibiting debts and liabilities amounting to £33,690, and assets of £6,237; but as the representatives of the press were excluded, no authenticated abstract of the discussion was furnished. The property set forth showed a dividend of about 3*s.* 8*d.* in the pound, but this amount would not be secured through a private arrangement or an appeal to bankruptcy. So dissatisfied were the creditors with the explanations furnished by the accountant and the partners, that after a lengthened discussion it was agreed that a declaration of insolvency should be signed, and that the firm should be forthwith placed in the *Gazette.* The old plea of the avoidance of bankruptcy on account of the expense was not listened to in this case, because it was felt, from the nice legal points and other difficulties, that a public investigation was essential.

STATEMENT OF AFFAIRS OF MESSRS. SKEEN AND FREEMAN, OF NO. 75, OLD BROAD STREET, JUNE 28, 1858.

DEBTOR.

To creditors unsecured—		
Open account	£12,463 17 6	
With acceptances	8,928 10 2	
		£21,392 7 8
To creditors with security—		
Value of securities—		
Bills of Exchange	£2,180 4 11	
Goods	13,901 6 0	
Debt with documents	1,134 1 7	
	£17,215 12 6	
Amount of claims	16,919 7 7	
Surplus, *per contra*	£296 4 11	
By liabilities—		
On bills receivable	£48,614 19 0	
Of which amount will rank on this estate...		£12,298 11 0
To creditors, viz.:—		
To be paid in full	100 0 0	
		£33,690 18 8

CREDITOR.

By debtors—		
Good	£4,671 7 6	
Doubtful, £455 1*s.* 8*d.*, estimated at	80 0 0	
	£4,751 7 6	
Less creditors to be paid in full	100 0 0	
		£4,651 7 6
By bills receivable	£1,161 17 3	
Valued at		250 0 0
By property available to estate		1,040 0 0
By surplus securities, *per contra*		296 4 11
		£6,237 12 5

This estate was wound-up in bankruptcy.

THE ESTATE OF MESSRS. HYDE, HODGE, AND CO.

A meeting of the creditors of Messrs. Hyde, Hodge, and Co., merchants, who stopped on the 4th, took place on the 14th August, 1858, when the following statement was presented by Messrs. Coleman, Turquand, Youngs, and Co., accountants:—

STATEMENT, AUGUST 14, 1858.

DEBTOR.

To creditors unsecured—									
Open account	£44,587	13	2						
Bills payable	55,651	19	0						
				£100,239	12	2			
To creditors in Honduras—									
To creditors partially secured—									
Claims	£103,213	0	11						
Estimated value of securities	44,100	0	0						
				59,113	0	11			
							£159,352	13	1
To creditors fully secured—									
Estimated value of security	£16,255	8	2						
Claims	12,325	3	11						
Surplus, *per contra*	£3,930	4	3						
To liabilities on bills payable, partially covered and good	7,550	10	1						
To liabilities on bills receivable	£209,533	10	10						
Considered good	£76,146	10	0						
Covered by credit balances	47,106	15	2						
The balance amounting to	86,280	5	8						
May rank against this estate							86,280	5	8
	£209,533	10	10						
							£245,632	18	9

CREDITOR.

By debtors—									
Considered good				£8,540	17	9			
Considered doubtful	£8,935	18	1						
Estimated at	1,350	0	0	1,350	0	0			
							£9,890	17	9
	£7,585	18	1						
Considered bad	4,774	3	0						
	£12,360	1	1						
By bills receivable—									
Estimated to produce							300	0	0
By property—									
Mahogany in docks, valued at	£400	0	0						
One-eighth share of ship "Maria Gray"	1,000	0	0						
							1,400	0	0
By surplus securities, *per contra*—									
Arising from goods estimated at	500	0	0						
Carried forward......	£500	0	0				£11,590	17	9

Brought forward	£500 0 0		£11,590 17 9
Arising from bills, good, estimated at	1,350 5 0		
	£1,850 5 0		1,850 5 0
Arising from bills, doubtful	2,079 19 3	2,079 19 3	
	£3,930 4 3		
By assets in Honduras			88,915 14 4
Being the balance appearing at the credit of the firm, February 28, 1858, as per balance-sheet, furnished by the agent there, $551,133 = ...	£110,226 12 0		
From which deduct subsequent remittances, and proceeds of goods sold and held as security £40,621 8*s.* 11*d.* Less draughts on London, £19,310 11*s.* 3*d.*.........	21,310 17 8		
	£88,915 14 4		
			£102,356 17 1
By freehold land in Honduras, 100 estates, containing 1,676 square miles = 1,072,640 acres..			
By partners' private estates—estimated surplus		£2,000 0 0	

The assets immediately available in this country were thus £13,441, in addition to which there were about 2,900,000 feet of mahogany coming forward, and a quantity of logwood, estimated to produce £51,290. Further shipments were also expected to the amount of £9,570. In the bills receivable the aggregate sum that could be proved upon the estate was £86,280, which would make the total liabilities about £245,000. In the event, however, of other houses continuing in safety, not more than £26,000 would come against the firm, reducing the liabilities to £185,000. It was stated that the freehold land in Honduras was good and productive, and that, at the lowest calculation, it ought to realize sufficient to show a considerable surplus. Emigration is going on from Jamaica to the colony, and sugar and corn are already being cultivated. Since 1855, losses have been sustained to the amount of £95,000. The capital, at the commencement of the year, was affirmed to have been £22,000. Some discussion took place upon the valuation of the timber, 5¾ per foot, when it appeared that the calculation was based upon the estimates of two persons well acquainted with the trade. A fear was also expressed that the creditors at Belize would contrive to secure payment in full, to the prejudice of the English claimants ; but it was stated that, contrary to the former practice, the settlement of colonial debts cannot now be preferentially enforced. It was ultimately resolved to wind-up the estate under inspection.

The dividend already paid is 5*s.* 6*d.* in the pound.

THE ESTATE OF MESSRS. JOHN CARMICHAEL AND CO

The following is the statement presented at the meeting of creditors of Messrs. John Carmichael and Co., on the 14th September, 1858, at Liverpool, and although it showed on the face of it 13*s.* 6*d.* in the pound, it was not anticipated that the estate would realize more than 5*s.*, as was originally intimated. The discussion was of a protracted nature, and the creditors unanimously were of opinion that a liquidation under inspectorship was the only course that could be adopted.

General Statement of the Affairs of Mr. John Carmichael, of Liverpool, Merchant, September 14, 1858.

LIABILITIES.

Acceptances unsecured		£154,616 8 6	
Expected to be retired by drawers		21,888 11 2	
			£132,727 17 4
Acceptances partially secured		£23,020 19 6	
Deduct securities held		6,704 8 1	
			16,316 11 5
Creditors on open account unsecured			56,932 9 4
Ditto, ditto, partially secured		£83,297 1 4	
Deduct securities held		58,467 18 9	
			24,829 2 7
Bills receivable, estimated bad		£108,500 0 0	
Deduct securities applicable		14,625 0 0	
			93,875 0 0
Total			£324,681 0 8
Liability to Glyn and Co. on promissory notes of A. Montgomery and Co., considered covered by remittances to come from Melbourne		£15,000 0 0	

ASSETS.

Good debts		£64,106 8 3	
Doubtful debts	£11,620 5 10		
Bad debts	47,690 7 1		
	£59,310 12 11		
Estimated to produce		6,000 2 1	
Cash balance		1,026 11 1	
Bills receivable		144 0 0	
Produce and goods on hand		1,730 0 0	
8-64th shares of barque "Guatemala Packet"	£316 13 4		
8-64th shares ditto ditto "Salvador Packet"	350 0 0		
		666 13 4	
Adventures to West Coast		1,258 0 0	
Shares in Canadian Steam Company, estimated		2,500 0 0	
Surplus securities in hands of creditors		50 0 0	
Household furniture, estimated		1,000 0 0	
Office furniture, ditto		50 0 0	
Property at Greytown, ditto		1,000 0 0	
		£79,531 14 9	
Honduras Accounts.			
Claim of former concern of John Carmichael and Co. upon Honduras government	£29,784 0 10		
Advances on account of Carmichael, Vidal, and Co.	46,831 9 4		
607 shares in the Anglo-French Company	60,913 11 7		
		137,529 1 9	
			217,060 16 6
Deficiency			£107,620 4 2

Of these shares 367 are applicable to claims of creditors as securities held by them, and when the value can be ascertained they will be so treated, and the statement altered accordingly.

Deficiency forward			£107,620 4 2
Accounted for as follows :—			
Bad and doubtful debts	£59,310 12 11		
Less estimated to produce	6,000 2 1		
		53,310 10 10	
Depreciation in shipping		3,402 5 11	
Sundry losses—			
On produce	7,417 9 7		
Property, Greytown	1,000 0 0		
Land at Oxton.	645 8 10		
Shares in Canadian Steam Company	385 0 0		
Adventures	133 14 10		
		9,581 13 3	
At debit of profit and loss		21,994 14 4	
Bills receivable	£108,500 0 0		
Less at credit of acceptors	89,169 0 2		
		19,330 19 10	
			£107,620 4 2

THE ESTATE OF MESSRS. A. MONTGOMERY AND CO.

At a meeting on the 15th September, 1858, of the creditors of Messrs. A. Montgomery and Co., engaged in trade as Australian and Honduras merchants, who failed on the 1st, the annexed accounts were presented by Mr. John Young, of the firm of Messrs. Coleman, Turquand, Youngs, and Co. The chair on the occasion was occupied by Mr. Caldecott, and among the parties interested were Mr. Morley, of Messrs. Morley and Co.'s, and Mr. Howell, of Messrs. Ellis and Everington's ; the legal representatives were Mr. Oliverson, of the firm of Messrs. Oliverson, Peachy, and Co. ; Mr. Murray, of Messrs. Murray, Son, and Hutchins ; and Mr. Parker.

Pro forma.—MESSRS. MONTGOMERY AND CO., NO. 2, GREAT WINCHESTER STREET. STATEMENT OF AFFAIRS, SEPTEMBER 1, 1858.

DEBTOR.

To creditors on open accounts			£15,432 8 2
To creditors on bills payable		£36,072 4 6	
To last amount which will be provided for by parties for whose account the bills were accepted		3,322 19 11	
Leaving to rank on this estate			32,749 4 7
To creditors who claim a lien on remittances on account of joint shipment [MF]			14,824 0 4
To liabilities on bills receivable	£35,410 10 8		
Of which there are expected to rank on the estate		£14,000 0 0	
To liability on joint guarantee		4,000 0 0	
			18,000 0 0
To liability under marriage settlement	8,000 0 0		
Liabilities			£81,005 13 1

CREDITOR.

By bills receivable on hand, good			£369 16 6
By debtors, good		£2,857 9 3	
„ doubtful	£26,844 8 6		
„ bad	7,047 13 6		
	£33,892 2 0		
Estimated to realize	8,217 5 6		
		8,217 5 6	
Carried to losses	£25,674 16 6		11,074 14 9
By counting-house furniture and lease of premises, estimated at			300 0 0
By Dun Copper Mining Company shares, estimated value			250 0 0
By household furniture and lease of house, estimated value			500 0 0
By adventures consigned to R. Bowden and Co., of Melbourne—			
Invoice price of goods on hand, 14th of July, 1858		£17,866 17 10	
Estimated to realize		16,859 5 7	16,859 5 7
Carried to losses		£1,007 12 3	
By joint adventure [MF] consigned to R. Bowden and Co., Melbourne—			
By invoice price of portion on hand, 14th of July, 1858			11,043 8 4
			£40,397 5 2
By creditors under £10			25 3 7
Assets			£40,372 1 7

STATEMENT SHOWING POSITION OF THE FIRM ON 1ST OF JANUARY, 1858, AND ACCOUNTING FOR THE PRESENT DEFICIENCY.

DEBTOR.

To capital, January 1, 1858, as per statement		£6,595 12 11
To profits	£1,679 10 6	
Less charges and interest	1,602 0 2	
		77 10 4
To difference in books		27 16 9
To deficiency, as per statement of affairs		40,633 11 6
		£47,334 11 6

CREDITOR.

By drawings, as per ledger			£727 11 7
By estimated losses, on shares—			
Dun Mountain Copper Mining Company	£557 0 4		
Berlin Waterworks	1,111 12 9		
		£1,668 13 1	
On adventures—			
Shipment H to Costa Rica	£90 1 1		
Costa Rica adventure	24 17 6		
No. 3, ditto	99 9 4		
Nos. 1 and 2, ditto	52 2 6		
Sundries, ditto	686 1 10		
Paper, ditto	13 11 4		
N—L, ditto, A. M. & Co.	156 7 3		
		1,122 10 10	
On counting-house furniture		140 19 6	
Carried forward		£2,932 3 5	£727 11 7

Brought forward	£2,932 3 5	£727 11 7
On bad and doubtful debts	25,674 16 6	
		28,606 19 11
Liabilities as per statement of affairs		18,000 0 0
		£47,334 11 6

It was explained that the assets nominally showed something under 10*s.* in the pound, but it was not supposed such an amount would be obtained by the creditors. The debtor, it was stated, had been engaged in business with a firm in Australia, Messrs. R. Bowden and Co., to whom he had made considerable consignments, and he had also been the London agent of Messrs. J. Carmichael and Co. In this latter capacity he was under liabilities to the extent of £18,000, the principal on accommodation bills. With respect to his operations in Australia, they were understood to represent a partnership; but the amount of assets which it was thought will be received from that side was limited. There were also some Honduras claims including a small island, the property of Mr. Montgomery, in the Bay of Honduras, but from these the returns were not expected to be large. A question was raised respecting advice lately sent out to Australia to make special appropriations from remittances derived from recent consignments, but it was understood that this and other points could be hereafter investigated, and eventually a resolution was passed agreeing to a liquidation, through the process of inspection, five of the principal creditors undertaking the management. In the course of the discussion it was elicited that the capital, according to the books, on the 1st of January, 1858, was £6,595; but that if proper allowance had been made for bad debts, etc., there would have been a slight deficiency. That sum had been subsequently increased through the liabilities incurred on account of Messrs. Carmichael and Co., Australian adventures, losses on shares and bad debts, to the amount representing the difference as exhibited by the present balance-sheet.

This estate was wound-up in bankruptcy.

THE ESTATE OF MESSRS. PARDOE, HOOMANS, AND PARDOE.

A meeting of the creditors of Messrs. Pardoe, Hoomans, and Pardoe, lately extensively engaged in trade as carpet and tapestry manufacturers, was held on the 17th September, 1858, at Kidderminster, Mr. J. Pitman, the Chairman of the Stourbridge and Kidderminster Banking Company, presiding, when the annexed account, prepared by Messrs. Kettle and Daniell, accountants, of Birmingham, was submitted. It will be noticed that it showed a discouraging state of affairs, and it was believed the prospects of liquidation were very unfavourable. After explanations and discussion, it was agreed that the estate should be immediately wound-up under the inspection of Mr. W. Grosvenor, Mr. G. B. Lea, of Kidderminster; and Mr. R. Waller, of Manchester, through the operation of the private arrangement clauses.

Statement of Affairs, September, 1858.

DEBTOR.

To amount due to creditors—	
On bonds, notes, etc., unsecured	£22,380 6 4
Ditto to trade creditors, on open accounts	6,521 17 3
Ditto on bills overdue and running	29,216 0 9
To amount due to London Life Association, for which they hold a mortgage upon properties and life policies, as *per contra*	18,450 0 0
To amount due to a creditor, for which she holds a second mortgage upon properties and life policies, as *per contra*	1,312 2 2
To amount due to bankers, unsecured	2,849 17 5
Carried forward	£80,730 3 11

		£	s.	d.
Brought forward		£80,730	3	11
To amount due to bankers, for which they hold a third mortgage upon properties and life policies, and a further security on the Caldwell Mill and Vicar Street properties, as *per contra*		20,906	13	7
To amount due for rent, taxes, and salaries, to be paid in full	£700 0 0			
		£101,636	17	6

No liability is likely to come on this estate in respect of bills of third parties endorsed and discounted; after careful consideration, it is believed the whole will be duly met at maturity.

CREDITOR.

		£	s.	d.
By stock of goods, materials, and stores, as per inventory		£14,439	10	1
By cash in hand		25	0	0
By amount due from debtors, supposed will realize		2,690	18	4
By thirty shares in public rooms, Kidderminster, say		200	0	0
By manufacturing plant and machinery, looms, patterns, etc., deducting steam-engines, boilers, and machinery, in the nature of fixtures, which are included in the mortgage securities, at the value appearing in Messrs. Pardoe and Co.'s accounts		18,740	0	0
By private effects of partners		1,500	0	0
		£37,595	8	5
Deduct rent, etc., to be paid in full		700	0	0
		£36,895	8	5
By freehold properties and life policies, subject to three mortgages, as *per contra*, at the valuation appearing in Messrs. Pardoe and Co.'s accounts—				
Old mill premises, and sling premises, the property of T. Pardoe, G. Hooman, and J. Hooman	£28,000 0 0			
Engine, boilers, and fixtures	2,260 0 0			
Old factory, Oxford Road, the property of T. Pardoe, G. Hooman, and J. Hooman	2,000 0 0			
Warehouse, offices, and dwelling-house, Worcester Street and Oxford Road, the property of T. Pardoe	4,000 0 0			
Scotch factory and dwelling-houses in Worcester Street, the property of G. Hooman and J. Hooman	1,500 0 0			
Lion Fields factory and land, the property of G. Hooman	1,200 0 0			
Life policies—Mr. Thomas Pardoe, £2,000; Mr. George Hooman, £3,000; Mr. James Hooman, £3,000; and Mr. James Pardoe, £3,000; estimated value	6,000 0 0			
		£44,960	0	0
By property held by bankers as further serity—				
Freehold land and house at Caldwell, the property of T. Pardoe, G. Hooman, and J. Hooman	300 0 0			
Caldwell mill and land, leasehold, the property of T. Pardoe and G. Hooman	1,291 0 0			
Premises in Vicar Street, leasehold, the property of G. Hooman and J. Hooman	600 0 0			
		£2,191	0	0
		£84,046	8	5

THE ESTATE OF MESSRS. J. PLOWES AND CO.

A meeting of creditors of Messrs. J. Plowes and Co., merchants, who stopped on the 7th, took place on the 28th September, 1858, when the following balance-sheet was submitted by Messrs. Coleman, Turquand, Youngs, and Co., accountants :—

STATEMENT OF THE AFFAIRS OF MESSRS. JOHN PLOWES AND CO., SEPTEMBER 28, 1858.

DEBTOR.

To creditors unsecured		£28,419 13 3
To creditors having recourse on Plowes, Son, and Co., Rio		7,678 9 2
To liabilities on bills payable, accepted as advances on shipments to Plowes, Son, and Co., Rio	£68,912 12 9	
Deduct, expected to be retired by drawers, who will then claim on the Rio house for net proceeds of their goods	27,300 19 10	
		41,611 12 11
Deduct liabilities on bills payable, drawn by Plowes, Son, and Co., but which are expected to be paid by them	4,382 11 2	
Deduct liabililities on bills receivable, none of which are expected to be claimed against the estate	5,031 4 5	
		£77,709 15 4

CREDITOR.

By cash at bankers'	£839 5 4	
Less amount considered specially remitted...	334 14 1	
		504 11 3
By bills receivable in hand, part of which is claimed as special remittances		6,255 14 7
By debtors, good		2,936 1 8
By produce in hand		981 8 7
By private property of J. H. Plowes		5,360 0 0
		£16,037 16 1

DOUBTFUL ASSETS.

By balance due from consigners of goods to Rio		£4,996 0 4
Plowes, Son, and Co., Rio, balance of general account	£24,276 3 5	
Estimated proceeds of goods against which advances have been made to parties who are unable to retire their draughts, estimated as sufficient to cover advances *per contra*	41,611 12 11	
		65,887 16 4
		£86,921 12 9

The estate thus shows a surplus contingent upon the realization of the doubtful assets. It appeared that there were two houses, one in London, consisting of Mr. J. Plowes, only, and the other in Rio de Janeiro, in which, besides that gentleman, there were three partners. The suspension was caused by one of the latter, Mr. Westwood, incurring losses by private speculation, and secretly giving acceptances in the name of the firm to the amount of £55,000. Against this sum he had cash and shares worth about £30,000, and it was therefore hoped that the firm would suffer from these transactions to the extent only of £25,000. The liabilities of the London house included advances by bill to consigners of goods to Rio equal in the aggregate to £68,912. Of this total, £27,310 were estimated to be retired by the drawers, who would then claim the net proceeds of the sale of their consignments in Brazil. The remaining £41,611 was covered by the merchandise on which the advances had been made. The position of the concern at Rio on the 30th June, after writing off all bad debts, was as follows:—Assets £170,000, and liabilities £145,000, leaving £25,000 to the credit of capital. In answer to questions, it was stated that the irregular acceptances above mentioned fell due in August, September, October, and November of that year. The house was in no way connected with one bearing a similar name in Buenos Ayres. After a long discussion, in the course of which a very favourable feeling was expressed towards Mr. Plowes, it was resolved to wind-up under inspection.

This estate has paid 4*s.* 6*d.* in the pound; about 14*s.* is anticipated.

THE ESTATE OF MESSRS. GEORGE CHAMBERS AND CO.

A meeting of the creditors of Messrs. G. Chambers and Co., fancy hardware manufacturers, who failed on the 11th September, 1858, was held on the 1st October, when the following statement was presented by Mr. G. H. Jay, the accountant:—

Statement of the Affairs of Messrs. George Chambers and Co., September 11, 1858.

DEBTOR.

To creditors unsecured—		
On open accounts		£2,794 2 11
On bills payable		7,571 19 7
To creditors holding security, fully covered—		
Value of security	£29,786 11 6	
Less amount of their claims	20,445 14 1	
Surplus, *per contra*	£9,340 17 5	
To liabilities on bills receivable, the acceptors of which have suspended payment		11,571 11 5
N.B.—Subject to any remittances that may be made in respect of goods consigned to, or in the hands of, the acceptors, and which amount at cost to £11,425 7*s.* These consignments may be so affected by the failure of the consignees, that, until advices are received, it is not possible to form any estimate of the extent to which this claim may be reduced.		
To liabilities on bills receivable	£27,946 12 4	
Not expected to rank against this estate.		
To liability on bill payable, accepted against goods shipped to the East Indies	£419 7 0	
Less estimated value of goods unsold	329 0 3	
		90 6 9
		£22,028 0 8

CREDITOR.

By cash—		
At bankers'	£10 1 6	
In hand	80 18 11	
		£91 0 5
By debtors—		
Good		462 19 2
Doubtful	£870 13 9	
Estimated at 10*s.* in the pound		435 6 10
Bad	59 6 4	
Carried forward		£989 6 5

Brought forward				£989	6	5
By stock—						
In Russia Row	£2,191	7	8			
At Studley	1,305	12	2			
	£3,496	19	10			
Estimated to realize				2,500	0	0
By implements in trade at Studley				73	0	0
By office fittings and fixtures				30	0	0
By consignments unsold				2,472	5	6
By surplus consignments, as *per contra*.........				9,340	17	5
By consignments in the hands of failed consignees, as *per contra*	11,452	7	0			
				£15,405	9	4
Less creditors to be paid in full, rents, salaries, etc.				436	6	8
				£14,969	2	8

It was explained by Mr. Lawrance, who represented the estate, that the suspension was entirely caused by the failure of Messrs. John Plowes and Co., between whom and Mr. Chambers business connections had for several years subsisted. The liabilities of the house were stated at £22,028, and the assets at £14,969, but the former included £11,571 on bills. Had it not been for this amount, a surplus of £4,500 would have been shown. It appeared that Mr. Chambers, having made consignments to the firm of Messrs. Plowes and Co. at Rio de Janeiro, obtained advances upon them, and in this manner the liability legitimately arose. The goods despatched were valued, at cost price, at £11,452, and if these should not have been appropriated, and were favourably realized, the estate would be relieved of any claims in this respect. The books had been well kept, and the transactions conducted in a regular and business-like manner. After some discussion and explanation, it was agreed to carry out a liquidation by inspection, the creditors generally being in favour of that course.

This estate is winding-up, with the sanction of bankruptcy, under inspection.

THE ESTATE OF MESSRS. WESTRUPP AND CO.

A meeting of the creditors of Messrs. Westrupp, millers, of Bromley, who recently suspended, was held on the 12th October, at the offices of Messrs. M'Leod and Stenning, when the following statement of affairs, prepared by Messrs. Quilter, Ball, Jay, and Co., was submitted and explained by Mr. W. Quilter. The figures showed that the assets were equal to a dividend of 7*s.* 6*d.* in the pound; and, after some discussion, it was agreed to accept a composition of that amount, provided it be paid in cash, so soon as the whole of the creditors should have agreed to the arrangement. It was expected that the realization of the estate would be anticipated by the friends of Messrs. Westrupp, and that they would consequently be enabled to give effect to the terms proposed. At the close of the proceedings, resolutions were passed in accordance with these arrangements.

Statement of the Affairs of Messrs. E. and E. F. Westrupp, September 28, 1858.

DEBTOR.

To creditors unsecured—						
On open accounts				£11,124	8	7
On bills payable				2,738	8	0
To creditor holding security—						
R. M. Last	£510	0	0			
Estimated value of security	510	0	0			
Carried forward..............				£13,862	16	7

Brought forward		£13,862 16 7
To liabilities on bills receivable	£489 6 9	
Expected to rank against this estate		40 7 6
		£13,903 4 1

CREDITOR.

By cash in hand		£375 0 0
By debtors—		
Good	£2,344 2 4	
Less 10 per cent. for collection	234 2 4	
		2,110 0 0
Doubtful	550 17 8	
Estimated at		208 0 0
Bad	227 0 6	
By stock		2,856 11 2
By lease and fixtures, held by creditors, as *per contra*	510 0 0	
By horses, implements in trade, etc.		220 0 0
		£5,769 11 2
Less creditors to be paid in full	447 2 0	
Estimated expenses of liquidation	100 0 0	
		547 2 0
		£5,222 9 2

THE ESTATE OF MESSRS. JAMES DAVIES AND SON.

A meeting of the creditors of Messrs. James Davies and Son, wholesale boot and shoe manufacturers, who suspended on the 27th November, was held on the 13th December, 1858. Mr. Mortimore, of the firm of Messrs. Streatfield, Laurence, and Co., presiding; when the annexed statement, prepared by Mr. G. H. Jay, of the firm of Messrs. Quilter, Ball, and Co., was submitted. The principal creditors were represented by Messrs. Wright and Bonar, and Mr. John Linklater appeared for some of the provincial creditors :—

STATEMENT OF AFFAIRS OF JAMES DAVIES AND SON, NOVEMBER 27, 1858.

DEBTOR.

To creditors unsecured		£84,959 16 7
To creditors for advances on goods consigned to them	£131,968 7 7	
To value of consignments	177,455 15 0	
Surplus, *per contra*	£45,487 7 5	
To creditors partially secured—		
Amount of claims	£1,130 0 0	
Value of security	150 0 0	
		980 0 0
To creditors fully secured—		
Value of security	553 13 6	
Amount of claims	381 13 6	
Surplus, *per contra*	£172 0 0	
To liabilities on bills receivable—		
Considered good	£82,299 5 7	
		£85,939 16 7

CREDITOR.

By cash at bankers'		£17 5 6		
In house		70 0 0		
				£87 5 6
By bills receivable on hand—Considered bad		£2,560 0 2		
By sundry debtors—Considered good		£5,166 18 6		
Ditto, doubtful	£2,917 8 2			
Estimated at		1,458 0 0		
				6,624 18 6
Considered bad		£1,137 14 2		
By stock in trade—				
London		£16,682 18 1		
Hackney Road		802 16 0		
Northampton		2,061 5 2		
Norwich		300 0 0		
				19,846 19 3
By surplus on goods consigned, as *per contra*				45,487 7 5
By surplus security in hands of creditors, *per contra*				172 0 0
By lease of premises, as per valuation		3,000 0 0		
Less amount borrowed on to pay wages since the date of suspension		750 0 0		
				2,250 0 0
By fixtures, machinery, utensils, etc., estimated at				1,285 9 0
By property				163 15 0
				£75,917 14 8
Deduct creditors under £10 each, rent, rates, salaries, and sundries				342 8 4
				£75,575 6 4

It was explained that the general figures in the balance-sheet showed a dividend of about 17*s.* or 18*s.* in the pound ; but, of course, it remained to be ascertained what would be the proceeds from the surplus of consignments in Australia, which stood at £45,487. Taking, however, the assets in this country at £30,000 against the £85,900 debts and liabilities, there was a prospect of a payment of 6*s.* 8*d.* With regard to the £82,200 liabilities on bills receivable, it was stated that not the least difficulty was anticipated, the several firms with which they were connected being considered solvent and able to provide for the required amounts. The losses of the firm were very heavy, the total of account current sales, received from Australia between June and September, showing a sacrifice of about £30,000. Soon after the turn of the half-year instructions were despatched to the colonies to advise caution in the realization of the consignments, and in the early part of that month the circumstance of the suspension was announced, accompanied by the fact that no more goods would be forwarded. Some of the creditors expressed dissatisfaction that the meeting could not be adjourned to allow of the statement of accounts being examined before a decision was arrived at, but it was intimated by Mr. Jay that it had been found necessary to consult thus early in order to prevent bankruptcy, proceedings having already commenced, which would mature in three days. Under these circumstances the accounts themselves could only be considered as presenting an approximate position, the details of deficiency not having been completed. The great question discussed was the advantage to be derived from the process of inspectorship, contrasted with bankruptcy in winding-up an estate like the present. The large majority of the creditors supported the views of the Chairman, Mr. J. Linklater, and Mr. Jay, in awarding preference to the former course, the arrangements necessary for dealing with the consignments rendering the exercise of prudence essential. Others advised that too great reliance should not be placed in the proceeds of the consignments, but to look to the assets here ; and these, estimated at 6*s.* 8*d.* in the pound, if carried into bankruptcy would, it was asserted, soon be reduced to 3*s.* It was mentioned, however, on behalf of Messrs. Davies and Son, that the transactions entered into were with the most respectable Australian houses, who would not force sales, and whose conduct would be perfectly regular. A creditor said he was pleased to hear this, but they would no doubt charge the colonial rate of interest—a question of consideration. Mr. F. Parbury, as representing the firm of Messrs. Parbury and Co., distinctly announced that he for one would not do so. It was eventually agreed to pass resolutions in favour of inspectorship, and Mr. Mortimore, of Messrs. Streatfield, Laurence, and Co.; Mr. Morris, of Messrs. Bevington and Morris; Mr. Somerville, of Messrs. Somerville and Co.; and Mr. H. Pound accepted the trust.

This estate has paid a first dividend of 3*s*. in the pound, and is winding-up under inspection with the sanction of bankruptcy.

THE ESTATE OF MESSRS. HICKS AND GADSDEN.

At a meeting of the creditors, on the 24th December, 1858, of Messrs. Hicks and Gadsden, who recently failed in the American trade, the following statement was presented by Mr. Jay, of the firm of Messrs. Quilter, Ball, and Co., the accountants:—

Statement of Affairs, December 6, 1858.

DEBTOR.

To sundry creditors		£9,045 4 2
To creditors fully secured, viz.:—		
Estimated value of securities	£4,400 13 2	
Less amount of claims	1,959 8 9	
Surplus, *per contra*	£2,441 4 5	
To liabilities on our acceptances, as per list	£60,488 18 11	
Against shipments of cotton, etc., in hand or *in transitu*, on which claims may arise on the sale of the cotton	54,198 3 2	
Unsecured	£6,290 15 9	
Of which it is estimated will prove claims		6,137 15 9
To liabilities on bills receivable, as per list	49,185 10 3	
Expected to rank as a claim		144 0 0
To liability on underwriting	100 0 0	
		£15,326 19 11

CREDITOR.

By cash at bankers' held as security against loan	£147 4 5	
By sundry debtors, as per list, considered good		£6,018 12 2
By doubtful	£9,368 19 11	
Estimated to produce		3,500 0 0
Considered bad	13 8 10	
By consignments outstanding and sundry property on hand, as per list	940 9 10	
Estimated to realize		650 0 0
By surplus securities held by creditors, *per contra*		2,441 4 5
		£12,609 16 7
Deduct creditors under £10 each, rent, etc., payable in full		70 0 0
		£12,539 16 7
By private estate of G. A. Hicks, estimated at		2,200 0 0
		£14,739 16 7

The assets, including £2,200, the separate property of Mr. Hicks, showed nominally a dividend of about 18*s*. in the pound, but no estimate was offered of the probable realization, though it was expected to be satisfactory. Of liabilities on bills for £60,488, £54,198 were covered by cotton, leaving little more than £6,000 to come against the estate. At present the sales on that account showed encouraging results, and the expectation seemed to be that the amount, as explained, would be provided for. The principal creditors were in Liverpool; and the cause of the suspension was attributed to the depreciation in cotton, and losses by the failure of Messrs. Francis and Frere, of London, and of Messrs. Reed and Nash, of New York. Of the £15,000 debts, £7,000 to £8,000 was represented by the Liverpool firms, but the relatives were also interested through loans to some extent. On the 1st of January, 1858, the house possessed a capital of about £10,000, the proportion due to Mr. Hicks being £7,408, and to Mr. Gadsden, £2,497. During the same period the partners' drawings did not exceed £1,700, the trade charges at the same time being limited to £275, law costs to £80, and interest and discount to £942. A lengthened conversation ensued on the course to be adopted for effecting the liquidation, Mr. Jay intimating that the debtors were prepared to follow any steps which might be suggested. They were anxious that the whole of their assets should be appropriated, even to the separate property of Mr. Hicks. Bankruptcy he considered not applicable to the realization in that case, as the creditors abroad, especially in New York and New Orleans, would be less likely to respond to the demands made upon them through that channel than if application were sought under the process of inspectorship, associated with the representations of the partners. This view was eventually concurred in, although meanwhile an attempt was made to obtain an adjournment; and at the conclusion of the proceedings a resolution was passed, authorizing the appointment of creditors to represent the general body by a deed of inspection.

This estate is winding-up under inspection, and has paid a first dividend of 4*s.* in the pound.

INDEX.

[*The Figures refer to the body of the work, the Numerals to the Appendix.*]

THE END.